Medicinal Plants - Chemical, Biochemical, and Pharmacological Approaches

Edited by Mozaniel Santana de Oliveira, Eloisa Helena de Aguiar Andrade, Ravendra Kumar and Suraj N. Mali

Published in London, United Kingdom

Medicinal Plants - Chemical, Biochemical, and Pharmacological Approaches
http://dx.doi.org/10.5772/intechopen.107682
Edited by Mozaniel Santana de Oliveira, Eloisa Helena de Aguiar Andrade, Ravendra Kumar and Suraj N. Mali

Contributors
Ruth T.S. Ofongo, Habeeb O Yusuf, Nurhan Kishali, Nitin Govindrao Dumore, Namita Girhepunje, Monali Dumore, Kishor Danao, Manoel Teixeira Souza Junior, Thalita Massaro Malheiros Ferreira, Fernanda Ferreira Salgado, Olga Costa Alves Souza, Rejane Valeriano Silva, Vivianny Nayse Belo Silva, Patrícia Abrão de Oliveira Molinari, Thales Lima Rocha, Dhouha Alimi, Azhar Hajri, Slimen Selmi, Hichem Sebai, Farah Al-Mamoori, Ashraf M.A. Qasem, Bamidele Sekinat Olayem, Akinbode Badiu Akinola, Origbemisoye Babawande Olaitan, Mostafa Fazeli, Ali S. Hassoon, Ahmed A. Kadhim, Madeha H. Hussein, Moleboheng Emily Binyane, Polo-Ma-Abiele Hildah Mfengwana, Sitheni Samson Mashele, Abdenaceur Reghmit, Farida Benzina-tihar, Fatma Sahir-Halouane, Mohammad Barati, Amir Modarresi Chahardehi, Muhammad Alamzeb, Behramand Khan, Ihsan Ullah, Muhammad Omer, Adnan, Sally Elnawasany, Vimbai Gobvu, Kudakwashe Chirigo, Takudzwa L. Charakupa, Clarice P. Mudzengi, Oladipupo A. A Lawal, Andy R. Opoku, Rebamang A. Mosa, Foluso O. Osunsanmi, Kehinde O. Amisu, Rakhi Mishra, Binit Dwivedi, Poonam Bansal, Sunayna Chaudhary, Bhupesh Gupta, Soniya Goyal, Raman Kumar, Pooja Sharma, Tanvi Taneja, Sonali Sangwan, Alda Pereira da Silva Oliveira, Manuel Pires Bicho, Maria do Céu Costa, Elliot Nyagumbo, Trust Nyirenda, Cephas Mawere, Donald Kapanga, William Pote, Michael Bhebhe, Lucy Mabaya, Fabian Maunganidze, Alfred Mutaramutswa, Godwins Ngorima, Ian Mutasa, Leroy Nhari, Emmanuel Kademeteme, Siham A. Salim

First published in London, United Kingdom, 2024 by IntechOpen
IntechOpen is the global imprint of INTECHOPEN LIMITED, registered in England and Wales, registration number: 11086078, 5 Princes Gate Court, London, SW7 2QJ, United Kingdom

British Library Cataloguing-in-Publication Data
A catalogue record for this book is available from the British Library

Additional hard and PDF copies can be obtained from orders@intechopen.com

Medicinal Plants - Chemical, Biochemical, and Pharmacological Approaches
Edited by Mozaniel Santana de Oliveira, Eloisa Helena de Aguiar Andrade, Ravendra Kumar and Suraj N. Mali
p. cm.
Print ISBN 978-1-80356-743-3
Online ISBN 978-1-80356-744-0
eBook (PDF) ISBN 978-1-80356-745-7

For EU product safety concerns:
IN TECH d.o.o., Prolaz Marije Krucifikse Kozulić 3, 51000 Rijeka, Croatia,
info@intechopen.com or visit our website at intechopen.com.

Meet the editors

Prof. Dr. Mozaniel Santana de Oliveira has a BS in Chemistry and an MSc and Ph.D. in Food Science and Technology. He has published eight-six articles, six books, forty-four international book chapters, and sixty-two conference abstracts. He has experience in engineering, food science and technology, phytochemistry, methods for prospecting bioactive molecules, chemistry and biotechnology of natural products, and allelopathy seeking new natural herbicides for the control of invasive plants. He is a professor of postgraduate programs in biological sciences, tropical botany, food science and technology, , and pharmaceutical sciences at the Federal University of Pará, Brazil. He was previously an ambassador for Bentham Science Publishers and an academy mentor at Web of Science. He has reviewed 350 scientific articles for 70 international journals and edited 138 scientific articles as a journal editorial board member. He is currently a postdoctoral researcher in the postgraduate program in biological sciences - tropical botany at Federal Rural University of the Amazon/ Museu Paraense Emílio Goeldi (UFRA/ MPEG).

Prof. Dr. Eloisa Helena de Aguiar Andrade has a BS in Pharmacy, a qualification in Biochemistry, an MSc in Chemistry of Natural Products, and a Ph.D. in Chemistry from the Federal University of Pará (UFPA), Brazil. She is currently Principal Researcher II of the Botany Coordination of the Museu Paraense Emílio Goeldi (MPEG), responsible for the Adolpho Ducke Multi-User Laboratory (LAD). She is also Associate Professor I and professor of the postgraduate programs in chemistry at UFPA; professor of postgraduate programs in biological sciences – tropical botany at the Federal Rural University of the Amazon/Museu Paraense Emilio Goeldi (UFRA/MPEG); and Professor of Biodiversity Biotechnology for the Bionorte Network, UFPA. She is the coordinator of the Pará State Hub of the Postgraduate Program in Biodiversity and Biotechnology (PPGBIONORTE-PA) of the Bionorte Network. Dr. Andrade is the author of more than 500 scientific contributions including articles, communications, book chapters, and books. She has experience in chemistry, with an emphasis on the chemistry of natural products, working on gas chromatography, gas chromatography/mass spectrometry on volatile and fixed (derivatized) chemical constituents, and more.

Dr. Ravendra Kumar obtained his MSc and Ph.D. in Agricultural Chemicals/Chemistry from G. B. Pant University of Agriculture and Technology, India, in 2006 and 2010, respectively. He qualified for ICAR NET in 2010. Dr. Kumar has industrial experience in the extraction, separation, and purification of nutraceutical and pharmaceutical components at India Glycols Limited, Dehradun. Since 2015, he has been an assistant professor in the Department of Chemistry, College of Basic Sciences and Humanities, G. B. Pant University of Agriculture and Technology. During this period, he published more than seventy-six research papers in reputed national and international journals and guided nine MSc and one Ph.D. students in natural product chemistry. Currently, he is guiding two MSc and two Ph.D. students in plant chemistry. He has also published twelve book chapters on pesticide chemistry/natural products and four review articles on plant chemistry. He received the 2018 Young Scientist Award at the first annual conference of the Society for Himalayan Action Research and Development (SHARAD) for his work in chemical and nanoscience. Dr. Kumar is a member of several chemical societies and journals. He is handling two national projects funded by the Uttarakhand Council for Biotechnology (UCB) and Uttarakhand State Council for Science and Technology (UCOST), Government of India.

Dr. Suraj N. Mali has a Ph.D. in Pharmacy. He is a former analytical scientist at Dr. Reddy's Laboratories, Hyderabad, India. He has an academic background in pharmaceutical science and technology from the Institute of Chemical Technology, Mumbai, India. He serves as a respected reviewer for multiple scientific journals and was designated as a Bentham Science Brand Ambassador for 2019–2020. He has more than ninety-five international journal publications to his credit. He received the Institute of Chemical Technology's (ICT) Masters Best Thesis Aditya Birla Award in 2019. His diverse expertise spans molecular modeling, synthetic chemistry, phytochemistry, pharmacology, and analytics, with a focus on drug design and synthesis. A recent publication in Nature Scientific Reports highlights his work in identifying antimycobacterial agents using computational tools. Dr. Mali was listed among the world's top 2% of scientists by Stanford University, USA, in 2023.

Contents

Preface

In the vast realm of medicine, the contributions of medicinal plants have been both timeless and invaluable. *Medicinal Plants – Chemical, Biochemical, and Pharmacological Approaches* is a comprehensive exploration of the multifaceted world of medicinal flora, shedding light on their diverse applications and therapeutic potential.

This book, skillfully curated by editors Mozaniel Santana de Oliveira, Eloisa Helena de Aguiar Andrade, Ravendra Kumar, and Suraj N. Mali, is organized into six sections, each offering a unique perspective on medicinal plants:

1. **Pharmacological Application**: Unveiling the therapeutic power of nature's pharmacy, this section explores the varied pharmacological applications of medicinal plants.

2. **Alkaloids**: A deep dive into alkaloids, the bioactive compounds found in plants, provides insights into their potential medicinal significance.

3. **Anti-Inflammatory and Antiviral**: Exploring the plant kingdom's natural anti-inflammatory and antiviral properties, this section uncovers nature's solutions to modern health challenges.

4. **Essential Oils**: The essential oils extracted from medicinal plants are highlighted for their diverse applications and aromatic healing properties.

5. **Medicinal Plants and Phytopathology**: Examining the intriguing relationship between medicinal plants and phytopathology, this section offers a unique perspective on plant health and healing.

6. **Ethnomedicine**: Bridging tradition and science, this section delves into the world of ethnomedicine, where cultural wisdom meets contemporary healthcare.

This book is a collaborative effort that brings together the expertise and insights of numerous scholars, making it an indispensable resource for researchers, students, and professionals in the fields of pharmacy, pharmacology, and botany. We hope that the content within these pages will inspire further exploration and appreciation of the remarkable healing potential offered by the world's diverse array of medicinal plants.

Dr. Mozaniel Santana de Oliveira
Postgraduate Program in Biological Sciences - Tropical Botany,
Research Campus,
Emílio Goeldi Paraense Museum,
Belém, PA, Brazil

Dr. Eloisa Helena de Aguiar Andrade
Research Campus,
Emílio Goeldi Paraense Museum,
Belém, PA, Brazil

Dr. Ravendra Kumar
Department of Chemistry,
College of Basic Science and Humanities,
G.B. Pant University of Agriculture and Technology,
Pantnagar, U.S. Nagar, Uttarakhand, India

Dr. Suraj N. Mali
Department of Pharmaceutical Sciences and Technology,
Birla Institute of Technology,
Mesra, India

Acknowledgment

The Editor Prof. Dr. Mozaniel Santana de Oliveira, would like to thank PDPG-POSDOC - Postgraduate Development Program (PDPG) Strategic Postdoctoral, as well as CAPES for the scholarship (process number: (88887.852405/2023-00)).

Dedication

The editor, Dr. Mozaniel Santana de Oliveira, dedicates this work to the entire scientific community of the Botanic coordination of the Museu Paraense Emilio Goeldi, and dedicates this work especially to his beloved wife Joyce Fontes and his beloved daughter Isabela de Oliveira.

Section 1

Pharmacological Application

Chapter 1

Natural Antioxidants: An Update

Muhammad Alamzeb, Behramand Khan, Ihsan Ullah, Muhammad Omer and Adnan

Abstract

Antioxidants are the body's defensive mechanism against reactive oxygen species damage, which is typically caused by the different physiological activities that take place within the body. These antioxidants can be obtained from a variety of sources, including the body's own endogenous antioxidants and exogenous dietary sources. Generally, food items and several types of medicinal plants are considered as the sources of natural antioxidants. Natural antioxidants possess wide variety of bioassay properties like anti-cancer, anti-aging, anti-inflammatory etc. The substitution of artificial dietary antioxidants with natural ones in recent decades has increased interest in low-cost raw materials, particularly agricultural-based products, for the discovery of new antioxidants. For both natural and synthetic antioxidants, reports of biological features such as anti-allergic, anti-mutation, anti-cancer and anti-aging activity have been reported. The most significant natural antioxidants come from regularly eating fruits and vegetables, although other plant materials and agricultural waste are also major sources of antioxidants.

Keywords: antioxidants, vegetables, plants, fruits, herbs

1. Introduction

Reactive nitrogen and oxygen species (RNS and ROS), including nitric oxide radicals, hydroxyl, and superoxide, can harm DNA in biological systems and cause oxidation of proteins and lipids in cells [1]. Free radicals can typically be scavenged by the body's antioxidant system, which helps to keep oxidation and anti-oxidation in the right proportion. But when the body produces too many ROS and RNS due to exposure to toxins from the environment, radiation, alcohol, or cigarette smoke, the body's natural balance of oxidation and anti-oxidation is disturbed, which can result in a number of chronic and degenerative illnesses [2, 3]. Intake of exogenous antioxidants could be increased to minimize the consequences of oxidative stress by scavenging free radicals, quenching singlet oxygen, and acting as reducing agents. These antioxidants also operate as scavengers of free radicals and quenchers of singlet oxygen [4].

Plants are the ultimate sources of natural antioxidants that are consumed or used medicinally. Antioxidants are obtained from vegetables, mushrooms, fruits, spices, cereals, flowers and herbs [5]. Additionally, antioxidants can also be obtained from businesses that deal with agricultural byproducts [6]. Flavonoids, lignans, stilbenes, anthocyanins and several other polyphenolic compounds, vitamins and carotenoids like carotenes and xanthophylls are obtained and derived from plants [7]. The natural antioxidants possess many pharmacological properties such as anti-cancer, anti-viral, anti-inflammatory and anti-bacterial [2, 7, 8].

Antioxidants can be divided into two major classes Antioxidants can be divided into two primary categories: Natural antioxidants and synthetic antioxidants. Free radical damage predominantly affects the cellular level of the body, and antioxidants protect against it there. As a result, enzymatic and nonenzymatic types of these antioxidants are also possible. The three primary enzyme-based antioxidants are glutathione peroxidase, catalase, and superoxide dismutase. The serum reflects the body's overall capacity for antioxidants, which is influenced by additional enzymes in the body [9]. The non-enzymatic class of antioxidants can be classified in several classes. Vitamins like vitamins C, E, A, peptides, enzyme co-factors (Q10) and a few minerals (selenium and zinc) usually serve as building blocks of the natural antioxidants [10]. The classification of natural antioxidants has been shown in **Figure 1**.

Some of the unfavourable or detrimental effects of synthetic antioxidant use have been uncovered by recent toxicological study. Researchers are now focusing their efforts on locating natural sources with sufficient antioxidant activity as a result of these studies. Furthermore, substantial concerns are raised regarding the cost and availability of these natural antioxidants. It is possible to define the numerous subcategories of natural antioxidants. Antioxidants, however, fall into two main categories: those that are present in frequently utilized or regular food items (such as

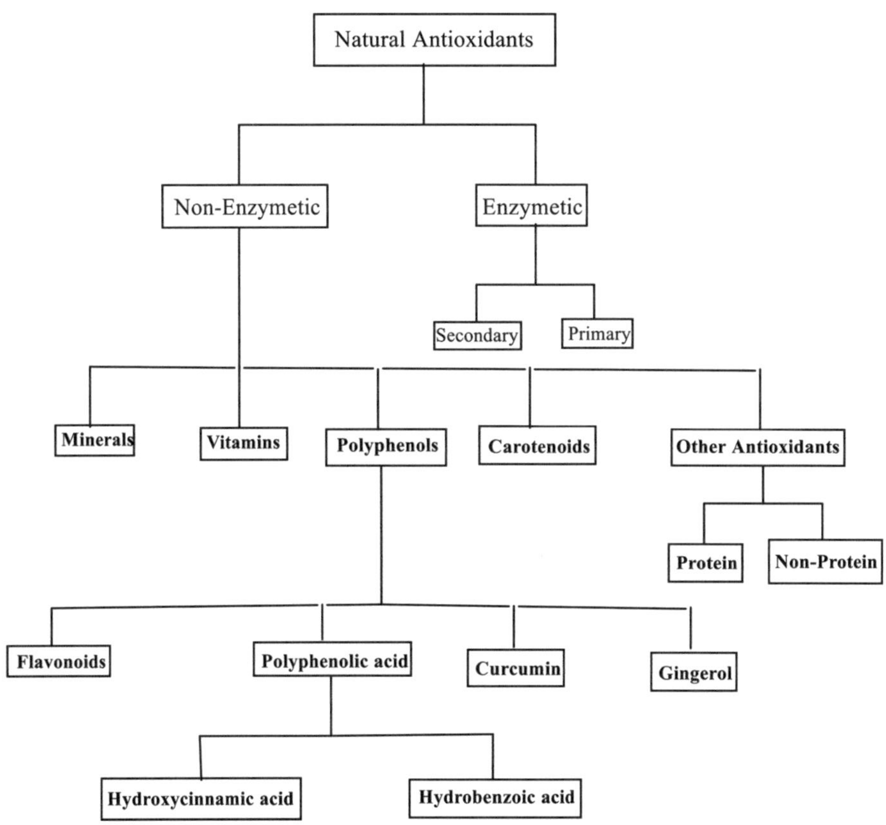

Figure 1.
Classifications of Antioxidants obtained from natural sources [11].

beans, fruits, vegetables and cereals) and those that are present in plants or herbs that have some antioxidant potential but are not frequently consumed (such as medicinal plants and wild herbs) [12, 13]. Until now, researchers from all around the world have concentrated on discovering inexpensive, more natural sources of antioxidants. These findings will be utilised by the food, pharmaceutical, and beauty industries as an improved option for produced supplements. Despite the fact that synthetic supplements have not yet been proven to have substantial detrimental effects, supplementing should generally emphasise getting back to nature. In the next ten years, natural products-based items are going to get more and more important, and research in the area of naturally occurring antioxidants will be more and more emphasized and pursued [14, 15].

The current chapter's major goal is to provide an overview and summary of the natural sources having antioxidant potential.

2. Oxidative stress

Oxygen, which is essential for sustaining cell viability and metabolism and is associated with aerobic living conditions, is also dangerous due to its paramagnetism. The paramagnetic nature of oxygen results in creation of very reactive intermediates chemicals. These chemicals are referred to as "reactive oxygen species" (ROS). These ROS are free radicals (FRs) in nature. Maximum natural stability refers to the coupling of the electrons in the corresponding molecular orbitals of stable neutral compounds. Because of this, when an orbital is having un-paired electrons, extremely reactive chemical entities are formed. The chemicals have the natural inclination to obtain an electron form neighboring molecules to account for their electron deficiency [16]. The main free radical is the triplet state of oxygen which possess two unpaired electrons. The rate of the reaction of triplet state is usually slow, however, due to metabolic transformation into one or more very reactive species it can dangerously interact with biological systems. This kind of metabolic activation is generally preferred in biological systems due to conversion of O_2 to H_2O during the phenomena of electron transport chain. During electron transport chain ROS and FRs are produced due to transfer of electrons [17]. External stimuli like sun radiation can induce free radicals to develop in biological systems since UV light exists. UV rays causes the homolytic bonds between molecules to disintegrate. As a disease worsens, FR can also manifest. For example, during heart attack, many FRs are produced when supply of glucose and oxygen are interrupted to cardiac muscles. Another outside factor which enhances the rate of formation of FR is known as chemical intoxication. The organism promotes FR release because it needs to convert toxic substances into less toxic ones. The toxicity of numerous numbers of drugs is due to their inclination to produce FRs and interference with processes for the formation of FR. Similarly, food contamination with herbicides and chemicals may also act a source of FR formation [18].

Inflammations are induced due to endogenous components which ultimately results in the promotion of FR. The FR occur in immune system's cleaning cells and are responsible for removing dangerous microbes. Tissue damage comes from excessive FR during this phase. Superoxide ions ($O_2^{\bullet-}$) are generated NADPH oxidase in the phagocytic cells. $O_2^{\bullet-}$ which is thought of as the main ROS. It may produce secondary ROS after interacting with other molecules through enzymatic processes. The protonation of $O_2^{\bullet-}$ may result in the formation of $H_2O^{\bullet}$ and H_2O_2. When water is exposed to UV light, molecular oxygen is exposed to cellular free radicals produced inside living cells, and water is photolyzed, $O_2^{\bullet -}$ is produced. as demonstrated by hemoproteins, NAD•, FpH•, semiquinone radicals, pyridinium cation radicals, etc. The phagocytic cells, during the process of respiration and oxygen ingestion, also produce $O_2^{\bullet -}$. The superoxide radical does not undergo rapid reaction with nucleic acids, polypeptides or carbohydrates [18].

Cells create •NO as a defense mechanism when nitric oxide synthase interacts with intracellular arginine. Lipid peroxidation in lipoproteins results from the creation of ONOO, which is created when O_2 and •NO mix. Autoimmune disorders which clearly demonstrate this phenomenon include vitiligo, Graves' disease, biliary cirrhosis, systemic lupus erythematosus, Hashimoto's disease, Rheumatoid arthritis, type 1 diabetes, inflammatory bowel syndrome, celiac disease, scleroderma, multiple sclerosis and psoriasis.

These chemical species are necessary for many of the chemical reactions that occur throughout metabolic activities, hence FR is required at all times. For instance, FR plays a role in the polymerization of glucose and amino acids to make glycogen and proteins.FR catalytically activates a variety of intermediary metabolic enzymes such as lipoxygenase, cyclooxygenase, and monoamine oxidase etc [18]. Antioxidant enzymes often successfully regulate these free radicals. Irreversible structural changes in essential macromolecules such as lipids, DNA and proteins also act as sources of ROS. Malonaldehyde and hydroperoxide, two substances that cause oxidative damage, are produced by these mechanisms. Neutral species like N_2O_3 and ONOOH as well as $NO^{\bullet}$, $ONOOCO^{2-}$ and $NO_2^{\bullet}$ $ONOO^-$ are all RNS. RNS are produced in small amounts during cellular growth, production of cellular energy, signaling, blood pressure modulation, relaxation of muscles, aggregation of platelets, neurotransmission and phagocytosis etc. [19, 20].

3. Importance of antioxidants

Both biochemical and biological defense systems have been developed by biological systems. in oxygenated circumstances. A microvascular system regulates the tissues' oxygen levels as far as physiological level is concerned, while at biochemical level, an enzymatic or non-enzymatic antioxidant defense system operates for the repair of the molecules.

3.1 Primary enzymatic type systems

Aerobic species have produced antioxidant enzymes such catalase, glutathione peroxidase, superoxide dismutase, and DT-diaphorase. SOD is responsible for the dismutation reaction that converts oxygen into hydrogen oxide, which is then converted back into oxygen and water in subsequent reactions that are catalyzed by catalase or GPx. The detoxification of a cell is performed by SOD. Because SOD requires a metal as a co-factor to detoxify a cell. Depending on the kind of metal ion required by SOD as a co-factor, different forms of the enzyme exist [21, 22]. CAT completes the detoxification process that SOD began by catalyzing the reduction of H_2O_2. Iron or manganese act as co-factor during the reduction of H_2O_2 and results in the production of water and oxygen molecules [23]. CAT is so efficient that a very large number of H_2O_2 molecules can be destroyed in a single second. Its main function is to eliminate the H_2O_2 created when fatty acids are oxidised. Peroxisomes are where CAT is mainly found. A vital intracellular enzyme called GPx breaks down lipid peroxides to corresponding alcohols and H_2O_2 in water; this predominantly occurs in the mitochondria and occasionally in the cytoplasm [24]. Selenium is necessary for the function of GPx. At least eight GPx enzymes (GPx1 to GPx8) are found in human beings [25].

Almost all cells contain GPx1, the most common selenoperoxidase among glutathione peroxidases. The enzyme is necessary to stop the oxidation of lipids, resulting in protecting the cells from oxidative stress [26]. When GPx activity is low, the functioning proteins and fatty acids in the cell membrane experience oxidative damage. The production and prevention of GPx, particularly GPx1, have been associated to a variety of diseases [27]. DT-diaphorase participates in the reduction of compounds with a quinone structure and catalyzes the conversion of quinone to quinol. Cells manufacture these enzymes under the direction of DNA [28].

3.2 Non-enzymatic type system

The antioxidants which capture FR, they constitute non-enzymatic type systems. They catch FR in order to prevent the radical initiation reaction. However, they are not as much reactive as that of the original FR. Antioxidants become free radicals in the process of neutralizing or trapping the radicals by donating electrons. The FR from antioxidants can be immediately and effectively neutralized by other antioxidants of this family. The cells utilize antioxidants and FR like α-tocopherol (vit E), ferritin, selenium, GSH, co-enzyme Q, zinc, bilirubin, cysteine, ascorbic acid (vit C), ubiquinone, melatonin and flavonoids. In some foods, the extracted flavonoids work with the ROS directly to form non-reactive or less reactive complexes, but in other foods, the flavonoids take part in the specific enzymatic catalysis as co-substrates [29].

4. Fruits and vegetables as sources of natural anti-oxidants

A class of chemicals with low and high molecular weights known as polyphenols is found in fruits and vegetables and has the ability to inhibit lipid oxidation. In addition to being functional derivatives like esters and methyl esters, most of them are the conjugates either mono or polysaccharides with one or more phenol linkages. This important class of natural antioxidants can be found in fruits like grapes, green and red teas, and other teas, especially those that are caffeine-free [30].

Yet, the polyphenols in teas are more important than those in fruits because of higher blood bioavailability. 15% to 20% of consumed polyphenols are absorbed by the human circulation (**Table 1**). This absorption is enhanced when no sugar molecules are present. Teas absorb polyphenols at a rate that is higher than that of fruits since fruits have a high sugar content [41, 42].

Flavonoids are also very rich sources of antioxidants. Food items like peaches, potatoes, berries, wheat and almonds are the richest sources of flavonoids [43, 44]. A subgroup of flavonoids called anthocyanin is found in berries and red wine. It is a potent antioxidant and has a lesser bioavailability than other flavonoids. Polyphenols can display their antioxidant properties and prevent the growth of plaque through low-density lipoprotein (LDL) oxidation [45]. Furthermore, it has been revealed that particular types of polyphenols can stop some important enzymes from oxidising, keeping their proper function. The family of carotenoids comes in second place to polyphenols as a significant class of phytochemical antioxidants present in fruits and vegetables. Veggies including potatoes, carrots, papayas, and apricots are the main sources of them [11].

S. no.	Common name	Antioxidant present	ORAC value (mmolTE/g)	Reference
1	Plum	Flavonoids, Phenolic Acids, Proanthocyanidins, Hydroxychalcones, isoprenoid glycosides.	94.8	[31]
2	Pomegranate	Polyphenols and Vitamin C and	1250	
3	Guava	Carotenoids, Lycophene, Vit C, anthocynin		[32]
4	Pears	Vit C, betalains, tauline, total carotenoids, flavonoids and total phenolics	140	[33]
5	Beet root	Carotenoids, Flavonoids Vit C and Vit E,	4100 (dry extract)	[34]
6	Apple	Flavonoids, Proanthocyanidins, Phenolic acids, Isoprenoid-glycosides, Flavanols, Hydroxychalcones etc.	17.0	[31]
7	Papaya	β-sitosterol, Quercetin	300	[35]
8	Pea	Carotenoids, Flavonoids, Vit C, Vit E, Thio compounds	0.019	[34]
9	Spinach	Carotenoids, Flavonoids, Vit C α-tocopherol,	0.152	[34]
10	Carrot	Carotenoids, Flavonoids, Vit C and E, Thio compounds	0.060	[34]
11	White onion	Carotenoids, Flavonoids, Vit C and E, Thio compounds	0.085	[34]
12	White cabbage	Carotenoids, Flavonoids, Vit C and E, Thio compounds	0.061	[34]
14	Tomato	Carotenoids, Flavonoids, Vit C and E, Thio compounds	0.067	[34]
15	Cauliflower	Carotenoids, Flavonoids, Vit C and E, Thio compounds	0.102	[34]
16	Grape juice	Anthocyanins	255.6–460	[36]
17	Coriander	Monoterpenoid,		[37]
18	Ginger	Phenols	1870.1	[38]
19	Nigella sativa	4-terpineol Thymoquinone, Carvacrol	1.0	[39]
20	Walnut	Phenolics	1320.6	[40]

Table 1.
Antioxidants obtained from various fruits, vegetables and natural sources.

The water-soluble antioxidant vitamin C, commonly referred to as ascorbic acid, is typically found in citrus fruits and vegetables including oranges, lemons, and tomatoes. It is a vitamin that is obtained from fruits and vegetables and serves as an antioxidant. It is advisable to consume vitamin C-containing fruits and vegetables in tiny, spaced-out portions rather than all at once because it demonstrates poor absorption when consumed in larger doses [46].

Vitamin E is reported to possess excellent antioxidant properties. It is a naturally occurring, nonpolar, fat-soluble vitamin that is present in lipid-rich foods including olives, almonds, and sunflower seeds. Vitamin E has a higher bioavailability than vitamin C because of its solubility in fat and potential for improvement when ingested with fatty meals [47].

5. Fruits and vegetable wastes as source of natural antioxidants

Producing, managing industrially, processing, preserving, and distributing fruits and vegetables all result in the generation of waste products. Over the past few decades, researchers have been experimenting with techniques to reuse these wastes in order to obtain medicinal benefits [48]. Vegetable and fruit wastes consists of the peelings, trimmings, seeds, shells, stems and pulp leftovers from juice extraction and starch or sugar processing. Between 25 to 30% of it is trash. These discarded scalps apparently contain more phenols and ascorbic acids than their pulp [49] likewise preferable in their unripe form than that of ripeness. Frequently, fruit peels have 2–27 times as much antioxidants as fruit pulp [50].

Only 25% as much phenolic compounds are present in banana pulp (232 mg/100 g) as there are in banana peels [51]. Cucumis sativus peel has been discovered to be an excellent source of flavonoids, which are regarded to have antioxidant potential. These wastes include a variety of bioactive components that can be gathered and used to create both culinary preparations and pharmaceutical preparations. The bioactive phytochemicals carotenes, tocopherols, terpenes, sterols, and polyphenols, all of which have strong antioxidant properties, are abundant in the tomato wastes. These natural antioxidants, which were derived from food waste, can be used to enhance food or make useful foods [52]. The mango peel is rich in dietary fibre, vitamin C, phenolic compounds, and carotenoids, among other antioxidants. These compounds have been proven to affect a variety of degenerative conditions, including cancer, Parkinson's disease, cataracts, and Alzheimer's disease [53]. Among the waste materials generated by the wine industry are degradable solids. These substances contain high levels of antioxidants, which have been shown to slow down a number of degenerative processes and have other advantageous impacts on health. Polyphenols make up around 6% of the waste produced by the coffee industry, whereas tannins make up about 4% [54, 55].

6. Important characteristics of antioxidants

A chemical or antioxidant system's main job is to stabilize the generated radical in order to prevent or detect a chain of oxidative propagation, which minimizes the body's exposure to oxidative damage [56]. Gordon categorized antioxidants according to that characteristic. Primary antioxidants (which halt a chain reaction and scavenge free radicals) and secondary, or preventive, antioxidants fall into two fundamental groups. A few examples of secondary antioxidant mechanisms include the deactivation of metals, stopping the formation of unfavorable volatiles, inhibiting lipid hydroperoxides, regenerating primary antioxidants, and removing singlet oxygen. Antioxidants are therefore "those substances that, in low quantities, act by preventing or greatly retarding the oxidation of easily oxidizable materials such as fats" [57].

7. Conclusion

Over the previous ten years, there has been an increase in interest in studying natural ingredients for usage in food and food products. Because natural sources are more useful and secure to use as dietary supplements than manufactured ones, researchers from all over the world are concentrating on them. Despite the fact that there have never been any cases of harm associated with the use of synthetic antioxidants, there is still a considerable desire from consumers for products that are close to nature meal artificial antioxidants and preservatives may also result in peroxidation of lipids and thus deterioration of quality and flavor of the food items.

Since ancient times, natural herbs, spices, and plant-based ingredients have been employed in traditional food preparation as flavorings, fragrances, and preservatives. A general overview of the possible benefits of several natural sources with respectable antioxidant capacity is what this chapter aims to deliver. The literature research gathered here will be useful to establish the relevance, active components, antioxidant potential and availability of various sources. This work will help the people to prioritize their daily requirements of natural antioxidants keeping in mind the cost-effectiveness and availability of natural sources because 70–80% of the world's population cannot afford current supplements and pharmaceuticals.

Author details

Muhammad Alamzeb[1*], Behramand Khan[2], Ihsan Ullah[3], Muhammad Omer[3] and Adnan[3]

1 Department of Chemistry, University of Kotli, Kotli, Azad Jammu & Kashmir, Pakistan

2 Department of Chemistry, Islamia College University, Peshawar, KPK, Pakistan

3 Institute of Chemical Sciences, University of Swat, Swat, Pakistan

*Address all correspondence to: zebchemist@gmail.com

References

[1] Li S, Tan HY, Wang N, Zhang ZJ, Lao L, Wong CW, et al. The role of oxidative stress and antioxidants in liver diseases. International Journal of Molecular Sciences. 2015;**16**:26087-26124 Epub 2015/11/06

[2] Wang F, Li Y, Zhang YJ, Zhou Y, Li S, Li HB. Natural products for the prevention and treatment of hangover and alcohol use disorder. Molecules (Basel, Switzerland). 2016;**21**:64 Epub 2016/01/12

[3] Zhou Y, Zheng J, Li S, Zhou T, Zhang P, Li HB. Alcoholic beverage consumption and chronic diseases. International Journal of Environmental Research and Public Health. 2016;**13**:522. Epub 2016/05/28

[4] Baiano A, Del Nobile MA. Antioxidant compounds from vegetable matrices: Biosynthesis, occurrence, and extraction systems. Critical Reviews in Food Science and Nutrition. 2016;**56**:2053-2068 Epub 2015/03/10

[5] Zhang JJ, Li Y, Zhou T, Xu DP, Zhang P, Li S, et al. Bioactivities and health benefits of mushrooms mainly from China. Molecules (Basel, Switzerland). 2016;**21**:938. Epub 2016/07/23

[6] Deng GF, Shen C, Xu XR, Kuang RD, Guo YJ, Zeng LS, et al. Potential of fruit wastes as natural resources of bioactive compounds. International Journal of Molecular Sciences. 2012;**13**:8308-8323 Epub 2012/09/04

[7] Manach C, Scalbert A, Morand C, Rémésy C, Jiménez L. Polyphenols: Food sources and bioavailability. The American Journal of Clinical Nutrition. 2004;**79**:727-747 Epub 2004/04/29

[8] Zheng J, Zhou Y, Li Y, Xu DP, Li S, Li HB. Spices for prevention and treatment of cancers. Nutrients. 2016;**8**:495. Epub 2016/08/17

[9] Anwar H, Rahman ZU, Javed I, Muhammad F. Effect of protein, probiotic, and symbiotic supplementation on serum biological health markers of molted layers. Poultry Science. 2012;**91**:2606-2613 Epub 2012/09/20

[10] Carocho M, Ferreira IC. A review on antioxidants, prooxidants and related controversy: Natural and synthetic compounds, screening and analysis methodologies and future perspectives. Food and Chemical Toxicology: An International Journal Published for the British Industrial Biological Research Association. 2013;**51**:15-25 Epub 2012/09/29

[11] Haseeb A, Ghulam H, Imtiaz M. Antioxidants from natural sources. In: Emad S, Ghada Mostafa A, editors. Antioxidants in Foods and Its Applications. Rijeka: IntechOpen; 2018. p. Ch. 1

[12] Asif M. Chemistry and antioxidant activity of plants containing some phenolic compounds. Chemistry International. 2015;**1**:35-52

[13] Ramalakshmi K, Rahath Kubra I, Jagan Mohan Rao L. Antioxidant potential of low-grade coffee beans. Food Research International. 2008;**41**:96-103

[14] Shebis Y, Iluz D, Kinel-Tahan Y, Dubinsky Z, Yehoshua Y. Natural antioxidants: Function and sources. Food and Nutrition Sciences. 2013;**04**:643-649

[15] Uddin G, Khan A, Alamzeb M, Ali S, Rashid MU, Alam M, et al. Biological

screening of ethyl acetate extract of Hedera nepalensis stem. African Journal of Pharmacy and Pharmacology. 2012;**6**:2934-2937

[16] Davies KJ. Oxidative stress: The paradox of aerobic life. Biochemical Society Symposium. 1995;**61**:1-31 Epub 1995/01/01

[17] Valko M, Leibfritz D, Moncol J, Cronin MT, Mazur M, Telser J. Free radicals and antioxidants in normal physiological functions and human disease. The International Journal of Biochemistry & Cell Biology. 2007;**39**:44-84 Epub 2006/09/19

[18] Fridovich I. The biology of oxygen radicals. Science. 1978;**201**:875-880 Epub 1978/09/08

[19] Limón-Pacheco J, Gonsebatt ME. The role of antioxidants and antioxidant-related enzymes in protective responses to environmentally induced oxidative stress. Mutation Research. 2009;**674**:137-147 Epub 2008/10/29

[20] Nakamura H, Nakamura K, Yodoi J. Redox regulation of cellular activation. Annual Review of Immunology. 1997;**15**:351-369 Epub 1997/01/01

[21] Dringen R, Pawlowski PG, Hirrlinger J. Peroxide detoxification by brain cells. Journal of Neuroscience Research. 2005;**79**:157-165 Epub 2004/12/02

[22] Fridovich I. Superoxide radical and superoxide dismutases. Annual Review of Biochemistry. 1995;**64**:97-112 Epub 1995/01/01

[23] Chelikani P, Fita I, Loewen PC. Diversity of structures and properties among catalases. Cellular and Molecular Life Sciences: CMLS. 2004;**61**:192-208 Epub 2004/01/28

[24] Góth L, Rass P, Páy A. Catalase enzyme mutations and their association with diseases. Molecular Diagnosis: A Journal Devoted to the Understanding of Human Disease through the Clinical Application of Molecular Biology. 2004;**8**:141-149 Epub 2005/03/18

[25] Ighodaro OM, Akinloye OA. First line defence antioxidants-superoxide dismutase (SOD), catalase (CAT) and glutathione peroxidase (GPX): Their fundamental role in the entire antioxidant defence grid. Alexandria Journal of Medicine. 2018;**54**:287-293

[26] Gill SS, Tuteja N. Reactive oxygen species and antioxidant machinery in abiotic stress tolerance in crop plants. Plant Physiology and Biochemistry: PPB. 2010;**48**:909-930 Epub 2010/09/28

[27] Rayman MP. Selenium in cancer prevention: A review of the evidence and mechanism of action. The Proceedings of the Nutrition Society. 2005;**64**:527-542 Epub 2005/11/30

[28] Chen S, Wu K, Knox R. Structure-function studies of DT-diaphorase (NQO1) and NRH: Quinone oxidoreductase (NQO2). Free Radical Biology & Medicine. 2000;**29**:276-284 Epub 2000/10/18

[29] Foti MC. Antioxidant properties of phenols. The Journal of Pharmacy and Pharmacology. 2010;**59**:1673-1685

[30] Carr AC, Zhu BZ, Frei B. Potential antiatherogenic mechanisms of ascorbate (vitamin C) and alpha-tocopherol (vitamin E). Circulation Research. 2000;**87**:349-354 Epub 2000/09/02

[31] Navarro M, Moreira I, Arnaez E, Quesada S, Azofeifa G, Vargas F, et al. Polyphenolic characterization and antioxidant activity of *Malus domestica* and *Prunus domestica* cultivars from

Costa Rica. Foods (Basel, Switzerland). 2018;7:15. Epub 2018/02/02

[32] Nantitanon W, Yotsawimonwat S, Okonogi S. Factors influencing antioxidant activities and total phenolic content of guava leaf extract. LWT – Food Science and Technology. 2010;**43**:1095-1103

[33] Fernández-López JA, Almela L, Obón JM, Castellar R. Determination of antioxidant constituents in cactus pear fruits. Plant Foods for Human Nutrition (Dordrecht, Netherlands). 2010;**65**:253-259 Epub 2010/09/03

[34] Ou B, Huang D, Hampsch-Woodill M, Flanagan JA, Deemer EK. Analysis of antioxidant activities of common vegetables employing oxygen radical absorbance capacity (ORAC) and ferric reducing antioxidant power (FRAP) assays: A comparative study. Journal of Agricultural and Food Chemistry. 2002;**50**:3122-3128 Epub 2002/05/16

[35] Oloyede O, Franco J, Roos D, Rocha JB, Linde Athayde M, Boligon A. Antioxidant properties of ethyl acetate fraction of unripe pulp of Carica papaya in mice. Journal of Microbiology, Biotechnology and Food Sciences. 2011;**1**:409-425

[36] Kim MJ, Jun JG, Park SY, Choi MJ, Park E, Kim JI, et al. Antioxidant activities of fresh grape juices prepared using various household processing methods. Food Science and Biotechnology. 2017;**26**:861-869 Epub 2017/07/12

[37] Wangensteen H, Samuelsen AB, Malterud KE. Antioxidant activity in extracts from coriander. Food Chemistry. 2004;**88**:293-297

[38] Stoilova I, Krastanov A, Stoyanova A, Denev P, Gargova S. Antioxidant activity of a ginger extract (*Zingiber officinale*). Food Chemistry. 2007;**102**:764-770

[39] Burits M, Bucar F. Antioxidant activity of Nigella sativa essential oil. Phytotherapy research: PTR. 2000;**14**:323-328 Epub 2000/08/05

[40] Oliveira I, Sousa A, Ferreira IC, Bento A, Estevinho L, Pereira JA. Total phenols, antioxidant potential and antimicrobial activity of walnut (*Juglans regia* L.) green husks. Food and Chemical Toxicology: An International Journal Published for the British Industrial Biological Research Association. 2008;**46**:2326-2331 Epub 2008/05/02

[41] Parr AJ, Bolwell GP. Phenols in the plant and in man. The potential for possible nutritional enhancement of the diet by modifying the phenols content or profile. Journal of the Science of Food and Agriculture. 2000;**80**:985-1012

[42] Quintavalla S, Vicini L. Antimicrobial food packaging in meat industry. Meat Science. 2002;**62**:373-380 Epub 2002/11/01

[43] Tain Y-L, Hsu C-N. Oxidative stress-induced hypertension of developmental origins. Preventive Aspects of Antioxidant Therapy. 2022;**11**:511

[44] Rashid MU, Alamzeb M, Ali S, Khan A, Semaan D, Igoli J, et al. A new ceramide along with eight known compounds from the roots of artemisia incisa Pamp. Records of Natural Products. 2015;**9**:3-294

[45] Farbstein D, Kozak-Blickstein A, Levy AP. Antioxidant vitamins and their use in preventing cardiovascular disease. Molecules. 2010;**15**:8098-8110

[46] McGhie TK, Walton MC. The bioavailability and absorption of anthocyanins: Towards a better

understanding. Molecular Nutrition & Food Research. 2007;**51**:702-713 Epub 2007/05/30

[47] Daniel JW. Metabolic aspects of antioxidants and preservatives. Xenobiotica: The Fate of Foreign Compounds in Biological Systems. 1986;**16**:1073-1078 Epub 1986/10/01

[48] Mirabella N, Castellani V, Sala S. Current options for the valorization of food manufacturing waste: A review. Journal of Cleaner Production. 2014;**65**:28-41

[49] Goulas V, Manganaris GA. Exploring the phytochemical content and the antioxidant potential of Citrus fruits grown in Cyprus. Food Chemistry. 2012;**131**:39-47

[50] Fatemeh S, Saifullah R, Abbas FM, Azhar ME. Total phenolics, flavonoids and antioxidant activity of banana pulp and peel flours: Influence of variety and stage of ripeness. International Food Research Journal. 2012;**19**:1041-1046

[51] Someya S, Yoshiki Y, Okubo K. Antioxidant compounds from bananas (Musa Cavendish). Food Chemistry. 2002;**79**:351-354

[52] Baiano A. Recovery of biomolecules from food wastes: A review. Molecules (Basel, Switzerland). 2014;**19**:14821-14842 Epub 2014/09/19

[53] Ayala-Zavala JF, Rosas-Domínguez C, Vega-Vega V, González-Aguilar GA. Antioxidant enrichment and antimicrobial protection of fresh-cut fruits using their own byproducts: Looking for integral exploitation. Journal of Food Science. 2010;**75**:R175-R181 Epub 2011/05/04

[54] Pujol D, Liu C, Gominho J, Olivella MÀ, Fiol N, Villaescusa I, et al. The chemical composition of exhausted coffee waste. Industrial Crops and Products. 2013;**50**:423-429

[55] Teixeira A, Baenas N, Dominguez-Perles R, Barros A, Rosa E, Moreno DA, et al. Natural bioactive compounds from winery by-products as health promoters: A review. International Journal of Molecular Sciences. 2014;**15**:15638-15678 Epub 2014/09/06

[56] Namiki M. Antioxidants/antimutagens in food. Critical Reviews in Food Science and Nutrition. 1990;**29**:273-300 Epub 1990/01/01

[57] Norma Francenia S-S, Raúl S-C, Claudia V-C, Beatriz H-C. Antioxidant compounds and their antioxidant mechanism. In: Emad S, editor. Antioxidants. Rijeka: IntechOpen; 2019. p. Ch. 2

Chapter 2

Antioxidant Fortification of Eggs through Nutrition of Laying Hens Administered Herbs/Medicinal Plants

Habeeb O. Yusuf and Ruth T.S. Ofongo

Abstract

The sole aim of raising pullet hens in the poultry industry is to produce eggs for human consumption in a large scale when they commence laying. Eggs are important dietary components to humans both adult and children and is classified as complete protein. However, certain quality of eggs produced by laying hens is further influenced by the diet consumed which in turn is determined by the quality of the feed ingredients making up the diet. Antibiotic residue in eggs and antimicrobial resistance are few concerns to consumers of poultry products. The current era of limiting antimicrobial utilization for livestock production has increased research into medicinal plants and herbs as suitable alternative. Antioxidant and anti-inflammatory activities reported in literature indicate the invaluable benefits of these plants both for humans and livestock. This book chapter attempts to present the 2,2-diphenyl-1-picrylhydrazyl (DPPH) antioxidant scavenging activity of eggs from laying hens fed medicinal plants – *Vernonia amygdalina* and *Ocimum gratissimum* as component of feed or administered orally as an aqueous extract. The DPPH antioxidant scavenging activity was present in eggs sampled but was better ($p < 0.05$) in eggs of laying hens administered aqueous *O. gratissimum* extract.

Keywords: antioxidants, antioxidant scavenging activity, medicinal plants, eggs, nutrition, laying hens

1. Introduction

Healthy food and healthy diet are a major concern to health-conscious individuals; since health lost is wealth lost. The concern of such individuals cannot be over emphasized. The genetic modification of plants and animals is an evolving development which is yet to be generally accepted by consumers of agricultural products. The current shift towards natural sources of antioxidants reported [1] can be attributed to the unfavorable effects of synthetic sources of antioxidants resulting from prolonged usage.

Reactive oxygen species (ROS) otherwise referred to as free radicals are by-products unavoidably produced in biological systems in the course of normal cellular energy function [1, 2]. They are also produced through exogenous sources such as environmental pollutants, radiation, pesticides [1, 3, 4] etc. just to mention a few. Free radicals are important for some biological functions physiologically by acting as cell signaling molecules which function against cellular responses as reported in literature [5, 6]. Oxidative stress elicits adverse effects on lipids, proteins and nucleic acids thereby resulting into a number of degenerative conditions [7–10]. Free radical scavengers or antioxidants delay/inhibit damage made to cells by converting ROS to non-reactive radical species [1]. Common antioxidants include vitamin E (α-tocopherol), vitamin C (ascorbic acid), and β-carotene [11, 12].

Generally speaking; consumption of antioxidant rich food by health-conscious individuals has more to do with preventing damage in tissues and privation of cellular functions resulting from free radicals' intermediates generated by cells during normal metabolism [2] or avoiding degenerative diseases. Antioxidant activities of plants are regarded as safe. This activity is attributed mainly to phenolic compounds [5, 13] which act as hydrogen donors, reducing agents, oxygen quenchers besides their metal chelating potential [3]. These properties play a significant role in neutralizing free radicals [3].

1.1 Eggs as quality food

Eggs are considered high quality protein which is readily digestible, however, with recent developments in human health, animal nutrition, it is gradually developing into a functional food for health and better wellbeing. The amino acid profile of eggs is adequate to meet both essential and non-essential amino acid needs of both adults and young children making it a complete protein. Eggs are also considered cheap, extremely nutritious, palatable and readily accessible across the globe [14, 15].

Egg yolk, egg white/albumen and egg shell with membrane respectively; accounts for approximately 27.5%, 63% and 9.5% of the whole egg [16]. The various constituent of egg nutrients are proteins and lipids 12% respectively; while the edible portion is made up of 74% water [17]. Eggs contain less than 1% carbohydrate along with vitamins and minerals [17–20]. Both high density and low-density lipoproteins in addition to livetin's is located in egg yolk. Ovalbumin, ovotransferrin, ovomucoid, ovomucin just to mention a few are protein fractions present in egg white [21]. The quality of egg protein is such that it is used as a golden standard for measuring the quality of other food proteins [18].

Greater part of lipids in egg yolk are present as triglycerides—almost 65%, however, carotenoids constitute less than 1% while phospholipids and cholesterol are 30% and 4% respectively [22]. Lipids from egg yolk have been used to supply long-chain polyunsaturated fatty acids, docosahexaenoic acid (DHA) and phospholipids incorporated into infant formula [23, 24]. They are also regarded as an excellent source of micronutrients—vitamins and minerals. According to published work [18] eggs contain approximately 16% required daily intake (RDI) 0f phosphorus. The RDI of eggs was reported for selenium (29%); iron (9%) and zinc (9%) respectively. In addition, eggs also provide 10% of the RDI for fat soluble vitamins, vitamin B2, B12, biotin and pantothenic acid [18]. As it is in spite of the nutritive quality of eggs, it appears the need to limit its cholesterol concentration, especially high-density lipoproteins are focused on meeting healthy outcomes for consumers of eggs. This has led to manipulating poultry diets to influence the nutritional constituents in eggs [25]. Several studies have been carried out in this regard and reported in literature [14, 15, 26–35]. Besides the fatty acid constituents of eggs, some minerals like selenium and iodine have been enriched in eggs via enrichment of feed [35, 36] and feed formulation [37].

Intake of antioxidants through diet is known to be important in reducing oxidative damage in cells and improving human health. Although eggs are known for their exceptional, nutritional quality, they are not generally considered as antioxidant foods [21].

1.2 Antioxidant content of eggs

Many of the compounds present in eggs—vitamin E, and A, selenium, phospholipids and carotenoids—show evidence of antioxidant properties; eggs are not normally considered as antioxidant foods [21]. Even though certain polyunsaturated fatty acids (PUFA) exhibit antioxidants properties; eggs are mostly consumed for their protein constituents and minerals which are components of egg.

Furthermore, it is important to note that with recent developments in enriching eggs via feed thereby making eggs as functional foods then consumption of eggs might just be a source of antioxidants for healthy living.

A recent report [38] testing the antioxidant property of a product containing pectic oligosaccharides, with prebiotic, chlorogenic as well as antioxidant effect on the possibility of enhancing egg laying performance and egg quality of laying hens. The results showed that the tested product enhances laying hens egg quality and performance, particularly by means of its antioxidant properties that play a part to sustain oxido-redox balance, consequently reducing the negative effects triggered by oxidation like degradation of egg quality [38]. By this, may the various nutrient components of egg which make up the quality of eggs can be improved upon by feed manipulation. The antioxidant properties of eggs are exhibited by the different components' present either in the egg yolk or egg white.

Ovalbumin, ovatransferrin, ovomucin, lysozyme, cystatin, in egg white reportedly have antioxidant properties. Phosvitin, phospholipids, carotenoids and vitamin E which are components of egg yolk.

Ovalbumin in egg white has free thiol groups that regulate redox status and bind metal ions thereby exerting antioxidant properties. In conjugation with saccharides increased antioxidant activity takes place [39, 40]. Ovomucin in egg white inhibit hydrogen peroxide H_2O_2-induced oxidative stress inhuman embryonic kidney [41]. Furthermore, lysozyme in egg white Suppress reactive oxygen species (ROS) and oxidative stress genes [42]. In the case of nutrient components with antioxidant properties in egg yolk; phospholipids are 10% of egg yolk dry matter. Hydrolyl amines in the side chains of egg yolk phospholipids play a role in radical scavenging with antioxidant properties [43]. The unsaturated backbone and aromatic rings of carotenoids present in egg yolk help in neutralizing singlet oxygen, free radicals what is more is protective against oxidative damage [44–46].

2. Antioxidant properties of medicinal plants

Current research thrust has exposed the numerous benefits of herbs and medicinal plants in the nutrition of farm animals. Needless to say, several herbs and medicinal plants formerly taken as traditional medicine have been used for animal feeding trials. In an earlier report [47, 48] a wide range of extracts from herbal plants have antioxidant properties.

In particular is extracts of herbs from the Labiatae family—oregano, thyme, basil, mint, rosemary, sage, savory, marjoram, etc. The antioxidant properties of hyssop and lavender were attributed to phenolic terpenoid compounds such as carvacrol, thymol, menthol and eugenol just to mention a few [2]. Evidence in literature regarding antioxidant properties of herbal plants [49] further revealed that both phenolic and nonphenolic compounds exhibit antioxidant properties in them. Example of nonphenolic compounds with antioxidant properties include glycosides [50]. A recent report [49] showed the antioxidant properties of phenols from *Vernonia amygdalina* Delile. (Asteraceae). High levels of phenolic content 54.61 ± 0.94 mg GAE/g in *V. amygdalina* [49] has also been reported from phytochemical analysis of *Vernonia amygdalina.* The methanolic extract indicated that phenolic compounds were highly detected (+++) while flavonoids were detected (++) [51]. The medicinal plant *V. amygdalina* is rich in flavonoids, tannins and saponins, which may possibly play a part in anti-oxidative effect [49, 51]. From literature *V. amygdalina* has a flavonoid content of 22.53 ± 0.91 mg QE/g [49]. Phenolics hydroxy groups present in the molecular structure of flavonoids earlier stated as exhibiting antioxidant properties are involved in antioxidant properties of flavonoids. The powerful antioxidant property exhibited by phenolic compounds [52] present in plants constituents can be attributed to the hydroxyl groups [53] present in them. The medicinal plant *Ocimum gratissimum* also has both antioxidant anti-inflammatory properties already attributed to its therapeutic benefits from literature [54, 55]. The presence of other phytochemicals - saponins, terpenoids, glycosides and alkaloids—in aqueous *O. gratissimum* besides flavonoids and phenols (Ofongo, 2023, unpublished data). These phytochemicals may further play a part in its anti-inflammatory and anti-oxidative activities [54, 55].

2.1 Medicinal plants as component of feed and their benefits to nutrition of laying hens

Medicinal as component of feeds is becoming a common practice in the livestock industry either as suitable alternative to antibiotics or for their growth performance and health promoting benefits to livestock and ultimately to consumers of livestock consumers. Other benefits of medicinal plants as components of feed include; antimicrobial; anti-inflammatory; immune modulatory effect and antioxidant effect.

As earlier stated, the efficacy of herbal extracts as antioxidant feed additives needs to be evaluated and correlated with their phenolic content [2]. The use of medicinal to feed laying hens is more or less an evolving practice in the poultry industry.

Firstly, from improving productive performance, modulation of cholesterol contents in eggs and egg quality have been reported [14, 15, 34, 38, 56, 57]. However, gradually, inclusion of medicinal plants in diets of laying hens is focusing at making eggs functional food by improving or fortifying the content of nutrients present in eggs with benefits to consumers of eggs such as antioxidant properties [34, 58–60].

This book chapter tries to illustrate possibility for antioxidant scavenging activity enhancement in eggs from laying hens by administering medicinal plants via feed or orally.

3. Materials and methods

3.1 Collection and identification of plant materials

Two medicinal plants (*Vernonia amygdalina* Ochile (Compositae) NDUP/21/14—**Figure 1** and *O. gratissimum* L. Lamiaceae) NDUP/12/13—**Figure 2**) were used in this study. Authentication of the plants were done at Herbarium Unit of the Department of Pharmacognosy, Faculty of Pharmacy, Niger Delta University. Fresh leaves of *V. amygdalina* and *O. gratissimum* were collected from Niger Delta University Teaching and Research Farm, Wilberforce Island, Bayelsa State, Nigeria.

Figure 1.
V. amygdalina.

Figure 2.
Ocimum gratissimum.

3.2 Preparation of extracts and chaff

A large amount of each leave was collected either first thing in the morning or late in the evening. The leaves were separated from the stalk and placed in a large container filled with clean drinkable water into which little salt to brine the water. This was done to remove durst and debris from the leaves. Thereafter, the leaves were placed in a sieve to remove the brine and rinsed in clean drinkable water without salt again to remove any brine water on the leaves. The leaves were afterwards placed in a sieve to drain out excess water. Each leave was chopped separately into fine particle size then one thousand gram (1000 g) of each leave was weighed separately for milling to obtain the aqueous extract. Seven hundred and fifty mills (750 ml) of clean drinkable water were used to mill 1000 g of each leave sample separately by means of an electric milling machine. The aqueous filtrate was obtained by passing the milled product through a cheese cloth. The obtained chaff was set aside, air dried then packaged in zip lock bag (**Figure 3A**) for use as component of feed (**Figure 3B**). The aqueous extract was administered to 22 weeks old bovan brown layer at an inclusion rate of 1 ml/bird administered twice a week. The obtained dried chaff was further milled to obtain fine particles and added to standard layers mash at an inclusion rate of 50 g/kg of complete feed. Feed mixed with chaff not for immediate use were stored (**Figure 4**).

3.3 Animal experiment

A feeding trial was carried out to access and determine improved DPPH concentration in eggs from laying hens administered medicinal plants either as component of feed of as aqueous extract. The experiment was arranged as a complete randomized

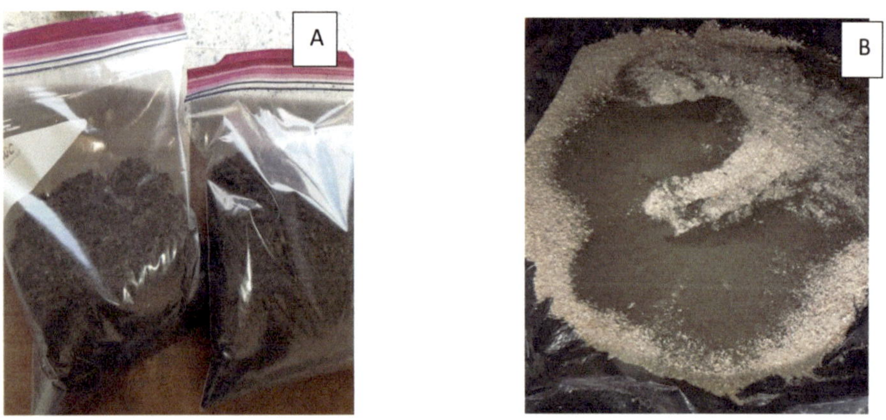

Figure 3.
Dried chaff of leaves in zip lock bag (A); fine milled chaff mixing into feed (B).

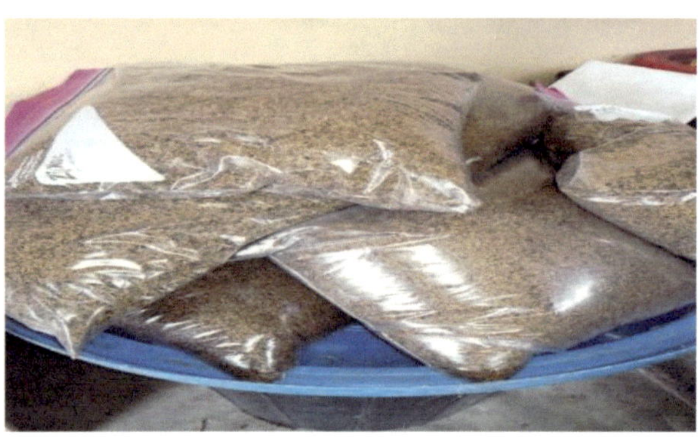

Figure 4.
Feed incorporated with chaff packaged in zip lock bag.

design having five (5) treatment groups of five (5) replicates and four (4) birds per replicate. A total of 100 bovan brown laying hens were purchased at 20 weeks of age randomly distributed to the above stated design then allowed to acclimatize for 2 weeks under the experimental treatments. Experimental collection of eggs for sampling commenced at week 22. The experiment was terminated after eight (8) weeks on day 56. Birds allocated to treatment 1 (T1) served as the control group. They were fed a standard layer's mash but where not administered aqueous plant extract neither was their feed supplemented with the chaff of *V. amygdalina* or *O. gratissimum*. Birds allocated to treatment 2 (T2) were fed diet supplemented with *V. amygdalina* chaff while birds in treatment 3 (T3) were administered 1 ml/bird of aqueous *V. amygdalina* twice a week. Birds assigned to treatment four (T4) had their diet supplemented with *O. gratissimum* chaff while birds in treatment five (T5) were administered 1 ml/bird of aqueous *O. gratissimum* extract twice a week.

3.4 Sample collection

On day 56, one egg per replicate was collected to determine antioxidant scavenging activity in inhibiting DPPH according to procedure illustrated below. The collected eggs were first cracked, homogenized then lyophilized to obtain a powder version of each egg before carrying out antioxidant scavenging activity in inhibiting DPPH using ascorbic acid as standard.

3.5 Proximate composition of feed and antioxidant scavenging activity in inhibiting DPPH

Samples of feed already supplemented with the respective chaff of *V. amygdalina* and *O. gratissimum* was collected into sample containers (100 g) for proximate analysis of feed as well as antioxidant activity in scavenging DPPH. The aqueous extract (30 ml) of either plant was also collected into sample bottles for antioxidant activity analysis in scavenging DPPH. Proximate composition of the experimental diets with and without each leave chaff as well as each respective chaff alone was carried out according to AOAC method.

Antioxidant activity of the plant extracts and feed supplemented with plant chaff in scavenging DPPH were evaluated on the basis of free radical scavenging effect of stable 2,2-diphenyl-1-picrylhydrazyl (DPPH). This was evaluated in comparison with Ascorbic acid standard, using a slightly modified method [61]. The compound DPPH is a synthetic compound not occurring in nature but it is utilized to evaluate the antioxidant activity of organic compounds such as vitamins, polyphenols and other phytochemicals.

Standard concentrations of Ascorbic acid standard were prepared at concentrations of 20, 40, 60, 80, and 100 ug/ml; respectively from a stock solution in triplicates using Methanol. Thereafter, a solution of DPPH was prepared using 0.1 mM of DPPH in methanol of which 2 ml of this solution was mixed with 3 ml of the test and standard solutions in test tubes. The solutions were shaken, then allowed to stand for 30 min in the dark before absorbance was measured at 517 nm using UV-VIS Spectrophotometer (Biomate 3, USA).

A standard control was prepared by mixing Methanol (3 ml) with 2 ml DPPH solution (0.1 mM, 1 ml). Methanol was used as a blank. The same procedure carried out with the Ascorbic acid standard was repeated with the test samples (leave chaff, aqueous extract and egg samples).

Percentage inhibition of DPPH was carried out using the formula below:

$$\text{Inhibition of DPPH} = (\text{Ac} - \text{Aa}) / \text{Ac} \times 100\%$$

Ac: absorbance of control sample.
Aa: absorbance of test samples or standard.

3.6 Statistical analysis

Collected data on DPPH scavenging activity in each sample (leaf chaff, aqueous extract, feed supplemented with leaf chaff and eggs) were subjected to analysis of variance (ANOVA). Statistically significant means were separated with Duncans Multiple Range test [62].

4. Results and discussions

4.1 Proximate composition of experimental diets

The proximate composition of layers mash fed to experimental birds as well as the proximate composition of *V. amygdalina* and *O. gratissimum* chaff is presented below in **Table 1**.

Crude protein concentration in *O. gratissimum* and *V. amygdalina* chaff was within the same range 12.00–13.00 g/kg DM. Ash concentration of *V. amygdalina* chaff was numerically higher (12.48 g/kg DM) than in *O. gratissimum* (9.17 g/kg DM). Dry matter concentration (943.15 g) was higher in *V. amygdalina* than in *O. gratissimum* (929.05 g). Proximate composition of feed showed an adequate crude protein concentration in the feed. Nitrogen free extract (carbohydrate) concentration was 61.61 g/kg DM in *O. gratissimum* which was numerically higher than value recorded in *V. amygdalina* and layers mash used in this study.

The synthetic compound 2,2-diphenyl-1-picrylhydrazyl (DPPH) is used as a reagent in antioxidant assays. It is used to evaluate the antioxidant activity of organic compounds such as vitamins, polyphenols and other phytochemicals. The addition of an organic compound to a solution of DPPH is used to measure the antioxidant activity of the compound. This activity is dependent on the ability of the compound to scavenge DPPH radical thereby limiting the ability of DPPH to absorb light at 517 nm. The lesser the absorbance of the solution, the higher the antioxidant scavenging

Nutrient	Layers mash	*Ocimum gratissimum* chaff	*V. amygdalina* chaff
Dry matter (gm)	926.80	929.05	943.15
Moisture (gm)	73.20	70.45	54.45
Crude protein	15.79	12.29	12.13
Crude fibre	14.88	12.01	12.33
Ash	10.76	9.17	12.48
Ether extract	5.79	4.90	5.86
NFE	52.78	61.63	57.20

NFE: nitrogen free extract.

Table 1.
Proximate composition of layers mash and leaf chaff (g/kg DM except otherwise stated).

activity of the test material. The DPPH scavenging activity is reported as percentage DPPH inhibition by the test compound. The lower the absorbance, the higher the percentage inhibition or free radical scavenging activity.

The antioxidant scavenging activity against 2,2-diphenyl-1-picrylhydrazyl (DPPH) by medicinal plants administered to layers or supplemented into complete feed is presented in **Table 2** below. Based on the DPPH scavenging activity (UV-absorbance at 517 nm); absorbance of DPPH was significantly ($p < 0.05$) low at 10 ug/ml concentration in *O. gratissimum* aqueous extract. The higher concentration resulted in significantly lower ($p < 0.05$) DPPH scavenging activity in *V. amygdalina* and *O. gratissimum* either in chaff or aqueous extract.

Table 3 shows the DPPH scavenging activity of *V. amygdalina* was high compared to *O. gratissimum* either as an aqueous extract or chaff. Values recorded was comparable to value obtained in Ascorbic acid standard. As indicated in **Table 3**: the least absorbance closest to value obtained using ascorbic acid was at a concentration of 200 ug/ml. The herb, *V. amygdalina* has antioxidant properties [63, 64].

The DPPH scavenging activity (% inhibition) is presented in **Figure 5**. *Vernonia amygdalina* chaff earlier reported in **Table 3** above elicited DPPH scavenging activity which was close to Ascorbic acid was significantly higher than values recorded for *O. gratissimum* leave chaff or aqueous extract. This value was also higher than that recorded for *V. amygdalina* aqueous extract in a dose dependent manner.

Expectedly, one will think the same trend will follow in DPPH scavenging activity for eggs collected from laying hens fed medicinal leaf chaff as component of feed or administered aqueous extract of either medicinal plant (**Table 4**). Rather; it was eggs from laying hens administered aqueous *O. gratissimum* that had numerically higher

Concentration (ug/ml)	*V. amygdalina* (AE)	*V. amygdalina* (chaff)	*Ocimum gratissimum* (AE)	*O. gratissimum* (chaff)	SEM
10	0.454^{b}	0.452^{c}	0.400^{d}	0.463^{a}	0.001
50	0.442^{a}	0.360^{d}	0.377^{c}	0.430^{b}	0.001
100	0.423^{a}	0.229^{d}	0.361^{c}	0.411^{b}	0.001
150	0.412^{a}	0.112^{d}	0.354^{c}	0.404^{b}	0.001
200	0.402^{a}	0.058^{d}	0.346^{c}	0.388^{b}	0.001

abcd: means along the same column with different superscripts are significantly different ($p < 0.05$)*; DPPH: 2,2-diphenyl-1-picrylhydrazyl; AE: aqueous extract; SEM: standard error of mean.*

Table 2.
DPPH scavenging activity (UV – Absorbance at 517 nm) in medicinal plants – Aqueous extract and chaff.

Concentration (ug/ml)	*V. amygdalina* (chaff)	Ascorbic acid
10	0.452	0.184
50	0.360	0.061
100	0.229	0.034
150	0.112	0.028
200	0.058	0.025

Table 3.
DPPH scavenging activity (UV-absorbance at 517 nm) in V. amygdalina *chaff compared to ascorbic acid standard.*

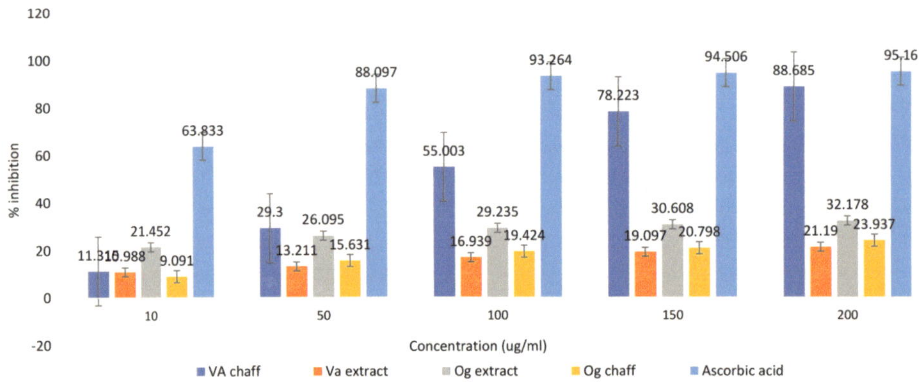

Figure 5.
DPPH scavenging activity (% inhibition). VA: Vernonia amygdalina; VA extract: Vernonia amygdalina aqueous extract; Og extract: Ocimum gratissimum *Aqueous extract; Og chaff:* O. gratissimum *Chaff.*

Treatment	Concentration (ug/ml)				
	10	50	100	150	200
Control group	8.044 ± 0.173	9.222 ± 0.065	10.922 ± 0.196	13.342 ± 0.285	16.285 ± 0.173
V. amygdalina chaff	8.829 ± 0.398	11.315 ± 0.113	12.230 ± 0.285	13.604 ± 0.173	13.996 ± 0.236
Ocimum gratissimum aqueous extract	9.287 ± 0.131	10.203 ± 0.131	12.165 ± 0.173	15.631 ± 0.196	17.332 ± 0.173
Ascorbic acid	63.833 ± 0.458	88.097 ± 0.429	93.264 ± 0.065	94.506 ± 0.227	95.160 ± 0.065

Table 4.
DPPH scavenging activity (% inhibition) in eggs from laying hens fed or administered medicinal plants.

DPPH scavenging activity which was better than that recorded for birds feed the layers mash control diet. Although values obtained in the study were lower than that recorded for Ascorbic acid; it can be said that may be the antioxidant components in *O. gratissimum* were better absorbed and incorporated into eggs of layers administered *O. gratissimum* aqueous extract than the chaff and *V. amygdalina* either as component of feed or aqueous extract for laying hens. The compound DPPH does not occur naturally and is not present in any organic compound; however, it is readily used to evaluate the antioxidant activity of organic compounds such as vitamins, polyphenols and other phytochemicals which can be present in medicinal plants and their extracts.

Natural products such as sesquiterpenoids and flavonoids were reported to be potential antioxidants [65–67]. These products are naturally obtained from food and use of medicinal plants. This fact further strengthens the antioxidative potential of *V. amygdalina*. Furthermore, results of earlier study added that not only flavonoids are responsible for any antioxidant effects of this plant species [68] but the presence of sesquiterpene lactones [69] might also play a part. Phytochemical screening of bitter leaf extract revealed a high concentration of flavonoids reported to be as the most abundant phytochemical present [63]. Antioxidant assay also indicated high levels of antioxidant activity and 2,2-diphenyl-1-picrylhydrazyl (DPPH) free radical scavenging activity which was concentration dependent.

The possibility of enhancing egg quality of laying hens by means of utilizing products exhibiting antioxidant properties can contribute to maintain oxido-redox balance thereby reducing the effect of oxidation on egg quality—degradation of eggs [38].

Possibly consumption of medicinal plants by laying hens either as feed component or aqueous extract can fortify the antioxidant properties of eggs. Poultry feed is composed mostly of plant ingredients which can be attributed to the antioxidant scavenging activity recorded in eggs from laying hens fed the control diet. Besides spoilage of eggs which is a concern to both consumers and producers of table eggs, the possibility of improving the not only the shelf life [38] but also improvement in good egg quality parameters [14] as well as designing eggs as functional foods.

Increased content of optimal ω-3 fatty acids, better ω-6/ω-3 fatty acids ratio, as well as good sensory profile of eggs optimal yolk color [14]. Furthermore, it has been shown in previous studies that fatty acid composition of eggs is dependent on fatty acid composition of feed given to laying hens which is subsequently transferred to the eggs [31].

A significant dose dependent DPPH scavenging activity (UV-absorbance at 517 nm) was observed in either medicinal plant consumed by laying hens and in eggs collected from the laying hens irrespective of treatment as presented in **Table 5**.

Table 6 further corroborated the possibility of plant to influence antioxidant scavenging activity of eggs from laying hens. The UV-absorbance at 517 nm for DPPH scavenging activity from eggs sampled showed significant improvement in antioxidant scavenging activity treatment 5 as earlier indicated in **Table 4**. Eggs from birds administered aqueous *O. gratissimum* significantly improved % inhibition of DPPH.

The possibility of producing functional eggs from laying hens for human consumption is wide. Apart from improving egg quality [38, 58], shelf life [38]; cholesterol content [57, 58], fatty acids profile and sensory profile [14, 15]. opportunities exist to improve antioxidant properties of eggs using medicinal plants. With increasing demand for enriched and functional foods which will provides various benefits to human health, eggs can be enriched with desirable nutrients [15] by means of dietary manipulation to achieve this goal [26, 27].

The all-important role of certain fatty acids such as linoleic and α-linolenic besides their long-chain (LC) n-6 and n-3 polyunsaturated fatty acids (PUFA) for humans have been reported some of the benefits ascribed to consumption of n-3 PUFA enriched eggs [15, 29]. Evidently this is made possible by feeding laying hens with

Concentration	Plants	Eggs
10ug/ml	0.443^{a}	0.467^{a}
50ug/ml	0.402^{b}	0.460^{b}
100ug/ml	0.356^{c}	0.454^{c}
150ug/ml	0.320^{d}	0.443^{d}
200ug/ml	0.298^{e}	0.433^{e}
SEM	0.001	0.001

abcde: means along the same column with different superscripts are significantly different ($p < 0.05$)*; DPPH: 2,2-diphenyl-1-picrylhydrazyl; SEM: standard error of mean.*

Table 5.
DPPH scavenging activity (UV – Absorbance at 517 nm in plants and eggs sampled.

Treatment	Eggs
T1	0.451^{bc}
T2	0.449^{c}
T3	0.456^{a}
T4	0.453^{b}
T5	0.443^{d}
SEM	0.001

abcd: means along the same column with different superscripts are significantly different ($p < 0.05$)*; DPPH: 2,2-diphenyl-1-picrylhydrazyl; SEM: standard error of mean; T1: control group; T2: treatment 2 –* V. amygdalina *chaff; T3: treatment 3 – 1 ml aqueous* V. amygdalina*; T4: Treatment 4 –* Ocimum gratissimum *chaff; T5: Treatment 5 – 1 ml* O. gratissimum *aqueous extract.*

Table 6.
Effect of treatment on DPPH scavenging activity (UV – Absorbance at 517 nm) in eggs from layers administered medicinal plant extract.

different by-products rich in PUFA [15, 27, 28, 31] such as flaxseed, rapeseed, microalgae, canola, chia (seed, meals or oils) etc.

Eggs for human consumption can have their fatty acid components enhanced by manipulating their feed by means of medicinal plants having antioxidant properties which may be transferred into the eggs. Furthermore, such eggs can serve as functional food for human consumption due to health benefits ascribe to eggs from laying hens fed with certain plants cum medicinal plants. In addition, the two medicinal plants reported here have antioxidant properties which can be of health benefit to consumers.

5. Conclusion

Although there are several variables for measuring antioxidant properties of medicinal plants; however, this study only looked at DPPH scavenging activity in eggs from laying hens fed *V. amygdalina* and *O. gratissimum* leaf chaff as components of the feed. It also evaluated DPPH scavenging activity when the aqueous extract of both plants is administered to laying hens. The DPPH scavenging activity of *V. amygdalina* chaff was high and comparable to ascorbic acid; but it was aqueous *O. gratissimum* administration to laying hens that yielded improved DPPH scavenging activity.

Acknowledgements

Miss Dressman, Mary Paul is gratefully acknowledged for collecting the eggs and participating in the study as part of her BSc project.

Conflict of interest

The authors declare no conflict of interest.

Author details

Habeeb O. Yusuf[1] and Ruth T.S. Ofongo[2]*

1 Faculty of Pharmacy, Department of Pharmacognosy, University of Lagos, Lagos State, Nigeria

2 Faculty of Agriculture, Department of Animal Science, Niger Delta University, Wilberforce Island, Bayelsa State, Nigeria

*Address all correspondence to: ruthofongo@ndu.edu.ng; ruthofongo@gmail.com

References

[1] Lourenço SC, Moldão-Martins M, Alves VD. Antioxidants of natural plant origins: From sources to food industry applications. Molecules (Basel, Switzerland). 2019;**24**(22):4132. DOI: 10.3390/molecules24224132

[2] Gholami-Ahangaran M, Ahmadi-Dastgerdi A, Azizi S, Basiratpour A, Zokaei M, Derakhshan M. Thymol and carvacrol supplementation in poultry health and performance. Veterinary Medicine and Science. 2022;**8**:267-288. DOI: 10.1002/vms3.663

[3] Fadzai B, Elaine C, Stanley M. Evaluation of nitrite radical scavenging properties of selected Zimbabwean plant extracts and their phytoconstituents. Journal of Food Processing. 2014. Article ID 918018:7. DOI: 10.1155/2014/918018

[4] Masoko P, Eloff JN. Screening of twenty-four south African *Combretum* and six *Terminalia* species (Combretaceae) for antioxidant activities. African Journal of Traditional, Complementary and Alternative Medicines. 2007;**4**:231-239. DOI: 10.4314/ajtcam.v4i2.31213

[5] Huyut Z, Beydemir S, Gülçin I. Antioxidant and antiradical properties of selected flavonoids and phenolic compounds. Biochemistry Research International. 2017;**2017**, Article ID 7616791:10. DOI: 10.1155/2017/7616791

[6] Comunian TA, Ravanfar R, de Castro IA, Dando R, Favaro-Trindade CS, Abbaspourrad A. Improving oxidative stability of echium oil emulsions fabricated by microfluidics: Effect of ionic gelation and phenolic compounds. Food Chemistry. 2017;**233**:125-134. DOI: 10.1016/j.foodchem.2017.04.085

[7] Sagar BK, Singh RP. Genesis and development of DPPH method of antioxidant assay. Journal of Food Science and Technology. 2011;**48**(4):412-422. DOI: 10.1007/s13197-011-0251-1

[8] Lobo V, Patil A, Phatak A, Chandra N. Free radicals, antioxidants and functional foods: Impact on human health. Pharmacognosy Reviews. 2010;**4**(8):118-126. DOI: 10.4103/0973-7847.70902

[9] Phaniendra A, Jestadi DB, Periyasamy L. Free radicals: Properties, sources, targets, and their implication in various diseases. Indian Journal of Clinical Biochemistry. 2015;**30**(1):11-26. DOI: 10.1007/s12291-014-0446-0

[10] Haseeb A, Ghulam H and Imtiaz M. Antioxidants from Natural Sources. In Antioxidants in Foods and Its Application. Emad Shalaby and Ghada Mostafa Azzam Eds. Intech. 2018. DOI: 10.5772/intechopen.75961

[11] Traber MG, Stevens JF. Vitamins C and E: Beneficial effects from a mechanistic perspective. Free Radical Biology & Medicine. 2011;**51**(5):1000-1013. DOI: 10.1016/j.freeradbiomed

[12] Idamokoro EM, Falowo AB, Oyeagu CE, Afolayan AJ. Multifunctional activity of vitamin E in animal and animal products: A review. Animal Science Journal. 2020;**91**:e13352. DOI: 10.1111/asj.13352

[13] Loi M, Paciolla C, Logrieco AF, Mulè G. Plant bioactive compounds in pre- and postharvest management for aflatoxins reduction. Frontiers in Microbiology. 2020;**11**:243. DOI: 10.3389/fmicb.2020.00243. PMID: 32226415; PMCID: PMC7080658

[14] Spasevski N, Peulić T, Banjac V, Rakita S, Puvača N, Kokić B, et al.

Effects of adding the functional co-extrudates and natural pigments in the diet of laying hens on egg quality. In: 26th World's Poultry Congress, book of abstracts. 2020. p. 445

[15] Vlaicu PA, Panaite TD, Turcu RP. Enriching laying hens' eggs by feeding diets with different fatty acid composition and antioxidants. Scientific Reports. 2021;**11**:20707. DOI: 10.1038/s41598-021-00343-1

[16] Cotterill OJ, Geiger GS. Egg product yield trends from shell eggs. Poultry Science. 1977;**56**:1027-1031

[17] Li-Chan ECY, Kim HO. Structure and chemical composition of eggs. In: Mine Y, editor. Egg Bioscience and Biotechnology. Hoboken, NJ, USA: John Wiley & Sons, Ltd; 2008. pp. 1-95

[18] Seuss-baum I. Nutritional evaluation of egg compounds. In: Huopalahti R, López-Fandiño R, Anton M, Schade R, editors. Bioactive Egg Compounds. Berlin, Heidelberg. Germany: Springer; 2007. pp. 117-144. DOI: 10.1007/978-3-540-37885-3_18

[19] Kovacs-Nolan J, Phillips M, Mine Y. Advances in the value of eggs and egg components for human health. Journal of Agricultural and Food Chemistry. 2005;**53**:8421-8431. DOI: 10.1021/jf050964f

[20] United States Department of Agriculture. United States Department of Agriculture: National Nutrient Database for standard reference Release 27. Available online: http://ndb.nal.usda.gov/ndb/ [Accessed: February 25, 2023]

[21] Nimalaratne C, Bandara N, Wu J. Purification and characterization of antioxidant peptides from enzymatically hydrolysed chicken egg white. Food Chemistry. 2015;**188**:467-472. DOI: 10.1016/j.foodchem.2015.05.014

[22] Hatta H, Kapoor M, Juneja L. Bioactive components in egg yolk. In: Mine Y, editor. Egg Bioscience and Biotechnology. Hoboken, NJ, USA: John Wiley & Sons, Ltd.; 2008. pp. 185-237

[23] Carlson S, Montalto M, Ponder D. Lower incidence of necrotizing enterocolitis in infants fed a preterm formula with egg phospholipids. Pediatric Research. 1998;**44**:491-498. DOI: 10.1203/00006450-199810000-00005

[24] Hoffman DR, Theuer RC, Castaneda YS, Wheaton DH, Bosworth RG, O'Connor AR, et al. Maturation of visual acuity is accelerated in breast-fed term infants fed baby food containing DHA-enriched egg yolk. Journal of Nutrition. 2004;**134**:2307-2313. DOI: 10.1093/jn/134.9.2307

[25] Surai PF. Effect of selenium and vitamin E content of the maternal diet on the antioxidant system of the yolk and the developing chick. British Poultry Science. 2000;**41**:235-243. DOI: 10.1080/713654909

[26] Ao T et al. Effects of supplementing microalgae in laying hen diets on productive performance fatty-acid profile and oxidative stability of eggs. Journal of Applied Poultry Research. 2015;**24**(3):394-400. DOI: 10.3382/japr/pfv042

[27] Gonzalez-Esquerra R, Leeson S. Alternatives for enrichment of eggs and chicken meat with omega-3 fatty acids. Canadian Journal of Animal Science. 2001;**81**(3):295-305. DOI: 10.4141/A00-092

[28] Carrillo S et al. Potential use of seaweeds in the laying hen ration to improve the quality of n-3 fatty acid enriched eggs. In: Nineteenth International Seaweed Symposium. Dordrecht: Springer; 2008. pp. 271-278

[29] FAO/WHO (Food and Agricultural Organization of the United Nations and World Health Organization). 2010 Fats and fatty acids in human nutrition. Report of an Extract Consultation Vol. 91 (FAO Food Nutrition Papers). 2010;**91**: 1-166. PMID: 21812367

[30] Hayat Z, Cherian G, Pasha TN, Khattak FM, Jabbar MA. Oxidative stability and lipid components of eggs from flax-fed hens: Effect of dietary antioxidants and storage. Poultry Science. 2010;**89**(6):1285-1292. DOI: 10.3382/ps.2009-00256

[31] Fraeye I, Bruneel C, Lemahieu C, Buyse J, Muylaert K, Foubert I. Dietary enrichment of eggs with omega-3 fatty acids: A review. Food Research International. 2012;**48**(2):961-969. DOI: 10.1016/j.foodres.2012.03.014

[32] Shahidi F, Ambigaipalan P. Omega-3 polyunsaturated fatty acids and their health benefits. Annual Review of Food Science and Technology. 2018;**9**:345-381

[33] Shinn S, Proctor A, Baum J. Egg yolk as means for providing essential and beneficial fatty acids. Journal of the American Oil Chemists' Society. 2018;**95**:5-11. DOI: 10.1146/annurev-food-111317-095850

[34] Kralik G, Kralik Z, Galovic O, Hanžek D. Cholesterol content and fatty acids profile in enriched n-3 PUFA and conventional eggs. In: World Poultry Congress 2022; Abstracts presented as webinar. Book of abstract page 57

[35] Charoensiriwatana W, Srijantr P, Teeyapant P, Wongvilairattana J. Consuming iodine enriched eggs to solve the iodine deficiency endemic for remote areas in Thailand. Nutrition Journal. 2010;**9**:68. DOI: 10.1186/1475-2891-9-68

[36] Bourre JM, Galea F. An important source of omega-3 fatty acids, vitamins D and E, carotenoids, iodine and selenium: A new natural multi-enriched egg. Journal of Nutrition, Health & Aging. 2006;**10**:371-376

[37] Naber EC. Modifying vitamin composition of eggs: A review. Journal of Applied Poultry Research. 1993;**2**:385-393

[38] Oueslati K, Ribeiro B, Chavatte D, Alleno C, Bouvet R. Positive impact of prebiotics and antioxidants on egg quality at the end of the laying hen production cycle. In: 26th World's Poultry Congress, Abstracts selected in 2020. p. 145

[39] Nakamura S, Kato A, Kobayashi K. Enhanced antioxidative effect of ovalbumin due to covalent binding of polysaccharides. Journal of Agricultural and Food Chemistry. 1992;**40**:2033-2037. DOI: 10.1021/jf00023a001

[40] Huang X, Tu Z, Xiao H, Wang H, Zhang L. Characteristics and antioxidant activities of ovalbumin glycated with different saccharides under heat moisture treatment. Food Research International. 2012;**48**:866-872. DOI: 10.1016/j.foodres.2012.06.036

[41] Chang O, Ha G, Han G. Novel antioxidant peptide derived from the ultrafiltrate of ovomucin hydrolysate. Journal of Agricultural and Food Chemistry. 2013;**61**:7294-7300. DOI: 10.1021/jf4013778

[42] Liu H, Zheng F, Cao Q, Ren B, Zhu L, Striker G, et al. Amelioration of oxidant stress by the defensin lysozyme. American Journal of Physiology. Endocrinology and Metabolism. 2006;**290**:E824-E832. DOI: 10.1152/ajpendo.00349.2005

[43] Sugino H, Ishikawa M, Nitoda T, Koketsu M, Juneja LR, Kim M, et al. Antioxidative activity of egg yolk phospholipids. Journal of Agricultural

and Food Chemistry. 1997;**45**:551-554. DOI: 10.1021/jf960416p

[44] Stahl W, Sies H. Antioxidant activity of carotenoids. Molecular Aspects of Medicine. 2003;**24**:345-351. DOI: 10.1016/s0098-2997(03)00030-x

[45] Ma L, Lin XM. Effects of lutein and zeaxanthin on aspects of eye health. Journal of the Science of Food and Agriculture. 2010;**90**:2-12. DOI: 10.1002/jsfa.3785

[46] Zhang LX, Cooney RV, Bertram JS. Carotenoids enhance gap junctional communication and inhibit lipid peroxidation in C3H/10T1/2 cells: Relationship to their cancer chemopreventive action. Carcinogenesis. 1991;**12**:2109-2114. DOI: 10.1093/carcin/12.11.2109

[47] Cuppett SL, Hall CA. Antioxidant activity of the Labiatae. In: Taylor S, editor. Advances in Food and Nutrition Research. Vol. 42. Academic Press Inc.; 1998. pp. 245-271. DOI: 10.1016/S1043-4526(08)60097-2. Available from: https://www.sciencedirect.com/science/article/pii/S1043452608600972

[48] Brenes A, Roura E. Essential oils in poultry nutrition: Main effects and modes of action. Animal Feed Science and Technology. 2010;**158**(1-2):1-14. DOI: 10.1016/j.anifeedsci.2010.03.007

[49] Harahap U, Dalimunthe A, Hertiani T, Mahatir MN, Satria D. Antioxidant and antibacterial activities of ethanol extract of *Vernonia amygdalina* Delile. Leaves. 2021. The International Conference on Chemical Science and Technology (ICCST – 2020) AIP Conf. Proc. 2342, 080011-1-080011-4; DOI: 10.1063/5.0045447

[50] Milos M, Mastelic J, Jerkovic I. Chemical composition and antioxidant effect of glycosidically bound volatile compounds from oregano (*Origanum vulgare* L. ssp. *hirtum*). Food Chemistry. 2000;*71*(1):79-83. DOI: 10.1016/S0308-8146(00)00144-8

[51] Ofongo RTS, Ohimain EI, Iyayi EA. Qualitative and quantitative phytochemical screening of bitter leaf and neem leaves and their potential as antimicrobial growth promoter in poultry feed. European Journal of Medicinal Plants. 2021;**32**(4):38-49. DOI: 10.9734/ejmp/2021/v32i430383

[52] Shahidi F, Janitha PK, Wanasundara PD. Phenolic antioxidants. Critical Reviews in Food Science and Nutrition. 1993;**32**:67-103. DOI: 10.1080/10408399209527581

[53] Hatano T, Edamatsu R, Mori A. Effects of the interaction of tannins with Co-existing substances. VI.: Effects of tannins and related polyphenols on superoxide anion radical, and on 1, 1-Diphenyl-2-picrylhydrazyl radical. Chemical & Pharmaceutical Bulletin. 1989;**37**:2016-2021. DOI: 10.1248/cpb.37.2016

[54] Olamilosoye KP, Akomolafe RO, Akinsomisoye OS, Adefisayo MA, Alabi QK. The aqueous extract of *Ocimum gratissimum* leaves ameliorates acetic acid induced colitis via improving antioxidant status and haematological parameters in male Wistar rats. Egyptian Journal of Basic and Applied Sciences. 2018;**5**(3):220-227. DOI: 10.1016/j.ejbas.2018.05.006

[55] Oyem JC, Chris-Ozoko LE, Enaohwo MT, Otabor FO, Okudayo VA, Udi OA. Antioxidative properties of *Ocimum gratissimum* alters Lead acetate induced oxidative damage in lymphoid tissues and haematological parameters of adult Wistar rats. Toxicology Reports. 2021;**8**:215-222. DOI: 10.1016/j.toxrep.2021.01.003

[56] Dauksiene A, Klementaviciute J, Gruzauskas R, Klupsaite D, Bartkiene E. Laying hens' production effectiveness increasing and quality improving by including to their diet sustainable plants (*Helianthus tuberosus* l.). In: 26th World's Poultry Congress, abstracts selected in 2020. p. 443

[57] Abdel-Wareth AAA, Lohakare JD. Productive performance, egg quality, nutrients digestibility, and physiological response of bovans brown hens fed various dietary inclusion levels of peppermint oil. Animal Feed Science and Technology. 2020;**267**:114554. DOI: 10.1016/j.anifeedsci.2020.114554

[58] Vlaicu PA, Panaite TD. Effect of dietary pumpkin (*Cucurbita moschata*) seed meal on layer performance and egg quality characteristics. Animal Bioscience. 2022;**35**(2):236-246. DOI: 10.5713/ab.21.0044

[59] Vakili R, Toroghian M, Torshizi ME. Saffron extract feed improves the antioxidant status of laying hens and the inhibitory effect on cancer cells (PC3 and MCF7) growth. Veterinary Medicine and Science. 2022;**8**:2494-2503. DOI: 10.1002/vms3.910

[60] Omri B, Alloui N, Durazzo A, Lucarini M, Aiello A, Romano R, et al. Egg yolk antioxidants profiles: Effect of diet supplementation with linseeds and tomato-red pepper mixture before and after storage. Foods. 2019;**8**:320. DOI: 10.3390/foods8080320

[61] Blois MS. Antioxidant determinations by the use of a stable free radical. Nature. 1958;**181**:1199-1200. DOI: 10.1038/1811199a0

[62] Steel RGD, Torrie JH. Principles and Procedures of Statistics. A Biometrical Approach. 2nd ed. New York, USA: McGraw-Hill; 1980. pp. 20-90

[63] Oriakhi K, Oikeh EI, Ezeugwu NO, Anoliefo O, Aguebor OES. Comparative antioxidant activities of extracts of *Vernonia amygdalina* and *Ocimum gratissimum* leaves. Journal of Agricultural Science. 2014;**6**:13-20. DOI: 10.5539/jas.v6n1p13

[64] Ekaluo UB, Ikpeme EV, Ekerette EE, Chukwu CI. *In vitro* antioxidant and free radical activity of some Nigerian medicinal plants: Bitter leaf (*Vernonia amygdalina* L.) and Guava (*Psidium guajava* Del.). Research Journal of Medicinal Plant. 2015;**9**(5):215-226. DOI: 10.3923/rjmp.2015.215.226

[65] Bork PM, Schimitz ML, Kuhnt M, Escher C, Heinrich M. Sesquiterpene lactone containing Mexican Indian medicinal plants and pure sesquiterpene lactones as potent inhibitors of transcription factor NF-*j*B. FEBS Letters. 1997;**402**:85-90. DOI: 10.1016/s0014-5793(96)01502-5

[66] Haraguchi H, Ishikawa H, Sanchez Y, Ogura T, Kubo Y, Kubo I. Antioxidative constituents of *Heterotheca inuloides*. Bio-organic & Medicinal Chemistry. 1997;**5**:865-871. DOI: 10.1016/S0968-0896(97)00029-1

[67] Jodynis-Liebert J, Murias M, Bloszyk E. Effect of sesquiterpene lactones on antioxidant enzymes and some drug-metabolizing enzymes in rat liver and kidney. Planta Medica. 2000;**66**:199-205. DOI: 10.1055/s-2000-8566

[68] Igile GO, Oleszek W, Jurzysta M, Burda S, Fanfunso M, Fasanmade AA. Flavonoids from Vernonia amygdalina and their antioxidant activities. Journal of Agricultural and Food Chemistry. 1994;**42**:2445-2448. DOI: 10.1021/jf00047a015

[69] Erasto P, Grierson DS, Afolayan AJ. Antioxidant constituents in *Vernonia amygdalina* leaves. Pharmaceutical Biology. 2007;**45**(3):195-199. DOI: 10.1080/13880200701213070

Chapter 3

The Endophytes: A New Resource for Vulnerable Plant Bioactive Compounds

Mostafa Fazeli

Abstract

Plant-associated microorganisms that live symbiotically in the plant body without causing disease symptoms are called endophytic microorganisms. Endophytes, including bacteria and fungi, can enhance the growth of the host plant and increase its resistance to pests, phytopathogens, and environmental stresses. In addition, endophytes can regulate the synthesis of plant secondary metabolites. Endophytes are a new reservoir for the discovery and production of valuable active substances. Some endophytic secondary metabolites are the same as host plants, such as paclitaxel. This finding has increased the importance of endophytes because the production of effective substances on an industrial scale in microorganisms is easier than in plants and has lower environmental costs. Therefore, endophytes need more attention in the pharmaceutical industry.

Keywords: endophyte, symbiosis, secondary metabolites, Taxol, endophytic fungi

1. Introduction

The rapid growth of human societies has increased the need to improve health standards and intensify food production. On the other hand, the emergence of drug resistance in pathogens and pests has become an increasing need to promote the search for new pharmaceutical and agricultural sources. Medicinal plants have been a valuable source of bioactive substances for a long time; however, environmental considerations, labor-intensive, high cost, and time-consuming have limited the use of these plant resources. On the other hand, the production of plant material in cell cultures faces technical challenges. The production of effective plant substances entered a new age with the discovery of the endophytic fungus *Taxomyces andreanea* in the yew, which could produce bioactive such as its host. Microorganisms are an attractive source of new biomaterials; also, they have the potential to increase the production of existing valuable materials. Plant-associated microorganisms called endophytes live in symbiosis with the tissues of their host plants. Many microorganisms, such as fungi, bacteria, and actinomycetes, have been discovered in endophytic relationships with plants [1].

The endophytes live asymptomatically in mutual association with plants. The endophytic lifestyle of microbes plays an important role in maintaining the health of plants by providing nutrients and defending plants against abiotic and abiotic stresses [2]. In addition, endophytes can produce many bioactive. Some of these substances are similar to the profile of the host plant's bioactive, which has increased the hope for cost-effective and environmentally friendly production. In the pharmaceutical and agricultural industries, bioactive compounds are known for their many applications. During the last two decades, endophytes have been recognized as important sources of bioactive compounds. Also, the proportion of new structures produced by endophyte isolates (51%) is significantly higher than that of soil isolates (38%), which has made endophytes one of the main natural product screening programs [3].

2. Endophytes

Microorganisms colonize many living plants in nature, and the degree of this microbial colonization varies by plant species. If the host plant tissue remains stable during this colonization, the relationship may vary from latent pathogenesis to mutual symbiosis. These microorganisms may be epiphytes, endophytes, or latent pathogens. Endophyte refers to microorganisms that are found under normal conditions in the tissues of living plants, without causing apparent diseases or visible symptoms of disease [4]. Endophytes are ubiquitous and spend a significant part of their life cycle without causing negative or obvious symptoms in the living tissues of the host plant. The word endophyte was first coined in 1866, where "endo" means "inside" and "phyte" means plant. They are mostly located in internal tissues such as roots, stems, leaves, flowers, and seeds. Endophytes may be transmitted horizontally or vertically [2], and some may even be seed-borne and passed on to the next generation [4]. A large community of endophytes lives inside the tissues of any plant. The diversity of endophytes is influenced by the host plant and its characteristics, including genotype, tissue, growth stage (age), and health status [5].

Endophytes have been isolated from all different parts of the plant. More than 200 genera from 16 bacterial phyla have been documented to be associated with endophytes [6]. It is also estimated that out of about 1.5 million species of fungi, one million of them are endophytic [7].

2.1 Endophytes: Plant interaction

Endophytes can provide benefits to their host plants. They mediate abiotic and biotic stress tolerance, reduce water consumption, and defend against pests and phytopathogens [8]. This interaction is controlled by endophyte and plant genes. The endophytic relationship is a novel and cost-effective plant-microbe evolutionary relationship that is driven by location and not defined by function [9]. Endophytic microbes are chemical synthesizers inside plants [10]. The imperceptible association of endophytes with the plant enables them to evolve [9]. It is the coevolution between endophytes and their host plant that determines the production of bioactive compounds. These compounds often play a role in the plant-microbe interaction in different ways and can bring different fitness benefits to the host plant [11, 12].

Plant compounds can be of plant origin or derived from endophytes or even can be produced by both. In the latter case, the endophyte may be involved in the entire pathway, but another scenario may be that only parts of the biosynthesis originate from the endophyte. In plant-endophyte interactions, significant changes appear in the secondary metabolism of symbionts, and these changes can be as a result of (i) induction of host metabolism by endophyte, (ii) induction of endophyte metabolism by the host, (iii) host and endophyte share part of a specific pathway, (iv) the host metabolizes endophyte products, and (v) the endophyte can metabolize host secondary compounds. [13]. Endophytes isolated from medicinal plants can produce bioactive metabolites and play a vital role in inducing secondary metabolite production by host plants [5, 14].

3. Secondary metabolites

Endophytes play a critical role in enhancing plant growth and are also known for their ability to produce bioactive with biotechnological applications. The use of herbal medicines is common in developing countries and up to 80% of people use this medicine. This traditional medicine has a long history. Medicinal plants are known for their rich sources of natural products. They are very valuable for disease prevention and treatment [15]. Endophytes communicate with their host plant through metabolic interactions [1, 16], which enable them to produce signaling molecules with interesting biological activities. In addition, the coevolution of endophytes with the host plant enables them to mimic the biological properties of the host and produce similar bioactive compounds [16].

Endophytes synthesize various bioactive compounds. However, compounds that have shown anticancer properties have attracted more attention, and in the meantime, the discovery of paclitaxel production by endophytic fungi has been a turning point in endophyte research.

3.1 Paclitaxel (Taxol)

Paclitaxel, with the brand name Taxol, is a terpenoid that was mainly obtained from the tissues of the yew plant; due to its amazing properties in binding to microtubules and inhibiting the division spindle, it is used in the treatment of various types of cancer, especially breast and ovarian cancer. It has been used a lot. However, extraction from plant sources due to the slow growth of the plant, the difficulty of purifying paclitaxel, and also its low amount in the plant tissues did not meet the needs of the market. Therefore, several methods, such as chemical synthesis, were also developed and commercialized. The scientists were also looking for alternative sources until the ability to synthesize it in the endophytic fungi of the host plant was discovered.

The discovery of *Taxomyces andreanea* from the Pacific yew, *Taxus brevifolia*, was undoubtedly a turning point in the field of bioprospecting for endophytes. This endophytic fungus demonstrated the ability to synthesize paclitaxel in the culture broth same as its host plant [17]. Microbial production of paclitaxel is very important for the development of the first billion-dollar anticancer drug business [17, 18]. Since this important discovery, several other endophytic fungi and bacteria showing paclitaxel production from yew and other plant species have been discovered (**Table 1**), including *Alternaria*, *Bartalinia*, *Fusarium*, *Lasiodiplodia*, *Metarhizium*, *Monochaetia*, *Pestalotiopsis*, *Penicillium*, *Phoma*, and *Spomatoichoanthermium* [13, 81–83]. The efficiency of paclitaxel among these fungal species varies [from nanograms to milligrams per liter], and their productivity is often lost during several generations of cultivation in laboratory conditions [84]. Microbial production of paclitaxel by endophytes has

Secondary metabolite	Host	Endophyte	Yield (μg/L)	Ref
Paclitaxel (Taxol)	*Taxus brevifolia*	*Taxomyces andreanae*	0.02–0.05	[17]
	T. wallichiana	*Pestalotiopsis microspora*	0.06–0.07	[19]
	Taxodium distichum	*Pestalotiopsis microspora* Cp-4	0.05–1.49	[20]
	Taxus cuspidata	*Alternaria* sp. Ja-69	0.16	[19]
	Taxus baccata	*Fusarium lateritium* Tbp-9	0.13	
	T. baccata	*Monochaetia* sp. Tbp-2	0.10	
	T. baccata	*Pestalotia bicilia* Tbx-2	1.08	
	T. cuspidata	*Pestalotiopsis microspora* Ja-73	0.27	
	T. wallachiana	*Pestalotiopsis microspora* Ne-32	0.5	
	T. sumatrana	*Pithomyces* sp. P-96	0.095	
	T. canadensis	*Erwinia taxi**	2.5–15	[21]
	Wollemia nobilis	*Pestalotiopsis guepinii* W-1f-2	0.49	[22]
	Torreya grandifolia	*Periconia* sp. No. 2026	0.03–0.83	[23]
	Ginkgo biloba	*Alternaria* sp.	0.12–0.26	[24]
	T. baccata	*Kitasatospora* sp. *	120	[25]
	T. baccata	*Penicillium* sp.	111	
	T. canadensis, T. brevifolia, T. hunnewelliana, T. baccata, T. cuspidata	*Bacillus cereus* ssp. *taxi*, *Bacillus megaterium* ssp. *taxi*, *Pantoea* sp., *Bacillus cereus*, *Bacillus subtilis* ssp. *taxi*, *Bacillus megaterium*, *Curtobacterium* sp., *Sphingomonas* ssp. *taxi**	1–25	[26]
	Tremacron mairei	*Tubercularia* sp. TF5	185.4	[27]
	T. yunnanensis	*Taxomyces* sp.	2.3	
	T. chinensis var. mairei	*Ozonium* sp. BT2	4–18	[28]
	T. cuspidata	*Botrytis* sp. HD181–23	206.34	[29]
	T. chinensis var. mairei	*Botrytis* sp. XT2	161.24	
	T. chinensis var. mairei	*Ectostroma* sp. XT5	276.75	
	T. chinensis var. mairei	*Papulaspora* sp. XT17	10.25	
	T. chinensis	*Alternaria alternata* TPF6	84.5	[30]
	T. chinensis var. mairei	*Fusarium mairei* Y1117	2.7	[31]
	T. chinensis var. mairei	*Ozonium* sp. EFY-21	21	[32]
	Aegle marmelos	*Bartalinia robillardoides* AMB9	187.6	[33]
	Cardiospermum helicacabum	*Pestalotiopsis pauciseta* CHP-11	113.3	[34]
	T. chinensis	*Fusarium mairei* UH23	286.4	[35]
	Citrus medica	*Phyllosticta citricarpa* No.598	265	[36]
	Podocarpus sp	*Aspergillus fumigatus* EPTP-1	557.8	[37]
	T. baccata	*Botryodiplodia theobromae* BT115	280.5	[38]
	T. cuspidata	*Fusarium arthrosporioides* F-40	131	[39]

Secondary metabolite	Host	Endophyte	Yield (μg/L)	Ref
	Cupressus sp	*Phyllosticta spinarum* No.625	235	[40]
	Wrightia tinctoria	*Phyllosticta tabernaemontanae*	461	[41]
	Thysanophrys celebica	*Fusarium solani*	1.6	[42]
	T. cuspidata	*Aspergillus niger* var. *taxi* HD86–9	273.6	[43]
	T. chinensis	*F. solanil* Tax- 3	163.35	[44]
	T. chinensis	*Metarhizium anisopliae* H- 27	846.1	[45]
	T. media	*Cladosporium cladosporioides* MD2	800	[46]
	T. media	*Aspergillus candidus* MD3	112	[47]
	T. chinensis	*Mucor rouxianus* DA10	ND	[48]
	Hibiscus rosa-sinensis	*Phyllosticta dioscoreae* No.605	298	[49]
	Terminalia arjuna	*Chaetomella raphigera* TAC15	79.6–211.1	[50]
	Terminalia arjuna	*Pestalotiopsis terminaliae*	211	[51]
	Taxus cuspidata	*Nodulisporium sylviforme*	450	[52]
	Morinda citrifolia	*Lasiodiplodia theobromae*	245	[53]
	Rhizosphere	*Pestalotiopsis malicola*	186	[54]
	Ginkgo biloba	*Phoma betae*	795	[55]
	T. baccata	*Stemphylium sedicola* SBU-16	6.9	[56]
	T. baccata	*F. redolens*	66	[57]
	Corylus avellana and T. baccata	*Penicillium aurantiogriseum* NRRL 62431	70	[58]
	T. wallichiana	*P. medicaginis*	1125	[59]
	T. chinensis var. mairei	*Aspergillus aculeatinus* Tax-6	334.92	[60]
	Podocarpus gracilior	*Aspergillus terreus* EFB108	20	[61]
	Rhizosphere	*Aspergillus flavipes*	185–850	[62]
	Rhizosphere	*Penicillium chrysogenum*	85	[62]
	Terminalia arjuna	*Alternaria brassicicola*	140.8	[63]
	Taxus sp.	*Aspergillus fumigatus* TPF-06	1590	[64]
	Catheranthus roseus	*Cladosporium cladosporioides*	700	[65]
	Tarenna asiatica	*Aspergillus oryzae*	95.04	[66]
	T. baccata	*Epicoccum nigrum* TXB502	61.35	[67]
	Sargassum polycystum	*Bacillus flexus* DMTMMB08*	ND	[68]
		Bacillus licheniformis DMTMMB10*		
	Acanthaphora specifera	*Oceanobacillus picturae* DMTMMB24*		
	Ginkgo biloba	*Penicillium polonicum* AUMC14487	90.53	[69]
	Mangifera indica	*Colletotrichum* sp. MIP-5	ND	[70]
	T. wallichiana	*Annulohypoxylon* sp. MUS1	282.05	[71]

Secondary metabolite	Host	Endophyte	Yield (μg/L)	Ref
	Persea americana	*Neopestalotiopsis clavispora* KY624416	100.6	[72]
	Moringa, Hibiscus	*Penicillium* sp. No.5	54.42–184.3	[73]
		Aspergillus niger No.10	43.95	
		Fusarium sp. No.8	26.8	
	Corylus avellana	*Stemphylium vesicarium* CA18	1400	[74]
	Corylus avellana	*Melanconium hedericola* CA12	1000	
	Calotropis procera, Catharanthus roseus	*Penicillium singorense*	13	[75]
	Millingtonia hortensis	*Cochliobolus hawaiiensis*	282	[76]
	T. wallichiana	*Aspergillus* sp. GBPI TWR F5	5450	[77]
baccatin III	*T. chinensis*	*Didmyostilbe* sp. DF110	ND	[78]
	T. wallichiana	*Diaporthe phaseolorum*	219	[79]
	T. wallichiana	*Trichoderma* sp. IRB54a	187.56	[80]
10-deacetyl baccatin III	*Corylus avellana*	*Melanconium hedericola* CA12	22,100	[74]
		Aspergillus microcysticus CA3	20,400	
		Arthrinium arundinis CA2	16,400	

Some strains that are only capable of producing precursors, such as baccatin III and 10-DAB, are listed separately.
**Reveals bacterial producers.*

Table 1.
Production of paclitaxel and some of its precursors by endophytic microorganisms; due to the multiplicity of different isolates from the same species, the name of the strain is also mentioned, as well as the amount of production in the strains noticed without subsequent manipulations and optimizations.

been observed mostly in fungal isolates. However, there are limited reports of the production of paclitaxel and some of its precursors by several strains of endophytic bacteria, such as *Erwinia taxi*, *Micromonospora* sp., *Streptomyces* sp., *Kitasatospora* sp., *Bacillus cereus*, *B. megaterium*, *Sphingomonas* ssp. *taxi*, *B. subtilis*, *Pantoea* sp., and *Curtobacterium* sp. [85]. Also, the discovery of paclitaxel-producing bacteria symbiotic with marine macroalgae *Sargassum polycystum* and *Acanthaphora specifera* showed that the search for endophytic sources of paclitaxel should not be limited to plants and terrestrials [68].

In addition to paclitaxel production, some endophytes can increase paclitaxel production in plants. Endophytic *Pseudodidymocyrtis lobariellae* fermentation broth can effectively increase paclitaxel accumulation in *T. chinensis* by regulating phytohormone metabolism and signal transduction and further regulating the expression of several key genes involved in paclitaxel biosynthesis [86]. The fermentation broth of *Kocuria* sp., *Micromonospora* sp., and *Sphingomonas* sp. also significantly increased the accumulation of taxanes in the stem cells of *T. yunnanensis* [87].

3.2 Vinca alkaloids

Vinblastine and vincristine are vinca alkaloids from *Catharanthus roseus* plant [88]. These compounds were the first herbal anticancer agents that were introduced to the clinical market. In the 1960s, vinblastine was used to treat breast cancer, testicular cancer, and Hodgkin's disease. Three years later, its oxidized derivative, vincristine, was introduced, which was widely used in the treatment of leukemia. These compounds inhibit the division spindle by irreversibly binding to microtubules and finally induce apoptosis. Vinblastine production from endophytic *Alternaria* was first described in 1998, followed by Lingqi et al. discovered an endophytic *Fusarium oxysporum* from *C. roseus* that successfully produced vincristine [89, 90]. These discoveries sparked a global hunt for new alternative sources of vinblastine and vincristine. Vincristine is most valuable as an anticancer agent. Endophytic *F. oxysporum* successfully biotransformed vinblastine to vincristine [91].

Palem et al. isolated an endophytic *Thalaromyces radicus* from *C. roseus* that could produce vinblastine and vincristine [92]. Ayob et al. isolated an endophyte *Nigrospora sphaerica* from *C. roseus* that can produce vinblastine. This fungus produced vinblastine with 10-fold better cytotoxicity to a breast cancer cell line compared to vinblastine extracted from *C. roseus* [93]. Endophytic fungal and bacterial species were found to have the ability to synthesize Vindoline —the precursor of vinca alkaloids— and have a high potential to be used as a biological elicitor in the production of vincristine [94, 95]. Also, a species of *Streptomyces* spp. was isolated from the rhizosphere soil of *C. roseus*, which can produce vinblastine and vincristine, (**Table 2**) [104].

Host plant	Endophyte	Vinca alkaloids	Strain	Ref
Catharanthus roseus	*Alternaria* sp.	Vinblastine	Fungal	[89]
	F. oxysporum	Vincristine	Fungal	[90]
	Mycelia sterilia 97CY-3	Vincristine	Fungal	[96]
	Fusarium solani	Vinblastine, Vincristine	Fungal	[97]
	F. oxysporum	Vinblastine, Vincristine	Fungal	[98]
	Talaromyces radicus	Vinblastine, Vincristine	Fungal	[92]
	Eutypella sp.	Vincristine	Fungal	[99]
	Nigrospora sphaerica	Vinblastine	Fungal	[93]
	Unidentified	Vinblastine, Vincristine	Fungal	[100]
	Microbacterium sp.	Vindoline	Bacterial	[94]
	Chaetomim globosum	Vinblastine	Fungal	[101]
	Curvularia verruculosa	Vinblastine	Fungal	[102]
	Botryosphaeria laricina	Vinblastine, Vincristine	Fungal	[103]
C. roseus Rhizospheric Soil	*Streptomyces* spp.	Vinblastine, Vincristine	Bacterial	[104]
C. roseus	*F. oxysporum*	Vinblastine, Vincristine	Fungal	[105]
	Alternaria sesami	Vindoline	Fungal	[95]
	Nigrospora zimmermanii	Vincristine	Fungal	[106]

Some of these isolates can biotransform vinblastine into vincristine. Also, some isolates only can synthesize vindoline as a valuable precursor of anticancer drugs.

Table 2.
Microbial production of vinca alkaloids by endophytic microorganisms; endophytic producers of vinca alkaloids have so far only been isolated from C. roseus *or its rhizosphere soil.*

3.3 Camptothecin

Camptothecin (CPT) is a pentacyclic quinoline alkaloid isolated from the wood of *Camptotheca acuminata* and the root of *Nothapodytes foetida*. Several reports show the therapeutic potential of CPT and its derivatives for the treatment of colon, cervical, uterine, lung, and ovarian cancer. Most of the two promising anticancer activities [107] are related to its main derivatives, 9-methoxycamptothecin and 10-hydroxycamptothecin, because CPT is not directly used as an anticancer drug due to its low solubility, short half-life, and toxicity [108–110]. These cytotoxic agents act by selectively inhibiting topoisomerase 1. and thereby disrupting the DNA replication process.

In 2005, the CPT-producing endophytic fungus *Entrophospora infrequens* was isolated from *N. foetida* [109]. Endophyte *Neurospora crassa* and *Nodulisporium* sp. isolated from *N. foetida* produces CPT in culture medium [111, 112]. There are also examples of endophytes that can produce hydroxylated CPT derivatives, for example, Mycelia sterilia XK001 can produce 10-hydroxycamptothecin, which is the clinically active derivative of CPT [107]. Most CPT-producing endophytes are fungi; however, there are also reports of bacterial producers (**Table 3**) [116, 121, 123, 126]. A CPT-producing endophytic fungus from the marine sponge *Cliona* sp. It has been isolated that unlike other endophytes isolated from soil and plant environments, and it has been isolated and identified from the marine environment and aquatic organisms [131].

Secondary metabolite	Host plant	Endophyte	Strain	Ref
CPT	*Nothapodytes foetida*	*Entrophospora infrequens*	Fungal	[109]
	Neanotis foetida	*Neurospora* sp.	Fungal	[111]
	Camptotheca acuminata	*F. solani*	Fungal	[108]
	N. foetida	*Nodulisporium* sp.	Fungal	[112]
	N. nimmoniana	*Botryosphaeria parva*	Fungal	[113]
	Apodytes dimidiata	*F. solani*	Fungal	[114]
	C. acuminata	*Trichoderma atroviride*	Fungal	[115]
	Miquelia dentata Bedd.	*Bacillus subtilis, Bacillus* sp., *Bacillus cereus, Lysinibacillus* sp.	Bacterial	[116]
	M. dentata Bedd.	*Fomitopsis* sp., *Alternaria alternata, Phomposis* sp.	Fungal	[117]
	C. acuminata	*Fusarium nematophilum, Alternaria Alternata, Phomopsis vaccinii*	Fungal	[118]
	N. foetida	*Fusarium oxysporum*	Fungal	[119]
	Catharanthus roseus	*F. solani*	Fungal	[120]
	C. acuminata	*Paenibacillus polymyxa*	Fungal	[121]
	N. nimmoniana	*Colletotrichm fructicola, Corynespora cassiicola*	Fungal	[122]
	N. nimmoniana	*F. solani*	Fungal	[110]
	Pyrenacantha volubilis	*Bacillus* sp., *B. subtilis, Bacillus amyloliquefaciens*	Fungal	[123]
	Piper betel L.	*Aspergillus niger*	Fungal	[124]

Secondary metabolite	Host plant	Endophyte	Strain	Ref
	Chonemorpha fragrans	*F. solani*	Fungal	[125]
	Ephedra foliata	*Kytococcus schroeter*	bacterial	[126]
	Ophiorrhiza mungos	*Meyerozyma* sp., *Talaromyces* sp.	Fungal	[127]
	N. nimmoniana	*Alternaria alstroemeriae, Alternaria burnsii*	Fungal	[128]
	Cipadessa baccifera	*Phyllosticta elongata*	Fungal	[129]
	Ficus elastica	*Aspergillus terreus, Aspergillus flavus*	Fungal	[130]
	sponge *Cliona sp.*	*Penicillium chrysogenum*	Fungal	[131]
	Cestrum parqui	*Aspergillus terreus*	Fungal	[132]
	No. nimmoniana	*Diaporthe* sp. F18	Fungal	[133]
10-hydroxyCPT	*No. nimmoniana*	Mycelia sterilia_ XK001	Fungal	[107]
Podophyllotoxin	*Podophyllum hexandrum*	*Alternaria* sp., *Penicillium* sp.	Fungal	[134]
	Diphylleia sinensis	*Penicillium* sp.	Fungal	
	Dysosma veitchii	*Monilia* sp., *Penicillium* sp.	Fungal	
	D. sinensis	*Penicillium implicatum*	Fungal	[135]
	D. veitchii	Penicillium implicatum	Fungal	[136]
	Juniperus vulgaris	*Alternaria* sp.	Fungal	[137]
	P. peltatum	*Phialocephala fortinii*	Fungal	[138]
	P. hexandrum	*Trametes hirsuta*	Fungal	[139]
	P. hexandrum	*Alternaria neesex*	Fungal	[140]
	Juniperus recurva	*Fusarium oxysporum*	Fungal	[141]
	P. hexandrum	*F. solani*	Fungal	[142]
	Solanum hexandrum	*Mucor fragilis*	Fungal	[143]
	Podophyllum emodi	*Alternaria tenuissima*	Fungal	[144]
	D. sinensis	*Penicillium* sp.	Fungal	[145]
	P. hexandrum	*Chaetomium globosum, Pseudallescheria* sp.	Fungal	[146]
	Dysosma versipellis	*Fusarium* sp.	Fungal	[147]
	P. hexandrum	*Penicillium* sp.	Fungal	[148]
	Dysosma difformis	*Penicillium* sp., *Trametes* sp., *Purpureocillium* sp., *Aspergillus* sp., *Ganoderma* sp.	Fungal	[149]
	D. difformis	*Fusarium proliferatum*	Fungal	[150]
Deoxypodophyllotoxin	*J. communis*	*Aspergillus fumigates*	Fungal	[151]
Huperzine A	*Huperzia serrata*	*Acremonium* sp.	Fungal	[152]
	Phlegmariurus cryptomerianus	*Blastomyces* sp.	Fungal	[153]
		Botrytis sp.	Fungal	

Secondary metabolite	Host plant	Endophyte	Strain	Ref
	Lycopodium serratum	*Penicillium chrysogenum*	Fungal	[154]
	H. serrata	*Shiraia* sp.	Fungal	[155]
		Cladosporium cladosporioides	Fungal	[156]
		Aspergillus flavus	Fungal	[157]
		Shiraia bambusicola	Fungal	[158]
		Colletotrichum gloeosporioides	Fungal	[159]
		Trichoderma sp.	Fungal	[160]
		Paecilomyces tenuis	Fungal	[161]
		Penicillium sp.	Fungal	[162]
	Phlegmariurus phlegmaria	*Ceriporia lacerate*	Fungal	[163]
	H. serrata	*Colletotrichum* sp., *Ascomycota* sp., *Sarcosomataceae* sp., *Dothideomycetes* sp.	Fungal	[164]
		Penicillium sp.	Fungal	[165]
		Alternaria brassicae	Fungal	[166]
		Penicillium polonicum, Colletotrichum gloeosporioides	Fungal	[167]
		Mucor racemosus, M. fragilis, Fusarium verticillioides, F. oxysporum, Trichoderma harzianum	Fungal	[168]
	Phlegmariurus taxifolius	*Fusarium* sp.	Fungal	[169]
	H. serrata	*Fusarium* sp.	Fungal	[170]

CPT and active derivative 10-hydroxyCPT, podophyllotoxin, deoxypodophyllotoxin act as anticancer, and Huperzine A approved for treatment of Alzheimer's disease.

Table 3.
Production of plant-derived secondary metabolites by endophytic microorganisms.

3.4 Podophyllotoxin

Podophyllotoxin is an aryltetralin lignin that uses in the synthesis of anticancer drugs. It is originally isolated from the resins of the *Podophyllum emodi*, which is traditionally used to treat genital warts [16]. Podophyllotoxin is a strong inhibitor of microtubules, while its derivatives inhibit topoisomerase 2. These derivatives are used to treat bronchial and testicular cancers. Podophyllotoxin production from endophytic fungi isolated from *Podophyllum* [syn. *Sinopodophyllum*] *hexandrum*, *Diphylleia sinensis*, and *Dysosma veitchii* were reported for the first time [134]. After that, two strains of the endophytic fungus *Phialocephala fortinii* from the rhizome of *P. peltatum*, which could produce podophyllotoxin under axenic culture conditions, were isolated and identified [138]. The fungus *Trametes* isolated from *P. hexandrum* is another endophyte capable of producing podophyllotoxin and podophyllotoxin glycosides [139]. In addition, *F. oxysporum* and *Aspergillus* endophytes isolated from

Juniperus recurva and *Juniperus communis* produced podophyllotoxin and deoxypadophyllotoxin, respectively [141, 151]. Podophyllotoxin production has also been reported from *Mucor fragilis*, and *Alternaria tenuissima* isolated from *P. emodi* was found to produce podophyllotoxin [143, 144]. Podophyllotoxin-producing endophytic fungi *Penicillium* sp., *Trametes* sp., *Purpureocillium* sp., *Aspergillus* sp. *Ganoderma* sp., and *Fusarium* spp. were isolated from plants of *Dysosma* spp. [149, 150]. Most podophyllotoxin-producing fungi belong to *Penicillium* sp., *Alternaria* sp., and *Fusarium* spp. genera, respectively, While there is no report of podophyllotoxin production among endophyte bacteria (**Table 3**).

Fungal production of podophyllotoxin is promising for mass production, and it is possible to provide affordable resources for commercial production by optimizing the cultivation methods and genetic changes of the producing microorganisms and reducing the pressure of harvesting from plant resources and giving the chance to producing plants for save from extinction.

3.5 Huperzine A

The lycopod *Huperzia serrata* is the main source of a natural lycopodium alkaloid called Huperzine A (HupA), which has attracted worldwide attention for its potential in the treatment of Alzheimer's disease. This compound is an acetylcholinesterase inhibitor that increases the availability of acetylcholine in central cholinergic synapses by highly selective and reversible inhibition of this enzyme and blocking its activity. The bulk of HupA is obtained from the *Huperziaceae* family. The *H. serrata* has a narrow geographical distribution, slow growth rate, and very low HupA content, which limits its natural harvest and HupA extraction. In the first report, the endophyte *Acremonium* sp. isolated from *H. serrata* has been capable of production of HupA [152]. Similarly, the endophyte *Shiraia* sp. Slf14 and *Cladosporium cladosporioides* isolated from *H. serrata* leaves also produced HupA [155, 156, 158]. In general, 32 endophytic fungi belonging to 15 genera were recorded to produce Hup A. These fungal endophytes were isolated from members of *Huperziaceae* family, including *H. serrata*, *Phlegmariurus phlegmaria*, and *Phlegmariurus taxifolius* (**Table 3**) [171].

Xia and colleagues isolated endophytic fungi *Mucor racemosus*, *M. fragilis*, *Fusarium verticillioides*, *F. oxysporum*, and *Trichoderma harzianum* from the *H. serrata*, which can inhibit acetylcholinesterase enzyme [168]. The endophyte *Ceriporia lacerate* successfully transformed HupA into five different compounds that showed potential acetylcholinesterase inhibitory activity [172]. Biotransformation, using fungal endophytes, is also a valuable approach to producing HupA derivatives. Microbial production of HupA has positive economic and environmental effects. This will be a practical strategy to meet the global market demand through microbial fermentation and genetic manipulation of the source fungi.

4. Industrial aspects

The role of plant compounds in the production of many clinically effective anticancer drugs is undeniable, but the production of herbal drugs is not always as expected. Because their production from plant resources faces serious challenges, many of these compounds are produced at a certain stage of plant growth or under certain environmental conditions, stress, or availability of nutrients. Also, the growth of plants is slow, and to collect and extract some products, they must reach acceptable growth. On the other hand, production in plant cell culture also faces technical challenges. Also, due to the extent and variety of bioactive in plants, the purification processes of the desired effective substances will be complicated and therefore expensive. Due to the limitations identified with the productivity and vulnerability of plant species as sources of new metabolites, microorganisms act as an available and inexhaustible resource of new pharmaceuticals [173].

Over many years, seasonal and climatic factors have caused failure in traditional methods of extracting bioactive from natural resources. The environmental issues that researchers face during the extraction of bioactive from plants make it necessary to adopt new approaches to obtain these compounds [174]. In the future, with the increase in population, the demand for pharmaceutical and agricultural products will increase day by day, and the future of endophytic fungi for the isolation of various beneficial compounds is bright. There is a great need to discover bioactive compounds from natural resources that can be used to treat various diseases. Recently, more attention has been paid to the production of bioactive from endophytic fungi because they are excellent for exploiting the biosynthetic pathway for the synthesis of bioactive. The main challenge is the low yield of desired active compounds obtained from endophytes. However, to meet the demand of pharmaceutical companies to increase the commercial production of drugs, genetic engineering technologies, drug design techniques, and microbial fermentation technology can be solutions to increase the rate of endophyte production [2]. In addition, the use of cell co-cultures of host plants and endophytes has improved the production rate. Some secondary metabolites may be produced by combined endophyte and host activity. Some endophytic bacteria produce secondary metabolites in medicinal plants. For example, *Bacillus altitudinis*, *Burkholderia* sp., and *Flavobacterium* sp. act as effective stimulators that increase ginsenoside concentrations by converting the major ginsenoside Rb1 to the minor ginsenoside Rg3 in the valuable medicinal plant ginseng [175–177]. Such biotransformations using endophytic bacteria have significant potential to intensify the accumulation of rare active substances in medicinal plants. The endophytic *Pseudomonas fluorescens* can increase the production of sesquiterpenoids in *Macrocephala Atractylodes* [178]. The endophyte *Bacillus subtilis* in the plant *Chuanxiong Ligusticum* enhances ligustrazine accumulation [179].

The interaction of endophytes with plant tissues asymptomatically increases the production of secondary metabolites. A double synthesis of podophyllotoxin was obtained from the interaction of endophytic fungi *Phialocephala fortinii* and rhizomes of *P. peltatum* [138]. Endophytic fungi *Stemphylium amaranthi* and *Gliomastix masseei* can be used as fungal stimulants to improve indole alkaloid production from *C. roseus* [180].

5. Conclusion

Throughout history, humans have used plants and plant-derived products to treat various ailments. Plant secondary metabolites or bioactive are known to be synthesized by plants. Microbes living inside host plant tissues are also known for their ability to synthesize substances similar to those synthesized by the host plant. Secondary metabolites, such as alkaloids, flavonoids, terpenoids, steroids, etc. synthesized by microbes, are known for their vital role as antioxidants and anticancer. The discovery of the ability to produce plant secondary metabolites in endophytes has raised many hopes for the production of these compounds on an industrial scale. Microorganisms reduce environmental concerns about the production of biological substances in plants because endophytic microbes have a high reproduction ability, the possibility of their genetic manipulation is easier, and the fermentation conditions for them are simpler, cheaper, and more diverse.

Author details

Mostafa Fazeli
Mehrdad Zist Tous Ltd, R&D Centers, Mashhad, Iran

*Address all correspondence to: fazelim921@mums.ac.ir

References

[1] Ezeobiora CE, Igbokwe NH, Amin DH, Mendie UE. Endophytic microbes from Nigerian ethnomedicinal plants: A potential source for bioactive secondary metabolites—A review. Bulletin of the National Research Centre. 2021;**45**(1):1-10

[2] Rana KL, Kour D, Kaur T, Devi R, Negi C, Yadav AN, et al. Endophytic fungi from medicinal plants: Biodiversity and biotechnological applications. In: Microbial Endophytes. Sawston, UK: Elsevier; 2020. pp. 273-305

[3] Segaran G, Sathiavelu M. Fungal endophytes: A potent biocontrol agent and a bioactive metabolites reservoir. Biocatalysis and Agricultural Biotechnology. 2019;**21**:101284

[4] Hassani M, Durán P, Hacquard S. Microbial interactions within the plant holobiont. Microbiome. 2018;**6**(1):1-17

[5] Wu W, Chen W, Liu S, Wu J, Zhu Y, Qin L, et al. Beneficial relationships between endophytic bacteria and medicinal plants. Frontiers in Plant Science. 2021;**12**:646146

[6] Gouda S, Das G, Sen SK, Shin H-S, Patra JK. Endophytes: A treasure house of bioactive compounds of medicinal importance. Frontiers in Microbiology. 2016;**7**:1538

[7] Priyadarshini MS, Panigrahi S, Rath C. Endophytes: Novel Microorganisms for Plant Growth Promotion. Tamil Nadu, India: Darshan publishers; 2022

[8] Hodkinson TR, Doohan FM, Saunders MJ, Murphy BR. Endophytes for a Growing World. Cambridge, UK: Cambridge University Press; 2019

[9] Kusari S, Spiteller M. Metabolomics of endophytic fungi producing associated plant secondary metabolites: Progress, challenges, and opportunities. In: Metabolomics. Rijeka, Croatia: InTechOpen; 2012. pp. 241-266

[10] Kaul S, Gupta S, Ahmed M, Dhar MK. Endophytic fungi from medicinal plants: A treasure hunt for bioactive metabolites. Phytochemistry Reviews. 2012;**11**:487-505

[11] Kusari S, Spiteller M. The promise of endophytic fungi as sustainable resource of biologically relevant pro-drugs: A focus on Cameroon. In: Fungi. Boca Raton, FL, USA: CRC Press; 2018. pp. 1-13

[12] Saxena S, Meshram V, Kapoor N. Muscodor tigerii sp. nov.-volatile antibiotic producing endophytic fungus from the Northeastern Himalayas. Annals of Microbiology. 2015;**65**(1):47-57

[13] Ludwig-Müller J. Plants and endophytes: Equal partners in secondary metabolite production? Biotechnology Letters. 2015;**37**:1325-1334

[14] Ek-Ramos MJ, Gomez-Flores R, Orozco-Flores AA, Rodríguez-Padilla C, González-Ochoa G, Tamez-Guerra P. Bioactive products from plant-endophytic gram-positive bacteria. Frontiers in Microbiology. 2019;**10**:463

[15] Pan S-Y, Zhou S-F, Gao S-H, Yu Z-L, Zhang S-F, Tang M-K, et al. New perspectives on how to discover drugs from herbal medicines: CAM's outstanding contribution to modern therapeutics. Evidence-Based Complementary and Alternative Medicine. 2013;**2013**:627375

[16] Meshram V, Gupta M. Endophytic fungi: A quintessential source of

potential bioactive compounds. Endophytes for a Growing World. 2019; **277**:277-309

[17] Stierle A, Strobel G, Stierle D. Taxol and taxane production by taxomyces andreanae, an endophytic fungus of Pacific yew. Science. 1993;**260**(5105): 214-216

[18] Stierle AA, Stierle DB. Bioactive secondary metabolites produced by the fungal endophytes of conifers. Natural Product Communications. 2015;**10**(10): 1671-1682

[19] Strobel G, Hess W, Ford E, Sidhu R, Yang X. Taxol from fungal endophytes and the issue of biodiversity. Journal of Industrial Microbiology. 1996;**17**:417-423

[20] Li J-y, Strobel G, Sidhu R, Hess W, Ford EJ. Endophytic Taxol-producing fungi from bald cypress, taxodium distichum. Microbiology. 1996;**142**(8): 2223-2226

[21] Landry N. Bacterial Mass Production of Taxanes with Erwinia. US5561055A: Google Patents; 1996

[22] Strobel GA, Hess W, Li J-Y, Ford E, Sears J, Sidhu RS, et al. Pestalotiopsis guepinii, a Taxol-producing endophyte of the Wollemi pine, Wollemia nobilis. Australian Journal of Botany. 1997;**45**(6): 1073-1082

[23] Li J, Sidhu R, Ford E, Long D, Hess W, Strobel G. The induction of Taxol production in the endophytic fungus—Periconia sp from Torreya grandifolia. Journal of Industrial Microbiology and Biotechnology. 1998; **20**:259-264

[24] Su K. Screening of Taxol-producing endophytic fungi from Ginkgo biloba and Taxus cuspidate in Korea. Agricultural Chemistry and Biotechnology. 1999;**42**:97-99

[25] Caruso M, Colombo A, Fedeli L, Pavesi A, Quaroni S, Saracchi M, et al. Isolation of endophytic fungi and actinomycetes taxane producers. Annals of Microbiology. 2000;**50**(1):3-14

[26] Page M, Landry N, Boissinot M, Helie M-C, Harvey M, Gagne M. Bacterial Mass Production of Taxanes and Paclitaxel. WO1999032651A1: Google Patents; 2000

[27] Wang B, Li A, Wang X. An endophytic fungus for producing Taxol. Science in China Series C. 2001;**31**:271-274

[28] Guo B, Wang Y, Zhou X, Hu K, Tan F, Miao Z, et al. An endophytic Taxol-producing fungus BT2 isolated from Taxus chinensis var. mairei. African Journal of Biotechnology. 2006; **5**(10):875-877

[29] Hu K, Tan F, Tang K, Zhu S, Wang W. Isolation and screening of endophytic fungi synthesizing Taxol from Taxus chinensis var. mairei. Journal of Southwest China Normal University (Natural Science Edition). 2006;**31**: 134-137

[30] Renpeng T, Qiao Y, Guoling Z, Jingquan T, Luozhen Z, Chengxiang F. Taxonomic study on a Taxol producing fungus isolated from bark of Taxus chinensis var. mairei. Wuhan zhi wu xue yan jiu= Wuhan Botanical Research. 2006;**24**(6):541-545

[31] Cheng L, Ma Q, Tao G, Tao W, Wang R, Yang J, et al. Systemic identification of a paclitaxel-producing endophytic fungus. Industrial Microbiology. 2007;**37**:23-30

[32] Zhou X, Wang Z, Jiang K, Wei Y, Lin J, Sun X, et al. Screening of Taxol-

producing endophytic fungi from Taxus chinensis var. mairei. Applied Biochemistry and Microbiology. 2007; **43**:439-443

[33] Gangadevi V, Muthumary J. Taxol, an anticancer drug produced by an endophytic fungus Bartalinia robillardoides Tassi, isolated from a medicinal plant, Aegle marmelos Correa ex Roxb. World Journal of Microbiology and Biotechnology. 2008;**24**:717-724

[34] Gangadevi V, Murugan M, Muthumary J. Taxol determination from Pestalotiopsis pauciseta, a fungal endophyte of a medicinal plant. Chinese Journal of Biotechnology. 2008;**24**(8): 1433-1438

[35] Dai W, Tao W. Preliminary study on fermentation conditions of Taxol-producing endophytic fungus. Chemical Industry and Engineering Progress. 2008;**27**(6):883-886

[36] Kumaran RS, Muthumary J, Hur B-K. Taxol from Phyllosticta citricarpa, a leaf spot fungus of the angiosperm Citrus medica. Journal of Bioscience and Bioengineering. 2008;**106**(1):103-106

[37] Sun D, Ran X, Wang J. Isolation and identification of a Taxol-producing endophytic fungus from Podocarpus. Wei sheng wu xue bao= Acta Microbiologica Sinica. 2008;**48**(5): 589-595

[38] Venkatachalam R, Subban K, Paul MJ. Taxol from Botryodiplodia theobromae (BT 115)—AN endophytic fungus of Taxus baccata. Journal of Biotechnology. 2008;**136**:S189-SS90

[39] Chang-Tian L, Yu L, Wang Q-J, Sung C-K. Taxol production by Fusarium arthrosporioides isolated from yew, Taxus cuspidata. Journal of Medical Biochemistry. 2008;**27**(4):454-458

[40] Senthil Kumaran R, Muthumary J, Hur B. Production of Taxol from Phyllosticta spinarum, an endophytic fungus of Cupressus sp. Engineering in Life Sciences. 2008;**8**(4):438-446

[41] Kumaran RS, Muthumary J, Hur B-K. Isolation and identification of an anticancer drug, Taxol from Phyllosticta tabernaemontanae, a leaf spot fungus of an angiosperm, wrightia tinctoria. The Journal of Microbiology. 2009;**47**(1): 40-49

[42] Chakravarthi B, Das P, Surendranath K, Karande AA, Jayabaskaran C. Production of paclitaxel by Fusarium solani isolated from Taxus celebica. Journal of Biosciences. 2008;**33**: 259-267

[43] Zhao K, Ping W, Li Q, Hao S, Zhao L, Gao T, et al. Aspergillus Niger var. taxi, a new species variant of Taxol-producing fungus isolated from Taxus cuspidata in China. Journal of Applied Microbiology. 2009;**107**(4):1202-1207

[44] Deng BW, Liu KH, Chen WQ, Ding XW, Xie XC. Fusarium solani, Tax-3, a new endophytic Taxol-producing fungus from Taxus chinensis. World Journal of Microbiology and Biotechnology. 2009;**25**:139-143

[45] Liu K, Ding X, Deng B, Chen W. Isolation and characterization of endophytic Taxol-producing fungi from Taxus chinensis. Journal of Industrial Microbiology and Biotechnology. 2009; **36**(9):1171

[46] Zhang P, Zhou P-P, Yu L-J. An endophytic Taxol-producing fungus from Taxus media, Cladosporium cladosporioides MD2. Current Microbiology. 2009;**59**:227-232

[47] Zhang P, Zhou P-P, Yu L-J. An endophytic Taxol-producing fungus

from Taxus x media, aspergillus candidus MD3. FEMS Microbiology Letters. 2009;**293**(2):155-159

[48] Miao Z, Wang Y, Yu X, Guo B, Tang K. A new endophytic taxane production fungus from Taxus chinensis. Applied Biochemistry and Microbiology. 2009;**45**:81-86

[49] Kumaran RS, Muthumary J, Kim E-K, Hur B-K. Production of Taxol from Phyllosticta dioscoreae, a leaf spot fungus isolated from Hibiscus rosa-sinensis. Biotechnology and Bioprocess Engineering. 2009;**14**:76-83

[50] Gangadevi V, Muthumary J. A novel endophytic Taxol-producing fungus Chaetomella raphigera isolated from a medicinal plant, Terminalia arjuna. Applied Biochemistry and Biotechnology. 2009;**158**:675-684

[51] Gangadevi V, Muthumary J. Taxol production by Pestalotiopsis terminaliae, an endophytic fungus of Terminalia arjuna (arjun tree). Biotechnology and Applied Biochemistry. 2009;**52**(1):9-15

[52] Zhao K, Sun L, Ma X, Li X, Wang X, Ping W, et al. Improved Taxol production in Nodulisporium sylviforme derived from inactivated protoplast fusion. African Journal of Biotechnology. 2011;**10**(20):4175-4182

[53] Pandi M, Kumaran RS, Choi Y-K, Kim HJ, Muthumary J. Isolation and detection of Taxol, an anticancer drug produced from Lasiodiplodia theobromae, an endophytic fungus of the medicinal plant Morinda citrifolia. African Journal of Biotechnology. 2011; **10**(8):1428-1435

[54] Bi J, Ji Y, Pan J, Yu Y, Chen H, Zhu X. A new Taxol-producing fungus (Pestalotiopsis malicola) and evidence for Taxol as a transient product in the culture. African Journal of Biotechnology. 2011;**10**(34):6647-6654

[55] Kumaran RS, Choi Y-K, Lee S, Jeon HJ, Jung H, Kim HJ. Isolation of Taxol, an anticancer drug produced by the endophytic fungus, Phoma betae. African Journal of Biotechnology. 2012; **11**(4):950-960

[56] Mirjalili MH, Farzaneh M, Bonfill M, Rezadoost H, Ghassempour A. Isolation and characterization of Stemphylium sedicola SBU-16 as a new endophytic Taxol-producing fungus from Taxus baccata grown in Iran. FEMS Microbiology Letters. 2012;**328**(2): 122-129

[57] Garyali S, Kumar A, Reddy MS. Taxol production by an endophytic fungus, Fusarium redolens, isolated from Himalayan yew. Journal of Microbiology and Biotechnology. 2013; **23**(10):1372-1380

[58] Yang Y, Zhao H, Barrero RA, Zhang B, Sun G, Wilson IW, et al. Genome sequencing and analysis of the paclitaxel-producing endophytic fungus Penicillium aurantiogriseum NRRL 62431. BMC Genomics. 2014;**15**(1):1-14

[59] Zaiyou J, Li M, Xiqiao H. An endophytic fungus efficiently producing paclitaxel isolated from Taxus wallichiana var. mairei. Medicine. 2017; **96**(27):e7406

[60] Qiao W, Ling F, Yu L, Huang Y, Wang T. Enhancing Taxol production in a novel endophytic fungus, Aspergillus aculeatinus Tax-6, isolated from Taxus chinensis var. mairei. Fungal Biology. 2017;**121**(12):1037-1044

[61] El-Sayed AS, Safan S, Mohamed NZ, Shaban L, Ali GS, Sitohy MZ. Induction of Taxol biosynthesis by Aspergillus terreus, endophyte of Podocarpus

gracilior Pilger, upon intimate interaction with the plant endogenous microbes. Process Biochemistry. 2018;**71**: 31-40

[62] El-Sayed AS, Ali DM, Yassin MA, Zayed RA, Ali GS. Sterol inhibitor "fluconazole" enhance the Taxol yield and molecular expression of its encoding genes cluster from Aspergillus flavipes. Process Biochemistry. 2019;**76**:55-67

[63] Gill H, Vasundhara M. Isolation of Taxol producing endophytic fungus Alternaria brassicicola from non-taxus medicinal plant Terminalia arjuna. World Journal of Microbiology and Biotechnology. 2019;**35**:1-8

[64] Kumar P, Singh B, Thakur V, Thakur A, Thakur N, Pandey D, et al. Hyper-production of Taxol from Aspergillus fumigatus, an endophytic fungus isolated from Taxus sp. of the Northern Himalayan region. Biotechnology Reports. 2019;**24**:e00395

[65] El-Sabbagh SM, Eissa OAE, Sallam MHE. Taxol production by an endophytic fungus cladosporioides isolated from Catheranthus roseus Cladosporium. Egyptian Journal of Experimental Biology (Botany). 2019; **15**(1):13-28

[66] Suresh G, Kokila D, Suresh TC, Kumaran S, Velmurugan P, Vedhanayakisri KA, et al. Mycosynthesis of anticancer drug Taxol by Aspergillus oryzae, an endophyte of Tarenna asiatica, characterization, and its activity against a human lung cancer cell line. Biocatalysis and Agricultural Biotechnology. 2020;**24**:101525

[67] El-Sayed E-SR, Zaki AG, Ahmed AS, Ismaiel AA. Production of the anticancer drug Taxol by the endophytic fungus Epicoccum nigrum TXB502: Enhanced production by gamma irradiation mutagenesis and immobilization technique. Applied Microbiology and Biotechnology. 2020;**104**(16):6991-7003

[68] Subramanian M, Marudhamuthu M. Hitherto unknown terpene synthase organization in Taxol-producing endophytic bacteria isolated from marine macroalgae. Current Microbiology. 2020;**77**:918-923

[69] Abdel-Fatah SS, El-Batal AI, El-Sherbiny GM, Khalaf MA, El-Sayed AS. Production, bioprocess optimization and γ-irradiation of Penicillium polonicum, as a new Taxol producing endophyte from Ginko biloba. Biotechnology Reports. 2021;**30**:e00623

[70] Jagan EG, Sharma P, Sureshkumar S, Pandi M. Isolation of Taxol and flavin-like fluorochrome from endophytic fungi of Mangifera indica. Journal of Pure & Applied Microbiology. 2021;**15** (4):2195-2208

[71] Gauchan DP, Vélëz H, Acharya A, Östman JR, Lundén K, Elfstrand M, et al. Annulohypoxylon sp. strain MUS1, an endophytic fungus isolated from Taxus wallichiana Zucc., produces Taxol and other bioactive metabolites. 3 Biotech. 2021;**11**(3):152

[72] Koutb M, Hassan E, El-Sokkary G, Saber S, Hussein N. Paclitaxel production by endophytic fungus, neopestalotiopsis clavispora KY624416 and subsequent extraction of chitosan from fungal biomass wastes. Global Nest Journal. 2021;**23**(3):370-380

[73] Abdel-Fatah SS, El-Sherbiny GM, Khalaf MA, El-Batal AI. Enhancement of Taxol production by endophytic fungi from Hibiscus and moringa plant using gamma irradiation. Egyptian Journal of Medical Microbiology. 2021; **30**(4):9-17

[74] Mohammadi Ballakuti N, Ghanati F, Zare-Maivan H, Alipour M, Moghaddam M, Abdolmaleki P. Taxoid profile in endophytic fungi isolated from Corylus avellana, introduces potential source for the production of Taxol in semi-synthetic approaches. Scientific Reports. 2022;**12**(1):9390

[75] Chowdhury DR, Chattopadhyay SK, Roy S. Isolation and partial characterization of bioactive components of Endophytic fungi Penicillium singorense, isolated from two Indian medicinal plants: Calotropis procera and Catharanthus roseus. American Journal of Microbiological Research. 2022;**10**(3):84-93

[76] Pandy R, Kumar SS, Suresh P, Annaraj J, Pandi M, Vellasamy S, et al. Screening and characterization of fungal Taxol-producing endophytic fungi for evaluation of antimicrobial and anticancer activities. Open Chemistry. 2023;**21**:1

[77] Adhikari P, Singh M, Pandey A. Production of Taxol by endophytic fungi isolated from roots of Himalayan yew (Taxus wallichiana Zucc.). Journal of Graphic Era University. 2022;**10**(2): 195-216

[78] Wang Y, Tang K. A new endophytic Taxol-and baccatin III-producing fungus isolated from Taxus chinensis var. mairei. African Journal of Biotechnology. 2011;**10**(72):16379-16386

[79] Zaiyou J, Li M, Guifang X, Xiuren Z. Isolation of an endophytic fungus producing baccatin III from Taxus wallichiana var. mairei. Journal of Industrial Microbiology and Biotechnology. 2013;**40**(11):1297-1302

[80] Li Y, Yang J, Zhou X, Zhao W, Jian Z. Isolation and identification of a 10-deacetyl baccatin-III-producing endophyte from Taxus wallichiana. Applied Biochemistry and Biotechnology. 2015;**175**:2224-2231

[81] Omeje EO, Ahomafor JE, Onyekaba TU, Monioro PO, Nneka I, Onyeloni S, et al. Endophytic fungi as alternative and reliable sources for potent anticancer agents. In: Natural Products and Cancer Drug Discovery. London, UK, Norderstedt, Germany: IntechOpen; 2017. pp. 52-60

[82] Vasundhara M, Kumar A, Reddy MS. Molecular approaches to screen bioactive compounds from endophytic fungi. Frontiers in Microbiology. 2016;**7**:1774

[83] Zhao J, Zhou L, Wang J, Shan T, Zhong L, Liu X, et al. Endophytic fungi for producing bioactive compounds originally from their host plants. Current Research, Technology and Education Topics in Applied Microbiology and Microbial Biotechnology. 2010;**1**:567-576

[84] Gond S, Kharwar R, White J Jr. Will fungi be the new source of the blockbuster drug Taxol? Fungal Biology Reviews. 2014;**28**(4):77-84

[85] Tejesvi MV, Pirttilä AM. Endophytic fungi, occurrence, and metabolites. In: Anke T, Schüffler A, editors. Physiology and Genetics: Selected Basic and Applied Aspects. Cham: Springer International Publishing; 2018. pp. 213-230

[86] Cao X, Xu L, Wang J, Dong M, Xu C, Kai G, et al. Endophytic fungus Pseudodidymocyrtis lobariellae KL27 promotes Taxol biosynthesis and accumulation in Taxus chinensis. BMC Plant Biology. 2022;**22**(1):1-18

[87] Liu Q, Li L, Chen Y, Wang S, Xue L, Meng W, et al. Diversity of endophytic microbes in Taxus yunnanensis and their potential for plant growth promotion

and taxane accumulation. Microorganisms. 2023;**11**(7):1645

[88] Cragg GM, Pezzuto JM. Natural products as a vital source for the discovery of cancer chemotherapeutic and chemopreventive agents. Medical Principles and Practice. 2016;**25**(Suppl. 2):41-59

[89] Bo G, Haiyan L, Lingqi Z. Isolation of an fungus producting vinbrastine. Journal of Yunnan University (Natural Sciences). 1998;**20**(3):214-215

[90] Lingqi Z, Bo G, Haiyan L, Songrong Z, Hua S, Su G, et al. Preliminary study on the isolation of endophytic fungus of Catharanthus roseus and its fermentation to produce products of therapeutic value. Zhong Cao Yao= Chinese Traditional and Herbal Drugs. 2000;**31**(11):805-807

[91] Kumar A, Ahmad A. Biotransformation of vinblastine to vincristine by the endophytic fungus Fusarium oxysporum isolated from Catharanthus roseus. Biocatalysis and Biotransformation. 2013;**31**(2):89-93

[92] Palem PP, Kuriakose GC, Jayabaskaran C. An endophytic fungus, Talaromyces radicus, isolated from Catharanthus roseus, produces vincristine and vinblastine, which induce apoptotic cell death. PLoS One. 2015;**10**(12):e0144476

[93] Ayob FW, Simarani K, Zainal Abidin N, Mohamad J. First report on a novel Nigrospora sphaerica isolated from Catharanthus roseus plant with anticarcinogenic properties. Microbial Biotechnology. 2017;**10**(4):926-932

[94] Anjum N, Chandra R. Endophytic bacteria of Catharanthus roseus as an alternative source of vindoline and application of response surface methodology to enhance its production. Archives of Biological Sciences. 2019; **71**(1):27-38

[95] Birat K, Siddiqi TO, Mir SR, Aslan J, Bansal R, Khan W, et al. Enhancement of vincristine under in vitro culture of Catharanthus roseus supplemented with Alternaria sesami endophytic fungal extract as a biotic elicitor. International Microbiology. 2022;**25**(2):275-284

[96] Xianzhi Y, Lingqi Z, Bo G, Shiping G. Preliminary study of a vincristine-proudcing endophytic fungus isolated from leaves of Catharanthus roseus. Zhong Cao Yao= Chinese Traditional and Herbal Drugs. 2004;**35**(1):79-81

[97] Kumar A, Abnave P, Ahmad A. Cultural, morphological and molecular characterization of vinca alkaloids producing endophytic fungus Fusarium solani isolated from Catharanthus roseus. International Journal of Botany and Research. 2013;**3**(2):2277-4815

[98] Kumar A, Patil D, Rajamohanan PR, Ahmad A. Isolation, purification and characterization of vinblastine and vincristine from endophytic fungus Fusarium oxysporum isolated from Catharanthus roseus. PLoS One. 2013; **8**(9):e71805

[99] Kuriakose GC, Palem PP, Jayabaskaran C. Fungal vincristine from Eutypella spp-CrP14 isolated from Catharanthus roseus induces apoptosis in human squamous carcinoma cell line-A431. BMC Complementary and Alternative Medicine. 2016;**16**(1):1-8

[100] Ashoka H, Hegde P, Manasa K, Madihalli C, Pradeep S, Shettihalli A. Isolation and detection of vinca alkaloids from endophytes isolated from Catharanthus roseus. European Journal

of Biomedical and Pharmaceutical Sciences. 2017;**10**:675-683

[101] Zafari D, Leylaiee S, Tajick MA. Isolation and identification of vinblastine from the fungus of Chaetomium globosum Cr95 isolated from Catharanthus roseus plant. Biological Journal of Microorganism. 2019;**8**(32):1-14

[102] Parthasarathy R, Shanmuganathan R, Pugazhendhi A. Vinblastine production by the endophytic fungus Curvularia verruculosa from the leaves of Catharanthus roseus and its in vitro cytotoxicity against HeLa cell line. Analytical Biochemistry. 2020;**593**: 113530

[103] Bandara CJ, Siriwardhana A, Karunaratne DN, Ratnayake Bandara BM, Wickramasinghe A, Krishnarajah SA, et al. Production of vincristine and vinblastine by the endophytic fungus Botryosphaeria laricina strain (CRS1) is dependent on stimulating factors present in Catharanthus roseus. The Natural Products Journal. 2021;**11**(2):221-230

[104] Andriambeloson OH, Noah RMA, Rigobert A, Jean-Marc C, Luciano R, Rado R. Isolation of Novel Vincristine and Vinblastine Producing Streptomyces Species from Catharanthus Roseus Rhizospheric Soil. Research Square. 2021. DOI: 10.21203/rs.3.rs-1082130/v1

[105] Ashraf J, Sharma MK, Biswas D. Separation, purification and characterization of vincristine and vinblastine from fusarium oxysporum, an endophytic fungus present in catharanthus roseus leaves. Journal of Advanced Scientific Research. 2021;**12** (01 Suppl 2):128-136

[106] Birat K, Binsuwaidan R, Siddiqi TO, Mir SR, Alshammari N, Adnan M, et al. Report on vincristine-producing endophytic fungus Nigrospora zimmermanii from leaves of Catharanthus roseus. Metabolites. 2022; **12**(11):1119

[107] Min C, Wang X. Isolation and identification of the 10-hydroxycamptothecin-producing endophytic fungi from Camptotheca acuminata decne. Acta Botanica Boreali-Occidentalia Sinica. 2009;**29**(3):614-617

[108] Kusari S, Zühlke S, Spiteller M. An endophytic fungus from Camptotheca acuminata that produces camptothecin and analogues. Journal of Natural Products. 2009;**72**(1):2-7

[109] Puri SC, Verma V, Amna T, Qazi GN, Spiteller M. An endophytic fungus from Nothapodytes f oetida that produces Camptothecin. Journal of Natural Products. 2005;**68**(12):1717-1719

[110] Ran X, Zhang G, Li S, Wang J. Characterization and antitumor activity of camptothecin from endophytic fungus Fusarium solani isolated from Camptotheca acuminate. African Health Sciences. 2017;**17**(2):566-574

[111] Rehman S, Shawl A, Kour A, Andrabi R, Sudan P, Sultan P, et al. An endophytic Neurospora sp. from Nothapodytes foetida producing camptothecin. Applied Biochemistry and Microbiology. 2008;**44**:203-209

[112] Rehman S, Shawl A, Kour A, Sultan P, Ahmad K, Khajuria R, et al. Comparative studies and identification of camptothecin produced by an endophyte at shake flask and bioreactor. Natural Product Research. 2009;**23**(11): 1050-1057

[113] Gurudatt P, Priti V, Shweta S, Ramesha B, Ravikanth G, Vasudeva R, et al. Attenuation of camptothecin

production and negative relation between hyphal biomass and camptothecin content in endophytic fungal strains isolated from Nothapodytes nimmoniana Grahm (Icacinaceae). Current Science. 2010; **98**(8):1006-1010

[114] Shweta S, Zuehlke S, Ramesha B, Priti V, Kumar PM, Ravikanth G, et al. Endophytic fungal strains of Fusarium solani, from Apodytes dimidiata E. Mey. ex Arn (Icacinaceae) produce camptothecin, 10-hydroxycamptothecin and 9-methoxycamptothecin. Phytochemistry. 2010;**71**(1):117-122

[115] Pu X, Qu X, Chen F, Bao J, Zhang G, Luo Y. Camptothecin-producing endophytic fungus Trichoderma atroviride LY357: Isolation, identification, and fermentation conditions optimization for camptothecin production. Applied Microbiology and Biotechnology. 2013; **97**:9365-9375

[116] Shweta S, Bindu JH, Raghu J, Suma H, Manjunatha B, Kumara PM, et al. Isolation of endophytic bacteria producing the anti-cancer alkaloid camptothecine from Miquelia dentata Bedd. (Icacinaceae). Phytomedicine. 2013;**20**(10):913-917

[117] Shweta S, Gurumurthy BR, Ravikanth G, Ramanan US, Shivanna MB. Endophytic fungi from Miquelia dentata Bedd., produce the anti-cancer alkaloid, camptothecine. Phytomedicine. 2013;**20**(3–4):337-342

[118] Su H, Kang J-c, Cao J, Mo L, Hyde KD. Medicinal plant endophytes produce analogous bioactive compounds. Chiang Mai Journal of Science. 2014;**41**(1):1-13

[119] Musavi SF, Dhavale A, Balakrishnan RM. Optimization and kinetic modeling of cell-associated camptothecin production from an endophytic Fusarium oxysporum NFX06. Preparative Biochemistry and Biotechnology. 2015; **45**(2):158-172

[120] Venugopalan A, Srivastava S. Enhanced camptothecin production by ethanol addition in the suspension culture of the endophyte, Fusarium solani. Bioresource Technology. 2015; **188**:251-257

[121] Pu X, Chen F, Yang Y, Qu X, Zhang G, Luo Y. Isolation and characterization of Paenibacillus polymyxa LY214, a camptothecin-producing endophytic bacterium from Camptotheca acuminata. Journal of Industrial Microbiology and Biotechnology. 2015;**42**(8):1197-1202

[122] Bhalkar BN, Patil SM, Govindwar SP. Camptothecine production by mixed fermentation of two endophytic fungi from Nothapodytes nimmoniana. Fungal Biology. 2016;**120**(6–7):873-883

[123] Soujanya KN, Siva R, Mohana Kumara P, Srimany A, Ravikanth G, Mulani FA, et al. Camptothecin-producing endophytic bacteria from Pyrenacantha volubilis Hook. (Icacinaceae): A possible role of a plasmid in the production of camptothecin. Phytomedicine. 2017;**36**: 160-167

[124] Aswini A, Soundhari C. Production of camptothecin from endophytic fungi and characterization by high-performance liquid chromatography and anticancer activity against colon cancer cell line. Asian Journal of Pharmaceutical and Clinical Research. 2018;**11**(3):166-170

[125] Clarance P, Lalitha J, Sales J, Khusro A, Agastian P. Anticancer

activity of camptothecin producing endophytes isolated from Chonemorpha fragrans (moon) Alston. (Apocynaceae). Research Journal of Biotechnology. 2019; **14**(5):74-82

[126] Ghiasvand M, Makhdoumi A, Matin MM, Vaezi J. Exploring the bioactive compounds from endophytic bacteria of a medicinal plant: Ephedra foliata (Ephedrales: Ephedraceae). Advances in Traditional Medicine. 2020; **20**:61-70

[127] Aswani R, Jasim B, Arun Vishnu R, Antony L, Remakanthan A, Aravindakumar CT, et al. Nanoelicitor based enhancement of camptothecin production in fungi isolated from Ophiorrhiza mungos. Biotechnology Progress. 2020;**36**(6):e3039

[128] Mohinudeen IAHK, Kanumuri R, Soujanya KN, Shaanker RU, Rayala SK, Srivastava S. Sustainable production of camptothecin from an Alternaria sp. isolated from Nothapodytes nimmoniana. Scientific Reports. 2021; **11**(1):1478

[129] Dhakshinamoorthy M, Ponnusamy SK, Nyayiru Kannaian UP, Srinivasan B, Shankar SN, Kilavan PK. Plant-microbe interactions implicated in the production of camptothecin – An anticancer biometabolite from Phyllosticta elongata MH458897 a novel endophytic strain isolated from medicinal plant of Western Ghats of India. Environmental Research. 2021; **201**:111564

[130] El-Sayed ASA, Khalaf SA, Azez HA, Hussein HA, El-Moslamy SH, Sitohy B, et al. Production, bioprocess optimization and anticancer activity of Camptothecin from aspergillus terreus and aspergillus flavus, endophytes of Ficus elastica. Process Biochemistry. 2021;**107**:59-73

[131] El-Sayed ASA, Hassan WHB, Sweilam SH, Alqarni MH, El Sayed ZI, Abdel-Aal MM, et al. Production, bioprocessing and anti-proliferative activity of Camptothecin from Penicillium chrysogenum, an endozoic of marine sponge, Cliona sp., as a metabolically stable Camptothecin producing isolate. Molecules. 2022;**27**:9

[132] El-Sayed ASA, George NM, Abou-Elnour A, El-Mekkawy RM, El-Demerdash MM. Production and bioprocessing of camptothecin from Aspergillus terreus, an endophyte of Cestrum parqui, restoring their biosynthetic potency by Citrus limonum peel extracts. Microbial Cell Factories. 2023;**22**(1):4

[133] Degambada KD, Kumara PAASP, Salim N, Abeysekera AM, Chandrika UG, Diaporthe sp. F18; a new source of camptothecin-producing endophytic fungus from Nothapodytes nimmoniana growing in Sri Lanka. Natural Product Research. 2023;**37**(1): 113-118

[134] Xianzhi Y, Shiping G, Lingqi Z, Hua S. Select of producing podophyllotoxin endophytic fungi from podophyllin plant. Natural Product Research and Development. 2003;**15**(5): 419-422

[135] Zeng S, Shao H, Zhang L. An endophytic fungus producing a substance analogous to podophyllotoxin isolated from Diphylleia sinensis. Journal of Microbiology. 2004;**24**:1-2

[136] Guo S, Jiang B, Su Y, Liu S, Zhang L. Podophyllotoxin and its analogues from the endophytic fungi derived from Dysosma veitchii. Biotechnology. 2004;**14**:55-57

[137] Lu L, He J, Yu X, Li G, Zhang X. Studies on isolation and identification of

endophytic fungi strain SC13 from harmaceutical plant Sabina vulgaris ant. and metabolites. Acta Agriculturae Boreali-occidentalis Sinica. 2006;**15**: 85-89

[138] Eyberger AL, Dondapati R, Porter JR. Endophyte fungal isolates from Podophyllum peltatum produce podophyllotoxin. Journal of Natural Products. 2006;**69**(8):1121-1124

[139] Puri SC, Nazir A, Chawla R, Arora R, Riyaz-ul-Hasan S, Amna T, et al. The endophytic fungus Trametes hirsuta as a novel alternative source of podophyllotoxin and related aryl tetralin lignans. Journal of Biotechnology. 2006; **122**(4):494-510

[140] Li C. Fermentation conditions of Sinopodophyllum hexandrum endophytic fungus on production of podophyllotoxin. Food and Fermentation Industries. 2007;**33**(9):28

[141] Kour A, Shawl AS, Rehman S, Sultan P, Qazi PH, Suden P, et al. Isolation and identification of an endophytic strain of Fusarium oxysporum producing podophyllotoxin from Juniperus recurva. World Journal of Microbiology and Biotechnology. 2008;**24**:1115-1121

[142] Nadeem M, Ram M, Alam P, Ahmad MM, Mohammad A, Al-Qurainy F, et al. Fusarium solani, P1, a new endophytic podophyllotoxin-producing fungus from roots of Podophyllum hexandrum. African Journal of Microbiology Research. 2012;**6**(10): 2493-2499

[143] Huang J-X, Zhang J, Zhang X-R, Zhang K, Zhang X, He X-R. Mucor fragilis as a novel source of the key pharmaceutical agents podophyllotoxin and kaempferol. Pharmaceutical Biology. 2014;**52**(10):1237-1243

[144] Liang Z, Zhang J, Zhang X, Li J, Zhang X, Zhao C. Endophytic fungus from Sinopodophyllum emodi (wall.) ying that produces Podophyllotoxin. Journal of Chromatographic Science. 2016;**54**(2):175-178

[145] Aharwal RP, Kumar S, Sandhu SS. Endophytic mycoflora as a source of biotherapeutic compounds for disease treatment. Journal of Applied Pharmaceutical Science. 2016;**6**(10): 242-254

[146] Wang T, Ma Y, Ye Y, Zheng H, Zhang B, Zhang E. Screening and identification of endophytic fungi producing podophyllotoxin compounds in Sinopodophyllum hexandrum stems. Chinese Journal of Experimental Traditional Medical Formulae. 2017;**39**: 402-408

[147] Tan X-m, Zhou Y-q, Zhou X-l, Xia X-h, Wei Y, He L-l, et al. Diversity and bioactive potential of culturable fungal endophytes of Dysosma versipellis; a rare medicinal plant endemic to China. Scientific Reports. 2018;**8**(1):5929

[148] Gohar UF, Attia Majeed BM, Mukhtar H. Optimum conditions for enhanced production of Podophyllotoxin from Penicillium sp. isolated from Khanspur, Pakistan. Pakistan Journal of Zoology. 2022;**54**(6):2775

[149] Thi Tran H, Thu Nguyen G, Thi Nguyen HH, Thi Tran H, Hong Tran Q, Ho Tran Q, et al. Isolation and cytotoxic potency of endophytic fungi associated with Dysosma difformis, a study for the novel resources of Podophyllotoxin. Mycobiology. 2022;**50**(5):389-398

[150] Nguyen GT, Nguyen HTH, Tran HT, Tran HT, Ho AN, Tran QH, et al. Enhanced podophyllotoxin production of endophyte Fusarium

proliferatum TQN5T by host extract and phenylalanine. Applied Microbiology and Biotechnology. 2023;**107**(17): 5367-5378

[151] Kusari S, Lamshöft M, Spiteller M. Aspergillus fumigatus Fresenius, an endophytic fungus from Juniperus communis L. Horstmann as a novel source of the anticancer pro-drug deoxypodophyllotoxin. Journal of Applied Microbiology. 2009;**107**(3): 1019-1030

[152] Li W, Zhou J, Lin Z, Hu Z. Study on fermentation condition for production of huperzine a from endophytic fungus 2F09P03B of Huperzia serrata. Chinese Medicinal Biotechnology. 2007;**2**(4): 254-259

[153] Zan J, Wang J, Pan S. Isolation and preliminary identification of the endophytic fungi which produce Hupzine a from four species in Hupziaceae and determination of Huperzine a by HPLC. Fudan University Journal of Medical Sciences. 2009;**36**(4): 445-449

[154] Zhou S, Yang F, Lan S, Xu N, Hong Y. Huperzine a producing conditions from endophytic fungus in SHB Huperzia serrata. Journal of Microbiology. 2009;**3**:32-36

[155] Zhu D, Wang J, Zeng Q, Zhang Z, Yan R. A novel endophytic Huperzine A–producing fungus, Shiraia sp. Slf14, isolated from Huperzia serrata. Journal of Applied Microbiology. 2010;**109**(4): 1469-1478

[156] Zhang ZB, Zeng QG, Yan RM, Wang Y, Zou ZR, Zhu D. Endophytic fungus Cladosporium cladosporioides LF70 from Huperzia serrata produces Huperzine A. World Journal of Microbiology and Biotechnology. 2011; **27**:479-486

[157] Wang Y, Yan R, Zeng Q, Zhang Z, Wang D, Zhu D. Producing huperzine a by an endophytic fungus from Huperzia serrata. Mycosystema. 2011;**30**(2): 255-262

[158] Wang Y, Zeng QG, Zhang ZB, Yan RM, Wang LY, Zhu D. Isolation and characterization of endophytic huperzine A-producing fungi from Huperzia serrata. Journal of Industrial Microbiology and Biotechnology. 2011; **38**(9):1267-1278

[159] Shu S, Zhao X, Wang W, Zhang G, Cosoveanu A, Ahn Y, et al. Identification of a novel endophytic fungus from Huperzia serrata which produces huperzine a. World Journal of Microbiology and Biotechnology. 2014; **30**:3101-3109

[160] Dong L-H, Fan S-W, Ling Q-Z, Huang B-B, Wei Z-J. Indentification of huperzine A-producing endophytic fungi isolated from Huperzia serrata. World Journal of Microbiology and Biotechnology. 2014;**30**:1011-1017

[161] Su J, Yang M. Huperzine a production by Paecilomyces tenuis YS-13, an endophytic fungus isolated from Huperzia serrata. Natural Product Research. 2015;**29**(11):1035-1041

[162] Han W, Song T, Yang S, Li X, Zhang H, Wu Y, et al. Identification of alkaloids and huperzine A-producing endophytic fungi isolated from wild Huperzia serrata. Journal of International Pharmaceutical Research. 2015;**6**:507-512

[163] Zhang F, Wang M, Zheng Y, Liu H, Zhang X, Wu S. Isolation and characterzation of endophytic Huperzine A-producing fungi from Phlegmariurus phlegmaria. Microbiology. 2015;**84**:701-709

[164] Wang Y, Lai Z, Li X-X, Yan R-M, Zhang Z-B, Yang H-L, et al. Isolation, diversity and acetylcholinesterase inhibitory activity of the culturable endophytic fungi harboured in Huperzia serrata from Jinggang Mountain, China. World Journal of Microbiology and Biotechnology. 2016;**32**:1-23

[165] Thi Minh Le T, Thi Hong Hoang A, Thi Bich Le T, Thi Bich Vo T, Van Quyen D, Hoang CH. Isolation of endophytic fungi and screening of Huperzine A–producing fungus from Huperzia serrata in Vietnam. Scientific Reports. 2019;**9**(1):16152

[166] Zaki AG, El-Shatoury EH, Ahmed AS, Al-Hagar OE. Production and enhancement of the acetylcholinesterase inhibitor, huperzine a, from an endophytic Alternaria brassicae AGF041. Applied Microbiology and Biotechnology. 2019;**103**:5867-5878

[167] Kang X, Liu C, Shen P, Hu L, Lin R, Ling J, et al. Genomic characterization provides new insights into the biosynthesis of the secondary metabolite huperzine a in the endophyte Colletotrichum gloeosporioides Cg01. Frontiers in Microbiology. 2019; **9**:3237

[168] Wen-Xia H, Zhong-Wen H, Min J, Han Z, Wei-Ze L, Li-Bin Y, et al. Five novel and highly efficient endophytic fungi isolated from Huperzia serrata expressing huperzine a for the treatment of Alzheimer's disease. Applied Microbiology and Biotechnology. 2020; **104**:9159-9177

[169] Cruz-Miranda OL, Folch-Mallol J, Martínez-Morales F, Gesto-Borroto R, Villarreal ML, Taketa AC. Identification of a Huperzine A-producing endophytic fungus from Phlegmariurus taxifolius. Molecular Biology Reports. 2020;**47**(1): 489-495

[170] Le TTM, Hoang ATH, Nguyen NP, Le TTB, Trinh HTT, Vo TTB, et al. A novel huperzine A-producing endophytic fungus Fusarium sp. Rsp5.2 isolated from Huperzia serrate. Biotechnology Letters. 2020;**42**(6): 987-995

[171] Putri NWPS, Ariantari NP. Production of huperzine a by fungal endophytes associated with huperziaceae plants. Journal Pharmaceutical Science and Application. 2023;**5**(1):45-52

[172] Ying Y-M, Shan W-G, Zhan Z-J. Biotransformation of Huperzine a by a fungal endophyte of Huperzia serrata furnished sesquiterpenoid–alkaloid hybrids. Journal of Natural Products. 2014;**77**(9):2054-2059

[173] Thirumalanadhuni V, Yerraguravagari LL, Palempalli UMD. Endophytic microflora: The fountainhead of anticancer metabolites —A systematic review. Recent Developments in Applied Microbiology and Biochemistry. 2021;**2**:13-20

[174] Madhusudhan CM, Bharathi RT, Prakash SH. Isolation and purification of bioactive metabolites from fungal endophytes–a review. Current Biochemical Engineering. 2015;**2**(2): 111-117

[175] Song X, Wu H, Yin Z, Lian M, Yin C. Endophytic bacteria isolated from Panax ginseng improves ginsenoside accumulation in adventitious ginseng root culture. Molecules. 2017;**22**(6):837

[176] Fu Y, Yin ZH, Yin CY. Biotransformation of ginsenoside Rb1 to ginsenoside Rg3 by endophytic bacterium Burkholderia sp. GE 17-7 isolated from Panax ginseng. Journal of Applied Microbiology. 2017;**122**(6): 1579-1585

[177] Fu Y. Biotransformation of ginsenoside Rb1 to gyp-XVII and minor ginsenoside Rg3 by endophytic bacterium Flavobacterium sp. GE 32 isolated from Panax ginseng. Letters in Applied Microbiology. 2019;**68**(2): 134-141

[178] Yang H-R, Yuan J, Liu L-H, Zhang W, Chen F, Dai C-C. Endophytic Pseudomonas fluorescens induced sesquiterpenoid accumulation mediated by gibberellic acid and jasmonic acid in Atractylodes macrocephala Koidz plantlets. Plant Cell, Tissue and Organ Culture (PCTOC). 2019;**138**:445-457

[179] Yin DD, Wang YL, Yang M, Yin DK, Wang GK, Xu F. Analysis of chuanxiong Rhizoma substrate on production of ligustrazine in endophytic Bacillus subtilis by ultra high performance liquid chromatography with quadrupole time-of-flight mass spectrometry. Journal of Separation Science. 2019;**42**(19):3067-3076

[180] Hemmati N, Azizi M, Spina R, Dupire F, Arouei H, Saeedi M, et al. Accumulation of ajmalicine and vinblastine in cell cultures is enhanced by endophytic fungi of Catharanthus roseus cv. icy pink. Industrial Crops and Products. 2020;**158**:112776

Chapter 4

A Review in Medical, Pharmacological and Industrial Importance of Roselle *Hibiscus sabdariffa* L.

Ali S. Hassoon, Ahmed A. Kadhim and Madeha H. Hussein

Abstract

Medicinal plants and plants occupy a major place in the world's agricultural and industrial production as the main source of pharmaceutical drugs or as a source of active substances that are used in the preparation of medicine. The Roselle plant *Hibiscus sabdariffa* L.(Malvaceae Family) is known in Arabic as "karkade" and "Roselle". The continental regions of Africa and Asia are the original home of the Roselle. The months of March and April and proceeds to reap the fruits during the months of October to the end of December. Recent researches and studies have pointed to the use of Roselle plant in the medical, food and industrial fields. It is used in the manufacture of jellies and jams. The syrup is added to some medical preparations. It is refreshing, moisturizing, and helps digestion. It is also a useful drink in the cases of bile. It causes urination and acts as an antifungal agent. and is used in the treatment of high blood pressure, It is also used in cancer treatment.

Keywords: rosella, medicinally plants, pharmcological uses, industrial uses of rosella, rosella as food

1. Introduction

The current interest is in the exploitation of the active substances of some pharmaceutical plants and their manufacture as important medical drugs for the treatment of many diseases and for the prevention, treatment and pain relief, and therefore these plants have taken a great deal of attention and studies as the main source of these drugs and the manufacture of safe medicines [1]. It has been used in many countries of the world to treat various viral and microbial diseases because they are more useful in different treatment stages and are safer and less expensive compared to laboratory-based chemical treatments [2].

Hibiscus sabdariffa L., which belongs to the Malvaceae family, is a medically important plant [3]. Its medicinal significance is concentrated in its leaves, which are a source of Hibiscus, which has a medical effect. Its leaves are rich in vitamin C, citric,

Tartaric, It also contains Protocatechuic Acid (PCA), an important antioxidant as well as its role in the treatment of certain cancers [4].

It is used as a refreshing beverage for the purpose of tempering the high temperatures and uses the red dye produced artificially as food nutritious natural food, and also enters the manufacture of jams, ice, candy, and food preservation, and the seeds contain a high proportion of oil up to the limit of 20–25% 30–35% protein [5].

2. Plant classification and naming

Hibiscus sabdariffa L. belongs to the Malvaceae family of Malvales, a Sabdariffa species of Hibiscus, a plant covered with Angiosperms dicotyledons seeds [6].

The origin of the name of the Rosella comes back to the area of Gujarat in India, and it is known by many names that differ from one region to another. The common name in Iraq is al-Gujarat tea. It is also called kaajah, red acid, kardipip, and gypsy. In Arab countries, it is called Karkade. In Brazil, it is called "Rosela" and is called commercially in America (Hibiscus). In England, it is called rozelle, sorrel, sour. In France, it is called "Bissap" and "Jamaica". In North and East Africa it is called "karkade or carcade" Senegal (bissap) and Nigeria (serrol) and in Iran is known as (Chaye -Torosh) [7, 8].

3. Plant description

Agglutaceous leafy plant with roots, spines, and legs, cylindrical with a smooth texture and greenish-red color, up to more than two meters depending on the variety. The leaves are simple, reciprocal, hand-shaped, finely edged, greenish-red, long-necked and long-necked. The leaves are white, yellow or red, consisting of five large, thick, succulent, thick leaves. The fruits are capsules containing a large number of kidney seeds. The texture is dark brown (**Figure 1**) [9, 10].

Figure 1.
Cultivation of rosella in my field of research.

4. Origin home and the appropriate environment

The original locality of the Gujarat plant is not yet clearly known, as it is believed that the tropical and subtropical regions of Africa are the original home of the Gujarat and others believe that India is its original home [11], That the Arab homeland is the original home of the Quraysh.

In the Arabian world, the cultivars are widely cultivated in semi-tropical regions of China, Malaysia, India, Indonesia, the Philippines, Vietnam, Nigeria, and Mexico. In the Arab world, they are grown in Saudi Arabia, Egypt, and Sudan [12]. Al-Qadisiyah Governorate, south of Iraq, especially in the Sunni area. It was cultivated as a summer crop in the southern and central regions. It was used as a refreshing drink and as a treatment for many diseases.

Cultivars are grown in most types of soils. However, they are grown in soft and fertile soil. They can withstand heavy soil conditions. The plant needs 4 to 8 months to complete its phytoplankton growth. The plant does not tolerate temperatures below 20° C during its growing stage. In areas where the temperature range is 28–35° C and air humidity is not more than 65% during the vegetative and syphilis stages, it is a long day plant. It requires about 13 hours of lighting per day to push the plant to bloom [13].

High humidity and high rainfall play a negative role during the harvest period. Drought conditions lead to a decrease in yield and a decrease in quality and quality of leaves. The plant needs about 130–200 mm of monthly rainfall rate during the first months of growth. The vegetative effect of the grits on the quality of the seeds, the prevailing environmental conditions, the date of harvest and post-harvest treatments, especially drying (**Figure 2**) [14].

5. Plant content of chemicals and effective compounds

Hibiscus species are different in their percentage of active food and chemical content depending on their genetic differences. Effective chemicals spread throughout the plant, which gives it high medical and pharmaceutical importance. It is a rich source of 25–35% protein, The most important of which are lysine, alkaline, and leucine [15].

Figure 2.
Roselle varieties (red, white and lined) in my research field.

It is also rich in calories. Each 100 g of calyxes contains 49 calories and contains large amounts of fiber up to 2.3 g per 100 g of juice [16].

Either containing fat or carbohydrates, it contains a good amount of up to 25% in dry sepals leaves, in addition to the presence of nutrients of high nutritional value in varying amounts such as iron, potassium, phosphorus, calcium, manganese, sodium, aluminum and chromium. Organic acids such as malic acid, citric, tartaric and ascorbic acid, which are responsible for the acid taste in the drink and some other organic acids such as Protocatechuic acid phenolic acid and also contain the aromatic acids aromatic acid, and that the leaves are rich in riboflavin acid Niacin riboflavin [3].

The sepals contain amino acids and the most common amino acids are aspartic acid [17]. It also contains the dye of anthocyanin, which is responsible for the red color of the plant, as well as some coloring substances such as carotene and thiamine. In addition, it contains many vitamins, most notably ascorbic acid (vitamin C). This is considered an important source of this vitamin in addition to containing vitamins A, B1, B2and B complex). The compounds also contain some of the classics such as Hibiscus hydrochloride with high physiological effect, which add high medicinal importance to the plant. These compounds are organic compounds that degrade acids by some enzymes to non-sugary substances [18].

In addition, the compounds also contain beta-compounds, which explains its dark red color when placed in acidic medium and contains some gels that may reach about 15% [19]. It is noted that the water extract of the saplings contains several important active compounds such as phenols, flavors, and tannins, which are different types of plant pigments that add various colors to the spores such as red, yellow and blue that are used in medicinal uses [20], and contains a chemical called Mucilage A refreshing, soothing substance is a simple, digestible sugar that adds flavor and sweetness to the drink of kora and also contains bacteriostatic substances [21].

Rosella seeds contain oil similar to the oil properties found in cotton seeds in terms of color, which amounts to about 17%. It can be edible, which is of high pharmacological importance. It is used for medical purposes in addition to containing a percentage of starch, carbohydrates, cellulose and cholesterol, And many organic acids such as oleic, formulaic, malvalic and citric acids [22]. Seeds are also an important source of many important nutrients such as potassium, phosphorus, magnesium, and micronutrients such as calcium, zinc, sulfur, sodium, and manganese [23]. In addition, the seeds contain a percentage of unsaturated fatty acids [24].

Leaves are also rich in chemicals active and important pharmaceutical, as they contain fats and fibers and many important compounds, such as Cliocosides and organic acids in addition to the food, including the most important calcium salts Calcium Oxalate. The roots of the Gujarat plant also contain Tartaric acid and Saponins [25].

6. The medical and pharmacological importance of rosella

In recent years, medicinal plants have received great attention from researchers for the importance of pharmaceuticals that have contributed effectively to the development of the medical aspect in all its fields. Rosella is one of the most important medicinal plants. The pharaohs used it in many medicinal recipes and used it as a disinfectant. And applied much scientific research where the compounds played by its parts played an important role in the treatment of many of the pathogens that confront human health [18].

Dahiru et al. [26] found that the oral administration of the mouse with the red leaf extract has significant effects against cancerous tumors because it contains the Protocatechuic Acid (PCA), which also helps to treat cirrhosis by removing the harmful effect of CCl4, And is a pharmacologically effective agent in minimizing the carcinogenic effect of diethylntro sardine in the liver, reducing the work of carcinogenic genes and thus leading to the anti-cancer function of anti-cancer [19]. It is also used for the extraction of blood viscosity by reducing systolic and diastolic blood pressure. It is used to calm the contractions in the muscles of the uterus, stomach, intestines, pain and antispasmodic pain, as well as the pain of the chest Pectoral and is an antidote to tapeworms and cylindrical [27]. As well as containing anthocyanin, which acts as an antioxidant because of its possession of phenols [28].

Rosella is also used as a Diuretic, tonic for heartbeat and for calming nerves. It is also important in the treatment of atherosclerosis, facilitating digestion of Dyspepsia, coughs, stimulation of bowel movement and the ability to stimulate the body to resist many intestinal diseases [29]. It also helps to heal ulcers and promote the growth of hair follicles. The ethanol extracts of leaves of Gujarat also contribute to the reduction of the accumulation of antilithiatic compounds formed in the kidneys, except for the oxalate compounds [30].

The researchers found that the extract of the leaves of the giraffes acts as an antibacterial agent, inhibiting the growth of sensitive and antibiotic-resistant bacteria. It is effective in killing the bacteria produced by the bacteriogenic bacteria of Bacillus and *E. coli* and the microbes that cause tuberculosis. It is used to nourish the muscles to prevent infection caused by Campylobacter bacteria and aerobes, as well as to prevent contamination caused by delay in fat metabolism by giving it to the cow embryo as well as for the treatment of fever and cholera diseases due to the acidity of algal tea (pH 3.5) [31].

Odigie et al. [32] found that injecting the human with a serum containing the extract of the leaf leaves of the Gujarat plant helped control the level of sugar in the blood and reduced the effect of atherosclerosis. Recent research by Essa et al. [33]. The red leaf extract has an effective effect on the levels of urea and ammonia by protecting the liver against the harmful effects of ammonium chloride and also protecting the liver from the levels of fat oxidation products such as HP (hydroperoxides) and AST (aspartate transaminase) radical.

Lin et al. [27] found that the use of sepals extracts resulted in lowering blood cholesterol (8.3–14.4%) after 1 month of study on a sample of 42 people at a rate of three times a day.

The findings of Mckay et al. [34] show that the almost daily intake of sepals reduces blood pressure in adults with high blood pressure, as it has proved to be beneficial in many dietary changes, including lowering cholesterol levels in the blood.

Research has also shown that eating Gujarat tea contributed to the treatment of anemia in humans Anemia [35] found that drinking this drink had multiple activities on certain biochemical properties in humans and was safe in reducing the disease, Is used in African folk medicine to this day as an antipyretic, anthelmintic, and oral infection as a local application of external wounds and in Iran is used to treat insomnia [36].

7. Food and industrial uses of rosella

The importance of the Gujarat plant lies in its many uses. It is an important food and industrial crop of high economic value in many countries of the world and is a good source of income. The different parts of the plant are used in many foods uses. Fresh and dried plant leaves are used in the preparation of hot and cold beverages.

Also in the manufacture of sweets and ice cream and ice cream, chocolate and ice mixtures for adding flavor and color to the food [37], and leaf leaves of the trees are also involved in several uses of food as the leaves of the green plant eaten fresh or cooked, It has a high nutritional value and is also used in the preparation of soups and salads, as well as in the preparation of spices as well as in some foods to add acid flavor to them [38]. On the other hand, cane seeds can be grounded or roasted to contain high protein, which is also a source of edible oil [39]. The seeds can also be used to prepare fodder crops for animals, especially sheep. The residues from the extraction of seed oil are also used to feed poultry and livestock, as well as to prepare fish food. The biological plant yield of leaves and fruits is also used with their contents in feeding the animals.

The plant of Gujarat is one of the most important industrial plants. The plant is used in many important industrial applications. It is grown as a source of natural fibers, which are used in the manufacture of plastic bags, clothing, fishing nets, and ropes, as well as the user's source in the preparation of industrial paper pulp [40]. The plant extracts are also used in the manufacture of cosmetics such as lipstick and food industries such as sweets and ice [41].

8. Conclusions

This study aims at understanding the importance of the Gujarat plant, it's propagation and planting, the appropriate environmental conditions and its most important medical, industrial and agricultural uses. In the light of this study, we can conclude that the Gujarat plant is one of the most important medicinal plants and it is necessary to expand its cultivation and production because of its medical and industrial importance. And to expand the extraction of medicinal compounds from it and its use in the pharmaceutical industries.

Author details

Ali S. Hassoon[1]*, Ahmed A. Kadhim[1] and Madeha H. Hussein[2]

1 Plant Production Techniques Department, Al-Musaib Tech. College, Al-Furat Al-AwsatTech. University, Iraq

2 Pharmacy Department, Al-Mansour Medical Instiute Tech, Midlle Tech. University, Iraq

*Address all correspondence to: com.hs.ali@atu.edu.iq

References

[1] Ayoubi MR. Part II. In: Alternative Medicine and Herbal Medicine. First ed. Cairo-Arab Republic of Egypt: Madbouli Press Publishing and Distribution; 2011 350 pages

[2] Borokini TI, Omotayo FO. Phytochemical and ethno botanical study of some selected medicinal plants from Nigeria. Journal of Medicinal Plants Research. 2012;**6**(7):1106-1118

[3] Abbas MK, Ali SA. Effect of foliar application of NPK on some growth characters of two cultivars of roselle (*Hibiscus sabdariffa* L.). American Journal of Plant Physiology. 2011;**6**:220-227

[4] Kılıç CS, Aslan S, Kartal M, Coskun M. Fatty acid composition of *Hibiscus trionum* L. (Malvaceae). Records of Natural Production. 2011;**5**(1):65-69

[5] Louis SJ, Kadams AM, Simon SY, Mohammed SG. Combining ability in Roselle cultivars for agronomic traits in Yola, Nigeria. Greener Journal of Agricultural Sciences. 2013;**3**(2):145-149

[6] Ajithadoss K, Pandian T, Rathinkumar S, Edwin R, Sekar T, Sakar P, et al. Botany Higher Secondary Second Year. 1st ed. Chennai: Government of Tamil Nadu Textbook Corporation College Road; 2006

[7] Anonymous. Exemplary description of 20 crop. Hibiscus Nederland e.v. In: Organic Farming in the Tropic and Subtropics. 1st ed. 2000

[8] Nasrallah AY, Hussam SS, Shamil IN. Effect of some plant growth regulators on field properties and antioxidant production of buckwheat leaves. Journal of Iraqi Agricultural Sciences. 2015;**46**(5):682-694

[9] Ross IA. Medicinal Plants of the World: Chemical Constituents, Traditional and Modern Medicinal Uses. Vol. Vol. 1. Humana Press Inc.; 2003

[10] Schippers RR. African Indigenous Vegetables: An Overview of Cultivated Species. UK: National resource institute. publisher Chatham; 2000

[11] Tounkara F, Amadou I, Wei LG, Hui HY. Effect of boiling on the physicochemical properties of Roselle seeds (*Hibiscus sabdariffa* L.) cultivated in Mali. African Journal of Biotechnology. 2011;**10**(79):18160-18166

[12] Eslaminejad T, Zakaria M. Morphological characteristics and pathogenicity of fungi associated with Roselle (*hibiscus sabdariffa* L.) diseases in Penang, Malaysia. Microbial Pathogenesis. 2011;**51**(5):325-337

[13] Sobhy D. Service and Cultivation of Hibiscus. The Egyptian Arabic Republic: Horticulture Research Institute, Medical and Aromatic Plants Research, Ministry of Agriculture and Land Reclamation-Agricultural Research Center, Central Administration of Agricultural Extension; 2005

[14] Plotto A. Hibiscus: Post-Production Management for Improved Market Access. Food and Agriculture Organization of the UN (FAO); 2004

[15] Hainida E, Ismail A, Hashim N, Zakiah A. Effects of defatted dried roselle (*Hibiscus sabdariffa* L.) seed powder on lipid profiles of hypercholesterolemia rats. Journal of the Science of Food and Agriculture. 2008a;**88**(6):1043-1050

[16] Awaji MN. Hibiscus tea Refreshing. First ed. House of civilization for publication, printing and distribution; 2006 99 pages

[17] Frimpong G. Investigating the Suitabity of (*Hibiscus sabdariffa* L.) Calyx Extract as Colouring Agent for Paediatric Syrup [Thesis]. Kumasi. Ghana: Department of Pharmaceutic. Kwame Nkrumah University of Science and Technology; 2008

[18] Saadi M. The Secrets and Secrets of Medicinal Plants and Drugs in Ancient and Modern Medicine. Oman Jourdan: Dar Al Yazouri Scientific Publishing and Distribution; 2006

[19] Dasouki HS. Fundamentals of Plant Physiology. The Egyptian Arabic Republic: Mansoura University; 2008

[20] Sheikh M. Effect of Number of Irrigation and Spray with the Extract of *Hibiscus sabdariffa* L. in the Growth and Yield of the Plant [Thesis]. Iraq: Faculty of Science, University of Babylon-Republic of Iraq; 2004

[21] Shaker KA, Rahman. Studying the chemical composition and technical characteristics of the flowers of the *Hibiscus sabdariffa* L. Journal of Iraqi Agricultural Sciences. 2002;**7**(8):171-177

[22] Mukhtar FB. Effect of some plant growth regulators on the growth and nutritional value of *Hibiscus sabdariffa* L. (red sorrel). International Journal of Pharmaceutical Sciences and Research. 2008;**2**(3):70-75

[23] Nzikou JM, Kalou GB, Matos L, Ganongo Po FB, Mboussi M, Moutoula FE, et al. Characteristic and nutritional evaluation of seed oil from Roselle (*Hibiscus sabdariffa* L.) in Gongo–Brazzaville. Current Research Journal of Biological Sciences. 2011;**3**(2):141-146

[24] Rao PU. Nutrient composition and biological evaluation of mesta (*Hibiscus sabdariffa* L.) seeds. Plant Foods for Human Nutrition. 1996;**49**(1):27-34

[25] Mahadevan N, Shivali, Pradeep K. *Hibiscus sadariffa* Linn-an overview natural product radiance. Natural Product Radiance. 2009;**8**(1):77-83

[26] Dahiru D, Obi OJ, Umaru H. Effect *of Hibiscus sabdariffa* L. calyx extract on carbon tetrachloride induced liver damage. Biology Biochemistry. 2003;**15**(1):27-33

[27] Lin T, Lin H, Chen C. *Hibiscus sabdariffa* L. extract reduces serum cholesterol in men and women. Nutrition Research. 2007a;**27**:140-145

[28] Tsai PJ, Mcintosh J, Pearce P, Caden B, Jordan TB. Anthocyanin and antioxidant capacity in roselle *Hibiscus sabdariffa* L. extract. Food Research International. 2002;**35**:351-356

[29] Ramadan AF, Jamil SM. Effect of spraying some nutrients on the growth and distribution of *Hibiscus sabdariffa* L. A. Natural and Total characteristics. Anbar Journal of Agricultural Sciences. 2010;**8**(4):323-336 Special number for the conference

[30] Betanabhatla KS, Christina AM, Sundar BS, Selvakumar S, Saravanan KS. Antilithiatic activity of *Hibiscus sabdariffa* Linn. On ethylene glycol-induced lithiasis in rats. Natural Product Radiance. 2009;**8**(1):43-47

[31] Yang L, Gou Y, Zhao T, Zhao J, Li F, Zhang B, et al. Antioxidant capacity of extracts from calyx fruits of roselle (*Hibiscus sabdariffa* L.). African Journal of Biotechnology. 2012;**11**(17):4063-4068

[32] Odigie I, Ettarh R, Adigun S. Chronic administration of aqueous extract

of *Hibiscus sabdariffa* attenuates hypertension and reverses cardiac hypertrophy in 2K-1C hypertensive rats. Journal of Ethnopharmacology. 2003;**86**:181-185

[33] Essa MM, Subramanian P, Suthakar G, Manivasagam T, Dakshayani KB, Sivaperumal R, et al. Influence of *Hibiscus sabdariffa* L. (Gongura) on the levels of circulatory lipid peroxidation products and liver marker enzymes in experimental hyperammonemia. Journal of Applied Biomedicine. 2006;**4**:53-58

[34] McKay DL, Chen O, Saltzman E, Blumberg JB. *Hibiscus Sabdariffa* L. tea (tisane) lowers blood pressure in pre hypertensive and mildly hypertensive adults. The Journal of Nutrition and Disease. 2010;**140**:298-303

[35] Ghislain MT, Giséle EL, Bertrand PMJ, Mathieu F, Onoré FK, Félicité TM, et al. Effect of "Foléré" juice (calyx of *Hibiscus sabdariffa* L.) on some biochemical parameters in humans. Pakistan Journal of Nutrition. 2011;**10**(8):755-759

[36] Tori HN. A Research Review on the Use of *Hibiscus sabdariffa* L. Oregon–USA: Professional Solutions (GAIA HERBS). Bastyr University; 2014

[37] Ali HM, Siddiqui MH, Basalah MO, Al-Whaibi MH, Sakran AM, Al-Amri A. Effects of gibberellic acid on growth and photosynthetic pigments of *Hibiscus sabdariffa* L. under salt stress. African Journal of Biotechnology. 2012;**11**(4):800-804

[38] Mungole A, Chaturvedi A. *Hibiscus sabdariffa* L. a rich source of secondary metabolites. International Journal of pharmaceutical sciences, review and research. 2011;**6**(1):83-87

[39] Atta S, Sarr B, Diallo AB, Bakasso Y, Lona I, Saadou M. Nutrients composition of calyces and seeds of three Roselle (*Hibiscus sabdariffa* L.) ecotypes from Niger. African Journal of Biotechnology. 2013;**12**(26):4174-4178

[40] Dutt D, Upadhyaya JS, Tyagi CH. Studies on Hibiscus cannabinus, *Hibiscus sabdariffa* L. and *Cannabinus sativa* pulp to be a substitute for softwood pulp - part 1: AS-AQ delignification process. Bio Resources. 2010;**5**(4):2123-2136

[41] Da-Costa-Rocha I, Bonnlaender B, Sievers H, Pischel I, Heinrich M. *Hibiscus sabdariffa* L. a phytochemical and pharmacological review. Food Chemistry. 2014;**165**:424-443

Section 2

Alkaloids

Chapter 5

Alkaloids as New Leads for Neurodegenerative Diseases

Farah Al-Mamoori and Ashraf M.A. Qasem

Abstract

Conventionally, diseases involving the selective loss of neurons are referred to as neurodegenerative diseases. Traditional and more recent compounds have been explored, but they only provide symptomatic benefits and have a large number of negative effects. It will be regarded as a modern vision if stronger molecules are found that can stop the pathophysiology of these diseases. In order to replace existing medications, natural compounds are being developed from plants and other sources. Natural products, including alkaloids that originate from plants, have emerged as potential protective agents against neurodegenerative disorders (e.g., Alzheimer's and Parkinson's), psychiatric conditions, and many more. They provided unique lead compounds for medicine. Alkaloids could be exploited as starting materials for novel drug synthesis or, to a lesser extent, used to manage neurodegenerative-related complications due to their diverse mechanistic effects. This chapter aims to highlight the importance of alkaloids as new leads for the development of potential clinical drug candidates for the management and treatment of neurodegenerative diseases.

Keywords: alkaloids, leads, neurodegenerative diseases, Parkinson's disease, Alzheimer's disease

1. Introduction

Neurodegenerative diseases are debilitating conditions that affect memory, cognition, mobility, and overall functioning. Although these diseases have diverse patterns of signs and symptoms, they have several characteristics: A high correlation with age, protein aggregation that is abnormal, and a natural history that is gradual and relentless. This group of illnesses is likewise distinguished by a slow beginning, with neuropathological alterations developing years before clinical manifestation [1]. Instances of neurodegenerative diseases include: Alzheimer's disease and Parkinson's disease.

Worldwide, dementia affects more than 25 million people, the majority of whom have Alzheimer's disease. It has had a significant influence on affected individuals, carers, and society in both developed and developing countries [2]. The abundance of experimental and clinical evidence suggests that Alzheimer's disease is a complicated disorder characterized by extensive neurodegeneration of the central nervous system with significant involvement of the cholinergic system, resulting in gradual cognitive decline and dementia [3]. New approaches, such as the detection of amyloid-beta (Aβ) and tumor necrosis factors (NFTs), lead to the amyloid and tau theories as potential causes of Alzheimer's development. Multi-target compounds that inhibit cholinesterases while also interfering with Aβ- aggregation and/or tau protein neuroinflammation may be effective in the treatment of Alzheimer's disease. Natural compounds, particularly plant alkaloids, have been a steady supply of innovative choices for the treatment of Alzheimer's disease. For example, the prototype of rivastigmine, physostigmine (*Physostigma venenosum*), is an cholinesterase enzyme inhibitors (IChE) inhibitor and allosteric modulator of the central nicotinic receptor. Galanthamine (*Galanthus woronowii*) is a selective acetylcholinesterase inhibitor, an allosteric modulator of the central nicotinic receptor, an inhibitor of Aβ aggregation, and an inducer of hippocampus neurogenesis [4]. Alkaloids have been one of the most appealing classes for searching for novel medications since the release of the Amaryllidaceae alkaloid as a drug in 2001 [3].

Parkinson's disease is the neurodegenerative disease with the second-highest prevalence. The age-adjusted prevalence was 205.89 per 100,000 people. The prevalence of advanced Parkinson's disease increased with age, from 3.77% in the 40–49 year age group to 25.86% in those over 89 years [5]. The development of pharmacotherapy for Parkinson's disease in terms of historically significant plant-derived substances, *Atropa belladonna* (deadly nightshade), *Hyoscyamus niger* (henbane), and *Datura stramonium* (thorn apple or jimsonweed), includes large amounts of pharmacologically potent anticholinergic tropane alkaloids (atropine, hyoscyamine, and hyoscine) [6].

Alkaloids, the main natural medicinal source, are a little-explored component of plant chemistry. They are cyclic organic compounds with at least one nitrogen atom [7]. The majority are biologically active. Alkaloids can be classified based on their ring chemistry or the amino acid from which they are formed [8]. The richness of alkaloid content varies between plant species, but most include a variety of these substances that are different in both their molecular structure and the biology of their effects. Individual plant alkaloid levels vary by component, life cycle, and season [9]. According to a review, 84% of medications licensed for central nervous system indications are derived from naturally occurring compounds [10].

Historically, the pharmaceutical industry originated from traditional plant medical knowledge. Natural scaffolds share molecular characteristics that can improve affinity with receptor binding sites. Natural structures confer more chirality (resulting in unique D- and L- stereoisomers) and more rigidity (due to ring conformations) than fully synthesized agents [10]. These chemicals can also go through the blood-brain barrier more easily. Because of evolutionary links between plants and mammals, natural products and natural-inspired medications may have a favorable influence on neurotransmitter systems [6]. This chapter summarizes the role of alkaloids in the management and treatment of neurodegenerative diseases and identifies them as lead compounds in the development of potential clinical drug candidates (**Table 1**).

2. Plant alkaloids as new leads for neurodegenerative diseases

2.1 Purine alkaloids: Theacrine

Some plants, like tea, coffee, chocolate, and mate, make purine alkaloids like caffeine, theobromine, theophylline, 7-methylxanthosine, and theacrine. The alkaloid theacrine is found in *Camellia kucha*, which is in the family Theaceae.

Class of alkaloids	Alkaloids	Plant source	Pharmacological activity	References
Purine alkaloid	Theacrine (**1**)	*Camellia kucha*	• In various animal models of Parkinson's disease, theacrine (**1**) has been shown to reverse dopaminergic cell loss and behavioral impairment –6-OHDA treated rats, and - 1-methyl-4-phenyl-1, 2, 3, 6-tetrahydropyridine (MPTP) treated mice, and zebrafish. Theacrine (**1**) alleviates apoptosis caused by oxidative damage and mitochondrial dysfunction.	[11]
Isoquinoline alkaloid	Berberine (**2**) Avicine (**3**) Chelerythrine (**4**) Sanguinarine (**5**) Aromoline (**6**)	*Berberis spp.* *Coptis chinensis, Phellodendron amurense, Hydrastis Canadensis, Zanthoxylum rigidum, Macleaya cordata* *Berberis vulgaris*	• Berberine (**2**) drastically reduced the level of the NLRP3 inflammasome, including the levels of NLRP3, the PYD and CARD domain-containing protein, cleaved caspase 1, and mature interleukin 1 beta in 1-Methyl-4-phenyl-1,2,3,6-tetrahydropyridine-induced Parkinson's disease mice. • Avicine (**3**) alkaloid had the highest cholinesterase inhibition, being more active against AChE than BChE It is a promising natural chemical and multifunctional candidate, serving as a good starting point for the creation of new Alzheimer's disease treatments. • Chelerythrine (**4**) and sanguinarine (**5**) demonstrated effective cholinesterase activity inhibition. • Aromoline (**6**) revealed significant hBuChE inhibitory action.	[12] [12] [13] [14]
Indole alkaloid	Geissoschizoline (**8**) *Conophylline* (**9**)	*Geissospermum vellosii* *Ervatamia microphylla*	• Geissoschizoline (**8**) inhibited both hAChE and hBChE via a mixed-type mechanism. Molecular docking experiments revealed geissoschizoline interactions with the active and peripheral anionic sites of hAChE and hBChE, indicating a dual site inhibitor profile. Furthermore, geissoschizoline demonstrated anti-inflammatory activity by lowering microglial NO and TNF- release. • In cultured brain cells, conophylline (**9**) was discovered to trigger autophagy. Autophagy activation improved cellular models of Parkinson's and Huntington's illnesses. As a result, conophylline (**9**) may prove effective in the development of chemotherapy for metabolic and neurological illnesses.	[15] [16]

Class of alkaloids	Alkaloids	Plant source	Pharmacological activity	References
Quinazoline alkaloid	Vasicinone (**10**) Dehydroevodiamine (**11**)	*Adhatoda vasica* *Evodiae fructus*	• The findings showed that vasicinone (**10**) promoted neuroprotection in SH-SY5Y cells via increasing autophagy and PINK-1/Parkin-mediated mitophagy. • Dehydroevodiamine (**11**) has a clear protective effect on the central nervous system. In recent years, a large number of studies have dehydroevodiamine reported that has a preventive effect on Alzheimer's disease induced by various models, and it exhibits good blood-brain barrier permeability.	[17] [18]
Protoalkaloid	Capsaicin (**12**)	*Capsicum annuum*	• Capsaicin (**12**) was found to inhibit apoptosis in a cell model of 6-OHDA-induced Parkinson's disease through regulating Actg1 and Gsta2. • Capsaicin (**12**) sped up the maturation of disintegrin and metalloproteinase, which stopped A from forming and changed the processing of Amyloid precursor protein towards cleavage. Capsaicin (**12**) also helped with other Alzheimer's disease -related diseases such as tau hyperphosphorylation, neuroinflammation, and neurodegeneration.	[19] [20]
Carbazole alkaloids	Clauselansiumines A (**13**) and B (**14**)	*Clausena lansium*	• Geranylated carbazole alkaloids displayed remarkable neuroprotective effects, with EC_{50} values ranging from 0.48 ± 0.04 to 12.36 ± 0.16 M. These geranylated carbazole alkaloids could be extremely important to the discovery of new agents for the treatment and prevention of Parkinson's disease.	[21]
Aporphine alkaloids	Compounds (**14**) and (**15**)	*Artabotrys spinosus*	• Two of isolated alkaloid exhibited the highest activity towards BChE and AChE	[22]
Norditerpenoid alkaloids	Lappaconitine (**16**) 3-*O*-acetylaconitine (**17**) Crassicauline A (**18**) Methyllycaconitine (**19**)	*Aconitum spp.* *Delphinium spp.*	• Activation of VGSC (3-*O*-acetylaconitine and crassicauline A) • Inhibition of VGSC (Lappaconitine) • Inhibition of nAChRS (methyllycaconitine)	[23] [24] [25] [26]

Table 1.
Summarize classes of alkaloids, their source, and neurodegenerative diseases they target.

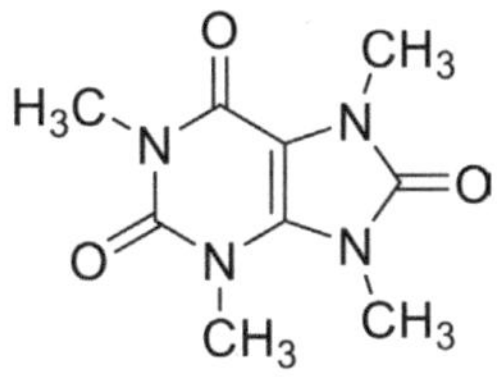

1

Figure 1.
Theacrine (1).

The leaves of *C. kucha* contain both theacrine **(1)** and caffeine. Theacrine **(1)** is thought to be made from caffeine through an N-methyltransferase process that uses *S*-adenosyl-L-methionine (SAM) as a methyl donor. Theacrine **(1)** is an adenosine receptor blocker that speeds up movement and makes people feel less tired. In animal models of Parkinson's disease, theacrine **(1)** stops the loss of dopaminergic cells and changes in behavior by reducing oxidative damage and mitochondrial dysfunction (**Figure 1**) [11].

2.2 Isoquinoline alkaloids: Berberine, avicine, chelerythrine, sanguinarine, and aromoline

The isoquinoline alkaloids include, most famously, the opiates morphine and codeine, as well as berberine. Berberine **(2)** is an alkaloid found in the roots, rhizomes, stems, and bark of several medicinal plants, including *Berberis*, *Coptis chinensis*, *Phellodendron amurense*, and *Hydrastis canadensis*. Berberine significantly reduced nucleotide-binding domain, leucine-rich–containing family, pyrin domain–containing-3(NLRP3) inflammasome levels in Parkinson's disease mic [27].

Moreover, avicine **(3)** is an alkaloid isolated from *Zanthoxylum rigidum* (Rutaceae family). It is the most effective dual cholinesterase inhibitor, with IC_{50} values of 0.15 and 0.88 M for both AChE and BuChE, respectively [12]. Chelerythrine **(4)** and sanguinarine **(5)** are the main active ingredients of *Macleaya cordata* (Papaveraceae family) [28]. These bioactives inhibit cholinesterase activity extremely well [13]. Aromoline **(6)**, an isoquinoline alkaloid, was extracted from the root bark of *Berberis vulgaris* (Berberidaceae family), where it showed a significant inhibitory activity against human Butyrylcholinesterase (BuChE) with an IC_{50} = 0.82 ± 0.10 μM [14]. The therapeutic potentiality of aromoline **(6)** is worthy of further investigation, as the computational analysis supports its high affinity and selectivity for the active site of human BuChE (**Figure 2**).

2.3 Indole alkaloids: Geissoschizoline and conophylline

Indole alkaloids have been found in many well-known plant groups, such as Apocynaceae, Rubiaceae, Nyssaceae, and Loganiaceae. Researchers think that indole alkaloids may have brain effects because they have the same structure as endogenous amines and neurotransmitters. Several substances with an indole group have been shown to bind to different serotonin receptors [29, 30].

Geissoschizoline **(7)** is an alkaloid isolated from *Geissospermum vellosii*emerges (Apocynaceae family) as a possible multi-target prototype that can be very useful in preventing neurodegeneration and restoring neurotransmission [15].

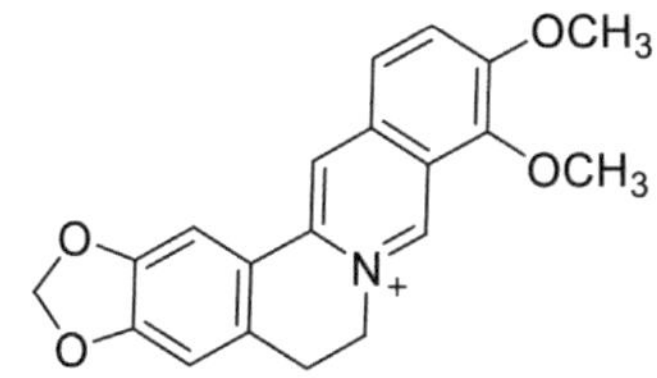

2

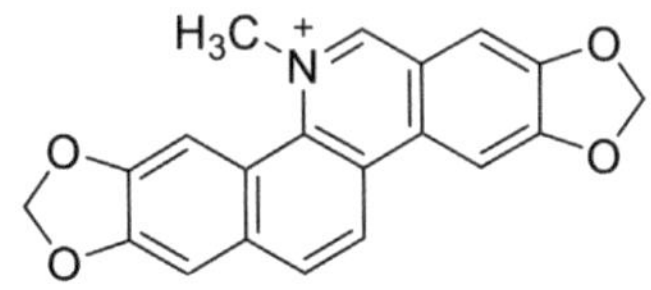

3

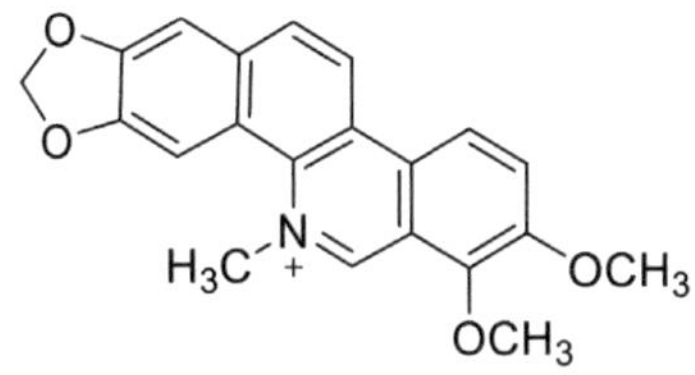

4

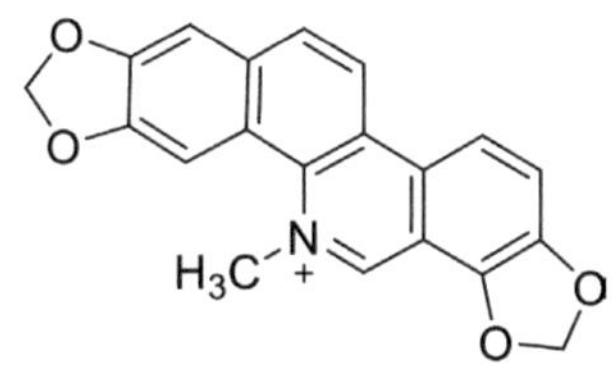

5

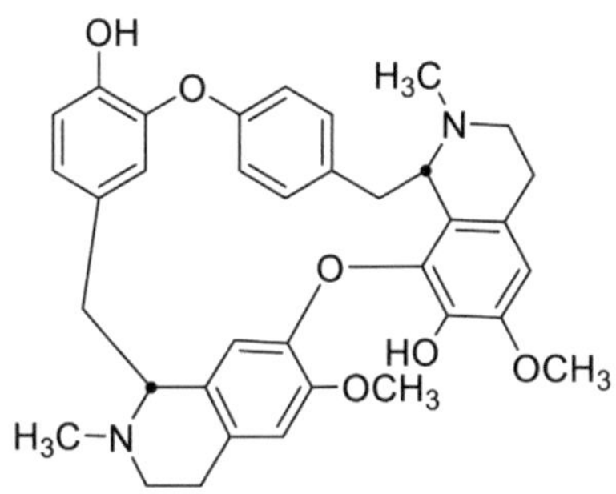

6

Figure 2.
Berberine (2), avicine (3), Chelerythrine (4), sanguinarine (5), and Aromoline (6).

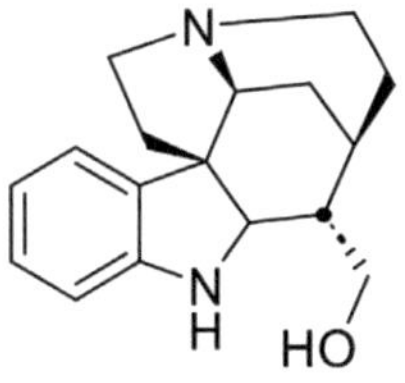

7

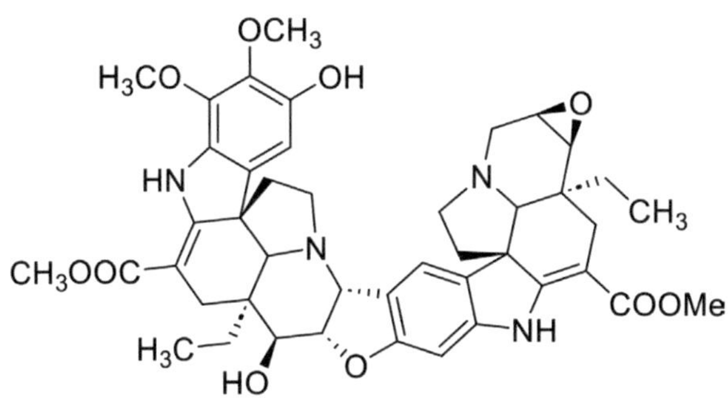

8

Figure 3.
Geissoschizoline (7) and Conophylline (8).

Neurodegenerative diseases are caused by nerve cell degeneration or death, and it was reported that autophagy is crucial for the prevention of such diseases. Conophylline **(8)**, isolated from *Ervatamia microphylla* (Apocynaceae family) leaves, was found to activate autophagy and suppress protein aggregation to protect the neural cells from cell death (**Figure 3**) [16].

2.4 Quinazoline alkaloids: Vasicinone and dehydroevodiamine

Quinazoline alkaloids belong to the N-based heterocyclic chemical class. To date, around 150 naturally occurring quinazoline alkaloids have been isolated from various plant species as well as animals and microbes; many are biogenetically generated from anthranilic acid. Vasicine was the first quinazoline alkaloid discovered, isolated from *Adhatoda vasica* (Acanthaceae family) and later from additional species [31]. Vasicinone **(9)**, is a vasicine autooxidation product. It demonstrates a neuroprotective mechanism in paraquat-induced Parkinsonian modalities in SH-SY5Y cells [17].

Dehydroevodiamine **(10)** is one of the bioactive components of *Evodiae Fructus* (Rutaceae family), which is widely used in traditional Chinese medicine. *Evodiae fructus* (Wuzhuyu in Chinese) is traditionally used for the treatment of various conditions, including migraine and central nervous system diseases [18]. Dehydroevodiamine **(10)** is the main component of *Evodiae fructus* for its neuroprotective action. Dehydroevodiamine **(10)** is highly permeable through the blood brain barrier and has a protective effect on Alzheimer's disease through its inhibitory effect on acetylcholine esterase (AChE). Clinical results on dehydroevodiamine **(10)** suggest that it's a potential drug candidate for stress-induced depression, neuronal death, and memory impairment [18].

In addition, chemical modification of dehydroevodiamine **(10)** results in carboxy-dehydroevodiamine. HCl (cx-DHE), which has a better water solubility, bioavailability, and effect on memory impairment. Through several clinical models in mice, the results suggested that cx-DHE is a promising drug candidate that could prevent the progression of Alzheimer's disease pathology (**Figure 4**) [18].

2.5 Protoalkaloids: Capsaicin

Capsaicn **(11)** is a pungent and irritant alkaloid isolated from *Capsicum annuum* (Solanaceae family). It's considered a protoalkaloid as it has

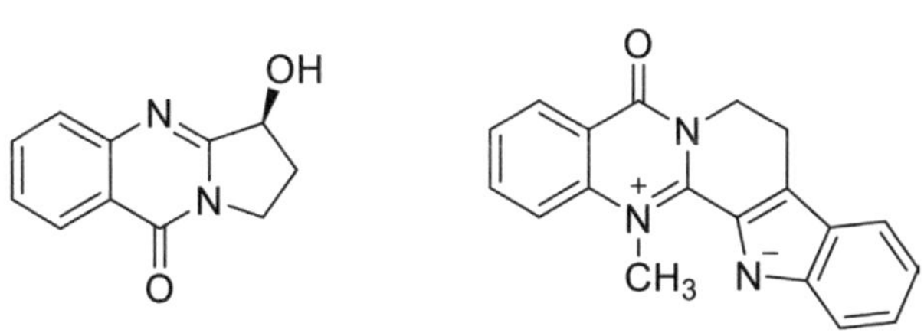

9 **10**

Figure 4.
Vasicinone (9) and Dehydroevodiamine (10).

non-hetercyclic nitrogen that comes from the amino acid precursors (phenylalanine and valine) [19].

It was previously reported that consumption of a capsaicin-rich diet was associated with better cognition. A recent study found that capsaicin **(11)** has a preventive effect on Alzheimer's disease by promoting the maturation of disintegrin and metalloproteinase 10 and also alleviating other Alzheimer's disease-type pathologies, such as neurodegeneration, tau hyperphosphorylation, and neuroinflammation [19]. These results suggest that supplementation with capsaicin **(11)** and chili peppers could be useful for the prevention and treatment of Alzheimer's disease. In addition, capsaicin **(11)** was found to protect the neural cells and reduce apoptosis by down-regulating Actg1 and up-regulating Gsta2 in the 6-6-hydroxydopamine (6-OHDA)-induced Parkinson's disease cell model (**Figure 5**) [20].

2.6 Carbazole alkaloids: Clauselansiumines A and B

Clauselansiumines A **(12)** and B **(13)** are two new geranylated carbazole alkaloids found in the stem and leaves of *Clausena lansium* (family Rutaceae). The alkaloids were unambiguously determined by spectral analysis, and their neuroprotective effect for Parkinson disease was tested against 6-hydroxydopamine induced cell death in human neuroblastoma and compared with curcumin as a positive control [21].

Clauselansiumines A **(12)** and B **(13)** displayed significant neuroprotective activity with an EC_{50} equal to 0.48 ± 0.04 μM and 0.98 ± 0.08 μM, respectively, which is more potent than the positive control that possessed an EC_{50} value of 6.03 ± 0.10 μM.

The structure activity relationship studies of clauselansiumines A **(12)** and B **(13)** and other geranylated carbazole alkaloids highlight the importance of the isopentenyl group at C-2′ and the methoxy group at positions 7 and 8 for the neuroptotective

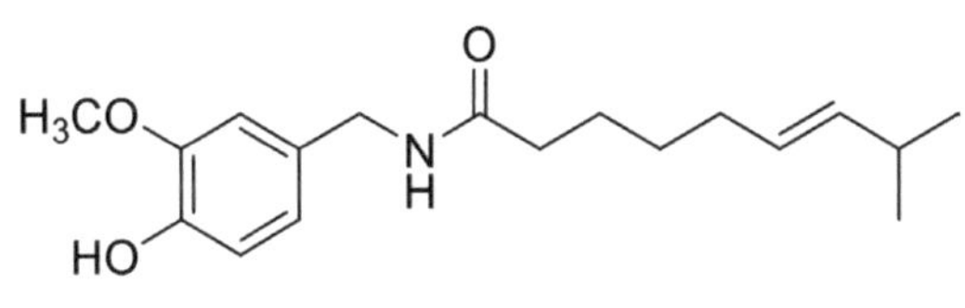

11

Figure 5.
Capsaicin (11).

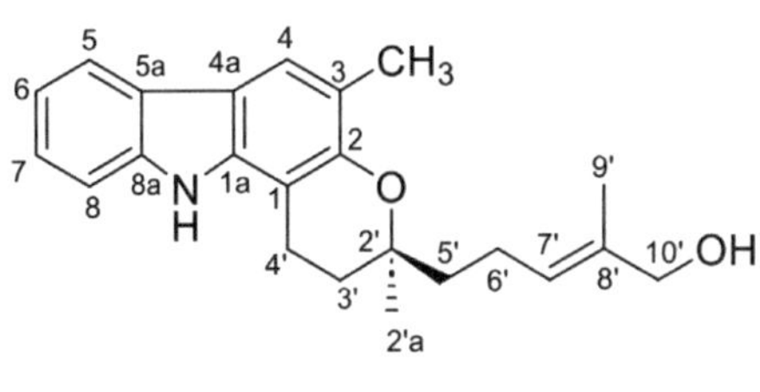

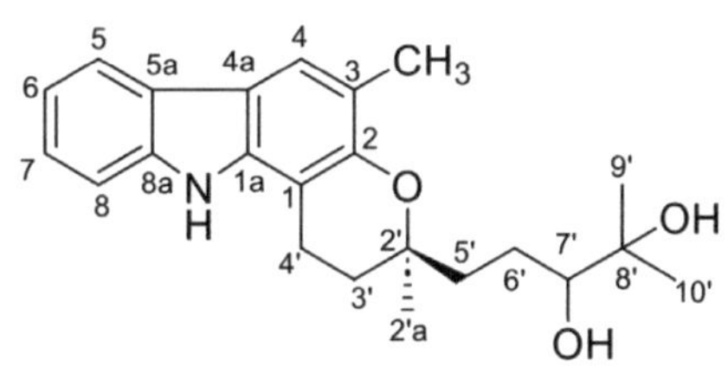

12 **13**

Figure 6.
Clauselansiumines A (12) and B (13).

activity [21]. In conclusion, the geranylated carbazole alkaloids separated from *C. lansium* could be considered promising candidates for therapeutic purposes in Parkinson's disease and other neural degenerative diseases (**Figure 6**).

2.7 Aporphine alkaloids

Aporphine alkaloids are a group of naturally occurring compounds with an aporphine nucleus isolated from several plant families such as Annonaceae, Papaveraceae, Ranunculaceae, and others. Recent work on the roots of *Artabotrys spinosus* (Annonaceae family) yielded the isolation of several aporphine alkaloids, of which two compounds **(14)** and **(15)** showed promising inhibitory activity towards AChE and BChE [22]. The *in silico* study confirmed the experimental results and supported the idea that compounds **(14)** and **(15)** are potential candidates for the treatment of Alzheimer's disease (**Figure** 7) [22].

2.8 Norditerpenoid alkaloids (C18 and C19): Lappaconitine, 3-O-acetylaconitine, bulleyaconitine A, and methyllycaconitine

The majority of norditerpenoid alkaloids (NDAs) are isolated from the genera Delphinium and Aconitum, and they are of pharmacological importance. NDAs have a complex hexacyclic structure (A, B, C, D, E, and F). Despite the chemical similarity between NDAs, they display various pharmacological actions, and such variety encourages researchers to work on their structure activity relationship [23]. Lappaconitine **(16)** is the first C_{18} NDA to be reported from *Aconitum septentrionale* Koelle in 1895 and the most successful NDA in terms of clinical application [23]. Lappaconitine **(16)** acts as a voltage gated sodium channel blocker and has a potent non-addictive analgesic effect that is comparable to morphine with an ED_{50} = 3.5 mg/kg [24]. The structure activity relationship studies highlight the importance of the benzoyl ester moiety and the amide group for the activity [23]. Based on the lappaconitine structure activity relationship, several lappaconitine **(16)** analogues were synthesized by replacing the amide moiety with different amides and sulphonamides; this strategy was successful in getting lead compounds as potential analgesics with comparable activity and lower toxicity to lappaconitine **(16)** [25].

3-*O*-acetylaconitine **(17)** and crassicauline A **(18)** are C_{19} NDAs isolated from *Aconitum* spp. They have a similar chemical structure to lappaconitine **(16)** but display an opposite pharmacological action as they keep VGSCs in their open state

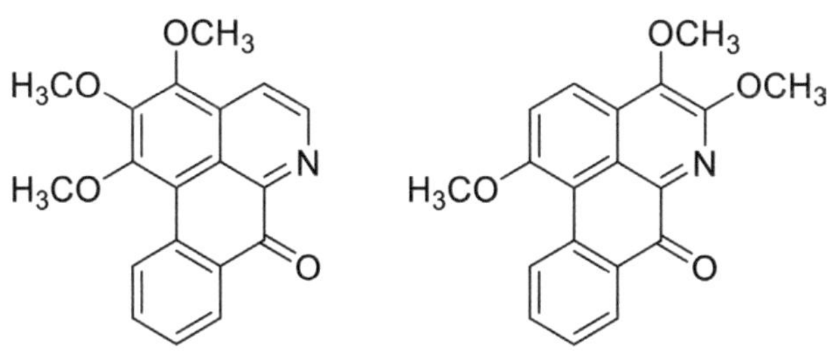

14 **15**

Figure 7.
Aporphine alkaloids (14) and (15).

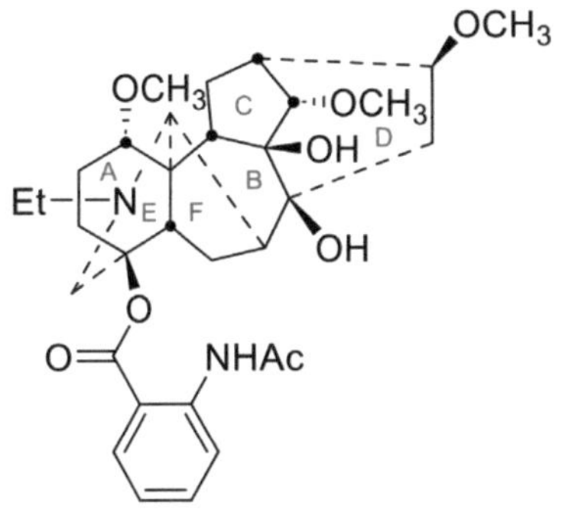

16

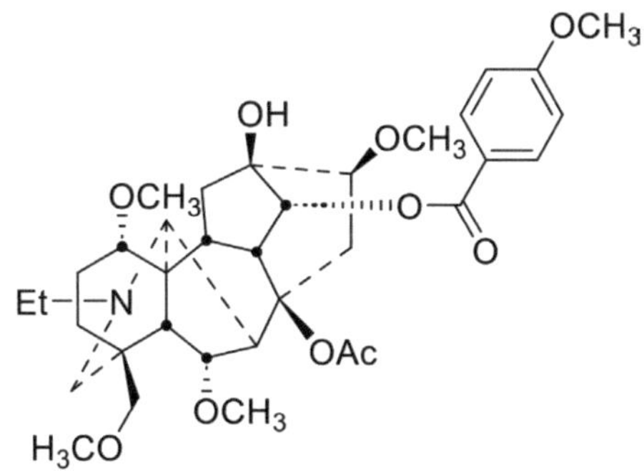

17

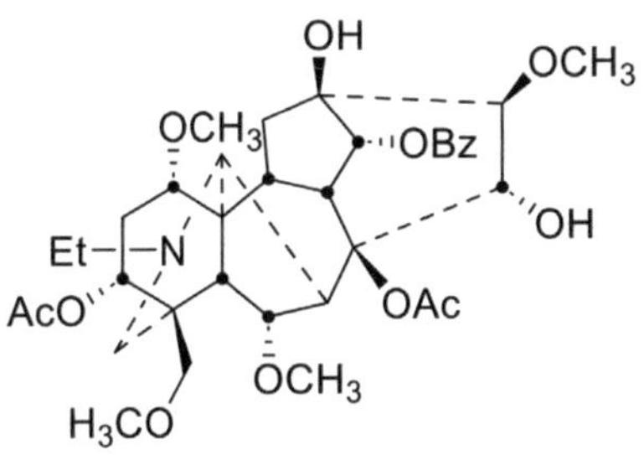

18

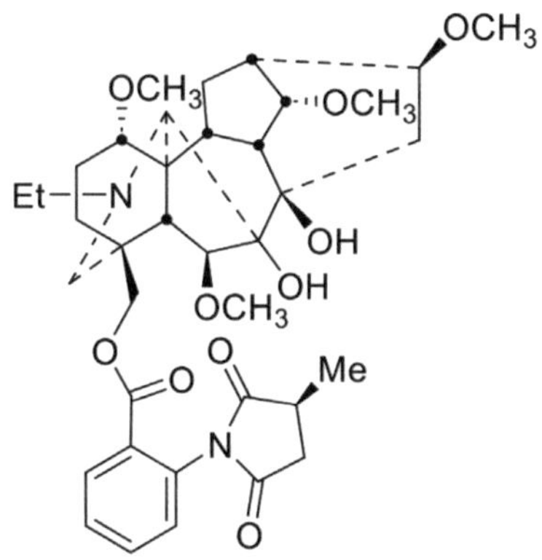

19

Figure 8.
Lappaconitine (16), 3-O-Acetylaconitine (17), Crassicauline A (18), and Methyllycaconitine (19).

conformation, and they exhibit a non-addictive potent analgesic activity that is comparable to morphine, where 3-*O*-acetylaconitine **(17)** and crassicauline A **(18)** have ED_{50} values of 0.16 and 0.087 mg/kg, respectively. 3-O-acetylaconitine **(17)** and crassicauline A **(18)** were introduced in China into clinical use in the 1980s as analgesic agents [23].

Methyllycaconitine (MLA) **(19)** is a C_{19} alkaloid that was first reported from *Delphinium brownii* Rydb by Manske in 1938. Methyllycaconitine **(19)** is one of the most potent competitive antagonists of α7-nicotinic acetylcholine receptors (nAChRs) with an IC_{50} value of 2 nM [23]. As the total synthesis of MLA **(19)** has not been achieved yet, the synthesis of simple small analogues could be useful to achieve better structure activity relationship understanding and possibly to identify potential candidates for the treatment of several neurodegenerative diseases, including Alzheimer's disease.

Structure activity relationship studies on MLA **(19)** showed that the neopentyl ester side-chain and the piperidine ring *N*-side chain are important features in MLA **(19)** activity [26]. The synthesis of several AE-bicyclic analogues of MLA **(19)** was reported recently, possessing different nitrogen and ester side chains. The antagonist effects of these analogues on human α7 nAChRs showed promising results that suggest that further optimization and research may enhance the activity of this analogue model (**Figure 8**) [26].

3. Conclusion and future perspective

There are many drugs that have been used to treat neurodegenerative illnesses, but none of them have been able to prevent the disease from getting worse. Instead, they have caused a lot of side effects. Several neurodegenerative illnesses can be treated with natural alkaloids that continue to grow stronger. Analysis of the physicochemical properties of alkaloids showed that most of them follow the Lipinski rules of drug likeness. But only a few alkaloids are widely used in clinical practice. Because natural alkaloids give patients hope that neurodegenerative diseases can be slowed down, it is very important to plan clinical trials for these kinds of compounds that have not even been tried in clinical trials yet. Also, the blood-brain barrier is a key component of keeping substances from going into the brain, and it needs more attention. Meanwhile, it's easy to see why the development of possible candidates into therapeutic leads has stalled because of problems with compounds that come from nature, such as low extraction yields and safety profiles. More studies have to be done on them before they can be used as therapeutics.

Conflict of interest

The authors declare no conflict of interest.

Author details

Farah Al-Mamoori* and Ashraf M.A. Qasem
Faculty of Pharmacy, Department of Pharmaceutical Sciences, Zarqa University, Zarqa, Jordan

*Address all correspondence to: fmamoori@zu.edu.jo

References

[1] Kromer R, Serbecic N, Hausner L, Froelich L, Aboul-Enein F, Beutelspacher SC. Detection of retinal nerve fiber layer defects in Alzheimer's disease using SD-OCT. Frontiers in Psychiatry. 2014;**5**:22. DOI: 10.3389/fpsyt.2014.00022

[2] Qiu C, Kivipelto M, von Strauss E. Epidemiology of Alzheimer's disease: Occurrence, determinants, and strategies toward intervention. Dialogues in Clinical Neuroscience. 2009;**11**(2):111-128. DOI: 10.31887/DCNS.2009.11.2/cqiu

[3] Vrabec R, Blunden G, Cahlíková L. Natural alkaloids as multi-target compounds towards factors implicated in Alzheimer's disease. International Journal of Molecular Sciences. 2023;**24**(5):4399. DOI: 10.3390/ijms24054399

[4] Lima JA, Hamerski L. Alkaloids as potential multi-target drugs to treat Alzheimer's disease. Studies in Natural Products Chemistry. 2019;**61**:301-334. DOI: 10.1016/B978-0-444-64183-0.00008-7

[5] Orozco JL, Valderrama-Chaparro JA, Pinilla-MonsalveGD,Molina-EcheverryMI, Pérez Castaño AM, Ariza-Araújo Y, et al. Parkinson's disease prevalence, age distribution and staging in Colombia. Neurology International. 2020;**12**(1):8401. DOI: 10.4081/ni.2020.8401

[6] Kempster P, Ma A. Parkinson's disease, dopaminergic drugs and the plant world. Fronties in Pharmacology. 2022;**13**:970714. DOI: 10.3389/fphar.2022.970714

[7] Pelletier SW. The nature and definition of an alkaloid. In: Pelletier SW, editor. Alkaloids: Chemical and Biological Perspectives. Vol. 1. New York: Wiley; 1983. pp. 1-31

[8] Aniszewski T. Alkaloids-Secrets of Life: Aklaloid Chemistry, Biological Significance, Applications and Ecological Role. Amsterdam: Elsevier; 2007

[9] Waller GR, Nowacki EK. Alkaloid Biology and Metabolism in Plants. London & New York: Plenum Press; 1978. pp. 121-141. DOI: 10.1007/978-1-4684-0772-3_4

[10] Bharate SS, Mignani S, Vishwakarma RA. Why are the majority of active compounds in the CNS domain natural products? A critical analysis. Journal of Medicinal Chemistry. 2018;**61**(23):10345-10374. DOI: 10.1021/acs.jmedchem.7b01922

[11] Duan WJ, Liang L, Pan MH, Lu DH, Wang TM, Li SB, et al. Theacrine, a purine alkaloid from kucha, protects against Parkinson's disease through SIRT3 activation. Phytomedicine. 2020;**77**:153281

[12] Plazas E, Hagenow S, Murillo MA, Stark H, Cuca LE, Plazas E, et al. Isoquinoline alkaloids from the roots of *Zanthoxylum rigidum* as multi-target inhibitors of cholinesterase, monoamine oxidase A and Aβ1-42 aggregation. Bioorganic Chemistry. 2020;**98**:103722. DOI: 10.1016/j.bioorg.2020.103722

[13] Tuzimski T, Petruczynik A, Szultka-Młyńska M, Sugajski M, Buszewski B. Isoquinoline alkaloid contents in *Macleaya cordata* extracts and their acetylcholinesterase and Butyrylcholinesterase inhibition. Molecules. 2022;**27**(11):3606. DOI: 10.3390/molecules27113606

[14] Hostalkova A, Marikova J, Opletal L, Korabecny J, Hulcova D, Kunes J, et al. Isoquinoline alkaloids from *Berberis vulgaris* as potential lead compounds for the treatment of Alzheimer's disease. Journal of Natural Products. 2019;**82**(2):239-248

[15] Lima JA, Costa R, et al. Geissoschizoline, a promising alkaloid for Alzheimer's disease: Inhibition of human cholinesterases, anti-inflammatory effects and molecular docking. Bioorganic Chemistry. 2020;**104**:104215. DOI: 10.1016/j.bioorg.2020.104215

[16] Umezawa K, Kojima I, Simizu S, Lin Y, Fukatsu H, Koide N, et al. Therapeutic activity of plant-derived alkaloid conophylline on metabolic syndrome and neurodegenerative disease models. Human Cell. 2018;**31**:95-101

[17] Huang CY, Sivalingam K, Shibu MA, Liao PH, Ho TJ, Kuo WW, et al. Induction of autophagy by vasicinone protects neural cells from mitochondrial dysfunction and attenuates paraquat-mediated Parkinson's disease associated α-synuclein levels. Nutrients. 2020;**12**(6):1707. DOI: 10.3390/nu12061707

[18] Fu S, Liao L, Yang Y, Bai Y, Zeng Y, Wang H, et al. The pharmacokinetics profiles, pharmacological properties, and toxicological risks of dehydroevodiamine: A review. Frontiers in Pharmacology. 2022;**13**:1040154

[19] Wang J, Sun BL, Xiang Y, Tian DY, Zhu C, Li WW, et al. Capsaicin consumption reduces brain amyloid-beta generation and attenuates Alzheimer's disease-type pathology and cognitive deficits in APP/PS1 mice. Translational Psychiatry. 2020;**10**(1):230

[20] Liu J, Liu H, Zhao Z, Wang J, Guo D, Liu Y. Regulation of Actg1 and Gsta2 is possible mechanism by which capsaicin alleviates apoptosis in cell model of 6-OHDA-induced Parkinson's disease. Bioscience Reports. 2020;**40**(6):BSR20191796

[21] Liu YP, Guo JM, Wang XP, Liu YY, Zhang W, Wang T, et al. Geranylated carbazole alkaloids with potential neuroprotective activities from the stems and leaves of *Clausena lansium*. Bioorganic Chemistry. 2019;**92**:103278

[22] Sichaem J, Tip-pyang S, Lugsanangarm K. Bioactive aporphine alkaloids from the Roots of *Artabotrys spinosus*: Cholinesterase inhibitory activity and molecular docking studies. Natural Product Communications. 2018;**13**(10):1934578X1801301011

[23] Qasem AMA, Zeng Z, Rowan MG, Blagbrough IS. Norditerpenoid alkaloids from *Aconitum* and *Delphinium*: Structural relevance in medicine, toxicology, and metabolism. Natural Product Reports. 2022;**39**(3):460-473

[24] Salehi A, Ghanadian M, Zolfaghari B, Jassbi AR, Fattahian M, Reisi P, et al. Neuropharmacological potential of diterpenoid alkaloids. Pharmaceuticals. 2023;**16**(5):747

[25] Li Y, Shang Y, Li X, Zhang Y, Xie J, Chen L, et al. Design, synthesis, and biological evaluation of low-toxic lappaconitine derivatives as potential analgesics. European Journal of Medicinal Chemistry. 2022;**243**:114776

[26] Qasem AMA, Rowan MG, Sanders VR, Millar NS, Blagbrough IS. Synthesis and antagonist activity of methyllycaconitine analogues on human α7 nicotinic acetylcholine receptors. ACS Biology & Medicinal Chemistry. 2023;**3**(2):147-157

[27] Huang S, Liu H, Lin Y, Liu M, Li Y, Mao H, et al. Berberine protects

against NLRP3 inflammasome via ameliorating autophagic impairment in MPTP-induced Parkinson's disease model. Frontiers in Pharmacology. 2021;**11**:618787

[28] Lin L, Liu YC, Huang JL, Liu XB, Qing ZX, Zeng JG, et al. Medicinal plants of the genus Macleaya (*Macleaya cordata*, Macleayamicrocarpa): A review of their phytochemistry, pharmacology, and toxicology. Phytothery Research. 2018;**32**(1):19-48. DOI: 10.1002/ptr.5952

[29] Kochanowska-Karamyan AJ, Hamann MT. Marine indole alkaloids: Potential new drug leads for the control of depression and anxiety. American Chemical Society. 2010;**110**(8):4489-4497. DOI: 10.1021/cr900211p

[30] Omar F, Tareq AM, Alqahtani AM, Dhama K, Sayeed MA, Emran TB, et al. Plant-based indole alkaloids: A comprehensive overview from a pharmacological perspective. Molecules. 2021;**26**(8):2297. DOI: 10.3390/molecules26082297

[31] Roja G, Vikrant BH, Sandur SK, Sharma A, Pushpa KK. Accumulation of vasicine and vasicinone in tissue cultures of *Adhatoda vasica* and evaluation of the free radical-scavenging activities of the various crude extracts. Food Chemistry. 2011;**126**(3):10338. DOI: 10.1016/j.foodchem.2010.11.115

Chapter 6

Methods of Alkaloids Synthesis

Nitin Dumore, Namita Girhepunje, Monali Dumore and Kishor Danao

Abstract

The investigation of plants used in traditional medicine in the early nineteenth century found alkaloids have developed into a group of natural products with exceptional structural and taxonomic diversity, as well as important chemical, biological, and medicinal importance. Since the early twentieth century, only a few routes have been thoroughly explored, and researchers have struggled to grasp their biogenesis and biosynthesis. Even for many pharmaceutically important alkaloids, there is still much to learn about how alkaloids are generated in nature, despite recent enzymatic efforts that have significantly advanced our understanding of this process. Certain aspects of empirically determined or speculated mechanistic routes of alkaloids creation are explored, with an emphasis on clinically relevant alkaloids.

Keywords: alkaloids, synthesis, plants, therapeutic, natural products, traditional medicines

1. Introduction

Alkaloids are organic compounds that are naturally occurring and are primarily found in plant sources, such as marine algae, and rarely in animals (e.g., in the toxic secretions of fire ants, ladybugs and toads). They are predominantly located in berries, bark, fruits, roots, and leaves of plants that produce seeds. Alkaloids often have a heterocyclic ring with at least one nitrogen atom [1].

Since their discovery and early isolation in the nineteenth century, alkaloids have attracted the attention of chemists' imaginations, creative energies, and very souls. They are still actively sought after because of their astounding and seemingly unlimited structural variety, the challenges their synthesis poses to even the most skilled and knowledgeable organic chemists, the range of biological reactions they supply, and the profusion of innovation and acrobatics in the routes of biosynthetic creation. In the past 200 years, no other group of natural compounds has stimulated both chemists and biologists [2]. The "principium somniferum" that Serturner isolated from opium and published in the Journal für Pharmazie in 1805 was most likely the first semi-purified alkaloid. However, it was because to the work of French scientists Pelletier and Caventou that alkaloids truly came of age. Following their successful separation of emetine in 1817, they isolated brucine, quinine, and strychnine between 1819 and 1821. Piperine, atropine, caffeine, solanine, chelidonine, coniine, nicotine, aconitine, and colchicine were all identified before 1833 as a result of other chemists taking on the challenge of researching the components of physiologically relevant plants [3]. By the time Berzelius' Lehrbuch der Chemie was published in 1837, the Swedish chemist had identified thirteen "Pflanzenbasen." Coke was discovered in 1860, and spartine in 1851. Thirteen "Pflanzenbasen" had already been recognised by Berzelius by the time his Lehrbuch der Chemie was published in 1837. However, it has remained difficult to define what an alkaloid is, thus no attempt will be made here to close the apparent gap. Despite not being polypeptides, proteins, or nucleic acids, they do contain nitrogen [4, 5]. A common reason why most alkaloids are optically active is that they contain tertiary nitrogen in their structural makeup. This leads to varied physical, chemical, and pharmacological properties in the various isomeric forms. For example, (+)-tubocurarine, which was isolated from *Chondrodendron tomentosum* (Bisset, 1992), has muscle-relaxant activity, whereas its leavo isomer has less activity [2].

It has given a very helpful overview of alkaloids and their role in biology and medicine [6]. Higher plants, particularly those with medical purposes or a reputation for being exceedingly deadly, provided the first isolates of alkaloids. In the late twentieth century, "alkaloids" were isolated from a wide variety of terrestrial and marine sources, including frogs, arthropods, mammals, insects, sponges, fish, fungus, and bacteria, as well as, of course, *Homo sapiens*, as the natural world was being chemically explored. The number of known alkaloids from higher plants alone has increased to at least 22,000, meaning that the sum from all sources is currently likely to be more than 30,000 [7]. Alkaloid isolations from the beginning were done before stereochemical concepts and the three-dimensional character of compounds were formed, before there was an understanding of the intricacy of molecular structure. The development of methods for figuring out the precise structures of these alkaloids—first via chemical analysis, then spectroscopy—posed one of the main difficulties over the following 160 years [8, 9]. In 1882, the structure of the first known alkaloid, xanthine, was discovered. The initial synthesis of alkaloids was published by Ladenburg, (+)-coniine, in 1886. Chemical deterioration under difficult circumstances was frequently engaged in structure determination. The result is, the core heterocyclic nucleus was occasionally the only one to survive, and this provided the foundation for the new field of organic chemistry known as heterocyclic nuclei. For instance, isolating quinoline from quinine by distilling it with KOH and indigo provided indole [10, 11]. Many of the alkaloids were extremely complicated structurally and stereochemically, making it difficult to determine their structure or synthesise them. Notwithstanding the fact that the indole alkaloid strychnine was discovered in 1818, it was not completely understood until 1947, and its synthesis was initially described by Woodward in a 1954 publication. On the other hand, Robinson successfully synthesised the crucial tropane nucleus in 1917 following what turned out to be biogenetic lines. However, it took a long time before this philosophical idea inspired the biogenetically-patterned synthesis of a wide variety of alkaloids [12, 13]. Synthesis and biogenesis advanced hand in hand in this setting, strengthened at key points by biosynthetic experiments, beginning with radioactive isotopes and afterwards the use of stable isotopes. Following these discoveries, an effort was made to identify, describe, and determine the genes and enzymes in charge of producing alkaloids inside the organism and within its cellular structure. However, testing is still far behind the ideas of spontaneous creation [14]. As a result, it is quite evident that "biogenesis" refers to theoretical ideas about the method of development of a natural product, whereas "biosynthesis" refers to the experimental confirmation of such pathway (feeding experiments with precursors, enzyme isolations and characterizations, etc.) [15]. For all but a few classes of alkaloid, experimentation has not yet surpassed theory, as will be shown in this chapter. As more and more alkaloid structures were revealed, it became necessary to divide "alkaloids" into different subgroups and analyse each of these groups separately. The "The Alkaloids, Chemistry and Pharmacology" series of publications, edited by R.H.F. Manske, was first released in 1950 and is still in print today. Alkaloids were categorised on the basis of their structural makeup, and groups of alkaloids were given names based on the heterocyclic nucleus of their parent compound, such as tropanes, indole alkaloids, isoquinoline alkaloids, benzylisoquinoline alkaloids, acridine alkaloids, steroidal alkaloids, etc. [16]. As it turned out, these categories also often—though not always—reflected a shared biosynthetic origin. For instance, the amino acid tryptophan, which has an indole nucleus, would provide an indole alkaloid. These alkaloid group names often still reflect a fundamental structural component as well as a shared biosynthetic source [17]. However, it is not suitable to designate them as "piperidine" or "quinoline" alkaloids because a large number of heterocyclic nuclei, such as the piperidine and quinoline nuclei, are known to have numerous biosynthetic origins. On the other hand, higher plants contain alkaloids at a rate of 14.2%, as indicated by plant genera. The 83 higher plant orders identified by Cronquist were examined by Cordell, Quinn, and Farnsworth and found to be the case (1730 of 7231). Though none have been isolated as of yet, alkaloids have been discovered in over 35 higher plant groupings. In addition, the alkaloids of 153 plant groups have never been examined [18]. Over 1870 alkaloid skeletal were recognised at that time, and over 21,120 alkaloids had their structural makeup established. These are the twenty most important: Apocynaceae, Asteraceae, Berberidaceae, Boraginaceae, Buxaceae, Celastraceae, Fabaceae, Lauraceae, Liliaceae, Loganiaceae, Menispermaceae, Papaveraceae, Piperaceae, Poaceae, Ranunculaceae, Rubiaceae, Rutaceae, and Solanace. There was a great deal of conjecture about the formation of alkaloids and the interactions between alkaloid groups before there was any actual evidence that alkaloids were derived from amino acids. Organic chemists took the lead when biosynthetic experimentation started in the early 1950s following the introduction of radio-labelled precursors, with groups led by Birch, Barton, Battersby, Arigoni, Scott, Spencer, and Leete who clarified many crucial fundamental alkaloid biosynthetic pathway elements [19]. When it became apparent, roughly 20 years ago, that research at the cellular and then enzyme levels was required, and from there to the cloning and expression of systems that could generate alkaloids ex situ, the groups of Zenk, Stöckigt, Kutchan, Robins, Yamada, and Verpoorte took the lead. The study of alkaloid biosynthesis from a regulatory and metabolic engineering approach has now entered a new phase. Leaders in this circumstance have included Kutchan, Facchini, and Yamada. Up until recently, Richard Herbert's heroic efforts allowed The Royal Society of Chemistry to publish Natural Products Reports, a review magazine that provided excellent coverage of this area of natural product chemistry and biology. Three significant reactions serve as the foundation for the biosynthesis of alkaloids: the Pictet-Spengler condensation, the Mannich reaction of a Schiff base with a nucleophile, and the phenolic coupling reaction. Before going over some of the amazing pathways that lead to the diversity of alkaloids, it is important to review these three reactions [20]. Alkaloids have long been thought to be crucial for humans, despite the fact that they are secondary metabolites, which might imply that they are useless. In very little quantities, alkaloids exhibited potent biological influences on human and animal species. Alkaloids are found in food and drink used by humans every day, as well as in several stimulant medications (**Figure 1**) [21].

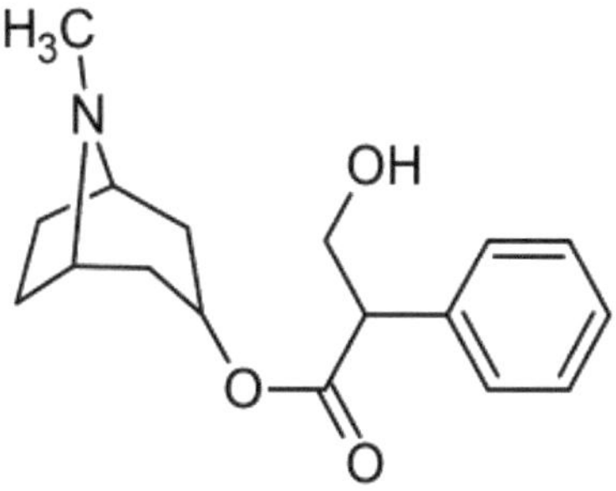

Figure 1.
Structure of atropine.

2. Occurrence

While alkaloids often present in all sections of a plant, they occasionally concentrate solely in one organ, leaving other parts of the plant free of them. For example, The potato plant's edible tubers are free of alkaloids, but its green parts are poisonous because they contain solanine. Alkaloids are not always synthesised in the organ in which they collect; for example, Tobacco's roots are where nicotine is produced before being carried to the leaves, where it is subsequently found. (Harborne and Herbert, 1995). In the epidermis of a human, about 300 alkaloids from over 24 classes have been discovered. The Phyllobates genus of frogs' skin was the source for the lethal neurotoxic alkaloids. Daly, (1993) separated different antibacterial alkaloids from reptile skin. Some isoquinoline and indole alkaloids were includes mammalian morphine but is segregated from them· The human diet includes several alkaloids in both food and beverages. The plants in the human diet in which alkaloids are present are not only coffee seeds, caffeine (**Figure 2**) [22].

3. Medicinal significance

Alkaloids have a variety of medicinal uses. Despite the fact that many of them exhibit local anaesthetic properties, they rarely have therapeutic uses. One of the most

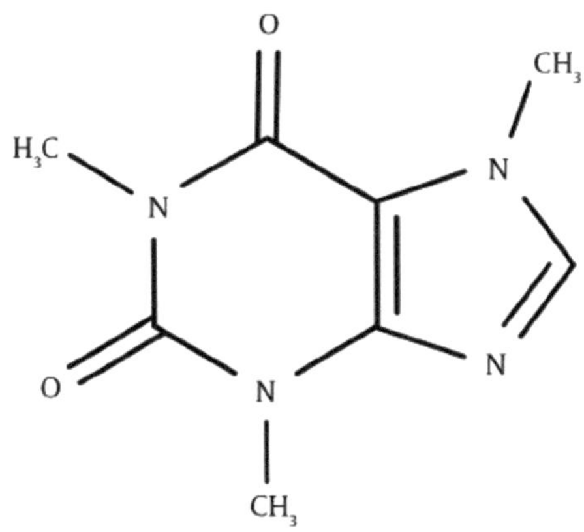

Figure 2.
Structure of caffeine.

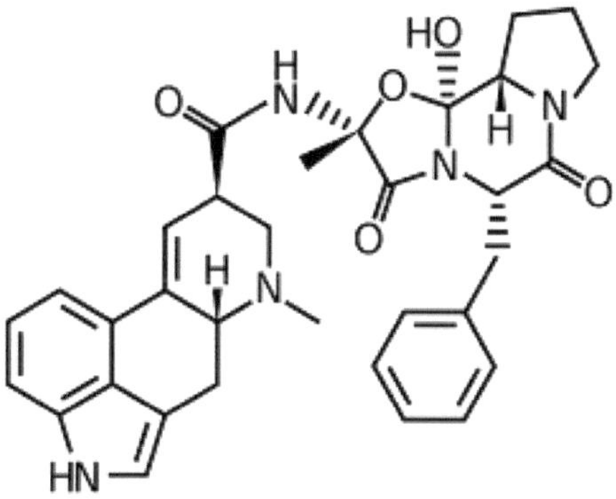

Figure 3.
Structure of ergotamine.

well-known alkaloids that has been utilised for medical purposes both historically and currently is morphine. This alkaloid is a strong narcotic that can only be used in small doses to treat pain due to its addictive properties [1].

Medicine has used alkaloids having hallucinogenic, narcotic, or analgesic characteristics, such as morphine, atropine, and quinine. While many alkaloids are abused as illegal substances, such as cocaine, For modern synthesised medications, several alkaloids served as model substances. Some alkaloids, such as strychnine and coniine, are too poisonous for any medicinal use. Additionally, new biologically active chemicals are continually being discovered in the plant. The drug atropine (Mann et al., 1994) relaxes smooth muscles and expands the eyeballs' pupils. Papaverine, a compound isolated from *Papaver somniferum*, has been shown by Pictet and Gums (1909) to have relaxing effects on blood vessel smooth muscle as well as intestinal and bronchial smooth muscle. Strong painkiller morphine is frequently prescribed to individuals with terminal illnesses. Although less strong, codeine performs similar pharmacological processes as morphine. Heroin is a highly addictive synthetic morphine derivative. As a particular analgesic, ergotamine (**Figure 3**) in the form of ergotamine tartrate coupled with caffeine is used to treat migraines [1].

4. Quinolines alkaloid

This particular type of quinolone-nucleus carrying alkaloid is only found in the bark of the Cinchona plant. However, a number of simple heteroaromatic quinolines have also been discovered in a number of marine sources. (2-heptyl-4-hydroxyquinoline from a marine pseudomonad and 4, 8-quinolinediol from cephalopod ink 2-heptyl-4-hydroxyquinoline from a marine pseudomonad). The primary alkaloids in this group include cinchonine, cinchonidine, quinine, and quinidine (**Figure 4**).

4.1 Quinoline alkaloid synthesis

The cross coupling of phenyl magnesium bromide with 2-chloroquinoline product I, which was catalysed by cobalt (II) acetylacetonate in dioxane at 50°C, resulted in the alkaloid in a 74% yield. By heating acetophenone with 2- aminobenzyl alcohol product (II) in dioxane in the presence of the catalyst for oxidation [Ru(DMSO)4]Cl2

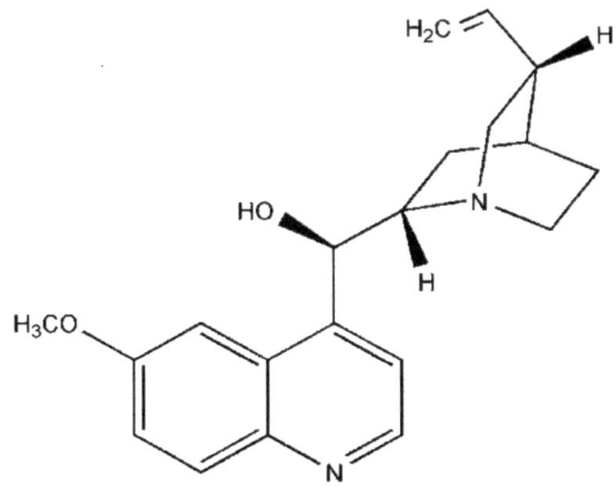

Figure 4.
Structure of quinine.

(2 mol%) and benzophenone as a hydrogen scavenger, the product was produced in 94% yield. In toluene at 100°C, molecular oxygen and hydrotalcite with ruthenium grafts, a multifunctional heterogeneous catalyst, have also been used to achieve this oxidative cyclisation in 89% yield. Reducing the titanium tetrachloride, zinc, and malononitrile-2-nitrochalcone adduct iii in boiling tetrahydrofuran resulted in the production of 2- phenylquinoline in a 78% yield. Finally, ytterbium (III) triflate in dichloromethane was used to achieve the three-component condensation of aniline with benzaldehyde and phenyl vinyl sulphide to yield 2-phenyl-4-phenylthio-1,2,3,4-tetrahydroquinoline 40 at room temperature .32 The alkaloid was subsequently produced in an overall yield of 23% by oxidising product iv with solid-supported periodate and thermolyzing the intermediate sulfoxide. These five adaptable methods were also used to create a variety of synthetic 2-phenylquinoline analogues, so they ought to work just as well when creating additional straightforward 2-arylquinoline alkaloids of related interest is the oxidation of 2-aryl-2,3-dihydroquinolin-4(1H)-ones such as product v with ferric chloride hexahydrate in boiling Methanol is used to produce 2-aryl-4-methoxyquinolines, including the naturally occurring compound 2-phenyl-4-methoxyquinoline vi, which was produced with a 78% yield [23].

4.2 Reagents and conditions in Quinoline alkaloid synthesis

i. PhMgBr (3 equiv.), Co(acac)2 (0.1 equiv.), dioxane, 50°C, 30 min (**Figure 5**);

ii. [Ru(DMSO)4]Cl2 (2 mol%), Ph2CO (1 equiv.), KOH (1 equiv.), dioxane, 80°C;

iii. Ru-grafted hydrotalcite–NEt3 (3 mol% in Ru), O2 (1 atm), PhMe, 100°C, 20 h;

iv. TiCl4 (4 equiv.), Zn (8 equiv.), THF, reflux, 3 h, then 39, rt., 4 h;

v. Yb (OTf)3 (0.05 equiv.), MgSO4, CH2Cl2, rt., 18 h;

vi. IO4 – on Amberlyst, dioxane– H2O, rt., 4 h, then 80°C, 18 h;

vii. FeCl3·6H_2O (2.5 equiv.), MeOH, reflux 2.5 h.

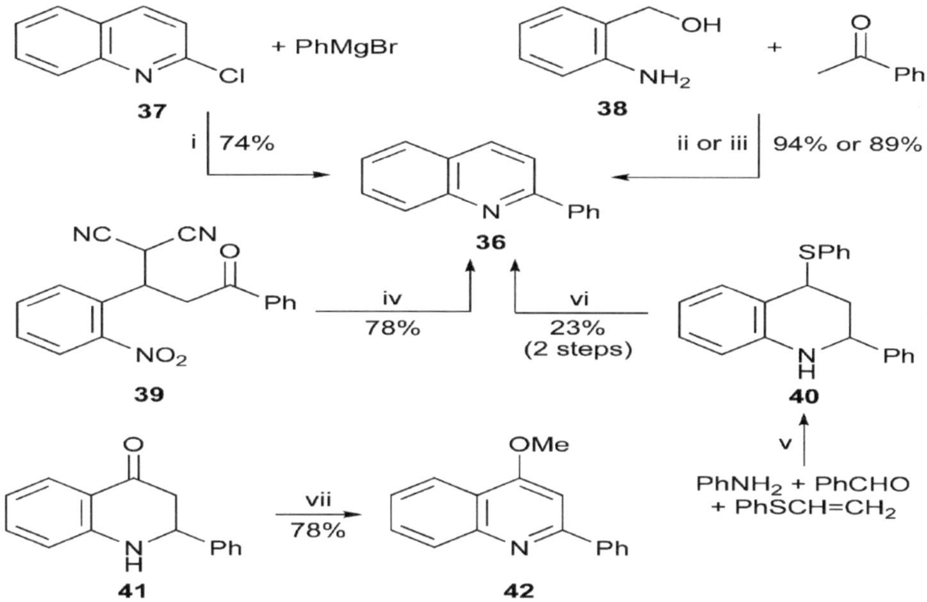

Figure 5.
Quinoline alkaloid synthesis.

5. Isoquinoline alkaloids

There are a few categories of isoquinolinoid marine alkaloids, however the majority of higher plants, isoquinoline alkaloids are present. The fundamental structural element is the isoquinoline nucleus. Numerous therapeutic effects, these particular alkaloids contain substances that have antiviral, antifungal, anticancer, antioxidant, antispasmodic, and enzyme inhibitory properties. Morphine and codeine are the two most significant and extensively studied isoquinoline alkaloids. They originate from tyrosine or phenylalanine.

They are produced using a ketone or an aldehyde and the precursor of dopamine (3, 4-dihydroxytryptamine). These alkaloids are further divided into the following categories: Simple isoquinoline alkaloids, such as salsoline and mimosamycin; benzylisoquinoline alkaloids, such as reticuline and imbricatine; bisbenzylisoquinoline alkaloids, such as fumaricine; manzamine alkaloids, such as manzamine a; pseudobenzylisoquinoline alkaloids, such as polycarpine and ledecorine; and secobisbenzylisoquinoline.

A vast family of naturally occurring substances known as isoquinoline alkaloids exhibits a wide variety of structural variation as well as biological and pharmacological activity. Much work has been done over the past 10 years to develop efficient synthetic methods to obtain these alkaloids in chiral nonracemic form. There have been a variety of approaches used that rely on diastereoselective or enantioselective catalytic processes For a very long time, isoquinoline alkaloids have been important targets for chemical synthesis, both as a source of intellectual challenge and as substances with potential medical value.

Recently, chiral N-sulfinyl β-arylethylamines have been employed as substrates for the asymmetric synthesis of isoquinoline alkaloids.72,73 The Mexican–Spanish team72 in a short and efficient synthesis of (+)-crispine A (177) applied sulfinamide

193, which was prepared from β-3,4-dimethoxyphenylethylamine and (S)-menthyl p-toluene sulfinate as the starting compounds [1].

5.1 Using chiral Oxazoloisoquinoline 794 and Arylmagnesium bromide as a substrate, (S)-(+)-Cryptostyline II (793) is synthesised

By cyclocondensing 4,5-dimethoxy-2-vinylbenzaldehyde with (R)-phenylglycinol in Asao's method, the main substrate, the oxazoloisoquinoline, was produced in 72% yield as a 93:7 combination of diastereomers and used for the synthesis of (S)-(+)-cryptostyline II (**Figure 5**). Thus, the reaction of 794 with 3, 4-dimethoxyphenylmagnesium bromide produced addition product 797, from which, after the chiral inductor was removed and N-methylation was performed, the target alkaloid 793 was recovered with 96% in 57% overall yield (**Figure 6**).

Amat's group employed it as a substrate for the synthesis of C-1-substituted tetrahydroisoquinoline derivatives (**Figure 6**), including alkaloids, starting with oxazolotetrahydroisoquinolone 795 made from aldehyde ester 796 and (R)-phenylglycinol in 58% yield (**Figure 5**). As a result, the reaction of 795 with the proper Grignard reagents produced addition products 798, which were separated as single isomers in a yield of 49–63%. (**Figure 7**) Removal of the N-chiral auxiliary led to lactam 799, in which reduction of the lactam carbonyl fulfilled the synthesis of five alkaloids: (−)-salsolidine (559), (−)-O,O-dimethylcoclaurine (ent-581), (−)-norcryptostyline II (ent-245), norcryptostyline III (558), and (−)-crispine A (ent-177) [24].

6. Indole alkaloids

Terpenoid indole alkaloids are present in a large number of plant species from the families Apocynaceae, Loganiaceae, Rubiaceae, and Nyssaceae (TIAs). TIAs are a broad class of structurally varied molecules that include substances with interesting pharmacological properties. The anti-neoplastic drugs vincristine and vinblastine, the anti-hypertensive drugs reserpine and ajmalicine, as well as the anti-arrhythmic drug ajmaline, are only a few of the terpenoid indole alkaloids used in modern medicine. The intermediate strictosidine, which is created by combining the amino acids

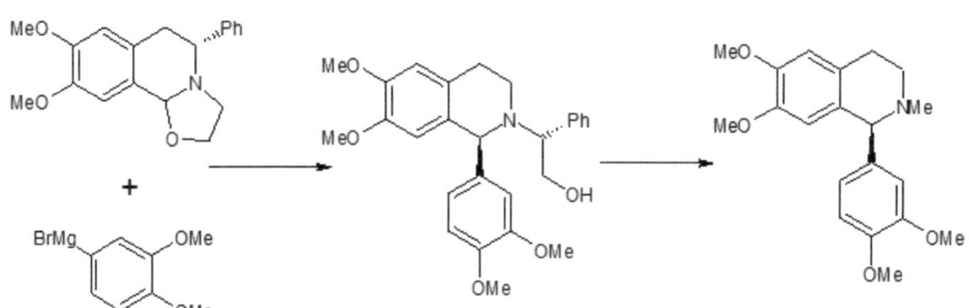

(S)-(+)-Cryptostyline II

793

Figure 6.
Synthesis of Cryptostyline II.

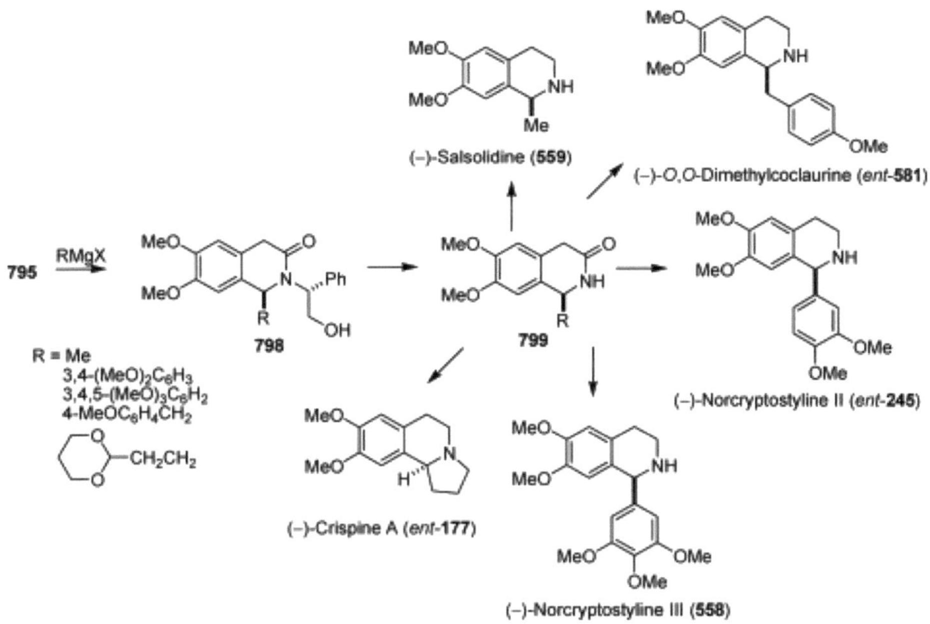

Figure 7.
Synthesis of Norcryptostyline. Grignard reagent reaction of Oxazolotetrahydroisoquinolone 795 with a series of 1-substituted Isoquinoline Alkaloids.

tryptamine and secologanin, which are respectively generated directly from the amino acid tryptophan and indirectly from (10-hydroxy-) geraniol, is essential to the biosynthesis of all terpenoid indole alkaloids [25].

7. Retrosynthetic analysis of arbornamine (monoterpene indole alkaloid)

Arbornamine (1) is a monoterpene indole alkaloid that was discovered in 2016 by Kam and co-unique from a Malayan Kopsia arborea. It has a unique 6/5/6/5/6 “arbornane” skeleton that is different from those of the eburnane and tacaman classes families (**Figure 8**).

Arbornamine’s (1) retrosynthesis calls for the worldwide Pentacyclic lactam 4 is reduced and produce the amino moiety and hydro Xymethine group. It was believed that the pentacyclic lactam 4 would result from a reductive Heck cyclization of vinyl iodide 5.4. The tetracyclic-lactam 6 would result from a The tryptamine derivative 75 and the dimethyl ester 8 of 2-ketoglutaric acid undergo Pictet-Spengler cyclization/ intramolecular ammonolysis. The unsaturated tetracyclic-lactam 5 may then created from there. The tryptamine Nb-nitrogen atom has a protecting group called a benzyl group. An undesirable regioisomeric -lactam 9 makes this design strategically important. In the foregoing one-pot reaction, it might have been produced if the free tryptamine had been employed.

We started our synthesis with a key cascade cyclization. According to **Figure 6**, The tetracyclic -lactam 6 was isolated in a yield of 73% by heating benzyl tryptamine 7 with dimethyl ester 8 and 2 equiv. of TFA in refluxing toluene. Using Pearlman’s catalyst, −lactam 6 was first hydrogenolyzed under atmospheric pressure to produce vinyl iodide under the circumstances required for the Nb-nitrogen atom’s

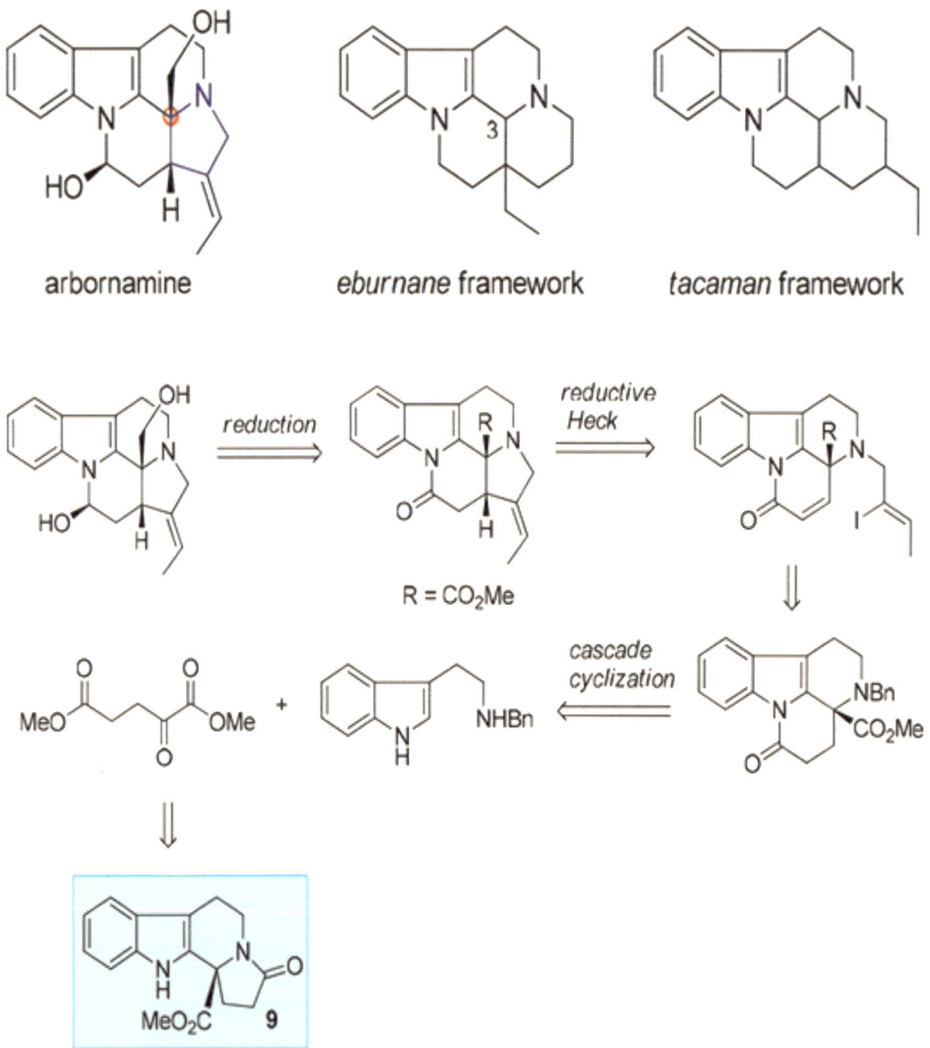

Figure 8.
Synthesis of monoterpene indole alkaloid.

benzyl protective group to be removed and transform it into 11. Thereafter, easily reachable The product was alkylated with (Z)-1- bromo-2-iodobut-2-ene7 to create vinyl iodide 11.

To produce arbornamine, it was globally reduced using lithium aluminium hydride (1). Both the methyl ester and the -lactam have been decreased during this process. The great facial selectivity of -lactam reduction was most likely related to the C-3 methyl ester's top-side shielding action, which allowed the hydride to approach from the bottom side. The synthetic sample's NMR results match those from the literature. As a result, we have created a brief initial method for arbornamine (1), a recently identified monoterpene indole alkaloid, to be completely synthesised. Six steps were all that it took to complete the synthesis, which had a 31% total yield. With the exception of the phase where the N-benzyl protecting group is cleaved, which is used deliberately to prevent the creation of the undesirable product, each step is effective in increasing molecular complexity.

Our synthesis got underway with a critical cascade cyclization. According to above Synthetic pathway, the tetracyclic -lactam 6 was isolated with a yield of 73% the benzyl tryptamine 7 is heated along with the dimethyl ester 8 and two equivalents of TFA in refluxing toluene. At room temperature and pressure, −lactam 6 was first hydrogenolyzed with Pearlman's catalyst to produce vinyl iodide, which was then applied to remove the benzyl protecting group from the Nb-nitrogen atom and transform it into 11. The result was then alkylated with easily available (Z)-1- bromo-2-iodobut-2-ene7 to create vinyl iodide 11. Combining vinyl iodide 11 with a common selenenylation/elimination procedure produced a conjugated tetracyclic -lactam 5. The scene was prepared for the last ring's completion with - lactam 5 in hand. Using a reductive Heck cyclization, Ni(cod)2 was used to mediate the creation of the necessary pentacyclic product 4 in 91% yield. The next product was a pentacyclic -lactam 4. Remaining solvent signals for CDCl3 were detected by 1H NMR (7.26) and 13C NMR (77.0). The following peak multiplicities were noted: Brs for electrospray ionisation (ESI), high resolution mass spectral (HRMS) data were acquired, and There were reported mass-to-charge ratios (m/z). Melting points were determined on a WRX- 5A melting point apparatus [25].

8. Tropane alkaloid

In general, but not always, the roots are where tropane alkaloid production takes place. Translocation then takes place through the aerial sections' xylem, where limited further metabolism may occur. For instance, many Datura species only convert hyoscyamine to hyoscine in the roots, so some synthesis may possibly take place in the aerial parts. For instance, despite the fact that A. Concentrations of mydriatic alkaloids are present in belladonna scions grown on foreign stocks and detached leaves of *A. belladonna* were discovered to have an increase in alkaloid content after 5 days, which was associated with a decrease in protein nitrogen (171).

From grafting studies, which typically used stocks and scions from plants of various solanaceous genera containing distinct alkaloids, such as Datura and Nicotine, it was possible to infer the root origins and subsequent migrations of alkaloids in a number of species (**Figure 9**).

8.1 Biosynthetic pathway of tropane alkaloid

Arginase is aAR. Ornithine decarboxylase, abbreviated ODC. Putrescine N-methyltransferase, or PMT. N-methylputrescine oxidase, or MPO. Tropineforming Reductase, abbreviated TRI. littorine mutase, CYP80F1. The hydroxylase of hyoscyamine 6 (H6H). Transferase for aromatic amino acids, AT4. Phenylpyruvic acid reductase, or PPAR.

8.2 Tropate ester biosynthesis

The (S)-tropic acid 7 ester moiety is a structural component of the alkaloids hyoscyamine 1 and scopolamine 2. Since even before the year 2000, there has been active discussion about the The biosynthesis of tropic acid 7, and the matter is still unresolved, at least in the finer points. We need to go back and look at Robinson's work from the years 1927 and 1955, when he first brought the topic of tropic acid biosynthesis to light and then went into greater detail about it in his book on the

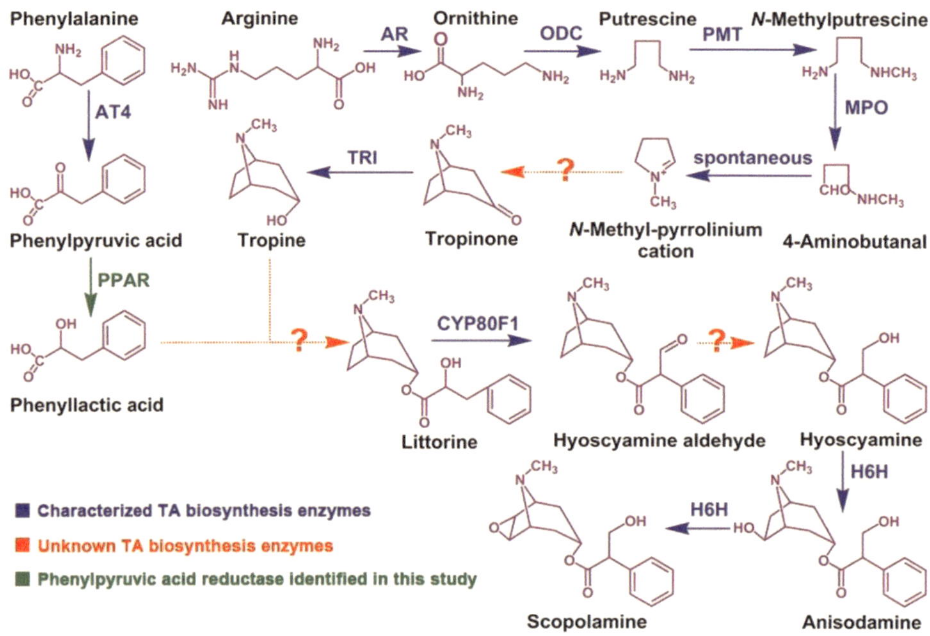

Figure 9.
Tropane alkaloid synthesis. Tropane alkaloids' proposed biosynthetic pathway in the Solanaceae.

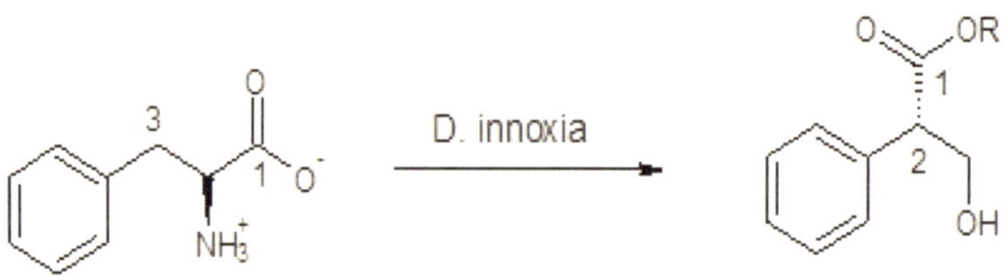

Figure 10.
Ester alkaloid synthesis.

structural relationships of natural products. The fact that tropic acid 7 has a branching carbon skeleton and must originate either from the isomerization of a phenylpropionoid moiety or through a unique synthesis was acknowledged. After feeding (1, 3-C_2) phenylalanine to Datura innoxia, Leete was able to conclusively demonstrate that it derived by isomerization. As indicated in the following scheme, the resulting hyoscyamine 1C now has the isotopes close to one another in positions C-1 and C-2 of the alkaloid (**Figure 10**) [26].

9. Xanthine alkaloids

Purine alkaloids, commonly referred to as xanthine alkaloids, consist of methylxanthines and methyluric acids and their structures are based on the xanthine and uric acid skeletons. Coffee (Coffea arabica), tea (*Camellia sinensis*), mate (*Ilex paraguariensis*), cocoa (*Theobroma cacao*), and guarana (*Paullinia cupana*), which are used to make popular non-alcoholic beverages, all contain caffeine

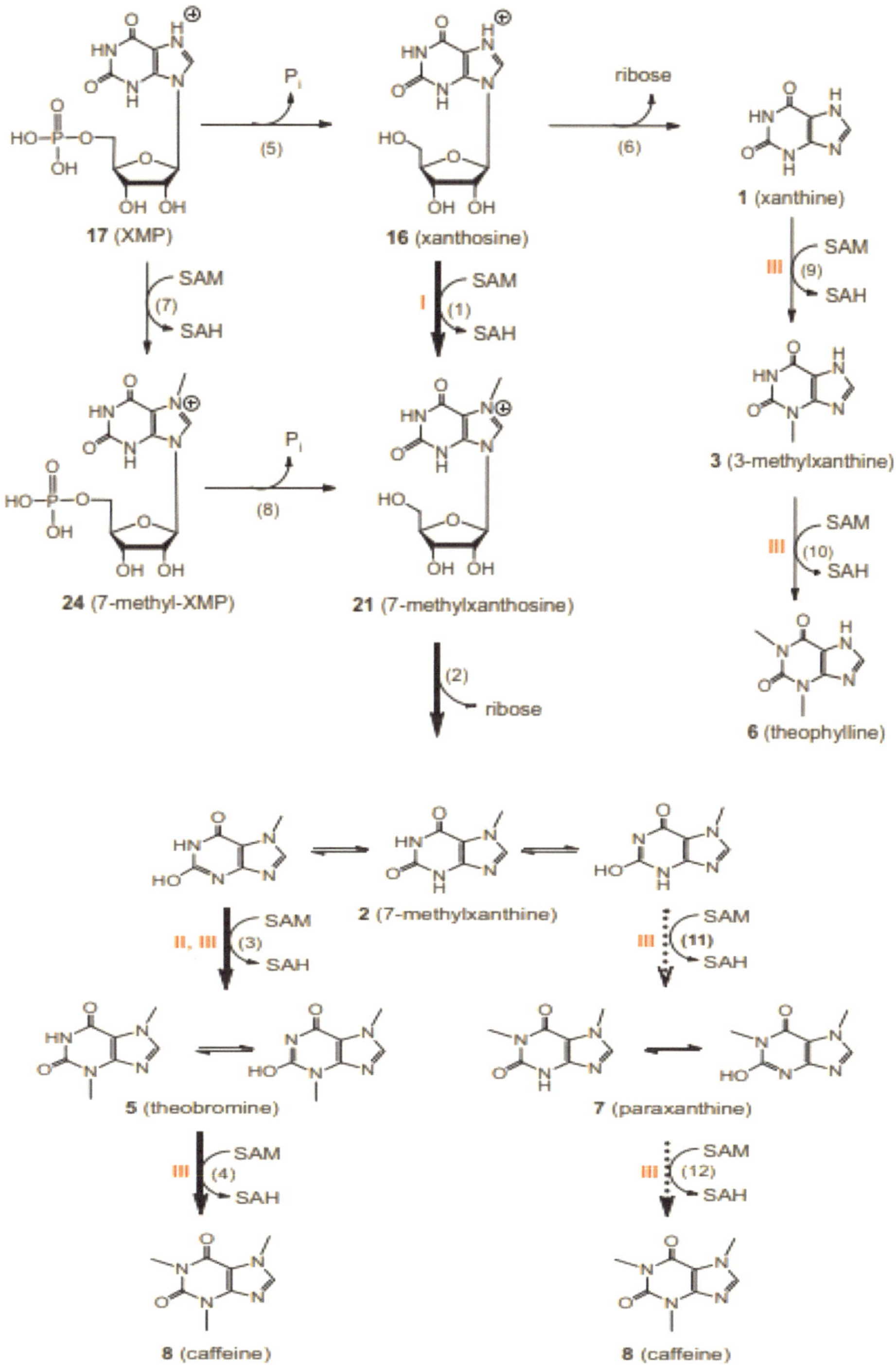

Figure 11.
Synthesis of caffeine.

(1, 3, 7-trimethylxanthine) and theobromine (3,7-dimethylxanthine). The isolation Caffeine of from coffee seeds was first reported independently in 1820 by the German researchers, Runge and von Giese. Caffeine was found as "thein" in tea leaves by Oudry in 1827. Daniell discovered it in kola nuts (*Cola acuminata*) in 1865, while Stenhouse discovered it in mate' in 1843. Woskresensky identified theobromine in cacao seeds in 1842. Salomon discovered paraxanthine (1,7-dimethylxanthine) from human urine in 1883, but Chou and Waller did not find it in coffee seeds until 1980. Fischer and Ach published the complete chemical synthesis of Caffeine in 1895. Studies on caffeine biosynthesis were initiated in the 1960s, while highly purified caffeine synthase was isolated by Kato et al. (**Figure 11**) [27].

Solid arrows represent the four steps that make up the main pathway (steps1–4). Three different N-methyltransferases are shown as I, II, and III: caffeine synthase, theobromine synthase, and 7-methylxanthosine synthase. N-methylnucleosidase is responsible for catalysing the second step, which converts 7-methylxanthosine to 7-methylxanthine. I, III. Due to the broad substrate specificities of caffeine synthase, minor routes, denoted by dotted arrows, may take place (III). The production of 7-methylxanthosine from XMP via 7-methyl-XMP (steps 7–8) was suggested by Schulthess et al., but no recombinant N-methyltransferases have been found to catalyse these conversions [27].

10. Pyrrole-imidazole alkaloids

As an example, consider the pyrrole-imidazole alkaloids (**Figure 1A**). Sponge natural products classified as pyrrole-imidazole alkaloids have approximately 150 congeners and are a broad and highly intricate class. Because of their chemical complexity, pyrrole-imidazole alkaloids have undergone a number of structural revisions, provided title compounds for organic synthesis, and maintained pharmaceutical interest due to their attractive bioactivity profiles. The enantioselective dimerization of three important monomeric building blocks, oroidin (1), hymenidin (2), and clathrodin (3), is suggested by retrosynthetic pathways for pyrrole-imidazole alkaloids. In fact, in vitro biomimetic research, these monomeric building blocks were dimerized utilising enzymes isolated from sponges that contained pyrrole-imidazole alkaloids. But there have not been many insights into the biosynthesis of 1–3 themselves; the only information we presently have came from observing the incorporation of radiolabeled amino acid precursors into 1 product (**Figure 12**) [28].

11. Synthesis of pyrrole-imidazole alkaloids is proposed

Retrobiosynthetic plan explaining how product 1 is produced from the building blocks of amino acids. 8 and 9 are hypothesised to be directly connected with pyrrole carboxylic acids in the chemical structures of pyrrole-imidazole alkaloids (B-E) EICs and MS^2 spectra mirror plots comparing the *Stylissa* metabolome's 7–9 and 11 identified metabolites to synthetic standards. We have highlighted important MS^2 ions (**Figure 13**).

Barbaleucamides A–B are shown in the metabolome BPC, which has superimposed EICs showing their existence. (B) EICs, produced within 1 ppm error tolerance for 7–12 across *Stylissa*, *Axinella*, *Agelas*, and *Dysidea* polar metabolomics LC/MS datasets. With the exception of EIC 7, which has the high abundance structural annotations of fragment ions, all EIC y-axes are similar as indicated. Each structural annotation's

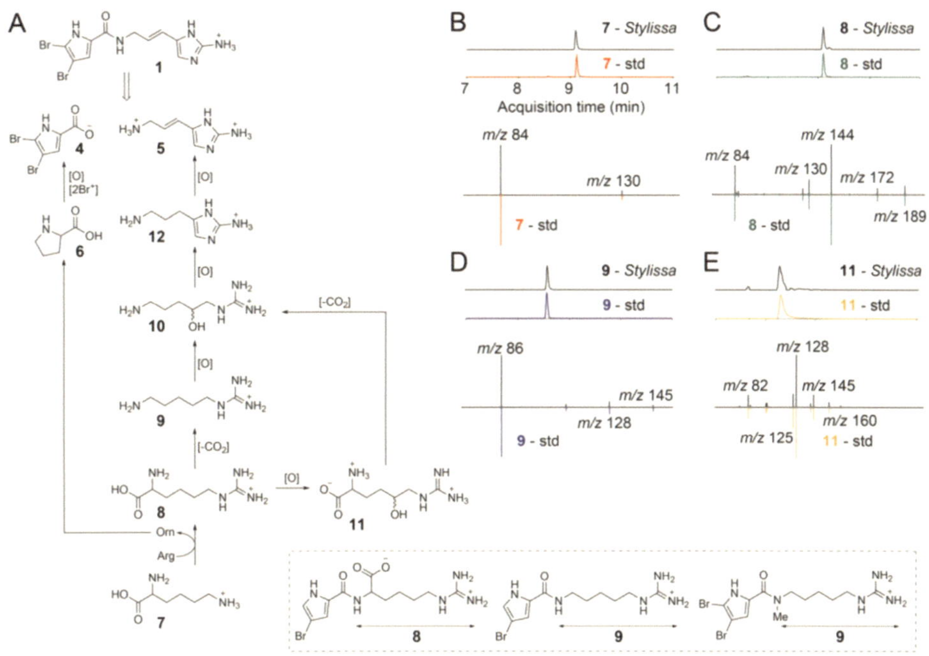

Figure 12.
Pyrrole imidazole alkaloid synthesis.

related ppm mistake is listed. The MS^2 spectra displayed here were obtained using an orbitrap mass spectrometer and extremely precise Fourier transform mass spectrometric fragmentation.

12. Piperidine alkaloids

This class of alkaloids' primary ring system is the piperidine nucleus. Monocycle molecules with the C5N nucleus are the primary defining feature of true piperidine alkaloids. The odour of piperidine alkaloids is one of their shared characteristics. They lead to long-term neurotoxicity. The majority of them come from plants. Even though the piperidine alkaloid is made from lysine, some piperidine alkaloids, like the straightforward pyrrolidine alkaloids, are also made from acetoacetate. Lobeline is a major alkaloid in this group [22].

13. Imidazole alkaloid

The imidazole ring structure of this form of alkaloid is what makes it unique. Since the imidazole ring was already formed during the precursor step, these alkaloids constitute an exception to the structure-transformation process. This type of structurally diverse alkaloids occurs in a variety of situations, particularly in marine and microbial alkaloids. They have a significant potential for therapeutic use and demonstrate a wide spectrum of biological activities [29].

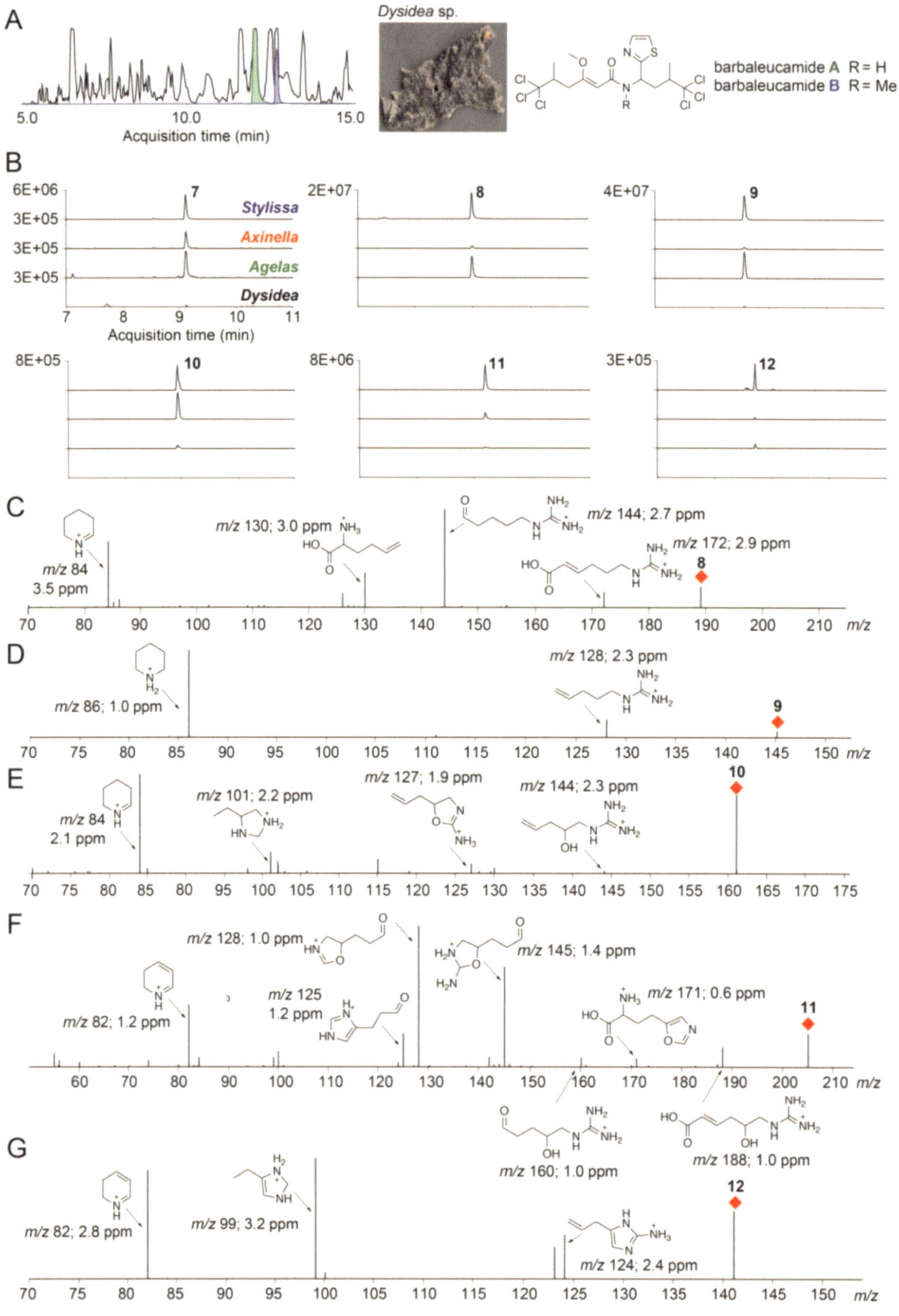

Figure 13.
List of pyrrole-imidazole alkaloid biosynthesis intermediates that have been rationalised. A Dysidea *species.*

14. Pyrrolizidine alkaloids

The pyrrolizidine nucleus is the defining characteristic of this class of alkaloids. Plants from the Fabaceae and Asteraceae families contain them. The bulk of pyrrolizidine alkaloids are present in plants as N-oxides, but when they are separated, they lose their functionality. A lot of research has been done on alkaloids because of their potentially harmful side effects, particularly liver damage. The animals that consume these alkaloids become antifeedants when they enter the food chain [30].

Acknowledgements

All authors are thankful to management and principal of Dadasaheb Balpande College of diploma in Pharmacy, Nagpur.

Author details

Nitin Dumore[1†], Namita Girhepunje[1*†], Monali Dumore[2†] and Kishor Danao[2†]

1 Dadasaheb Balpande College of Diploma in Pharmacy, Nagpur, Maharashtra, India

2 Dadasaheb Balpande College of Pharmacy, Nagpur, Maharashtra, India

*Address all correspondence to: namita.tilgule71@gmail.com

† All authors are equally contributed for the preparation of chapter.

References

[1] Kurek J. Introductory Chapter: Alkaloids-their Importance in Nature and for Human Life. London, UK: InTech; 2019

[2] Hesse M. Alkaloids, Nature's Curse or Blessing, Weinheim, Germany: Wiley-VCH, A very interesting book of the history, biological significance, and synthesis of alkaloids 2003, 42, 40, 4852-4854

[3] Geoffrey A. Cordell, Alkaloids and their Biosynthesis - Introduction to Alkaloids. A Biogenetic Approach, Phytochemistry and Pharmacognosy, Natural Products Inc., Evanston, 1981, 21, 3, 1055

[4] Dewick PM. Medicinal Natural Products. A Biosynthetic Approach. Second ed. Chichester, UK: John Wiley & Sons. [The alkaloid chapter in this book offers a useful and succinct overview of alkaloids as medicinal agents]; 1997. p. 466

[5] Cordell GA, Quinn-Beattie ML, Farnsworth NR. The potential of alkaloids in drug discovery. Phytotherapy Research [A review of the occurrence of alkaloids and their biological potential]. 2001;**15**:183-205

[6] Wink M. Molecular modes of action of cytotoxic alkaloids: From DNA intercalation, spindle poisoning, topoisomerase inhibition to apoptosis and multiple drug resistance. In: The Alkaloids, Chemistry and Biology. Vol. 64. Elsevier Publishers, A review of the relationships of the structure of alkaloids and their interactions with cell systems; 2007. pp. 1-47

[7] Herbert RB. The biosynthesis of plant alkaloids and nitrogenous microbial metabolites. Natural Product Reports, A series of reviews on the biosynthesis of alkaloids in plants, fungi, bacteria, and marine organisms. 2003;**20**(5):494-508

[8] Kutchan TM. Alkaloid biosynthesis – The basis for metabolic engineering of medicinal plants. The Plant Cell, American Society of Plant Physiologists, A good introduction to the relevance of metabolic engineering in developing alkaloids. 1995;7(7):1059-1070

[9] Zenk MH, Juenger M. Evolution and current status of the phytochemistry of nitrogenous compounds. Phytochemistry, Why the study and the continuous development of alkaloid biosynthesis using metabolic engineering is important. 2007;**68**(22–24):2757-2772

[10] Usera AR, O'Connor SE. Mechanistic advances in plant natural product enzymes. Current Opinion in Chemical Biology, A good introduction to the relevance of metabolic engineering in developing alkaloids. 2009;**13**:492-498

[11] Roberts M, Strack D, Wink M. Biosynthesis of alkaloids and betains. Annual Plant Reviews, An overview of alkaloid biosynthesis from chemical, enzymatic and gene perspectives. 2010; **40**:20-91

[12] Suzuki K-I, Yamada Y, Hashimoto T. Expression of Atropa belladonna putrescine N-methyl transferase gene in root pericycle. Plant Cell Physiology, The cDNAs for putrescine N-methyl transferase are described. 1999;**40**: 289-297

[13] Heim WG, Sykes KA, Hildreth SB, Sun J, Lu RH, Jelesko JG. Cloning and characterization of a Nicotiana tabacum methyl putrescine oxidase transcript. Phytochemistry, One of the key enzymes

in tropane alkaloid biosynthesis is described. 2007;**68**:454-463

[14] Lounasmaa M, Tamminen T. The tropane alkaloids. In: Cordell GA, editor. The Alkaloids, Chemistry and Pharmacology. Vol. 44. San Diego, California: Academic Press, A review of the tropane alkaloids; 1993. pp. 1-114

[15] Robins RJ, Walton NJ. The biosynthesis of tropane alkaloids. In: Cordell GA, editor. The Alkaloids, Chemistry and Pharmacology. Vol. 44. San Diego, California: Academic Press. An overview of tropane alkaloid biosynthesis; 1993. pp. 115-187

[16] Robins RJ, Abraham TW, Parr AJ, Eagles J, Walton NJ. The biosynthesis of tropane alkaloids in Datura stramonium: The identity of the intermediates between N-methylpyrrolinium salt and tropinine. Journal of the American Chemical Society, Acetoacetate is incorporated intact into the tropane nucleus. 1997;**119**:10929-10934

[17] Sandala GM, Smith DM, Radom L. The carbon skeleton rearrangement in tropane alkaloid biosynthesis. Journal of the American Chemical Society, Quantum chemistry calculations suggest a concerted carbocation rearrangement in hyoscyamine biosynthesis. 2008;**130**: 10684-10690

[18] Humphrey A.J. and O'Hagan D. Tropane alkaloid biosynthesis. A century old problem unresolved. Natural Product Reports, An historical overview of the complexities of tropane alkaloid biosynthesis, 2001, 18, 494-502

[19] Stöckigt J, Panjikar S. Structural biology in plant natural product biosynthesis – Architecture of enzymes from monoterpenoid alkaloid and tropane alkaloid biosynthesis. Natural Product Reports, A contemporary view of the importance of studying the enzymes of tropane and indole alkaloid biosynthesis. 2007;**24**:1382-1400

[20] Khadem S, Marles RJ. Chromone and flavonoid Alkaloids: Occurrence and bioactivity. Molecules. 2012;**17**(12): 191-206

[21] Dey P, Kundu A, Kumar A, Gupta M, Lee BM, Bhakta T, et al. Recent Advances in Natural Products Analysis. Elsevier; 2020

[22] Dey P, Kundu A, Kumar A, Gupta M, Lee BM, Bhakta T, et al. Analysis of alkaloids (indole alkaloids, isoquinoline alkaloids, tropane alkaloids). In: Recent Advances In Natural Products Analysis. Elsevier; 2020. pp. 505-567

[23] Michael OP. Quinoline, quinazoline and acridone alkaloids. Natural Product Reports. 2007;**24**(1):223

[24] Chrzanowska M, Grajewska A, Rozwadowska MD. Asymmetric synthesis of Isoquinoline alkaloids. Chemical Reviews. 2016;**116**(19): 12369-12465

[25] Collu G, Unver N, Peltenburg-Looman AMG, van der Heijden R, Verpoorte R, Memelink J, et al. Geraniol 10-hydroxylase1, a cytochrome P450 enzyme involved in terpenoid indole alkaloid biosynthesis. FEBS Letters. 2001;**508**(2):215-220

[26] Andrew J. Humphreya, David O'Haganb, a century old problem unresolved. Natural Product Reports. 2001;**18**(5):494-502

[27] Zheng Y, Yue B-B, Wei K, Yang Y-R. Total synthesis of (−)-Geissoschizol through Ir-catalyzed allylic Amidation as the key step. Organic Letters. 2017; **19**(23):6460-6462

[28] Mohanty I, Moore SG, Yi D, Biggs JS, Gaul DA, Garg N. Vinayak Agarwal precursor-guided mining of marine sponge metabolomes lends insight into biosynthesis of pyrrole-imidazole alkaloids. ACS Chemical Biology. 2020; **15**(8):2185-2194

[29] Kohnen-Johannsen KL, Kayser O. The imidazole alkaloid with the greatest medicinal significance is pilocarpine. Tropane Alkaloids: Chemistry, Pharmacology, Biosynthesis and Production, National Liabrary of Medicine. 2019;**24**(4):796

[30] Schramm S, Köhler N, Rozhon W, Alkaloids P. Biosynthesis, biological activities and occurrence in crop plants. Molecules. 2019;**24**(3):498

Chapter 7

Alkaloids: The Potential of Their Antimicrobial Activities of Medicinal Plants

Mohammad Barati and Amir Modarresi Chahardehi

Abstract

Given the potential adverse effects of chemical drugs, utilizing natural products with diverse therapeutic and antimicrobial compounds is advisable. Countries can use indigenous flora from their regions in vegetation for medicinal purposes. Several nations exhibit distinctive indigenous flora owing to their geographic positioning and climatic conditions. These plants have been the subject of our research, which has explored their antimicrobial properties against fungi, parasites, bacteria, and viruses. Studies have investigated the therapeutic and antimicrobial effects of plants and their bioactive compounds, such as alkaloids, flavonoids, and terpenoids. Among them are alkaloids, a diverse class of naturally occurring chemicals, such as tropanes, terpenoids, and steroids. Some of these medicinal plants have been found to possess antioxidant and anti-inflammatory properties in addition to their antimicrobial effects. This chapter explores the antimicrobial potential of alkaloids found in medicinal plants.

Keywords: alkaloids, medicinal plants, antimicrobial activity, secondary metabolite, bioactive compounds

1. Introduction

Researchers are discovering infectious diseases are a major threat to world health [1, 2]. For millennia, medicinal plants have yielded an abundance of therapeutic compounds, which have been incorporated into traditional pharmacological practices across the globe [3, 4]. Since the dawn of time, people have known that plants have healing properties, making botanic medicine one of the first forms of therapy [5–7]. Antibacterial medications have traditionally been derived from natural materials. This avenue of inquiry declined in the 1980s as scientists shifted their focus to synthetic compound libraries because of their greater flexibility [8]. Antibiotic and antifungal medication discovery are crucial in the face of the rise of multidrug-resistant (MDR) fungi and bacteria [9]. The emergence of multidrug-resistant organisms is a significant worldwide health concern [10]. The incorrect use of antibiotics in human and animal health care is largely responsible for the rise of MDR strains. Consequently, the search for alternative, nonantibiotic-dependent solutions to this critical issue has become an urgent and imperative challenge [11]. In particular, each plant produces small quantities of secondary metabolites—tiny compounds like terpenoids, polyphenols, phenolics, alkaloids, essential oils, etc. [12]. The discovery of novel pharmacological compounds that can treat serious ailments has greatly benefited from research into medicinal plants [13]. Some plants, such as mustard, ginger, basil, garlic, cinnamon, sage, curry, and many other crude extracts, for instance, have antibacterial activity against many different forms of bacteria, including gram-positive and gram-negative [14]. Medicinal plants include phytochemicals, often responsible for their biological activity, commonly found in these plant sources [1]. Secondary metabolites found in plants include active chemical molecules with potential therapeutic applications for various diseases [15]. Isolated secondary metabolites in plants are thought to account for fewer than 10% of the total. Metabolites are commonly employed to safeguard against insects, herbivores, and microbes. The diverse range of aromatic substances and their oxygen-substituted derivatives plants synthesize accounts for the extensive variety observed [16]. Recently, drug resistance has emerged as a major issue in healthcare; the rate at which drug-resistant diseases are increasing is far higher than the rate at which new medications are being tested and authorized for human use. Thus, it is crucial to create new antimicrobial drugs [17–19].

Infectious illnesses caused by microorganisms significantly contribute to human suffering and death. About 60% of the biomass on Earth is thought to be composed of microbial species. This, together with their tremendous genetic, metabolic, and physiological variables, renders them a danger to the well-being and progress of human communities everywhere [20]. Hence, nature is the source of a significant proportion of the drugs currently used, derived from microorganisms, flora, or fauna. Identifying and synthesizing novel compounds possessing pharmacological properties depends on the natural environment's biodiversity [12, 21]. Many plant components are available without a prescription from herbal distributors and natural-food stores, and self-administration of these drugs is common even though their purity is often questionable [22]. Chemical analysis of medicinal plants has uncovered various bioactive chemicals, including saponins, tannins, and alkaloids [1, 23]. Also, flavonoids, terpenoids, and alkaloids are the primary constituents of phytochemicals in the plant kingdom [24]. The pharmacologically active compounds encompass a variety of alkaloids that can be categorized into several classes, such as piperidines, pyrrolizidines, quinolizidines, imidazoles, tropanes, pyrrolidines, indoles, isoquinolines, and purines [15]. They belong to a vast group of naturally occurring chemical compounds that include at least one nitrogen atom (particularly in the form of an amino or amido group). The nitrogen atoms often form a ring shape [25]. Alkaloids are plant-derived bioactive compounds typically exhibiting alkaline properties due to their nitrogen atoms [26].

With many plants still waiting to be discovered and examined for their phytochemical compositions, the future of therapeutic plants seems bright. Synthetic medicine design and development have benefited from learning about medicinal plants [1]. Thus, alkaloids are the subject of intensive study because they may constitute a novel class of naturally occurring antibiotics with a broad antibacterial range, few side effects, and a low propensity to result in drug resistance. The present chapter centers on investigating the antimicrobial potential of alkaloids obtained from medicinal plants against human pathogenic microorganisms, specifically emphasizing multi-drug-resistant clinical strains. The chapter elucidates the mechanism of action of these alkaloids when available and underscores their concentrations and usage.

2. Plant products as an antimicrobial agent

Pathogenic bacteria create dangerous and potentially fatal infectious diseases that affect humans [27, 28]. On the other hand, antibiotic resistance is a significant issue in the twenty-first century, and infectious illnesses are still the second-greatest cause of mortality globally despite the success of antibiotic discoveries [1]. The growing incidence of antimicrobials-microbes resistance is causing growing alarm among scientists. The advent of drug-resistant bacteria has increased the difficulty and expense of creating newer antimicrobials from novel chemical compounds [15, 28]. Despite the approval of synthetic antimicrobial agents in numerous countries, using natural compounds derived from microbial, animal, or plant sources has garnered significant interest among researchers [29]. Numerous researchers are currently engaged in the investigation of plants to identify potential antimicrobial agents [15]. The quest for compounds possessing antimicrobial properties is common, and scholars have shown interest in medicinal plants due to their widespread use in traditional medicine as a treatment for various infectious ailments [30]. Hence, the demand for and research into plant-based pharmaceuticals and nutritional aids has increased rapidly in recent years [31]. Studies conducted on plants utilized in traditional medicine have been performed *in vitro* within the realm of microbiology, with a particular focus on the proliferation of infectious bacteria [30]. Betoni et al. found that plant compounds can either act as antimicrobial agents that complement antibiotics or increase a pathogen's susceptibility to an antibiotic that would have otherwise been ineffective [30].

Researchers from fields as diverse as ethnopharmacology, botany, microbiology, and natural products chemistry scour the planet in search of phytochemicals and "leads" that might be refined into effective antimicrobial drugs [31]. New medications can be developed by optimizing the structural makeup of phytochemicals present in plants [1]. Phytochemicals and other substances derived from plants have been used to treat a wide range of infectious diseases because they exhibit good antibacterial action against many human infections [29, 32]. However, it is widely established that several extracts and components of plants have antibacterial activity. Unfractionated extracts are typically used in these studies, despite their low *in vitro* antimicrobial activity. *In vivo* tests were rarely used to verify the results of these investigations [12]. Phytochemicals, which are bioactive organic chemical compounds, are present in medicinal plants [33, 34]. These compounds protect against chronic diseases, including those caused by metabolic or genetic disorders and infectious diseases. They are present in various foods made from plants, including cereals, veggies, and fruits [1]. There are several classes of phytochemicals, including carotenoids, alkaloids, phenolics, organosulfur compounds, and nitrogen-containing compounds [5].

3. Alkaloids

Alkaloids are naturally occurring compounds sourced from various organisms, including plants (which comprise approximately 300 plant families), bacteria, fungi, and animals [12]. The compounds and biomolecules exhibit significant diversity, yet all these chemicals are byproducts of the amino acid biosynthesis process or the transamination reaction [35]. Alkaloids are predominantly solid compounds that are commonly found in higher plants. The aforementioned botanical families, namely Leguminoceae, Papaveraceae, Solanaceae, Ranunculaceae, Annonaceae,

Amaryllidaceae, Liliaceae, Apocynaceae, Boraginaceae, Loganiaceae, Magnoliaceae, Berberidaceae, Piperaceae, Gnetaceae, Rutaceae, Lauraceae, Menispermaceae, and Rubiaceae, are known to exhibit a high prevalence of the subject matter [36]. Certain plant species employ naturally occurring insecticides or pesticides to protect themselves against the harmful effects of select insect species. The synthesis of vegetal alkaloids primarily occurs in herbaceous and vascular plants [12]. The Arabic word al-qali designates the source of soda. German scientist Carl F. W. Meissner developed the term "alkaloid" in 1819 to describe this compound [36]. One of the biggest groups of secondary metabolites in plants, alkaloids are present in some economically relevant plant families [37]. As mentioned, they are present in various kingdoms. However, their distribution is restricted within each domain [8]. Alkaloids are classified into multiple categories. The categorization is founded upon the compounds' heterocyclic ring structure and biosynthetic forerunners. The abovementioned compounds comprise indoles, pyrrolizidines, quinolizidines, pyrrolidines, piperidines, tropanes, isoquinoline, purines, and imidazoles [15]. The amino acids nicotinic acid, L-histidine, L-ornithine, L-tryptophan, L-lysine, L-tyrosine, acetate, L-phenylalanine, anthranilic acid, and L-phenylalanine are all precursors to the alkaloid phenylpropanoid [35]. Alkaloids also exhibit various pharmacological and biological properties and may be found in many herbal treatments [38]. Alkaloids have been the fundamental framework for advancing multiple antibiotics showing a broad activity spectrum [16]. Nicotine, caffeine, and cocaine are just a few examples of alkaloids incorporated into popular culture as drugs used for entertainment or abuse. Certain alkaloids have been identified as possessing high toxicity levels, resulting in numerous instances of human poisoning [16].

Alkaloids have a wide array of pharmacological activities, including antibacterial activity [12]. Most alkaloids exert their effects via efflux pump inhibitor (EPI) activity, which is considered a potential mechanism of antibacterial action [29]. In addition to their use as stimulant medications, alkaloids may be found in many of the foods and drinks we consume regularly. They have shown several pharmacological effects, including those of local anesthetic, anticancer, analgesic, pain-relieving, antifungal, anti-inflammatory, neuropharmacological, and antimicrobial, [25], antimalarial action, oxytocic and vasoconstrictor activity (ergometrine), activity against the central nervous system (brucine), and activity against the cholinergic system (atropine) [16]. Alkaloids, which derive their name from their resemblance to alkalis, can undergo salt formation upon reaction with acids, similar to inorganic alkalis. The nitrogen atoms exhibit basic properties in acid-base responses [25]. Alkaloids are characterized by a nitrogen atom that accepts protons and multiple amine hydrogens that donate protons. Hence, the biological activity of biomolecules is primarily attributed to their ability to establish hydrogen bonds with other biomolecules such as enzymes, receptors, and proteins [12, 24]. Thus, alkaloids can be used for a variety of pharmacological purposes [24]. Several antibiotics have been developed from alkaloids: the quinolones were discovered by accident during the production of quinine; the structure of metronidazole was altered from that of azomycin; and the quinoline scaffold was utilized to create bedaquiline [8]. Alkaloids can also be found in other medications like linezolid and trimethoprim scaffolding. Academic institutions, private companies, and public-private partnerships continue studying alkaloids to create effective antibacterial drugs [8].

A straightforward quantitative approach for identifying alkaloids in plants was developed by Li et al. [39]. Using tetrahydrofurfuryl methacrylate as the monomer, in situ radical polymerization was used to construct a polymer-based chromatographic monolithic column. Based on the results of the technique validation, the accuracy of the spiking recovery measures is between 98.89 and 102.06%. These findings demonstrate the constructed monolithic column's viability for avoiding the lengthy analysis time required by conventionally packed C18 columns in quantitatively analyzing alkaloids from actual medicinal and culinary plant foods [39]. Alkaloids are used internally to improve health, physical performance, and the immune system. These entities are common in daily dietary intake, drinks, and supplementary products. Several compounds present in plants exhibit advantageous characteristics. Compounds such as caffeine, guaranine, and mateine, found in various plants, including coffee, have been observed to possess anti-inflammatory, antioxidant, and stimulatory properties. Additionally, cocoa contains theobromine and paraxanthine, which act as antioxidants. Ginger, conversely, contains gingerol and shogaols, which are phenolic alkenones that possess antioxidant, anti-inflammatory, antimicrobial, and antitumoral properties [37]. However, we provide a brief overview of the class of alkaloids concerning antimicrobial activity.

3.1 Alkaloids classification

At present, the number of identified alkaloids exceeds 18,000 [15]. Natural antibacterial alkaloids have been the subject of research since the 1940s, although most of the earliest studies did not go far enough to determine minimum inhibitory concentrations (MICs). Despite this class's large number of chemicals, only a fraction of their biosynthesis routes have been determined [40]. The chemical makeup or inherent biological source of these entities determines their classification [16]. Chemical structure and characteristics are used to divide alkaloids into several classes. The feasibility of classifying alkaloids based on their natural origin arises because certain alkaloids are limited to specific sources [16]. The chemical structure or biological origin of alkaloids allows for two broad categories:

1. The initial category comprises three types: protoalkaloids, or biological amines, nonheterocyclic or unconventional alkaloids. These alkaloids contain nitrogen in their side chains. The following category includes the heterocyclic or conventional alkaloids, also known as true alkaloids, which possess nitrogen within the heterocycle, and pseudoalkaloids [36]. The basic carbon skeleton of pseudoalkaloids is not directly formed from amino acids. Still, it is connected to amino acid processes and is derived via an amination or transamination process from amino acid precursors or postcursors. Common pseudoalkaloids include capsaicin, caffeine, and ephedrine [36].

2. The second division may be subsequently classified into 14 subgroups based on the ring shape due to its deep structural complexity [16, 24].

As mentioned above, there are primarily three classes of alkaloids [36]:

3.1.1 Protoalkaloids

Alkaloids having a closed ring structure are protoalkaloids; they are chemically perfect but have a straightforward molecular structure. Among the alkaloids, they are in the minority [35]. The most notable examples of these alkaloids include yohimbine, mescaline, and hordenine (a phenethylamine) (**Figure 1**). Hordenine, a Tyr-derived

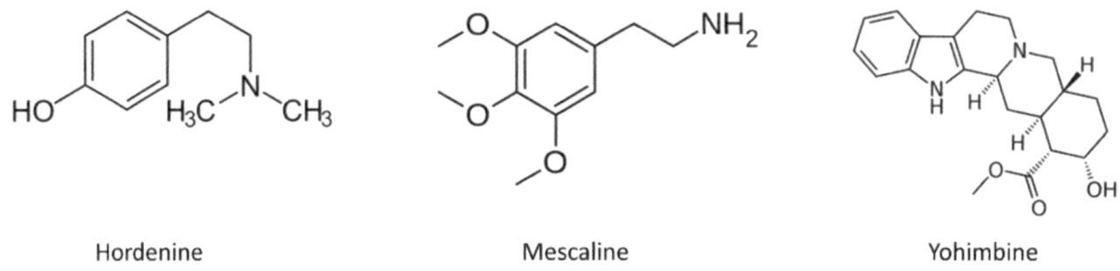

Figure 1.
Some examples of protoalkaloids.

phenylethylamine alkaloid, was initially discovered in *Hordeum vulgare* (barley) [41]. They are prescribed for various conditions, from mental illness to chronic pain to neuralgia. The nitrogen atom in these alkaloids comes from a source other than the heterocyclic ring structure; instead, it is generated from an amino acid. Typically, L-tryptophan and L-tyrosine are the precursors to these alkaloids. Simple alkaloids make form the framework of this minor class [36]. Protoalkaloids are compounds where the heterocyclic bond does not include the N atom from an amino acid. One type of alkaloid consists of compounds derived from the amino acids L-tryptophan and L-tyrosine [35].

3.1.2 True alkaloids

These alkaloids and their precursor amino acids both have nitrogen in a heterocyclic ring. These entities exhibit high reactivity and possess significant biological efficacy [36]. These compounds can dissolve in water and form salts soluble in water. Additionally, many of these compounds exhibit a crystalline structure and can undergo conjugation with acids to form salts. Most authentic alkaloids are characterized by their solid state and bitter flavor, except nicotine, a brown liquid. Common true alkaloids include cocaine, morphine, and quinine [36]. Morphine, an alkaloid generated from tyrosine, has a nitrogen-containing heterocyclic ring and is used as a painkiller. It exhibits potent analgesic effects and is widely used as a painkiller in clinical settings [42]. Not all alkaloids show significant biological efficacy; some have no known pharmacological activity [43].

These subgroups have unique properties and uses, making them essential modern medicine and research components. Understanding the classification of alkaloids is an important step in understanding their potential therapeutic applications. For example, various pharmacological effects are associated with indole alkaloids found in plants, many of which are thought to be attributable to the indole nucleus [44]. Common plant families proven to contain indole alkaloids include Loganiaceae, Rubiaceae, Apocynaceae, and Nyssaceae. Preclinical and clinical research has shown that several of the discovered indole alkaloid compounds are particularly effective [44]. According to their antimicrobial activity, the most critical phytocompounds across all alkaloid chemical groups are shown in **Table 1**.

Monoterpenoid indole alkaloids are a class of widely recognized alkaloids that are derived from tryptamine and secologanin. Numerous alkaloids exhibit intricate structures and significant biological properties, rendering them intriguing. Various species belonging to the Apocynaceae family, including *Tabernanthe iboga*, *Voacanga africana*, and multiple *Tabernaemontana* species, synthesize alkaloids, including the ibogan type [116]. Antibiotic and well-known alkaloid tryptanthrin (TRYP) (indolo[2,1-b] quinazolin-6,12-dione) is found in *Candida lypolica*, higher plants, and numerous

Types of alkaloids	Formula	Plant family	Targeted microorganism	Concentrations range (μg/mL)	Reference
Indole alkaloids					
Brassicaceous indoles	$C_{13}H_9N_3O_2S$	Brassicaceae			
Caulilexin A	$C_{10}H_9NOS_2$	Brassicaceae	*Sclerotinia sclerotiorum*, *Leptosphaeria maculans*, *Rhizoctonia solani*	5×10^{-4} M	[45]
Camalexin (3-thiazol-2′-yl-*indole*)	$C_{11}H_8N_2S$	Brassicaceae	*Alternaria brassicae*	80 μg/mL	[46]
β-Carbolines	$C_{11}H_8N_2$				
Borrerine	$C_{16}H_{20}N_2$	Rubiaceae	*Staphylococcus aureus*, *Vibrio cholerae*	50 and 6 μg/mL, respectively	[47]
Borreverine	$C_{32}H_{40}N_4$	Rubiaceae	*S. aureus*, *V. cholerae*		[47]
Canthin-6-one (canthinone)	$C_{14}H_8N_2O$	Simaroubaceae	*S. aureus*, *Mycobacterium* sp.	8–32 μg/mL	[48]
Rhetsinine	$C_{19}H_{17}N_3O_2$	Rutaceae	*Xanthomonas oryxae* pv *oryzae*, *Xanthomonas oryxae* pv *oryzicola*	1 and 4.5 μg/mL	[49]
Carbazoles	$C_{12}H_9N$				
Glycozolidol	$C_{14}H_{13}NO_2$	Nitrariaceae and Rutaceae	*Proteus vulgaris*, *Bacillus firmis*, *S. lutea*, *S. aureus*, *Agrobacterium tumefaciens*	200 μg/mL/well	[50]
Benzoisofuranone	$C_8H_6O_2$	Rutaceae	*S. aureus*, *B. subtilis*, *Escherichia coli*, *P. vulgaris*, *Aspergillus niger*, *Candida albicans*	3.13–100 μg/mL	[51]
Harmane	$C_{12}H_{10}N_2$	Nitrariaceae	*V. anguillarum*	3.1 μg/mL	[52]
			Cryptococcus neoformans, *A. niger*, *Cryptococcus gattii*, *C. albicans*	Very weak inhibited	[53]
Koenigine	$C_{19}H_{19}NO_3$	Nitrariaceae and Rutaceae	*Candida* sp.	MIC_{90}: 12.5–100 μg/mL	[54]
3,3′-[Oxybis(methylene)]bis (9-methoxy-9H-carbazole)	$C_{28}H_{24}N_2O_3$	Nitrariaceae and Rutaceae	*P. vulgaris* and *C. albicans*	6.2 and 25 μg/mL, respectively	[51]
Monoterpenoid indole alkaloids					
Scholarisine	$C_{19}H_{18}N_2O_2$	Apocynaceae	*Gibberella pulicaris* and *Cercospora nicotianae*	MIC: 1.37–1.91 μM	[55]

Types of alkaloids	Formula	Plant family	Targeted microorganism	Concentrations range (μg/mL)	Reference
Kopsiflorine	$C_{23}H_{28}N_2O_5$	Apocynaceae	*S. aureus*	IZ: 9.7 mm	[56]
Erchinines A and B	—	Apocynaceae	*B. subtilis*, *Trichophyton rubrum*	0.78 and 0.78, 12.5 and 6.25, respectively	[57]
Melokhanine A	$C_{19}H_{26}N_2O_3$	Apocynaceae	*Pseudomonas aeruginosa*, *Enterococcus faecalis*	2–5 μM	[58]
Ibogaine	$C_{20}H_{26}N_{2O}$	Apocynaceae	*E. coli, B. subtilis, A. flavus, A. niger, Rhizoctonia phaseoli, K. pneumoniae, S. aureus, S. pneumoniae, A. flavus, C. albicans, and R. phaseoli*	50–60 μg/mL	[59]
Vobasine	$C_{21}H_{24}N_2O_3$	Apocynaceae	*A. niger* and *A. flavus*	50–60 μg/mL	[59]
Voacamine	$C_{43}H_{52}N_4O_5$	Apocynaceae	*R. phaseoli*, *P. chrysogenum*, and *C. albicans*	50–60 μg/mL	[59]
Cadambine	$C_{27}H_{32}N_2O_{10}$	Rubiaceae	Weakly against: *Staphylococcus epidermidis*, *S. aureus*, *B. cereus*, *B. subtilis*, and *C. albicans*	3.3–164 μg/mL	[60]
Strictosidine	$C_{27}H_{34}N_2O_9$	Rubiaceae	*K. pneumoniae, Providencia smaitii,* and *E. coli*	12.5, 25, and 50 μg/mL, respectively	[61]
Tubotaiwine	$C_{20}H_{24}N_2O_2$		*Mycobacterium tuberculosis*	100 μg/mL	[62]
Diterpene alkaloids	$C_{24}H_{39}NO_4$	Ranunculaceae			
Vilmorrianone, panicutine, 8-acetylheteropyllisine	$C_{23}H_{29}NO_4$	Ranunculaceae	*Allescheria boydii*, *A. niger*, *E. floccosum*, *Pleurotus ostreatus*	–	[63]
Miscellaneous					
Tryptanthrin	$C_{15}H_8N_2O2$		*E. floccosum*, *T. mentagrophytes*, *Trichophyton rubrum*, *Trichophyton tonsurans*, *M. gypseum*, and *Microsporum canis*	3.1–6.3 μg/mL	[64]
			C. neoformans, and *Cryptococcus deuterogattii*	MIC/MFC: 2/ > 64 and 8/ 32 μg/mL	[65]
Dehydroevodiamine	$C_{19}H_{15}N_3O$	Rutaceae	*X. oryxae* pv *oryzae*	1.4 μg/mL	[49]
Piperidine Alkaloids					

Types of alkaloids	Formula	Plant family	Targeted microorganism	Concentrations range (μg/mL)	Reference
Piperlongumine	$C_{17}H_{19}NO_5$	Piperaceae	*C. albicans*, *S. aureus* and *P. aeruginosa*	MIC: 3.9 μg/mL	[66]
Piperine	$C_{17}H_{19}NO_3$	Piperaceae	*C. albicans*, *R. solani*, *Fusarium gramineum*, *Alternaria tenuissima*, *Gloeosporium theae-sinensis*, *Phytophthora capsici*, and *Phomopsis adianticola*	100 μg/mL	[66]
Quinolizidine					
Quinolizidine	$C_9H_{17}N$	Nymphaeaceae, Fabaceae	*E. faecalis*, *Enterococcus faecium*, *S. aureus*, and Methicillin-resistant Staphylococcus aureus (MRSA)	2–16 μg/mL	[67]
Phenanthroindolizidine		Lauraceae, Moraceae, Asclepiadaceae			
7-Demethoxytylophorine	$C_{23}H_{25}NO_3$	Apocynaceae	*Penicillium italicum, Penicillium digitatum*	MIC/MFC: 1.5/6.2 and 1.5/12.5 μg/mL, respectively	[68, 69]
Tylophorinine	$C_{23}H_{25}NO_4$	Apocynaceae	*C. albicans*, *Candida krusei*, *Candida glabrata*, and *A. fumigatus*	0.6–5 μg/mL	[68]
Tylophorinidine	$C_{22}H_{23}NO_4$	Apocynaceae	*C. albicans*, *Candida krusei*, *Candida glabrata*, and *A. fumigatus*	2–8 μg/mL	[68]
Securinega alkaloids					
viroallosecurinine	$C_{13}H_{15}NO_2$	Phyllanthaceae	*P. aeruginosa* and *S. aureus*	MIC: 0.4 μg/mL	[70]
Securinine, Allosecurinine	$C_{13}H_{15}NO_2$	Phyllanthaceae	*P. aeruginosa*, *S. aureus*, and *M. smegmatis*	Weak activity	[70]
norsecurinine	$C_{12}H_{13}NO_2$	Phyllanthaceae	filamentous fungi	Inhibited at Low concentrations	[71]
Miscellaneous					
Dihydrodioscorine	$C_{13}H_{21}NO_2$	Dioscoreaceae	*Sclerotium rolfsii*, *C. lunata*, *F. moniliforme*, *Botryodiplodia theobromae*, and *Macrophomina phaseolina*	Inhibited the mycelial growth	[72]
Pandamarilactone-1	$C_{18}H_{23}NO_4$	Dioscoreaceae	*E. coli*, *P. aeruginosa*, and *S. aureus*	Weak activity	[73]

Types of alkaloids	Formula	Plant family	Targeted microorganism	Concentrations range (μg/mL)	Reference
Haloxyline B	—	Chenopodiaceae	*M. tuberculosis* H37Rv	50 μg/mL	[74]
Quinoline Alkaloids					
Simple Quinolines					
4-Methylquinoline	$C_{10}H_9N$	—	*S. aureus*	MIC/MBC values of 12.2/ 50 μg/mL	[75]
4-methoxy-2-phenylquinoline	$C_{16}H_{13}NO$	Rutaceae	*M. tuberculosis* H37Rv	16 μg/mL	[75]
Dictamine	$C_{12}H_9NO_2$	Rutaceae	*Micrococcus luteus* (TISTR 884) and *B. cereus* (TISTR 688)	26 and 64 μg/mL, respectively	[76]
γ-Fagarine	$C_{13}H_{11}NO_3$	Rutaceae	broad-spectrum antibacterial	Moderate activity	[77]
Robustine	$C_{12}H_9NO_3$	Rutaceae	broad-spectrum antibacterial	Moderate activity	[77]
Benzylisoquinolines					
Reticuline	$C_{19}H_{23}NO_4$				
Fuyuziphine	—	Papaveraceae	*Alternaria brassicicola*, *A. solani*, *Alternaria melongenae*, *C. maculans*, *Erysiphe cichoracearum*, and *Helminthosporium pennisetti*	500 ppm	[78]
Bisbenzylisoquinolines					
Tetrandrine	$C_{38}H_{42}N_2O_6$	Menispermaceae	*S. aureus* and MRSA	weakly bactericidal	[79]
Tiliacorinine	$C_{36}H_{36}N2O_5$	Menispermaceae	*M. tuberculosis*	6.2 μg/mL	[80]
2′-nortiliacorinine	$C_{35}H_{34}N_2O_5$	Menispermaceae	*M. tuberculosis*	3.1 μg/mL	[80]
Tiliacorine	$C_{36}H_{36}N_2O_5$	Menispermaceae	*M. tuberculosis* and *A. tenuissima*	3.1 and 100 μg/mL, respectively	[80, 81]
Aporphines					
Aporphine	$C_{17}H_{17}N$	Illiciaceae, Trimeniaceae	bacteria and fungus in plants	suppressed a wide variety of bacteria and fungus	[82]

DOI: http://dx.doi.org/10.5772/intechopen.112364

Types of alkaloids	Formula	Plant family	Targeted microorganism	Concentrations range (μg/mL)	Reference
Liriodenine	$C_{17}H_9NO_3$	Illiciaceae, Trimeniaceae	bacteria and fungus in plants	suppressed a wide variety of bacteria and fungus	[82]
Anonaine	$C_{17}H_{15}NO_2$	Magnoliaceae, Annonaceae	*B. cereus*, *E. coli*, *S. aureus*, and *S. epidermidis*	diameters of 20, 8, 14, and 12 mm, respectively	[83]
Lysicamine	$C_{18}H_{13}NO_3$	Annonaceae	*L. monocytogenes*, Methicillin-resistant Staphylococcus aureus (MSSA), *S. pneumoniae*, *Actinobacillus* sp., and *K. pneumoniae*	1.4–20 μg/mL	[84]
O-methylmoschatoline	$C_{19}H_{15}NO_4$	Annonaceae	*B. subtilis*, *E. coli*, and *Salmonella typhi*	64 μg/mL	[82]
Artabotrine	$C_{20}H_{23}NO_4$	Annonaceae	*K. pneumoniae*	MIC/MBC: 2.5/2.5 μg/mL	[84]
Azaoxoaporphine sampangine	$C_{15}H_8N_{2O}$	Annonaceae	*C. albicans*, *C. glabrata*, *C. kruseii*, *A. fumigatus*, and *C. neoformans*	3.1, 3.1, 6.2, 6.2, and 0.05 μg/mL, respectively	[85]
Lanuginosine	$C_{18}H_{11}NO_4$	Annonaceae	*B. cereus*, *S. aureus*, *E. coli*, *K. pneumoniae*, and *P. aeruginosa*	IZ: 12, 14, 10, 14, and 12 mm, respectively	[86]
Nordicentrine	$C_{19}H_{19}NO_4$	Menispermaceae	*M. tuberculosis*	12.5 μg/mL	[87]
Dicentrinone	$C_{19}H_{13}NO_5$	Menispermaceae	*M. tuberculosis*	Moderate antimycobacterial	[88]
Oxoaporphine thailandine	$C_{39}H_{62}O_{14}$	Menispermaceae	*S. pneumoniae*, *S. aureus*, *E. faecalis*, and *M. tuberculosis*	30, 30, 60, and 6.2 μg/mL, respectively	[89]
Isoboldine	$C_{19}H_{21}NO_4$	Ranunculaceae	*A. baumanii*, *B. subtilis*, *C. albicans*, *P. aeruginosa*, *E. coli*, *P. mirabilis*, *K. pneumoniae*, and *S. aureus*	Moderate activity	[90]
Roemerine	$C_{18}H_{17}NO_2$	Lauraceae	MRSA, *A. fumigatus*, *C. albicans*, *C. glabrata*, *C. krusei*, *Candida tropicalis*, *Candida parapsilosis*, and *S. aureus*	10 μg/mL for *C. albicans*	[91, 92]
Magnoflorine	$C_{20}H_{24}NO_{4+}$	Menispermaceae	*C. albicans*, *C. parapsilosis* var. *parapsilosis*, *T. rubrum*, and *T. mentagrophytes*	Moderate activity	[93]
Protopines					
Protopine	$C_{20}H_{19}NO_5$	Papaveraceae	*C. albicans*	4 μg/mL	[90]
Allocryptopine	$C_{21}H_{23}NO_5$	Papaveraceae	*P. aeruginosa*, *S. aureus*, *E. coli*, and *S. agalactiae*	Weak activity	[94]

Types of alkaloids	Formula	Plant family	Targeted microorganism	Concentrations range (µg/mL)	Reference
Protoberberines					
Pendulamine A	—	Annonaceae	*B. subtilis*, *P. aeruginosa*, *S. aureus*, *Corynebacterium hoffmanii*, *K. pneumoniae*, *S. typhi*, *Micrococcus lysodickycus*, and *S. paratyphi A*	0.02–2 µg/mL	[95]
Pendulamine B	—	Annonaceae	*Corynebacterium hoffmanii*, *S. faecalis*, *S. aureus*, *S. typhi*, *S. viridans*, *M. lysodickycus*, *P. aeruginosa*, *K. pneumoniae*, and *S. paratyphi* A	0.02–2 µg/mL	[95]
Spirobenzylisoquinolines					
Parfumine	$C_{20}H_{19}NO_5$	Papaveraceae	*A. baumanii*, *B. subtilis*, *K. pneumoniae*, *E. coli*, *P. aeruginosa*, *P. mirabilis*, and *S. aureus*	Moderate activity	[90]
Fumarophycine	$C_{22}H_{23}NO_6$	Papaveraceae	*A. baumanii*, *B. subtilis*, *K. pneumoniae*, *E. coli*, *P. aeruginosa*, *P. mirabilis*, and *S. aureus*	Moderate activity	[90]
Fumariline	$C_{20}H_{17}NO_5$	Papaveraceae	*A. baumanii*, *B. subtilis*, *K. pneumoniae*, *E. coli*, *P. aeruginosa*, *P. mirabilis*, and *S. aureus*	Moderate activity	[90]
Benzophenanthridines					
Stylopine or sanguinarine	$C_{20}H_{14}NO_{4+}$	Papaveraceae	*A. baumanii*, *B. subtilis*, *K. pneumoniae*, *E. coli*, *P. aeruginosa*, *P. mirabilis*, and *S. aureus*	Moderate activity	[90]
Dihydrosanguinarine	$C_{20}H_{15}NO_4$	Papaveraceae	*S. mutans, S. aureus, P. aeruginosa, E. coli*, and *S. agalactiae*	32, 31.3, 250, 62.5, 15.6 µg/mL	[94]
6-Methoxydihydrosanguinarine	$C_{21}H_{17}NO_5$	Papaveraceae	*S. aureus* and MRSA	IZ: 17 mm	[96]
			E. faecalis and *S. aureus*	MIC/MBC: 5/10, 2.5/5 µg/mL	[97]
8-Hydroxydihydrosanguinarine		Papaveraceae	MRSA	MIC range: 0.4 to 7.8 µg/mL, and MBC range: 1.9 to 31.2 µg/mL	[81]
Norsanguinarine	$C_{19}H_{11}NO_4$	Papaveraceae	*A. baumanii*, *B. subtilis*, *K. pneumoniae*, *E. coli*, *P. aeruginosa*, *P. mirabilis*, and *S. aureus*	Moderate activity	[90]

Types of alkaloids	Formula	Plant family	Targeted microorganism	Concentrations range (μg/mL)	Reference
Allocryptopine	$C_{21}H_{23}NO_5$	Papaveraceae	*S. epidermidis*, *S. aureus*, *S. pyogenes*, *B. subtilis*, *K. pneumoniae*, and *E. coli*	6.2/12.5, 12.5/50, 12.5/50, 25/50, 12.5/25, 25/25 μg/mL, respectively	[98]
8-Hydroxydihydrochelerythrine	$C_{21}H_{19}NO_5$	Papaveraceae	MRSA	MIC: 0.9–15.6 μg/mL, MBC: 7.8–62.5 μg/mL	[99]
Dihydrochelerythrine	$C_{21}H_{19}NO_4$	Papaveraceae	MRSA, *E. coli*	8–128 5 μg/mL	[77]
Chelerythrine	$C_{21}H_{18}NO_4$	Papaveraceae	*C. albicans*, *S. cerevisae*, and *C. neoformans*	MIC/MBC: 3.1/3.1, 6.2/6.2, and 3.1/6.2 μg/mL, respectively	[98]
Corynoline	$C_{21}H_{21}NO_5$	Papaveraceae	*Cladosporium herbarum*	3 μg/spot	[100]
Acetylcorynoline	—	Papaveraceae	*C. herbarum*	3 μg/spot	[100]
Norchelerythrine	$C_{20}H_{15}NO_4$	Rutaceae	*M. tuberculosis*	25 μg/mL	[101]
Avicine	$C_{20}H_{14}NO_{4+}$	Rutaceae	*S. epidermidis*, *S. aureus*, *S. pyogenes*, *B. subtilis*, *K. pneumoniae*, and *E. coli*	3.1/12.5, 1.5/25, 1.5/12.5, 1.5/6.2, and 6.2/12.5 μg/mL, respectively	[98]
Rhoifoline B	$C_{21}H_{17}NO_5$	Rutaceae	*S. aureus*, *S. epidermidis*, *E. coli*, *E. cloacae*, *K. pneumoniae*, *P. aeruginosa*, and *S. dysenteriae*	Moderate activity	[102]
Nitidine	$C_{21}H_{18}NO_{4+}$	Rutaceae	*M. luteus*, *S. aureus*, and *M. smegmatis*	Weak activity	[103]
Protoberberines					
Berberine	$C_{20}H_{18}NO_{4+}$	Berberidaceae	*K. pneumonia* and *A. baumanii*	8 μg/mL	[90]
Palmatine	$C_{21}H_{24}NO_{4+}$	Berberidaceae, Papaveraceae, Ranunculaceae, and Menispermaceae	*A. baumanii*, *E. coli*, *P. mirabilis*, *P. aeruginosa*, *K. pneumoniae*, *S. aureus*, and *B. subtilis*	Moderate activity	[90]
Phthalides					
Bicuculline	$C_{20}H_{17}NO_6$	Fumariaceae	*A. brassicae*, *F. udum*, and *Curvularia lanata*	200 ppm	[90]
Hasubanans					

Types of alkaloids	Formula	Plant family	Targeted microorganism	Concentrations range (µg/mL)	Reference
Glabradine	$C_{19}H_{19}NO_7$	Menispermaceae	*S. aureus* and *S. mutans*	50 µg/mL	[104]
Amaryllidaceae Alkaloids		Amaryllidoideae			
Crinamine	$C_{17}H_{19}NO_4$	Amaryllidoideae	Some bacteria	—	[105]
Lycorine	$C_{16}H_{17}NO_4$	Amaryllidoideae	*C. glabrata*, *Candida dubium*, *C. albicans*, *Lodderomyces elongisporus*, and *S. cerevisae*	512, 39, 32, 64, and 97.3 µg/mL	[78]
			Alternaria oleracea, *C. gloeosporioides*, *F. graminearum*, *Colletotrichum ophiopogonis*, and *Pleospora lycopersici*	100 µg/mL	[106]
Lycoricidine		Amaryllidoideae		IZ: 12 mm	[78]
Narciclasine	$C_{14}H_{13}NO_7$	Amaryllidoideae	*Corynebacterium fascians* and *C. neoformans*	Highly growth inhibition	[107]
Tazettine	$C_{18}H_{21}NO_5$	Amaryllidoideae	*L. elongisporus* and *C. dubliniensis*	Weak activity	[108]
Miscellaneous					
Quinolinones					
Antidesmone	$C_{19}H_{29}NO_3$	Euphorbiaceae	Carbendazim-sensitive strains of *S. sclerotiorum*, and Carbendazim-resistant strains of *S. sclerotiorum* *Botryosphaeria dothidea*, *Pestalotipsis guepinii*, *Colletotrichum musae*, *Colletotrichum orbiculare*, *Pestalotiopsis longiseta Phylophthora nicotianae*	50 µg/mL	[109]
Waltherione C	$C_{22}H_{21}NO_3$	Malvaceae	*B. dothidea*, *Colletotrichum orbiculare*, *Colletotrichum musae*, *Pestalotiopsis longiseta*, *Pestalotipsis guepinii*, *Phylophthora nicotianae*, carbendazim-sensitive strains of *S. sclerotiorum*, and carbendazim-resistant strains of *S. sclerotiorum*	50 µg/mL	[109]
Evocarpine	$C_{23}H_{33}NO$	Rutaceae	MRSA and *S. aureus*	8 µg/mL	[110]
Acridanones					
1-hydroxy-3,4-dimethoxy-10-methylacridan-9-one	$C_{16}H_{15}NO_4$	Rutaceae	*E. coli*	Growth inhibition	[51]

Types of alkaloids	Formula	Plant family	Targeted microorganism	Concentrations range (μg/mL)	Reference
Phenanthrene Alkaloids					
Aristolochic acid	$C_{17}H_{11}NO_7$	Aristolochiaceae	*Moraxella catarrhalis*	MIC and MBC: 25/50 μg/mL	[111]
1-N-monomethylcarbamate-argentinine-3-O-β-D-glucoside	—	Menispermaceae	MRSA	500 μg/disk, IZ: 8 mm	[112]
Pyrrolidines and Imidazole Alkaloids		Piperaceae			
Pyrrolidines		Piperaceae			
Brachyamide B	$C_{20}H_{25}NO_3$	Piperaceae	*C. albicans*	IC_{50}: 41.8 μg/mL	[113]
Pandanus lactones			*C. neoformans*	IC_{50}: 7.1 μg/mL	[114]
Pandamarilactonine A	$C_{18}H_{23}NO_4$	Pandanaceae	*E. coli*, *P. aeruginosa*, and *S. aureus*	Moderate activity	[73]
Diterpene Alkaloids		Ranunculaceae			
8-acetylheterophyllisine	$C_{24}H_{35}NO_5$	Ranunculaceae	*Pleurotus ostreatus*, *Allescheria boydii*, *A. niger*, and *E. floccosum*	Growth inhibition	[63]
Vilmorrianone	$C_{23}H_{27}NO_5$	Ranunculaceae	*Pleurotus ostreatus*, *Allescheria boydii*, *A. niger*, and *E. floccosum*	Growth inhibition	[63]
Panicutine	$C_{23}H_{29}NO_4$	Ranunculaceae	*Pleurotus ostreatus*, *Allescheria boydii*, *A. niger*, and *E. floccosum*	Growth inhibition	[63]
Steroidal Alkaloids					
N-formylconessimine	—	Apocynaceae	MSSA	32 μg/mL	[115]
Conimine	$C_{22}H_{36}N_2$	Apocynaceae	MRSA	128 μg/mL	[115]
Isoconkuressine	—	Apocynaceae	MSSA and MRSA	Growth inhibition	[115]

IZ: inhibition zone; and MIC: minimum inhibitory concentration.

Table 1.
Classification of alkaloids in plants family based on their antimicrobial activity.

marine microbes [117]. Various biological and pharmacological qualities are related to the several structural scaffolds, and a wide variety of functional group modifications is found in the broad class of plant-specific metabolites known as benzylisoquinoline alkaloids. N-Methylation is a widely used modification technique that forms intermediates and final products in the tertiary and quaternary metabolic pathways [118].

3.2 Some selected alkaloids with antimicrobial activity

Various alkaloids found in nature have been shown to have antimicrobial effects against a wide range of diseases [15]. Some selected alkaloids with potent antimicrobial activity include berberine, quinine, and vincristine. The potential for these particular alkaloids' antibacterial action to expand therapy choices for infectious disorders caused by drug-resistant microbes or those not responding to conventional therapies has been widely discussed [16]. Hence, this review focuses on alkaloids with antibacterial activity against MDR microorganisms. Also, this article describes the most influential alkaloids with potent antibacterial properties. Here are some selected examples of these compounds:

3.2.1 Berberine

The natural isoquinoline alkaloid berberine has been shown to have minimal toxicity [119]. Berberine, derived from *Berberis* spp., is a prominent quaternary ammonium salt of protoberberines. It exhibits various antimicrobial properties, particularly against Gram-negative bacteria [24]. *Berberis vulgaris*, *Coptis chinensis*, *Hydrastis canadensis*, *Coptidis rhizoma*, *Xanthoriza simplicissima*, *Phellodendron amurense*, and *Chelidonium majus* all contain it, among many others, making them useful as therapeutic herbs [119]. Berberine is an effective antibacterial agent that may one day replace conventional antibiotics and help combat the problems caused by antibiotic resistance. Methanol extract of *Pancratium illyricum* L. bulbs yielded the isoquinoline alkaloid ungeremine. Its antimicrobial qualities have been well-praised. As mentioned earlier, the compound can induce a significant augmentation in DNA cleavage through its selective targeting and inhibition of bacterial topoisomerase IA [29]. Herpes, influenza, and respiratory syncytial viruses are susceptible to berberine's antiviral actions [34, 119]. Berberine's mechanism of action against *V. cholerae* and *E. coli*-induced diarrhea has been thoroughly investigated. The effects of *E. coli* and *V. cholerae* enterotoxins were found to be directly inhibited by berberine *in vitro* as early as 1982 [120]. Berberine's antibacterial activity against *S. aureus* has been shown in *in vitro* investigations [121]. As reported in reference, berberine and CinA can undergo self-assembly, forming nanoparticles (NPs) that exhibit bacteriostatic properties against MRSA and potentially eliminate biofilms [40]. Cinnamaldehyde (CinA) is a principal constituent of the *Cinnamomi cortex*, a traditional spice that finds extensive usage in everyday routines [122].

The alkaloid berberine sulfate is harvested from the bark and roots of several plants. It exhibits antibacterial, antifungal, and antiprotozoal properties. Berberine sulfate disrupts fimbrial formation in *Streptococcus pyogenes*, impeding bacterial attachment to mucosal or epithelial surfaces [123]. On the other hand, L-Tyr is widely recognized as the biosynthesis precursor of berberine. 13 different enzymatic processes are involved in the production of berberine from L-Tyr. Notably, biochemical analysis has been performed on all of the enzymes in this pathway [24].

3.2.2 Caffeine

Numerous plant species derive caffeine (1,3,7-trimethyl xanthine) from methylated alkaloids. It is structurally related to uric acid [124]. However, recent studies have shown that caffeine also has antimicrobial properties, which has led to increased interest in its potential use as an alternative to traditional antibiotics. Understanding caffeine's antimicrobial activity is crucial in developing new treatments for drug-resistant infections, making it an important area of research. Another study by Ibrahim et al. found that growth inhibition was most noticeable at concentrations of 0.50% and above against *E. coli* [124]. Also, caffeine concentrations in coffee extracts are high enough to concern human health, with 50% antibacterial activity against *S. enterica* [125].

3.2.3 Capsaicin (CAP)

The berries of virtually all peppers in the genus *Capsicum* contain capsaicin, also known as 8-methyl-N-vanillyl-6-nonenamide [12]. Peppers, especially chili peppers, are members of the Solanaceae plant family, responsible for their distinctive flavor [11]. *Capsicum annuum* powder is a commonly utilized seasoning in various culinary traditions across the globe. Apart from its gastronomic application, CAP is employed for analgesic purposes in different severe and persistent medical conditions [12]. Pepper fruits may contain capsaicin at a rate of up to 1% of their total weight. It is naturally produced in the epidermal cells of the placenta, which are located close to the seeds. The compound tends to accumulate in the form of "blisters" on the surface of the placenta. The molecule is a potent agonist of the transient receptor potential vanilloid ion-channel receptor 1 (TRPV1), eliciting its characteristic hot, burning sensation. However, the beneficial effects of capsaicin and the TRPV1 receptor cannot be attributed primarily to this interaction [11]. In an *in vitro* investigation [126], six capsaicin derivatives were developed, each possessing phenolic hydroxyl, a benzene ring, and amide structures. These derivatives were subsequently evaluated for their antibacterial properties against *E. coli* and *S. aureus*. Two powerful chemicals found in *Capsicum* species were shown to have antimicrobial capabilities, and Cichewicz and Thrope identified them. The experiment results showed that the plain and heated extracts displayed different levels of inhibition against *Streptococcus pyogenes*, *B. subtilis*, *B. cereus*, *Clostridium tetani*, and *Clostridium sporogenes* [127].

3.2.4 Colchicine

Colchicine has been around longer than most other pharmaceuticals [128]. The use of colchicine as a pharmacological agent in humans has been permitted by the Food and Drug Administration (FDA). It is a safe and productive anti-inflammatory medication derived from the *Colchicum* and *Gloriosa* plant species. Colchicine has been utilized in treating cardiovascular ailments due to its distinctive effectiveness as an anti-inflammatory agent [24]. The chemical origins of colchicine have been the subject of extensive research, facilitated by numerous feeding studies utilizing isotope-labeled substrates in Colchicum plants. Furthermore, a well-defined biosynthetic hypothesis has been established thanks to structural study of colchicine-related alkaloids isolated from several members of the Colchicaceae family [24]. The first biosynthetic studies on colchicine were performed by Leete in 1960 [129].

The medical application of colchicine in cancer chemotherapy is restricted due to its comparatively high toxicity, despite its potency as an anticancer agent. Nevertheless, colchicine is currently utilized in therapy [130]. Colchicine's potential anticancer impact on hypo-pharyngeal carcinoma was studied. Colchicine dose-dependently suppressed hypo-pharyngeal human cell proliferation [128]. Colchicine inhibited adhesion, migration, and cell invasion via decreasing expression of MMP9, uPA, and FAK/SRC [128]. Researchers have shown that colchicine inhibits the reproduction of the Flaviviridae family of viruses by blocking microtubule polymerization. Researchers believe colchicine, a well-known anti-inflammatory medication, can cure COVID-19 by decreasing inflammation [131].

3.2.5 Piperine

Piperine has been extracted from various species of the Piperaceae botanical family [132], as shown chemically in **Figure 2** [132]. Piperine is a major compound of black pepper (*Piper nigrum*) and long pepper (*Piper longum*), two species of the Piperaceae family. Studies suggest piperine exhibits bioavailability-enhancing properties for select nutritional substances [133]. The biting quality that is distinct from black pepper is attributed to piperine. Piperine exhibits numerous pharmacological properties and confers various health advantages, particularly for chronic ailments. These benefits include mitigation of anti-inflammatory effects, insulin resistance, amelioration of hepatic steatosis [134], anti-aging, antidiabetic, cardioprotective, antimicrobial, and anti-obesity [132]. When ciprofloxacin and a piperidine-type alkaloid from

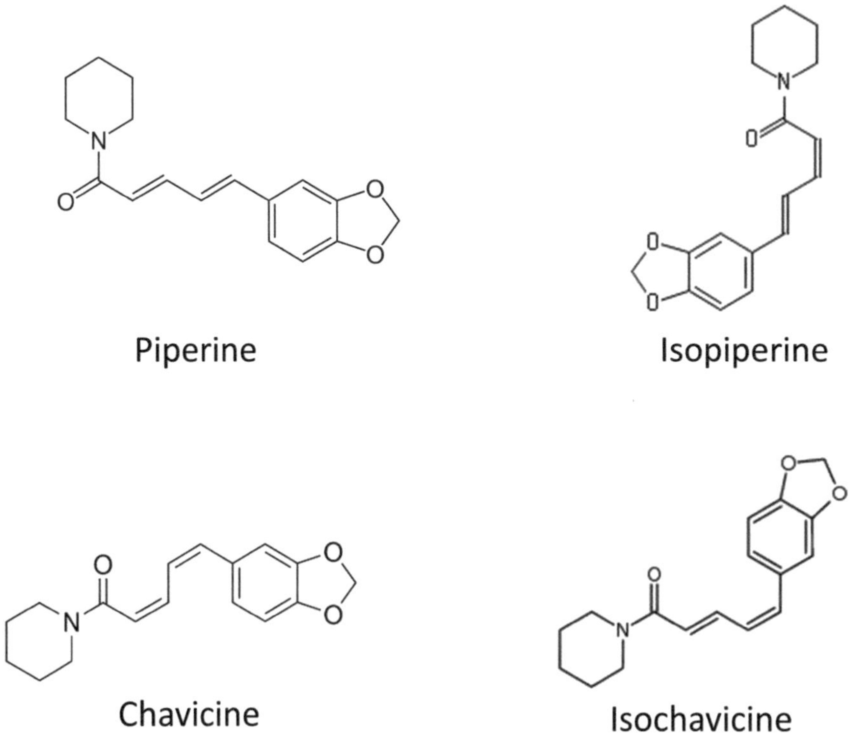

Figure 2.
Piperine and its structural isomers (adapted from Ul-Haq et al. [132]).

the plants. Together, *P. longum* and *P. nigrum* were able to inhibit the development of a mutant *S. aureus* and considerably reduce MIC values for *S. aureus* [135].

In the case of absorption, it is noteworthy that piperine exhibits no metabolic transformations upon absorption, as evidenced by its presence in both intestinal tissues and serosal fluid. This suggests that piperine remains unaltered throughout the absorption process [132].

3.2.6 Reserpine

Reserpine, an indole alkaloid extracted from the plant Rauwolfia serpentina, is well-known for its potent EPI action. The co-administration of reserpine has improved the antibiotic susceptibility of various bacterial species, such as *Micrococcus* spp., *Streptococcus* spp., and *Staphylococcus* spp. [29]. Combining reserpine with other commercially available antibiotics has been shown to improve the antibiofilm response and eradicate a sizable amount of bacterial biofilm in a urinary catheterization model, as reported by Parai et al. [136]. In another study, many acyl reserpine derivatives were made and tested for their antimycobacterial and antioxidant activities against *Mycobacterium* TB, strain H (37) Rv. This was done because reserpine is thought to have therapeutic benefits. According to the findings, 10 of 18 derivatives exhibited more significant suppression of antimycobacterial activity than reserpine [137]. On the other hand, reserpine inhibits AcrB. Acriflavine resistance protein B (AcrB) is an MDR efflux transporter that belongs to the Resistance-nodulation-division (RND) superfamily [138].

3.2.7 Tomatidine

Steroid alkaloid tomatidine is harvested from nightshade plants, including tomatoes, potatoes, and eggplant. As monotherapy or in combination with aminoglycosides, there is evidence that it is highly effective as an antibacterial agent against *S. aureus* [29]. Tomatoes and tomatidine, as found by Silva-Beltrán et al., have great promise as a source of several bioactive chemicals, antioxidants, and antibacterial agents [139]. Tomatidine exhibited bacteriostatic activity against small-colony variants linked to their impaired electron transport system. The electron transport inhibitor 4-hydroxy-2-heptylquinoline-N-oxide (HQNO) increased the sensitivity of typical *S. aureus* strains to tomatidine [140].

3.2.8 Conessine

Holarrhena antidysenterica, a member of the Apocynaceae family, has a long history of medical usage for treating dysentery, diarrhea, fever, and bacterial infections [141]. Conessine is a steroidal alkaloid. The therapeutic actions of *H. antidysenterica* barks are due to the presence of alkaloids, specifically the steroidal alkaloid conessine. There is preliminary evidence that this compound can kill gram-positive and gram-negative bacteria [141]. Based on the existing evidence, it can be inferred that the steroidal crude extract of *H. antidysenterica* and conessine exhibit properties of efflux pump inhibitors (EPIs). Recently, it has been reported that the steroidal extract and alkaloid conessine can augment the efficacy of antibiotics by impeding the AdeIJK efflux pump in *A. baumannii* [142].

Other alkaloid classes, namely indolizidine, pyrrole-imidazole alkaloid, quinoline, aaptamine, indole, isoquinoline, piperazine, polyamine, bisindole, quinolone, indole-quinoline, agelasine, aaptamine-indole, pyridoacridine, and bispyrrole have been reported to exhibit antibacterial activity [37].

4. Alkaloids derived from medicinal plants and their antimicrobial activities

The distribution of alkaloids within plant tissues is heterogeneous, as mentioned previously, with varying concentrations observed across plant parts such as roots, seeds, leaves, fruits, and bark. Distinct alkaloid types may exist in various parts of a single plant [12]. The alkaloids are the most abundant secondary metabolites in the *Zanthoxylum* genus, and they exhibit a wide variety of biological functions due to their structural diversity [143]. A study by Farouk et al. indicated that *Eurycoma longifolia* leaf extracts were tested for antibacterial efficacy against *Pseudomonas aeruginosa* and *S. aureus* bacteria. The extracts were prepared using various solvents, including acetone, ethanol, phosphate buffer, and methanol at 5–100 mg/mL concentrations. Several extracts inhibited bacterial growth, with the widths of the inhibition zones ranging from 7 to 25 mm [144]. In addition to causing serious side effects, treating fungal infections with antifungal drugs often leads to drug-resistant strains of the fungus. This highlights the critical need to investigate potential new antifungal medicines. It has been shown that alkaloids isolated from the leaves of *Ruta graveolens* L. are fungi toxic [145]. Flavonoids and quinoline alkaloids isolated from the roots of *Waltheria indica* L. showed that to have antifungal activity against *Candida albicans* [146]. **Table 2** summarizes some selected medicinal plants that possess alkaloids with antimicrobial properties.

In a study by Erdemoglu et al. [154], capillary GC-MS identified 15 alkaloids. 13α-hydroxylupanine (50.78%) and lupanine (23.55%) were assessed to be the significant alkaloids in the aerial parts of *L. angustifolius*. Ammodendrine, tetrahydrorhombifoline, isoangustifoline, α-isolupanine, 5,6-dehydrolupanine, 11,12-dehydrolupanine, 13α-tigloyloxylupanine, 13α-acetoxylupanine, angustifoline, 13α-isovaleroyloxylupanine, 13α-valeroyloxylupanine, 13α-*cis*-cinnamoyloxylupanine, and 13α-*cis*-cinnamoyloxy-17-oxolupanine were analyzed as the minor alkaloids of the substances in this plant. The alkaloid extract showed modest effectiveness against *E. coli,* while a strong point against *B. subtilis*, *S. aureus*, and *P. aeruginosa*. The extract was only moderately effective against *Candida albicans* and *C. krusei* [154]. Although native to the Middle East and Mediterranean regions, *Peganum harmala* has been introduced to Australia and the United States [155]. The alkaloids of *P. harmala* are concentrated in its roots and seeds. All 13 Gram-positive (*S. pyogenes*, *S. epidermidis*, *S. aureus*, *L. monocytogenes B. pumilus*, *B. cereus*, and *B. anthracis*) and Gram-negative (*Brucella melitensis*, *P. aeruginosa*, *Salmonella typhi*, *Klebsiela pneumoniae*, *E. coli*, and *P. mirabilis*) bacteria tested showed inhibition by methanol extract [155]. *Papaver somniferum,* belonging to the Papaveraceae botanical family, has been the subject of extensive research due to its benzylisoquinoline alkaloids (BIAs), which have been utilized for medicinal purposes since ancient times. It is notable for being the sole commercial source of morphine and codeine and is regarded as the model plant for BIA research. *P. somniferum* synthesizes vital alkaloids, such as sanguinarine, papaverine, and noscapine [162].

Native to Oman, *Ficus sycomorus* has had its leaf extracts investigated for their ability to eradicate *Haemophilus influenzae*, *S. aureus*, *E. coli*, and *Proteus* spp. [152]. *Ficus sycomorus* is abundant in flavonoids, alkaloids, tannins, and phenolic compounds.

Plant and family	Common name	Part of plant	Extraction solvent	Method of detection	Bioactive compound	Ref.
Antibacterial						
Alchornea laxiflora / Euphorbiaceae	Three-veined bead string, Lowveld bead string, Venda bead string,	Leaf	Methanol and distilled water	—	—	[147]
Amaryllis belladonna/ Amaryllidaceae	Jersey lily	Bulb	Chloroform, Ethanol, and n-butanol	HPTLC	(-)-Amarbellisine, (-)-lycorine, (-)-pancracine, (+)-vittatine, (+)-11-hydroxyvittatine, and (+)-hippeastrine	[148]
Stephania glabra/ Menispermaceae	Hairless tape vine	Tuber	Ethanol	—	gindarine, gindaricine, gindarinine, columbamine, jatrorrhizine and magnoflorine	[149]
Zanthoxylum spp./Rutaceae	Pricklyash	—	—	—	Quinoline, isoquinoline, indole, quinazoline, indolopyridoquinazoline	[143]
Eurycoma longifolia/ Simaroubaceae	Tongkat Ali	Leaf	Acetone, methanol, and ethanol			[144]
Morus alba/Moraceae	Mulberry	Root	Water extract	NMR	piperidine	[150]
Glycyrrhiza glabra L./Fabaceae	Licorise	Aerial parts	Methanol	—	—	[151]
Ficus sycomorus/Moraceae	Mulberry Fig, Sycamore Fig	Leaf	Methanol	—	—	[152]
Telosma (Pergularia) pallida/ Apocynaceae	Telosma vine	Air-dried roots	—	—	pergularinine and tylophorinidine	[153]
Lupinus angustifolius L./ Fabaceae	Lupine	Aerial parts	Dichloromethane	GC-MS	13α-Hydroxylupanine (50.78%) and lupanine (23.55%)	[154]
Murraya koenigii (L) Spreng/ Rutaceae	Curry tree	The stem barks	Petroleum ether	UV, IR, MS, and a series of 1D and 2D NMR	Benzoisofuranone and carbazole	[51]
Peganum harmala/Nitrariaceae	Wild rue, Syrian rue, esfand, espand, harmel	Root and seed	Methanolic extract	TLC	Pegamine, vasicine, harmine, harmane, harmaline, harmalol, and vasicinon	[155]

Plant and family	Common name	Part of plant	Extraction solvent	Method of detection	Bioactive compound	Ref.
Phoenix dactylifera L./Arecaceae	Date palm	Leaf and pit	Methanol and acetone	—	—	[156]
Antifungal						
Ruta graveolens L./Rutaceae	Rue, common rue, herb-of-grace	Leaf	Hexane	(1)H and (13)C NMR	1-methyl-2-[6'-(3″,4″-methylenedioxyphenyl) hexyl]-4-quinolone	[145]
Waltheria indica/Malvaceae	Sleepy morning	Aerial parts	Dichloromethane	COSY, HSQC, HMBC, NOESY NMR, UV, IR, and HRESIMS	Waltheriones and 5®-vanessine	[146]
Antiviral						
Phellodendron amurense/ Rutaceae	Amur cork tree	Bark	Aqueous and ethanol	—	Berberine	[157]
Moringa oleifera/Moringaceae	ben oil tree, drumstick tree, horseradish tree, and benzolive tree	Leaf	Water extract	LC-MS	Gentiatibetine	[158]
Nuphar lutea/Nymphaeaceae					thiobinupharidines and thiobinuphlutidines	[159]
Antiparasitic						
Argemone Mexicana/ Papaveraceae	Mexican poppy	Leaves and stems	Methanolic extract	Dragendorff's reagent	Berberine	[160]
Spondias mombin/ Anacardiaceae	Yellow mombin	Bark and leaves	Aqueous and ethanol	—	—	[161]

Table 2.
Selected medicinal plants possess antimicrobial activity based on their alkaloids as components.

The leaves were subjected to methanol extraction, and subsequent extraction with various solvents. The disk diffusion technique results showed that at concentrations of 0.22–2.02 mg/mL, the crude leaf extracts showed antibacterial activity against *E. coli*, with inhibition diameters ranging from 0 to 9 mm [152].

The Apocynaceae plants, *Catharanthus roseus*, and *Rauwolfia serpentina* are known for their production of significant alkaloids, including serpentine, vinblastine, vincristine, ajmalicine, reserpine, and ajmaline. These plants are role models for understanding how monoterpene indole alkaloids (MIA) are synthesized. Considerable knowledge exists regarding the physiological and ecological factors producing MIA in *C. roseus* [37].

The date palm is widely distributed throughout the Arabian Peninsula and is recognized as a significant economic crop. Date palms possess various chemical compounds such as vitamins, flavonoids, steroids, alkaloids, tannins, and carbohydrates. Except for *E. faecalis*, both the methanol and acetone extracts showed potent antibacterial activity [156].

5. Alkaloids' antibacterial mechanism of action

Alkaloids have been observed to affect various metabolic systems in animals, and their toxic mechanism of action can display considerable variability. Toxicity may present itself via enzymatic alterations that affect physiological functions, obstruction of DNA synthesis and repair mechanisms by intercalating with nucleic acids, or modulation of the nervous system. Various alkaloids can exert an influence on different physiological processes [37]. However, bactericidal drugs are those that, in the absence of confounding variables, result in a 99.9% reduction in bacterial viability at doses no higher than four times the MIC [96]. Most research shows that alkaloids are antibacterial, not bacteriostatic, though this might vary depending on the species of specific alkaloids (such as chelerythrine and prosopilosidine) [8, 15]. The MIC values of squalamine have been demonstrated to be bactericidal within 1–2 hours, killing 99.99% or more of gram-positive and gram-negative bacteria [8]. Their primary antibacterial methods involve blocking bacterial metabolism, altering membrane permeability, and blocking the creation of nucleic acids and proteins [17]. Techniques involving the controlled introduction of pathogens or herbivores, the physical or chemical stimulation of their presence, and the subsequent monitoring of gene expression, enzyme activity, and concentrations of precursors and the alkaloid itself have proven effective [37]. The distinct classes of alkaloids exhibit varying mechanisms of action as antibacterial agents [37]. The antibacterial properties of pergularinine and tylophorinidine, which belong to the indolizine class of alkaloids, are attributed to their ability to inhibit the dihydrofolate reductase enzyme, thereby impeding the synthesis of nucleic acids [153]. Agelasines alkaloids affect bacterial hemostasis by inhibiting the dioxygenase enzyme BCG 3185c, contributing to their antibacterial action. Agelasine D is an alkaloid with antimycobacterial activities, and its overexpression and binding affinity in studies led to the result mentioned above [163]. The respiratory inhibition effects of synthetic quinolone alkaloids, as well as the cell division inhibition effects of isoquinolines, including protoberberine, berberine, benzophenanthridine, and sanguinarine through perturbation of the Z-ring, have been documented. Additionally, the phenanthridine isoquinoline alkaloid ungeremine has been found to inhibit nucleic acid synthesis. In contrast, the indolizidine alkaloids pergularinine and tylophorinidine have been shown to suppress nucleic acid synthesis by inhibiting dihydrofolate reductase [37]. The mechanisms of action about antibacterial activity exhibit variation across distinct alkaloids. The following examples are being examined [16]:

1. Disruption of the bacterial membrane.

    Several alkaloids from herbal plants have been discovered to exhibit antimicrobial activity by disrupting the bacterial membrane. For example, herbal alkaloids like berberine and palmatine have been proven to cause bacterial cell death by rupturing their membrane [164, 165]. Additionally, squalamine is a polyamine alkaloid with a detergent-like mode of action, depolarizing Gram-positive bacteria membranes and disrupting Gram-negative bacteria's outer membranes [16]. The cytoplasmic membrane is disturbed by phenanthroindolizidine alkaloids [166]. For instance, berberine attacked the mitochondrial membrane of fungi and resulted in cytoplasmic damage in *Streptococcus agalactiae* (CVCC 1886 strain, obtained from the Microbiological Lab of Sichuan Agricultural University, Ya'an, China), whereas liriodenine caused cytoplasmic changes and cell wall destruction in *Paracoccidioides brasiliensis* [9].

2. Interfering with cell division.

    Pergularinine and tylophorinidine, two phenanthroindolizidine plant alkaloids, can block the production of nucleic acids. Protein, RNA, and DNA synthesis rely on pyrimidine and purine precursors, produced by the crucial enzyme dihydrofolate reductase [16]. DNA-protein cross-linking and DNA cross-linking are two mechanisms through which certain alkaloids, such as aristolochic acids, can cause mutations [167]. Interaction with DNA is thought to be the primary mechanism by which quinoline alkaloids exert their antibacterial and antifungal effects [9]. Another example is berberine, which was effective against *Actinobacillus pleuropneumoniae* and *Streptococcus agalactiae* (CVCC 1886) by inhibiting DNA synthesis and preventing synthesis [168].

3. Bacterial enzyme and respiratory system inhibition:

    Alkaloids from herbal plants have been reported to inhibit bacterial enzymes and respiratory systems. For example, inhibiting the respiratory system of bacteria, including *S. aureus*, has been demonstrated for the alkaloid tetrandrine, which is present in several medicinal plants [15]. Additionally, berberine can inhibit bacterial enzymes like DNA gyrase leading to cell death [15]. Also, the alkyl methyl quinolone alkaloids exhibit potent and selective antibacterial properties against *H. pylori* using respiratory inhibition [169].

4. Modulating the expression of virulence genes.

    The regulatory protein ToxT has been identified in *V. cholerae.* It plays a crucial role in activating various virulence determinants, including the genes responsible for encoding virulence factors. Additionally, Yang et al. report that cholera toxin and ToxT co-regulated pilus [170]. The isoquinoline alkaloid known as virstatin has been found to effectively inhibit ToxT, which subsequently results in the inhibition of virulence factors. The research showed that it prevented *V. cholerae* from colonizing the intestines of newborn mice models [16].

On the other hand, the majority of quinoline and indole-based antifungal and antibacterial alkaloids discovered in Asian angiosperms, respectively, target DNA, topoisomerases, and the cytoplasmic membrane as their primary sites of action [9].

6. Conclusions and future

Alkaloids comprise a vast and heterogeneous category of compounds that exhibit a broad-spectrum of biological functions that hold immense significance for plants, animals, and humans. These compounds possess remarkable pharmacological properties. The advent of antibiotic-resistant microorganisms has substantially compromised antibiotic effectiveness. To date, a new approach to tackling antibiotic resistance is urgently needed. In the coming years, bioactive compounds will likely be discovered using phytochemicals, which exhibit a variety of chemical structures and methods of action. Alkaloids exhibit varying primary functions across different plant species, and their metabolic profiles are often associated with distinct environmental factors and developmental cues, thereby providing evident adaptive advantages. Concerning potential toxicity to other organisms or the production of bioactive metabolites for therapeutic applications, the variation in plant alkaloid metabolism and accumulation is crucial. Alkaloids are effective in this review report as an alternate therapy for combating the emergence and spread of multidrug-resistant infections and the harmful effects of some antibiotics. The following compounds have been identified as primary candidates due to their MIC of less than 1 μg/mL: 8-Acetylnorchelerythrine, cryptolepine, sampangine, 8-hydroxydihydrochelerythrine, 6-methoxydihydrosanguinarine, 2′-nortiliacorinine, tiliacorine, rhetsisine, pendulamine A and B, tylophorinine, tryptanthrin, viroallosecurinine, and vallesamine.

Author details

Mohammad Barati and Amir Modarresi Chahardehi*
Infectious Diseases Research Center, AJA University of Medical Sciences, Tehran, Iran

*Address all correspondence to: amirmch@gmail.com

References

[1] Ugboko HU et al. Antimicrobial importance of medicinal plants in Nigeria. ScientificWorldJournal. 2020; **2020**:7059323

[2] Barati M, Sharifi I, SHarififar F. In vitro evaluation of anti-leishmanial activities of Zataria multiflora Boiss, Peganum Harmala and Myrtus Communis by colorimetric assay. Journal of Kerman University of Medical Sciences. 2010;**16**(1):32-42

[3] Gong X et al. Plant pharmacophylogeny: Review and future directions. Chinese Journal of Integrative Medicine. 2022;**28**(6):567-574

[4] Cruz Martinez C, Diaz Gomez M, Oh MS. Use of traditional herbal medicine as an alternative in dental treatment in Mexican dentistry: A review. Pharmaceutical Biology. 2017; **55**(1):1992-1998

[5] Aiyegoro O, Okoh A. Use of bioactive plant products in combination with standard antibiotics: Implications in antimicrobial chemotherapy. Journal of Medicinal Plants Research. 2009;**3**(13): 1147-1152

[6] Ghaderi A et al. Evaluation of antileishmanial effect of the plant extract of alpha-pinene (Pistacia atlantica) in vitro and in vivo. Scientific Journal of Kurdistan University of Medical Sciences. 2018;**23**(5):32-44

[7] Modarresi Chahardehi A et al. Antidepressant-like effects of selected crude extracts of Pilea microphylla in mice model of depression. American Journal of Agricultural and Biological Sciences. 2013;**8**(1):75-81

[8] Cushnie TPT, Cushnie B, Lamb AJ. Alkaloids: An overview of their antibacterial, antibiotic-enhancing and antivirulence activities. International Journal of Antimicrobial Agents. 2014; **44**(5):377-386

[9] Sulaiman M et al. Antibacterial and antifungal alkaloids from Asian angiosperms: Distribution, mechanisms of action, structure-activity, and clinical potentials. Antibiotics. 2022;**11**(9):1146

[10] Shin J, Prabhakaran VS, Kim KS. The multi-faceted potential of plant-derived metabolites as antimicrobial agents against multidrug-resistant pathogens. Microbial Pathogenesis. 2018;**116**: 209-214

[11] Fuchtbauer S et al. Antibacterial properties of capsaicin and its derivatives and their potential to fight antibiotic resistance-A literature survey. European Journal of Microbiology and Immunology (Bp). 2021;**11**(1):10-17

[12] Alibi S, Crespo D, Navas J. Plant-derivatives small molecules with antibacterial activity. Antibiotics (Basel). 2021;**10**(3):231

[13] El-Saber Batiha G et al. Traditional uses, bioactive chemical constituents, and pharmacological and toxicological activities of Glycyrrhiza glabra L. (Fabaceae). Biomolecules. 2020; **10**(3):352

[14] Gonelimali FD et al. Antimicrobial properties and mechanism of action of some plant extracts against food pathogens and spoilage microorganisms. Frontiers in Microbiology. 2018;**9**:1639

[15] Thawabteh A et al. The biological activity of natural alkaloids against herbivores, cancerous cells and

pathogens. Toxins (Basel). 2019; **11**(11):656

[16] Othman L, Sleiman A, Abdel-Massih RM. Antimicrobial activity of polyphenols and alkaloids in Middle Eastern plants. Frontiers in Microbiology. 2019;**10**:911

[17] Yan Y et al. Research Progress on antibacterial activities and mechanisms of natural alkaloids: A review. Antibiotics (Basel). 2021;**10**(3):318

[18] Modarresi-Chahardehi A et al. Screening antimicrobial activity of various extracts of Urtica dioica. Revista de biologia tropical. 2012;**60**(4): 1567-1576

[19] Modarresi Chahardehi A. Infectious Diseases; Along with a Set of Questions and Explanations of Key Words. Tehran, Iran: Royan Pazhouh Publication; 2023

[20] Radulovic NS et al. Antimicrobial plant metabolites: Structural diversity and mechanism of action. Current Medicinal Chemistry. 2013;**20**(7): 932-952

[21] Modarresi, Chahardehi A et al. Effects of ethyl acetate extract of Urtica dioica on Bacillus subtilis strain ATCC 6633: Structural degeneration study. In: National Postgraduate Seminar (NPS 2014). Fostering Collaborative for the Advancement of Microbiology. Malaysia: Universiti Putra Malaysia; 2014

[22] Cowan MM. Plant products as antimicrobial agents. Clinical Microbiology Reviews. 1999;**12**(4): 564-582

[23] Modarresi Chahardehi A, et al. Cytotoxicity activity of Elatostema umbellatum against cancer cell lines. In: The 2nd Annual International Conference in Conjunction with the 8th IMT-GT UNINET Bioscience Conference; Darussalam, Banda Aceh, Indonesia. Banda Aceh: Universitas Syiah Kuala; 2012

[24] Huang W et al. Biosynthesis investigations of terpenoid, alkaloid, and flavonoid antimicrobial agents derived from medicinal plants. Antibiotics. 2022; **11**(10):1380

[25] Joanna K. Chapter 1, Introductory chapter: Alkaloids-their importance in nature and for human life. In: Joanna K, editor. Alkaloids. Rijeka: IntechOpen; 2019

[26] Ti H et al. Progress of plant medicine derived extracts and alkaloids on modulating viral infections and inflammation. Drug Design, Development and Therapy. 2021;**15**: 1385-1408

[27] Zhao Y et al. Antimicrobial effects of chemical compounds isolated from traditional Chinese herbal medicine (TCHM) against drug-resistant bacteria: A review paper. Mini Reviews in Medicinal Chemistry. 2019;**19**(2): 125-137

[28] Hashemi A et al. Antibacterial effects of methanolic extracts of Zataria multiflora, Myrtus communis and Peganum harmala on Pseudomonas aeruginosa producing ESBL. Journal of Arak University of Medical Sciences. 2011;**14**(4):104-112

[29] Khameneh B et al. Review on plant antimicrobials: A mechanistic viewpoint. Antimicrobial Resistance & Infection Control. 2019;**8**(1):118

[30] Betoni JE et al. Synergism between plant extract and antimicrobial drugs used on Staphylococcus aureus diseases. Memórias do Instituto Oswaldo Cruz. 2006;**101**(4):387-390

[31] Evans SM, Cowan MM. Plant products as antimicrobial agents. In: Cosmetic and Drug Microbiology. U.S.A.: CRC Press; 2016. pp. 227-254

[32] Barati M et al. Anti-leishmanial activity of Gossypium hirsutum L., Ferula assa-foetida L. and Artemisia aucheri Boiss. Extracts by colorimetric assay. Anti-Infective Agents. 2014;**12**(2): 159-164

[33] Chahardehi AM et al. Baja citotoxicidad, y actividad antiproliferativa sobre las celulas cancerosas, de la planta Senna alata (Fabaceae). Revista de Biología Tropical. 2021;**69**(1):317-331

[34] Ghaffari H et al. Inhibition of herpes simplex virus type 1 infection by Sambucus ebulus extract in vitro. Medical Journal of the Islamic Republic of Iran. 2021;**35**:9

[35] Aniszewski T. Chapter 1-definition, typology, and occurrence of alkaloids. In: Aniszewski T, editor. Alkaloids. 2nd ed. Boston: Elsevier; 2015. pp. 1-97

[36] Dey P et al. Analysis of alkaloids (indole alkaloids, isoquinoline alkaloids, tropane alkaloids). Recent Advances in Natural Products Analysis. 2020;**2020**: 505-567. DOI: 10.1016/B978-0-12-816455-6.00015-9

[37] Matsuura HN, Fett-Neto AG. Plant alkaloids: Main features, toxicity, and mechanisms of action. In: Gopalakrishnakone P, Carlini CR, Ligabue-Braun R, editors. Plant Toxins. Dordrecht: Springer Netherlands; 2015. pp. 1-15

[38] Yu X et al. An innovative extraction strategy for herbal medicine by adopting p-sulphonatocalix[6]/[8]arenes. Phytochemical Analysis. 2022;**33**(7): 1068-1085

[39] Li M et al. Simple quantitative analytical methods for the determination of alkaloids from medicinal and edible plant foods using a homemade chromatographic monolithic column. Journal of Chromatography. B, Analytical Technologies in the Biomedical and Life Sciences. 2019;**1128**: 121784

[40] Huang X et al. Self-assemblies based on traditional medicine berberine and cinnamic acid for adhesion-induced inhibition multidrug-resistant Staphylococcus aureus. ACS Applied Materials & Interfaces. 2019;**12**(1): 227-237

[41] Schenck CA, Maeda HA. Tyrosine biosynthesis, metabolism, and catabolism in plants. Phytochemistry. 2018;**149**:82-102

[42] Szántay C, Dörnyei G, Blaskó G. Chapter 2 the morphine alkaloids. In: Cordell GA, Brossi A, editors. The Alkaloids: Chemistry and Pharmacology. Maryland, U.S.A.: National Institutes of Health Bethesda; Academic Press; 1994. pp. 127-232

[43] Ding Y et al. Phytochemical and biological investigations of Amaryllidaceae alkaloids: A review. Journal of Asian Natural Products Research. 2017;**19**(1):53-100

[44] Omar F et al. Plant-based indole alkaloids: A comprehensive overview from a pharmacological perspective. Molecules. 2021;**26**(8):2297

[45] Pedras MSC et al. The phytoalexins from cauliflower, caulilexins A, B and C: Isolation, structure determination, syntheses and antifungal activity. Phytochemistry. 2006;**67**(14):1503-1509

[46] Jimenez LD, Ayer WA, Tewari JP. Phytoalexins produced in the leaves of

Capsella bursa-pastoris (shepherd's purse). Phytoprotection. 1997;**78**(3): 99-103

[47] Maynart G et al. Antibacterial effect of borreverine, an alkaloid isolated from Borreria verticillata (Rubiaceae). Comptes Rendus des Seances de la Societe de Biologie et de Ses Filiales. 1980;**174**(5):925-928

[48] O'Donnell G, Gibbons S. Antibacterial activity of two canthin-6-one alkaloids from Allium neapolitanum. Phytotherapy Research: An International Journal Devoted to Pharmacological and Toxicological Evaluation of Natural Product Derivatives. 2007;**21**(7):653-657

[49] Su X-L et al. Three new quinazolines from Evodia rutaecarpa and their biological activity. Fitoterapia. 2018;**127**: 186-192

[50] Bhattacharyya P, Chakrabartty P, Chowdhury B. Glycozolidol, an antibacterial carbazole alkaloid from Glycosmis pentaphylla. Phytochemistry. 1985;**24**(4):882-883

[51] Rahman MM, Gray AI. A benzoisofuranone derivative and carbazole alkaloids from Murraya koenigii and their antimicrobial activity. Phytochemistry. 2005;**66**(13):1601-1606

[52] Aassila H et al. Identification of harman as the antibiotic compound produced by a tunicate-associated bacterium. Marine Biotechnology. 2003; **5**:163-166

[53] Cruz KS et al. Screening and antifungal activity of a β-Carboline derivative against Cryptococcus neoformans and C. gattii. International Journal of Microbiology. 2019;**2019**: 7157845

[54] Joshi T et al. Pyranocarbazoles from Murraya koenigii (L.) Spreng. as antimicrobial agents. Natural Product Research. 2018;**32**(4):430-434

[55] Wang W, Cheng MH, Wang XH. Monoterpenoid indole alkaloids from Alstonia rupestris with cytotoxic, anti-inflammatory and antifungal activities. Molecules. 2013;**18**(6):7309-7322

[56] Xu S, Bian R, Chen X. Methods of Pharmacology Experiment. Beijing, China: People's Sanitation Press; 2003. pp. 1651-1653

[57] Yu HF et al. Nepenthe-like indole alkaloids with antimicrobial activity from Ervatamia chinensis. Organic Letters. 2018;**20**(13):4116-4120

[58] Cheng GG et al. Bioactive monoterpenoid indole alkaloids with diverse skeletons from Melodinus khasianus. Journal of Natural Products. 2016;**79**(9):2158-2166

[59] Singh B, Sharma RA, Vyas GK. Antimicrobial, antineoplastic and cytotoxic activities of indole alkaloids from Tabernaemontana divaricata (L.) R. Br. Current Pharmaceutical Analysis. 2011;**7**(2):125-132

[60] Karaket N et al. Chemical and bioactivity evaluation of the bark of Neonauclea purpurea. Natural Product Communications. 2012;**7**(2):169-170

[61] Qin X-J et al. Indole alkaloids with antibacterial activity from aqueous fraction of Alstonia scholaris. Tetrahedron. 2015;**71**(25):4372-4378

[62] Kawakami J et al. Antibacterial and antifungal activities of tryptanthrin derivatives. Transactions of the Materials Research Society of Japan. 2011;**36**(4):603-606

[63] Atta ur R, et al. Antifungal diterpenoid alkaloids from Delphinium

denudatum. Journal of Natural Products. 1997;**60**(5):472-474

[64] Hao Y et al. Discovery of tryptanthrins as novel antiviral and anti-phytopathogenic-fungus agents. Journal of Agricultural and Food Chemistry. 2020;**68**(20):5586-5595

[65] Wu JY et al. Topoisomerase I inhibitor evodiamine acts as an antibacterial agent against drug-resistant Klebsiella pneumoniae. Planta Medica. 2013;**79**(1):27-29

[66] Wang J et al. Natural phenolic derivatives based on piperine scaffold as potential antifungal agents. BMC Chem. 2020;**14**(1):24

[67] Silva Teles MMR et al. Alkaloids of the Lauraceae. The Alkaloids. Chemistry and Biology. 2019;**82**:147-304

[68] Xin Z et al. Isolation of antofine from Cynanchum atratum BUNGE (Asclepiadaceae) and its antifungal activity against Penicillium digitatum. Postharvest Biology and Technology. 2019;**157**:110961

[69] Peng L et al. Antibacterial activity and mechanism of berberine against Streptococcus agalactiae. International Journal of Clinical and Experimental Pathology. 2015;**8**(5):5217-5223

[70] Mensah JL et al. Antibacterial activity of the leaves of Phyllanthus discoid us. Journal of Ethnopharmacology. 1990; **28**(1):129-133

[71] Singh AK, Pandey MB, Singh UP. Antifungal activity of an alkaloid Allosecurinine against some fungi. Mycobiology. 2007;**35**(2):62-64

[72] Adeleye A, Ikotun T. Antifungal activity of dihydrodioscorine extracted from a wild variety of Dioscorea bulbifera L. Journal of Basic Microbiology. 1989;**29**(5):265-267

[73] Laluces HMC et al. Antimicrobial alkaloids from the leaves of Pandanus amaryllifolius. Journal of Applied Pharmaceutical Science. 2015;**5**(10): 151-153

[74] Bibi N et al. In vitro antituberculosis activities of the constituents isolated from Haloxylon salicornicum. Bioorganic & Medicinal Chemistry Letters. 2010;**20**(14):4173-4176

[75] Kim MG et al. Antimicrobial potentials of active component isolated from Citrullus colocynthis fruits and structure-activity relationships of its analogues against foodborne bacteria. Journal of the Science of Food and Agriculture. 2014;**94**(12):2529-2533

[76] Aguinaldo AM et al. Quinoline alkaloids from Lunasia amara inhibit Mycobacterium tuberculosis H37Rv in vitro. International Journal of Antimicrobial Agents. 2007;**29**(6): 744-746

[77] Tantapakul C et al. Antibacterial compounds from Glycosmis puberula twigs. Natural Product Communications. 2014;**9**(12):1705-1707

[78] Pandey MB et al. Inhibitive effect of Fuyuziphine isolated from plant (Pittapapra) (Fumaria indica) on spore germination of some fungi. Mycobiology. 2007;**35**(3):157-158

[79] Zhang H et al. Synergistic anti-candidal activity of tetrandrine on ketoconazole: An experimental study. Planta Medica. 2010;**76**(1):53-61

[80] Sureram S et al. Antimycobacterial activity of bisbenzylisoquinoline alkaloids from Tiliacora triandra against multidrug-resistant isolates of

Mycobacterium tuberculosis. Bioorganic & Medicinal Chemistry Letters. 2012;**22**(8):2902-2905

[81] Singh K et al. Tiliacorinine, a new systemic fungicide effective against Alternaria blight of pigeon pea (Cajanus cajan)/Tiliacorinine, ein neues systemisches Fungizid mit Wirkung gegen die Alternaria-Blattfleckenkrankheit an Taubenerbsen (Cajanus cajan). Zeitschrift für Pflanzenkrankheiten und Pflanzenschutz/Journal of Plant Diseases and Protection. 1991;**98**(2):213-219

[82] Rahman MM et al. Antibacterial and cytotoxic compounds from the bark of Cananga odorata. Fitoterapia. 2005; **76**(7–8):758-761

[83] Paulo Mde Q et al. Antimicrobial activity of benzylisoquinoline alkaloids from Annona salzmanii D.C. Journal of Ethnopharmacology. 1992; **36**(1):39-41

[84] Tan KK et al. Antibacterial alkaloids from Artabotrys crassifolius Hook.f. & Thomson. Natural Product Research. 2015;**29**(24):2346-2349

[85] Agarwal AK et al. Role of heme in the antifungal activity of the azaoxoaporphine alkaloid sampangine. Eukaryotic Cell. 2008;**7**(2):387-400

[86] Khan M, Kihara M, Omoloso A. Antimicrobial activity of the alkaloidal constituents of the root bark of Eupomatia laurina. Pharmaceutical Biology. 2003;**41**(4):277-280

[87] Lekphrom R, Kanokmedhakul S, Kanokmedhakul K. Bioactive styryllactones and alkaloid from flowers of Goniothalamus laoticus. Journal of Ethnopharmacology. 2009;**125**(1): 47-50

[88] Camacho-Corona MdR et al. Evaluation of some plant-derived secondary metabolites against sensitive and multidrug-resistant Mycobacterium tuberculosis. Journal of the Mexican Chemical Society. 2009;**53**(2):71-75

[89] Makarasen A et al. Cytotoxic and antimicrobial activities of aporphine alkaloids isolated from Stephania venosa (Blume) Spreng. Planta Medica. 2011; **77**(13):1519-1524

[90] Orhana I et al. Antiviral and antimicrobial profiles of selected isoquinoline alkaloids from Fumaria and Corydalis species. Zeitschrift fur Naturforschung-Section C Journal of Biosciences. 2007;**62**(1–2):19-26

[91] Ma C et al. Potent activities of Roemerine against Candida albicans and the underlying mechanisms. Molecules. 2015;**20**(10):17913-17928

[92] Agnihotri VK et al. Constituents of Nelumbo nucifera leaves and their antimalarial and antifungal activity. Phytochemistry Letters. 2008;**1**(2):89-93

[93] Kim J et al. Antifungal activity of magnoflorine against Candida strains. World Journal of Microbiology and Biotechnology. 2018;**34**(11):167

[94] Kosina P et al. Phytochemical and antimicrobial characterization of Macleaya cordata herb. Fitoterapia. 2010;**81**(8):1006-1012

[95] Faizi S et al. New antimicrobial alkaloids from the roots of Polyalthia longifolia var. pendula. Planta Medica. 2003;**69**(4):350-355

[96] Choi JG et al. Antibacterial activity of Hylomecon hylomeconoides against methicillin-resistant Staphylococcus aureus. Applied Biochemistry and Biotechnology. 2010;**160**(8):2467-2474

[97] Xue X et al. TLC bioautography-guided isolation and antimicrobial, antifungal effects of 12 alkaloids from Hylomecon japonica roots§. Natural Product Communications. 2017;**12**(9): 1439-1442

[98] Tavares Lde C et al. Structure-activity relationship of benzophenanthridine alkaloids from Zanthoxylum rhoifolium having antimicrobial activity. PLoS One. 2014; **9**(5):e97000

[99] Zuo GY et al. Synergistic antibacterial and antibiotic effects of bisbenzylisoquinoline alkaloids on clinical isolates of methicillin-resistant Staphylococcus aureus (MRSA). Molecules. 2011;**16**(12):9819-9826

[100] Guang Ma W, Fukushi Y, Tahara S. Fungitoxic alkaloids from Hokkaido corydalis species. Fitoterapia. 1999; **70**(3):258-265

[101] Phatchana R, Yenjai C. Cytotoxic coumarins from Toddalia asiatica. Planta Medica. 2014;**80**(8–9):719-722

[102] Hu J et al. Alkaloids from Toddalia asiatica and their cytotoxic, antimicrobial and antifungal activities. Food Chemistry. 2014;**148**:437-444

[103] Gu JQ et al. Cytotoxic and antimicrobial constituents of the bark of Diospyros maritima collected in two geographical locations in Indonesia. Journal of Natural Products. 2004;**67**(7): 1156-1161

[104] Semwal DK et al. The genus Stephania (Menispermaceae): Chemical and pharmacological perspectives. Journal of Ethnopharmacology. 2010; **132**(2):369-383

[105] Adesanya SA et al. Stilbene derivatives from Cissus quadrangularis. Journal of Natural Products. 1999; **62**(12):1694-1695

[106] Shen JW et al. Lycorine: A potential broad-spectrum agent against crop pathogenic fungi. Journal of Microbiology and Biotechnology. 2014; **24**(3):354-358

[107] Pettit GR, Melody N, Herald DL. Antineoplastic agents. 450. Synthesis of (+)-pancratistatin from (+)-narciclasine as relay(1a). The Journal of Organic Chemistry. 2001;**66**(8):2583-2587

[108] Nair JJ, van Staden J. Antifungal constituents of the plant family Amaryllidaceae. Phytotherapy Research. 2018;**32**(6):976-984

[109] Liang C et al. Broad-spectrum antifungal activity of dichloromethane extract of Waltheria indica stems and isolated compounds. Industrial Crops and Products. 2019;**142**:111855

[110] Adams M et al. Cytotoxicity and p-glycoprotein modulating effects of quinolones and indoloquinazolines from the Chinese herb Evodia rutaecarpa. Planta Medica. 2007;**73**(15):1554-1557

[111] Suliman Mohamed M et al. Activity of Aristolochia bracteolata against Moraxella catarrhalis. International Journal of Bacteriology. 2014;**2014**:481686

[112] Zeng YB et al. Antimicrobial glycoalkaloids from the tubers of Stephania succifera. Archives of Pharmacal Research. 2017;**40**(4): 429-434

[113] Tuntiwachwuttikul P et al. Chemical constituents of the roots of Piper sarmentosum. Chem Pharm Bull (Tokyo). 2006;**54**(2):149-151

[114] Shi YN et al. Antifungal amide alkaloids from the aerial parts of Piper

flaviflorum and Piper sarmentosum. Planta Medica. 2017;**83**(1–02):143-150

[115] Zhou LN et al. Antibacterial steroidal alkaloids from Holarrhena antidysenteriaca. Chinese Journal of Natural Medicines. 2017;**15**(7):540-545

[116] de Lourdes FD et al. Chapter 9-Biological activity and 13C NMR spectral data of iboga-type skeleton alkaloids. In: Atta ur R, editor. Studies in Natural Products Chemistry. Vol. 72. Karachi, Pakistan: Center for Molecular Medicine and Drug Research University of Karachi, Elsevier; 2022. pp. 287-369

[117] Kirpotina LN et al. Therapeutic effects of Tryptanthrin and Tryptanthrin-6-Oxime in models of rheumatoid arthritis. Frontiers in Pharmacology. 2020;**11**

[118] Morris JS, Facchini PJ. Isolation and characterization of reticuline N-methyltransferase involved in biosynthesis of the aporphine alkaloid magnoflorine in opium poppy. The Journal of Biological Chemistry. 2016; **291**(45):23416-23427

[119] Warowicka A, Nawrot R, Goździcka-Józefiak A. Antiviral activity of berberine. Archives of Virology. 2020; **165**(9):1935-1945

[120] Sack RB, Froehlich JL. Berberine inhibits intestinal secretory response of Vibrio cholerae and Escherichia coli enterotoxins. Infection and Immunity. 1982;**35**(2):471-475

[121] Wang D et al. Global transcriptional profiles of Staphylococcus aureus treated with berberine chloride. FEMS Microbiology Letters. 2008;**279**(2): 217-225

[122] Guzman JD. Natural cinnamic acids, synthetic derivatives and hybrids with antimicrobial activity. Molecules. 2014;**19**(12):19292-19349

[123] Sun D, Courtney HS, Beachey EH. Berberine sulfate blocks adherence of Streptococcus pyogenes to epithelial cells, fibronectin, and hexadecane. Antimicrobial Agents and Chemotherapy. 1988;**32**(9):1370-1374

[124] Ibrahim SA et al. Application of caffeine, 1,3,7-trimethylxanthine, to control Escherichia coli O157:H7. Food Chemistry. 2006;**99**(4):645-650

[125] Almeida AAP et al. Antibacterial activity of coffee extracts and selected coffee chemical compounds against Enterobacteria. Journal of Agricultural and Food Chemistry. 2006;**54**(23): 8738-8743

[126] Wang X et al. Synthesis of amide derivatives containing capsaicin and their antioxidant and antibacterial activities. Journal of Food Biochemistry. 2019;**43**(12):e13061

[127] Cichewicz RH, Thorpe PA. The antimicrobial properties of Chile peppers (capsicum species) and their uses in Mayan medicine. Journal of Ethnopharmacology. 1996;**52**(2):61-70

[128] Dhyani P et al. Anticancer potential of alkaloids: A key emphasis to colchicine, vinblastine, vincristine, vindesine, vinorelbine and vincamine. Cancer Cell International. 2022; **22**(1):206

[129] Leete E, Nemeth PE. The biogenesis of the alkaloids of colchicum. I. The incorporation of phenylalanine into colchicine1. Journal of the American Chemical Society. 1960;**82**(23): 6055-6057

[130] Huczyński A et al. Synthesis, antiproliferative and antibacterial

evaluation of C-ring modified colchicine analogues. European Journal of Medicinal Chemistry. 2015;**90**:296-301

[131] Golpour M et al. The effectiveness of colchicine as an anti-inflammatory drug in the treatment of coronavirus disease 2019: Meta-analysis. International Journal of Immunopathology and Pharmacology. 2021;**35**

[132] Haq IU et al. Piperine: A review of its biological effects. Phytotherapy Research. 2021;**35**(2):680-700

[133] Atal CK, Dubey RK, Singh J. Biochemical basis of enhanced drug bioavailability by piperine: Evidence that piperine is a potent inhibitor of drug metabolism. The Journal of Pharmacology and Experimental Therapeutics. 1985;**232**(1):258-262

[134] Derosa G, Maffioli P, Sahebkar A. Piperine and its role in chronic diseases. Advances in Experimental Medicine and Biology. 2016;**928**:173-184

[135] Khan IA et al. Piperine, a phytochemical potentiator of ciprofloxacin against Staphylococcus aureus. Antimicrobial Agents and Chemotherapy. 2006;**50**(2):810-812

[136] Parai D et al. Reserpine attenuates biofilm formation and virulence of Staphylococcus aureus. Microbial Pathogenesis. 2020;**138**:103790

[137] Begum S et al. Antimycobacterial and antioxidant activities of reserpine and its derivatives. Natural Product Research. 2012;**26**(22):2084-2088

[138] Shaheen A et al. Reserpine is the new addition into the repertoire of AcrB efflux pump inhibitors. Molekuliarnaia Biologiia (Mosk). 2019;**53**(4): 674-684

[139] Silva-Beltrán NP et al. Total phenolic, flavonoid, tomatine, and tomatidine contents and antioxidant and antimicrobial activities of extracts of tomato plant. International Journal of Analytical Chemistry. 2015;**2015**:284071

[140] Mitchell G et al. Tomatidine inhibits replication of Staphylococcus aureus small-colony variants in cystic fibrosis airway epithelial cells. Antimicrobial Agents and Chemotherapy. 2011;**55**(5):1937-1945

[141] Siriyong T, Voravuthikunchai SP, Coote PJ. Steroidal alkaloids and conessine from the medicinal plant Holarrhena antidysenterica restore antibiotic efficacy in a Galleria mellonella model of multidrug-resistant Pseudomonas aeruginosa infection. BMC Complementary and Alternative Medicine. 2018;**18**(1):285

[142] Siriyong T et al. Holarrhena antidysenterica extract and its steroidal alkaloid, conessine, as resistance-modifying agents against extensively drug-resistant Acinetobacter baumannii. Microbial Drug Resistance. 2016;**22**(4): 273-282

[143] Wei WJ et al. A review on classification and biological activities of alkaloids from the genus Zanthoxylum species. Mini Reviews in Medicinal Chemistry. 2021;**21**(3):336-361

[144] Farouk A, Nawi M, Hassan S. Antibacterial peptides from Euycoma longifolia (Tongkat Ali) and Labisia pumila (Kacip Fatimah) leaves in Malaysia. Science Brun. 2008;**9**: 55-63

[145] Oliva A et al. Natural fungicides from Ruta graveolens L. leaves, including a new quinolone alkaloid. Journal of Agricultural and Food Chemistry. 2003;**51**(4):890-896

[146] Cretton S et al. Antifungal quinoline alkaloids from Waltheria indica. Journal of Natural Products. 2016;**79**(2):300-307

[147] Akinpelu DA et al. Evaluation of antibacterial and antifungal properties of Alchornea laxiflora (Benth.) Pax. & Hoffman. Evidence-based Complementary and Alternative Medicine. 2015;**2015**:684839

[148] Evidente A et al. (−)-Amarbellisine, a lycorine-type alkaloid from Amaryllis belladonna L. growing in Egypt. Phytochemistry. 2004;**65**(14): 2113-2118

[149] Semwal DK, Semwal RB. Efficacy and safety of Stephania glabra: An alkaloid-rich traditional medicinal plant. Natural Product Research. 2015;**29**(5): 396-410

[150] Asano N et al. N-containing sugars from Morus alba and their glycosidase inhibitory activities. Carbohydrate Research. 1994;**259**(2): 243-255

[151] Sultana S et al. Antimicrobial, cytotoxic and antioxidant activity of methanolic extract of Glycyrrhiza glabra. Agriculture and Biology Journal of North America. 2010;**1**(5):957-960

[152] Al-Matani SK, Al-Wahaibi RNS, Hossain MA. Total flavonoids content and antimicrobial activity of crude extract from leaves of Ficus sycomorus native to Sultanate of Oman. Karbala International Journal of Modern Science. 2015;**1**(3):166-171

[153] Rao KN, Venkatachalam SR. Inhibition of dihydrofolate reductase and cell growth activity by the phenanthroindolizidine alkaloids pergularinine and tylophorinidine: The in vitro cytotoxicity of these plant alkaloids and their potential as antimicrobial and anticancer agents. Toxicology In Vitro. 2000;**14**(1):53-59

[154] Erdemoglu N, Ozkan S, Tosun F. Alkaloid profile and antimicrobial activity of Lupinus angustifolius L. alkaloid extract. Phytochemistry Reviews. 2007;**6**(1):197-201

[155] Darabpour E et al. Antibacterial activity of different parts of Peganum harmala L. growing in Iran against multi-drug resistant bacteria. EXCLI Journal. 2011;**10**:252-263

[156] Perveen K, Bokhari NA, Soliman DA. Antibacterial activity of Phoenix dactylifera L. leaf and pit extracts against selected Gram negative and Gram positive pathogenic bacteria. Journal of Medicinal Plants Research. 2012;**6**(2):296-300

[157] Wang W et al. In vitro antioxidant, antimicrobial and anti-herpes simplex virus type 1 activity of Phellodendron amurense Rupr. From China. The American Journal of Chinese Medicine. 2009;**37**(1):195-203

[158] Rahayu I, Timotius KH, Analysis P. Antimutagenic and antiviral activity of Moringa oleifera L. leaf infusion: In vitro and in silico studies. Molecules. 2022;**27** (13):4017

[159] Weiss S et al. In vitro and in vivo therapeutic potential of 6,6'-Dihydroxythiobinupharidine (DTBN) from Nuphar lutea on cells and K18-hACE2 mice infected with SARS-CoV-2. International Journal of Molecular Sciences. 2023;**24**(9):8327

[160] Elizondo-Luévano JH et al. Berberine: A nematocidal alkaloid from Argemone mexicana against Strongyloides venezuelensis.

Experimental Parasitology. 2021;**220**: 108043

[161] Agbaje EO, Onabanjo AO. Toxicological study of the extracts of anti-malarial medicinal plant Enantia chlorantha. The Central African Journal of Medicine. 1994;**40**(3):71-73

[162] Hagel JM, Facchini PJ. Benzylisoquinoline alkaloid metabolism: A century of discovery and a brave new world. Plant & Cell Physiology. 2013; **54**(5):647-672

[163] Arai M et al. Identification of the target protein of agelasine D, a marine sponge diterpene alkaloid, as an anti-dormant mycobacterial substance. Chembiochem. 2014;**15**(1):117-123

[164] Luo Y et al. Berberine prevents non-alcoholic steatohepatitis-derived hepatocellular carcinoma by inhibiting inflammation and angiogenesis in mice. American Journal of Translational Research. 2019;**11**(5): 2668-2682

[165] Brahma U et al. Antimicrobial and anti-biofilm activity of hexadentated macrocyclic complex of copper (II) derived from thiosemicarbazide against Staphylococcus aureus. Scientific Reports. 2018;**8**(1):8050

[166] Chen C et al. Inhibitory effect of 7-Demethoxytylophorine on Penicillium italicum and its possible mechanism. Microorganisms. 2019;**7**(2):36

[167] Kuete V et al. Antimycobacterial, antibacterial and antifungal activities of the methanol extract and compounds from Thecacoris annobonae (Euphorbiaceae). South African Journal of Botany. 2010;**76**(3):536-542

[168] Kang S et al. The antibacterial mechanism of berberine against Actinobacillus pleuropneumoniae. Natural Product Research. 2015;**29**(23): 2203-2206

[169] Tominaga K et al. In vivo action of novel alkyl methyl quinolone alkaloids against Helicobacter pylori. The Journal of Antimicrobial Chemotherapy. 2002; **50**(4):547-552

[170] Yang J, Tauschek M, Robins-Browne RM. Control of bacterial virulence by AraC-like regulators that respond to chemical signals. Trends in Microbiology. 2011;**19**(3):128-135

Chapter 8

Synthetic Alkaloids: Cantharidin Derivatives

Nurhan Kishali

Abstract

Cantharidin is a naturally occurring cyclic anhydride found in many insect species, particularly Lytta vesicatoria, known as the Spanish fly. Although highly poisonous, dried Spanish fly has been used as an aphrodisiac since ancient Greeks and Romans. Spanish fly has been used in eastern medicine for many years as a natural anticancer agent, especially in the treatment of hepatoma and esophageal carcinoma. Over time, its stotoxicity was determined to be high and its use was limited. Later, alkaloid derivatives with no stotoxic effect were produced synthetically and evaluated as anticancer agents. Since cantharidin obtained from insects is not an alkaloid but its derivatives with lower stotoxicity, cantharimide and norcantharimide are cyclic imides, they can be evaluated in the class of alkaloids. Cantharimide and norcantharimide compounds have gained importance in terms of their stotoxic effect on many cancer cell lines. Many studies have been done on their synthesis and anticancer properties for many years.

Keywords: alkaloids, cantharidin, cantharimide, norcantharimide, anticancer

1. Introduction

Alkaloids are chemical compounds that contain basic Nitrogen atoms and are produced naturally by various organisms. Alkaloids can contain some groups with neutral [1] or acidic properties [2]. Alkaloids are usually organic bases. They form salts when reacted with acid and form alkaline solutions when dissolved. Primary sources of alkaloids are flowering plants. Plants use alkaloids for defense against herbivores and pathogens. It has been determined that 20% of plants contain alkaloids. Alkaloids are cyclic compounds that contain nitrogen (**Figure 1**) [3].

Plenty of alkaloids have been used in medicine for ages and even nowadays, they are prominent medical compounds. Since primitive times, alkaloids obtained from plant extracts have been used in medicines and poisons. In ancient times, plant extracts containing alkaloids were used to treat numerous ailments, including snake-bite, fever, and insanity. Generally, alkaloids are extremely toxic at low concentrations, even if they have a therapeutic effect. Defense chemicals of plants against microorganisms, insects, and herbivores, as well as other plants using allelopathic active chemicals [4]. Their taste is bitter. They are usually colorless crystals at room temperature and are optically active [5]. Purely isolated plant alkaloids and their

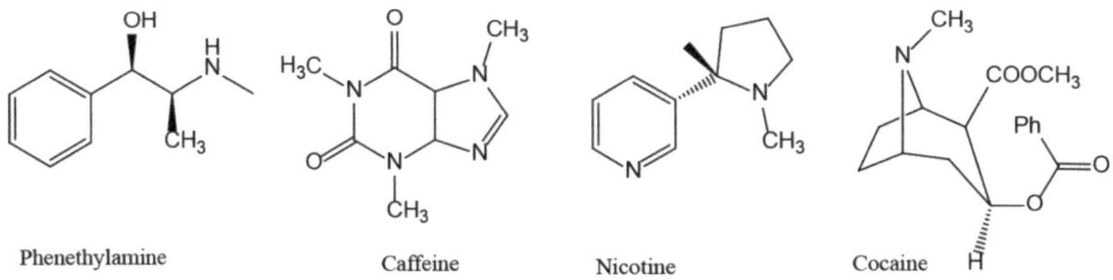

Figure 1.
Examples of alkaloids.

synthetic derivatives are used as basic medicinal agents due to analgesic, antispasmodic, and bactericidal effects [6]. Alkaloids generally affect the nervous system in humans (especially acetylcholine, epinephrine, norepinephrine, gamma-aminobutyric acid, dopamine, and serotonin) [7]. Alkaloids such as berberine (in eye medications) and sanguinarine (in toothpaste) are used as antiseptics (**Figures 2** and **3**) [8].

While the search for new anticancer drugs continue, old drugs are viewed as new options. The dried body of Mylabris, the Chinese bubble beetle, has been the focus of attention for its anticancer properties, as it is known for traditional medicine in China for more than two thousand years, where it has been used as a traditional medicine. The oldest data in China on the use of Mylabris as a medicine dates back to 300–168 BC. In Europe, it was found about 77 AD in a medical article published in Materia Medica. The active ingredient of Mylabris has been identified as cantharidin. Later, it

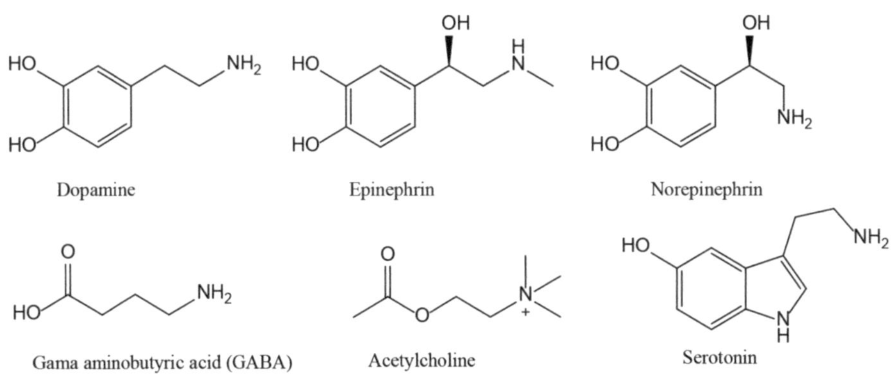

Figure 2.
Examples of alkaloids acting on the nervous system.

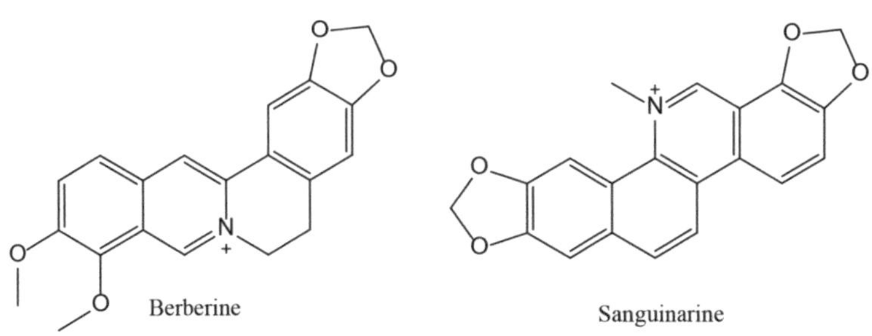

Figure 3.
Chemical formula of berberine and sanguinarine alkaloid.

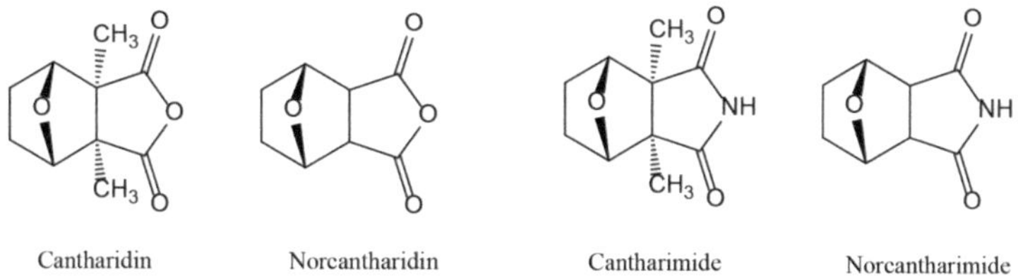

Figure 4.
The chemical formula of cantharidin, norcantharidin, cantharimide, and norcantharimide.

was determined that cantharidin has both anticancer activity and leukocytosis and hemorrhagic cystitis properties [9]. When the natural product cantharidin was purified and used for a long time, its cytotoxic effects began to be observed, and norcantharidin, a demethylated derivative, was synthesized. Subsequently, nitrogen-containing derivatives cantharimide and norcantharimide were synthesized. Later, derivatives of these four analogs were synthesized both in the amide ring and in the cyclohexyl ring, and biological activity studies were carried out (**Figure 4**).

Norcantharidin slows the proliferation of tumors such as HeLa, CHO, CaEs-17, BEL-7402, SMMC7721 human hematoma, HEP-2, and human epidermoid larynx carcinoma [10–12]. Norcantharidin has been found to have fewer nephrotoxic and inflammatory effects [13–17]. Based on the structure–activity relationship, more analogs have been synthesized. For this reason, based on the structure–activity relationship, the researchers synthesized more analogs, and each synthesis was supported by biological activity and anticancer studies [18–28].

Disodium cantharidate and norcantharidin derivatives, among the compounds synthesized analogously, are among the derivatives synthesized in the earliest period by Wang et al. [10]. Because of their found stronger antihepatoma activities, these derivatives were more popular than cantharidin itself. However, they also cause minimal urinary irritation (**Figure 5**) [10].

Cantharidin has many derivatives besides disodium cantharidat and norcantharidin, some of which are hydrocantharidimite, methyl cantharidimite, and dehydronorkantharidin (**Figure 5**). Chinese scientists are collecting their medical literature by researching cantharidin and its derivatives, but there are still large gaps in researchers' knowledge about these drugs and their effects [29–32]. According to current information, norcantharidin, a demethylated analog of cantharidin, is stated to slow down the proliferation of tumors such as HeLa, CHO, esophageal carcinoma (CaEs-17), hepatoma (BEL-7402 and SMMC-7721), epidermoid larynx carcinoma (HEP2), and human epidermoid larynx carcinoma [10–12].

Two main methods were used in the synthesis of cantharimide and norcantharimides. One of them is the addition of furan and (2,3-dimethyl)maleic anhydride and then its conversion to the imide derivative. The other (2,3-

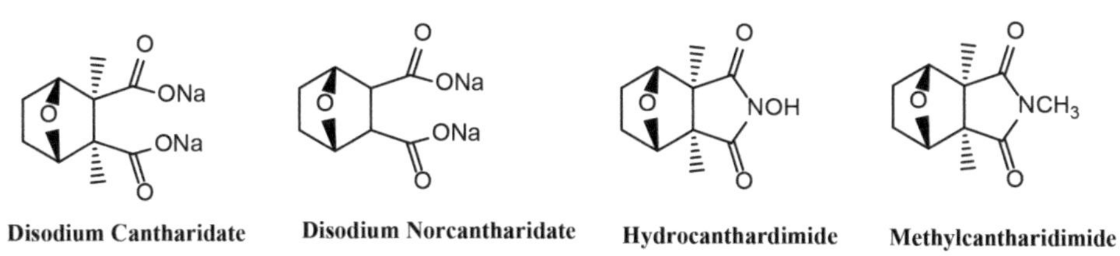

Figure 5.
Examples of early synthesized derivatives of cantharidin.

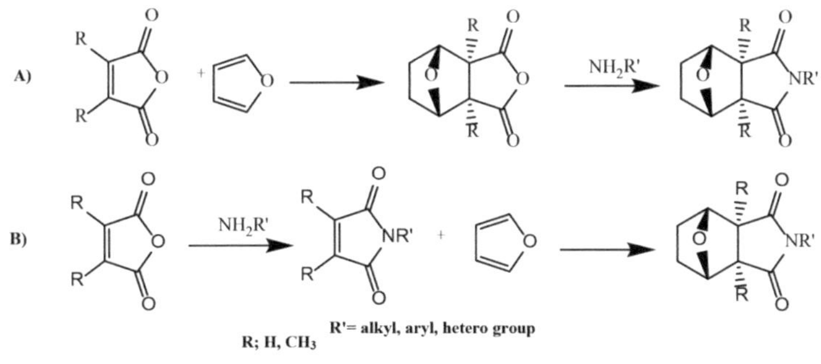

Figure 6.
General synthesis methods of cantharimide and norcantharimides (R′: Primary amine containing the desired derivative to be synthesized).

dimethylmaleimide) is made by initially synthesizing maleimides and adding Diels-Alder with furan (**Figure 6**).

2. Cantharimides

Cantharidine is found in *Cantharis vesicatory*, *Lytta caraganae*, *Mylabris phalerata*, *Meloidae*, *Oedemeridae*, and various other insect species. Derivatives of cantharidin are synthesized to reduce their blistering effects on the skin, reduce their toxic properties, and benefit from their antitumor activity. Since the 1950s, it has been determined that various anhydride and imide derivatives exhibit a wide range of biological activities (antidepressant, anticonvulsant) as well as antitumor properties [33–43]. For this reason, cantharimide, which has an imide structure, attracted the attention of Guang-Sheng WANG reported that he synthesized cantharimide and N-methyl cantharimide in an article he prepared in 1989 and conducted an activity study in KB cell culture. As a result of his study, he stated that both compounds showed antitumor activity [10].

Pen-Yuan Lin and his group synthesized 10 different cantharimide derivatives using various tryptamine, indolyl, naphthyl, and pyridyl amines [34]. Lin used Zhang's method in his synthesis. In this method, in the presence of triethyl amine at high pressure, the related amine compound gives cantharimides as a result of an addition reaction under pressure at 200°C (**Figures** 7 and **8**) [44].

Another of the cantharimide derivatives is the derivatives obtained by the addition of heterocyclic groups. These derivatives were synthesized for cytotoxicity tests against human hepatocellular carcinoma cells. In this synthesis, N-thiazolyl and N-thiadiazolyl cantharimides were synthesized by the method used above. According to the results of the study, they observed that the side groups attached to N-Thiazolyl and N-Thiadiazolyl amine compounds showed higher cytotoxicity than cantharidin in

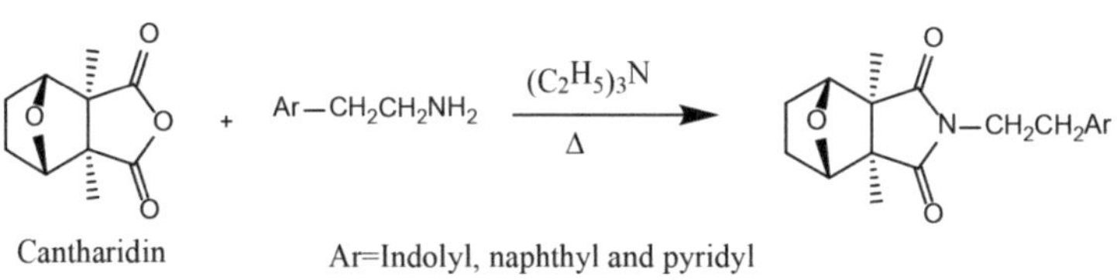

Figure 7.
Synthesis of tryptamine, indolyl, naphthyl, and pyridyl canthrimidines [44].

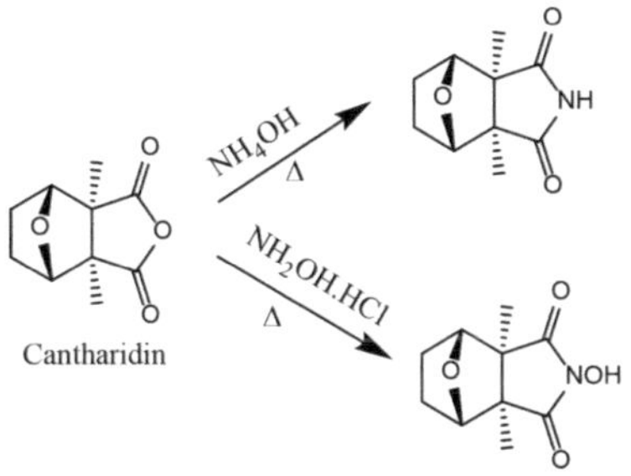

Figure 8.
Synthesis of cantharimides at high pressure [44].

the area of electron-withdrawing (such as -NO_2), whereas a decrease in cytotoxic activity was observed in methyl-substituted compounds (**Figure 9**) [45].

In 2007, a new cantharimide derivative was synthesized by Chan et al. Its structure was elucidated and its cytotoxicity on SK-Hep-1 hepatoma cells was determined. In this study, two cantharimides were synthesized with 2-amino benzothiazole derivatives. Cantharidin and 2-amino benzothiazole derivatives were added to a tube containing dry toluene and triethylamine (TEA), and related derivatives were synthesized at 200°C [45]. As a result of cytotoxic studies of the study group, it was determined that the compounds showed inhibition on SK-Hep-1 hepatoma. Using the results from the study, the group is also designing new (**Figure 10**) molecules [46].

Cantharimide derivatives were also synthesized using aliphatic primary amines, phenethyl amines, aniline, and pyridine components by high-pressure addition reaction method. In the synthesis, the product was obtained with a yield of 29—96%. It is thought that the primary factor affecting the efficiency is the inductive effect of electron-negative groups. Another factor affecting the yield is the nucleophilic and basicity strength of the amines. Again, cytotoxicity studies of the compounds obtained in this synthesis were performed in human hepatocellular carcinoma (Hep G2) and myeloid leukemia cells (HL-60). In the evaluation, it was determined that compounds (**10** and **16**) with electron-withdrawing NO_2 groups in the pyridyl and benzene rings showed strong inhibitory effects in both cell lines. Methyl-containing compounds

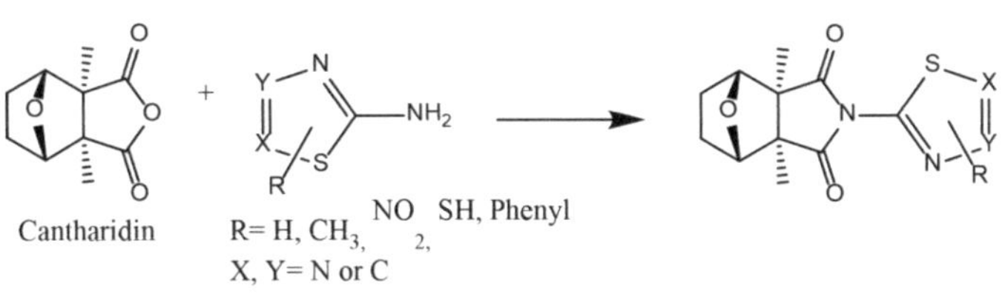

Figure 9.
The synthesis of the N-thiazolyl- and N-thiadiazolyl cantharimides [45].

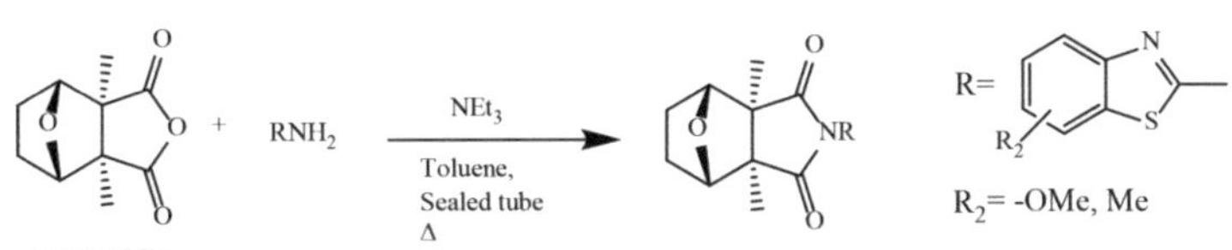

Figure 10.
The synthesis of the (Me/-OMe)-2-amino benzothiazole cantharimides [46].

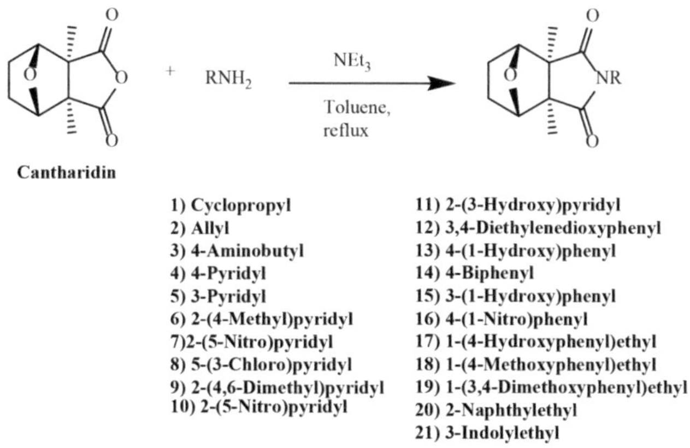

Figure 11.
Cantharimides are synthesized by the high-pressure addition reaction method [33].

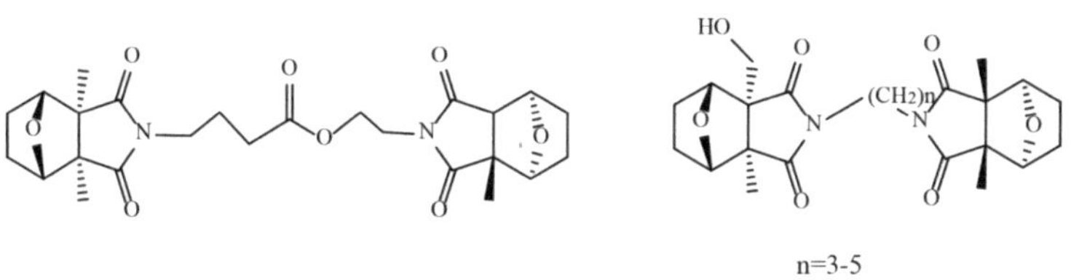

Figure 12.
Dimeric-canthraimides isolated from M. Phaletata [34].

(**6**, **7**, and **9**) have less effect, while halogen-containing compound (**8**) has a moderate effect. Compound **10** with the 2-(3-nitro pyridyl) group showed stronger cytotoxicity. Para-Pyridyl imide **4** showed greater potency than ortho- or meta-pyridyl imide. 3-Pyridyl (**5**) and N-phenyl imides (**14** and **16**) also showed strong cytotoxicity. Compounds with planar side chains (**12**, **14**, and **20**) and N-azaethyl or N-aryl compounds (**17**, **18**, **19**, and **21**) showed moderate cytotoxicity on both Hep G2 and HL-60. However, aliphatic chain imides (**1**, **2**, and **3**) were found to have very low effects in the studied cell lines (Hep G2, HL-60) (**Figure 11**) [33].

Until 2017, about thirty cantharidin derivativeshave been isolated from insects of the genus *Mylabris* and *Hycleus*. As cantharimide compounds are known to exist in different insects from *Mylabris* and *Hycleus* species, thirteen new cantharidin derivatives have been isolated from the whole body of *Mythicomyia phalerata* as part of studies to discover new potential antitumor agents [34]. During this purification, dimeric cantharimides were also obtained. The cytotoxic effects of the isolated derivatives against HepG2, MDA-MB-231, and A-549 cell lines were investigated, and it was determined that all isolated compounds showed high activity, except for the compound called Canthaminomide F (**Figure 12**) [34].

Due to the presence of dimeric products in cantharimides obtained from natural sources, some researchers have included these compounds in their synthesis. In a study conducted in this way, cantharimide dimers were synthesized and the structure was determined. The dimers synthesized in the same study were also isolated from natural sources. Along with these dimers, two different cantharimides were purified (**Figure 13**) [35].

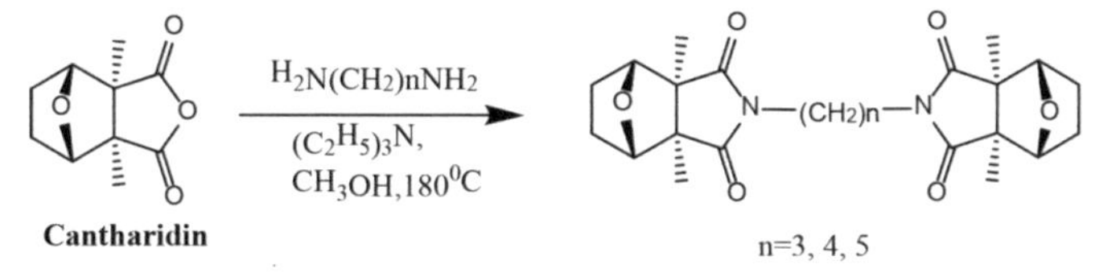

Figure 13.
Cantharimide dimers from the Chinese Blister Beetle *[35].*

3. Norcantharimides

Studies on norcantharimides are examined by dividing them into two groups distribution methods and medicinal chemicals. Compounds such as norcantharimide [36] and norcantharidin dimer [37] belong to the class of medicinal chemicals. Norcantharimides are made from either the imide ring [38] or the cyclohexyl ring [39] in their medicinal chemical derivative synthesis. Then, new derivatives [40] were obtained by opening the imide ring in these derivatives (**Figures 14** and **15**).

In addition, when dimer products were detected in cantharimide derivatives purified in natural sources, dimer structures in norcantharimide derivatives were included in the synthesis [36]. Norcantharidin-dimer analogs as analogs of norcantharidin, norcantharidin with lactose acid, norcantharidin with amantadine, water-soluble norcantharidin with chitosan analogs, esterification of norcantharidin, amino acid norcantharimides, N-substituted dehydronortharimid analogs, immune liposomes, and modifications of norcantharidin show potential in the anticancer field [36].

Some norcantharidin analogs are known to have very good PP1 and PP2A inhibitory activity [22, 41–43]. In addition, studies by McCluskey et al. reported that some norcantharidin analogs, which have no toxic effect on human cell lines, kill trichostrongylid nematode *Trichostrongylus vitrinus* and *Haemonchus contortus larvae* [47].

McCluskey synthesizes a large number of norcantharimide in his studies and conducts various activity studies. In one of these studies, he synthesized 54 norcantharimide compounds and conducted a toxicity study against H. Contortus, which showed serine–threonine phosphatases (PP1 and PP2A) [48]. As a result of the study, it was determined that three of the 54 analogs synthesized were almost completely lethal against H. contortus and showed at least five times more inhibition

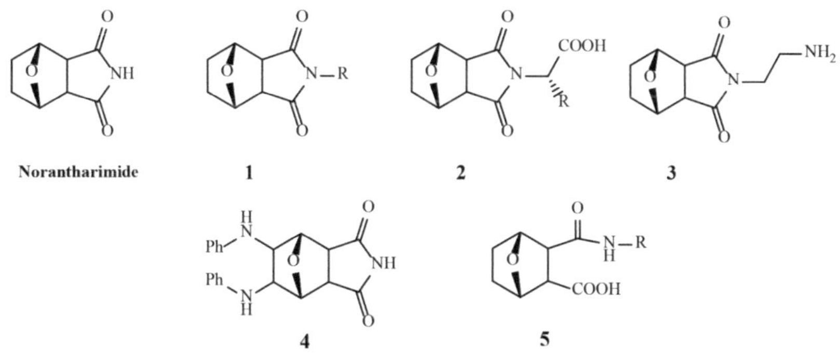

Figure 14.
Examples of norcantharimide as a medicinal chemical [36–40].

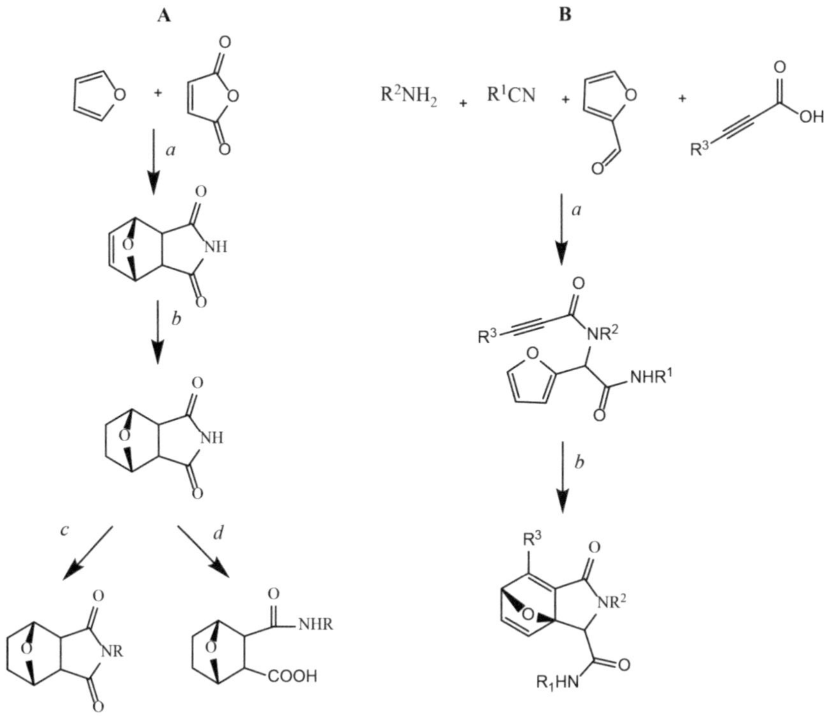

Figure 15.
(A) Reagents and conditions: (a) Et_2O, rt., 48 h; (b) acetone, 10% Pd–C, $H_2(g)$ 50 psi, 18 h; (c) RNH_2, $PhCH_3$, reflux, 24–36 h (d) RNH_2, THF, rt [41]. (B) Reagents and conditions: (a) CH_3OH, rt. 30 min; (b) $PhCH_3$, sealed tube 200°C, 36 h.

than the control compounds. McCluskey' reported that he synthesized 54 compounds by the methods given below **Figure 14** [47].

In another study published by McCluskey in 2011, eighteen phosphate esters of the side group attached to the imide nitrogen were synthesized. It has bioscreened them in nine human cancer cell lines (HT29 ve SW480, MCF-7, A2780, H460, A431, DU145, BE2-C, SJ-G2). As a result of the study, he stated that he obtained a new series of norcantharidin analogs with broad-spectrum antiproliferative activity [41]. Another important finding from the study is the relative ease of Phosphate ester hydrolysis. Of the phosphates studied, diphenyl and bis-trichloroethyl analogs showed the highest level of cell death (**Figure 16**) [49].

McCluskey also conducted biological activity studies by synthesizing norcantharidin-dimer analogs. McCluskey explained the relationship between heterocyclic substituted (nor)cantharidin analogs and PP1 and PP2A as a result of his studies. McCluskey et al. synthesized many norcantharidin analogs, including two bis-norcantharimides. Among the compounds it synthesized were compounds containing 10, 12, and 14 alkyl chains attached to the imide nitrogen, 1,2-diol units, and two norcantharimide attached to dodecyl. Among these groups, two norcantharimide (Bis-Norcantharimide) bound to dodecyl showed the highest activity in the cell line [42].

Evaluating the results obtained from this study, Cuifang Cai showed that the compound containing N-$C_{14}H_{29}$ (**Figure 17**) side group has a longer half-life and

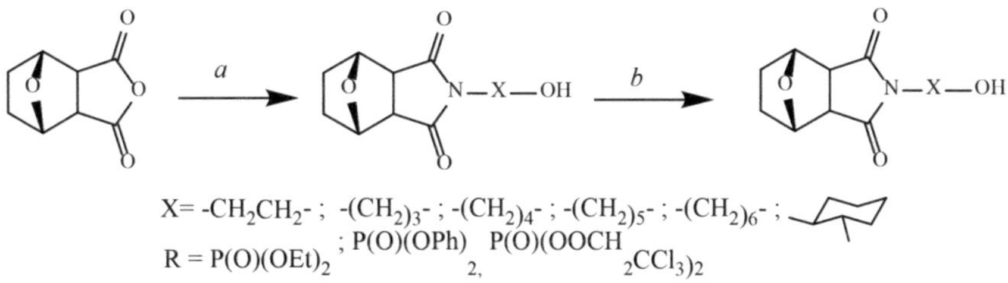

Figure 16.
Reagents and conditions: (a) H_2N–X–OH, D, 18 h; (b) ClP(O)(OR)$_2$, n-Bu_2O, Et_3N, rt [49].

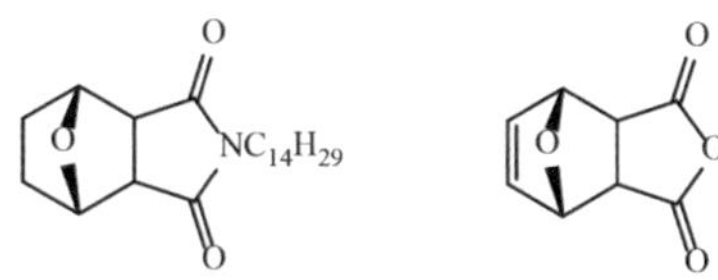

N-$C_{14}H_{29}$ Norcantharidin

Figure 17.
N-$C_{14}H_{29}$ and norcantharimide, whose physical and chemical stability was determined by Cuifang Cai [50].

higher volume of distribution (Vss) compared to norcantharidin, as a result of the pharmacokinetic study performed by injecting it into rats. The results obtained from this study suggested that the formulation to be prepared with N-$C_{14}H_{29}$ is a promising alternative with its high encapsulation efficiency, significant physical and chemical stability, and long half-life (**Figure 17**) [50].

Dimeric compounds norcantharidin compounds have been synthesized and studied by many researchers for the treatment of cancer, HIV, Alzheimer's, malaria, and various parasitic diseases. Some of the scientists who carried out these studies are Gervais [51], McCluskey [42], T. Nakatani [35], Tang [52], and Tan [38].

When Xue-Jie Tan and his group realized that good results were obtained both in extract analysis and in many synthesis and activity studies, they used four dimeric norcantharimide directly synthesized in their studies. Two of these compounds were synthesized for the first time by Xue-Jie Tan and his group, while the remaining two were synthesized by Dominic V. McGrath's group. However, Xue-Jie Tan and his group presented the crystal structure, spectroscopic properties, and anticancer activities of these four unsaturated dimeric norcantharimides to the knowledge of researchers (**Figure 18**) [38].

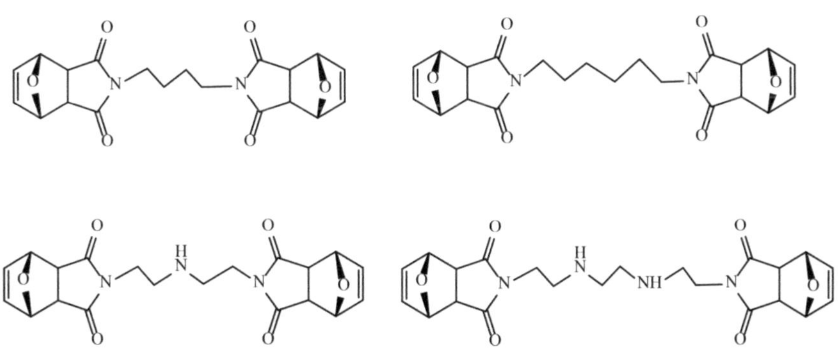

Figure 18.
Dimer-norcantharimides synthesized by Tan et al. [38].

4. Conclusion

Alkaloids are defined as amine compounds naturally produced by plants. In addition, alkaloids are defined as secondary metabolites with important biological properties. It is also known that some alkaloids are beneficial for some diseases. In light of this information, we can evaluate cantharimide and norcantharimide derivatives in the alkaloid class, even if they are not of plant origin. Considering the information, we have briefly compiled above, the starting compound is a product of natural origin. Later, all of the synthesized derivatives contained nitrogen atoms, and positive test results were obtained on many disease-causing agents. After obtaining a large number of alkaloid cantharimide derivatives by extraction, it has also begun to be synthesized as a medicinal chemical. Researchers at the time sought to improve previous methods. In this review, it was concluded that medicinal chemical-synthesized derivatives of an insect-derived alkaloid can be used for pharmaceutical purposes.

Acknowledgements

We would like to thank Atatürk University (Project number: 2012/470 and FBG-2018-6476) for its financial support of this work.

Author details

Nurhan Kishali
Faculty of Sciences, Department of Chemistry, Ataturk University, Erzurum, Turkey

*Address all correspondence to: nhorasan@atauni.edu.tr

References

[1] McNaught AD, Wilkinson A. Iupac. Compendium of chemical terminology. 1997;**2**(3). Available from: http://publications.iupac.org/compendium/A.html

[2] Manske RHF. The Alkaloids. Chemistry and Physiology. New York, NY: Academic Press; 1965

[3] Pelletier SW. The nature and definition of an alkaloid, Alkaloids: Chemical and Biological Perspectives. 1983;**1**:1-31

[4] Molyneux RJ, Nash RJ, Asano N. In: Pelletier SW, editor. Alkaloids: Chemical and Biological Perspectives. Vol. 11. Oxford: Pergamon; 1996

[5] Harbone JP. Phytochemical methods, a guide to modern technique of plant analysis. London: Chapmann and hall; 1973. pp. 1-271

[6] Stary F. The Natural Guide to Medicinal Herbs and Plants. Barnes & Noble Inc; 1996

[7] Roberts MF, Wink M. Biochemistry, ecology, and medicinal applications. Alkaloids. New York, London: Plenum Press; 1998. pp. 1-7

[8] Cordell GA. Introduction to alkaloids: A biogenetic approach. John Wiley & Sons; 1981

[9] Mack P, Ha X-F, Cheng L-Y. HPB Surgery. Efficacy of Intra-arterial Norcantharidin in S-uppressing Tumour 4C-labelled Glucose Oxidative Metabolism in rat Morris Hepatoma. 1996;**10**:65-72

[10] Wang GS. Medical uses of mylabris in ancient China and recent studies. Journal of Ethnopharmacology. 1989;**26**: 147-162

[11] Wang KR, et al. Biological effect of sodium norcantharidate on several mammalian cell lines. In: Abstracts, 4th International Congress of Cell Biology. Ottawa, Canada: National Research Council Canada; 1988. p. 76

[12] Liao HF, Su SL, Chen YN, et al. Norcantharidin preferentially induces apoptosis in human leukemic Jurkat cells without affecting viability of normal blood mononuclear cells. Food and Chemical Toxicology. 2007;**45**:1678-1687

[13] Liu XH, Blazsek I, Comisso M, et al. Effects of norcantharidin, a protein phosphatase type-2A inhibitor, on the growth of normal and malignant haemopoietic cells. European Journal of Cancer. 1995;**31**:953-963

[14] Deng LP, Liu FM, Wang HY. 1,3-dipolar cycloaddition reaction of novel 5,6-dehydronorcantharidin derivafives of substituted aromatic amines withpotential antitumor activities. Journal of Heterocyclic Chemistry. 2005;**42**:13-l8

[15] Purevsuren G, Koblicova Z, Trojanek J. Cantharidinimide, a novel substance from Mylabris mongolica dokht. Ceskoslovenska Pharmacie. 1987;**36**:32-34

[16] Lee JY, Chung YJ, Bae YS, et al. Synthesis of hexapeptide and tetrapeptide analogs of the immunomodulating peptides. Journal of the Chemical Society, Perkin Transactions. 1998;**2**:359-366

[17] Berson JA, Reynolds RD, Jones WM. The stereochemistry and mechamsm of the Diels-Alder reaction. An “intenal” mechanism for the interconveresion of endo-exo isomers. Journal of the American Chemical Society. 1956;**78**: 6094-6097

[18] McCluskey A, Keane MA, Walkom CC, et al. The first two cantharidin analogues displaying PP1 selectivity. Bioorganic & Medicinal Chemistry Letters. 2002;**12**:391-393

[19] McCluskey A, Taylor C, Quinn RJ, et al. Inhibition of protein phosphatase 2A by cantharidin analogues. Bioorganic & Medicinal Chemistry Letters. 1996;**6**:1025-1028

[20] McCluskey A, Keane MA, Mudgee LM, et al. Norcantharimide analogues possessing terminal phosphate esters and their anti-cancer activity. Bioorganic & Medicinal Chemistry. 2000;**35**:957-964

[21] McCluskey A, Bowyer MC, Collins E, et al. Anhydride modified cantharidin analogues: Synthesis, inhibition of protein phosphatases 1 and 2A and anticancer activity. Bioorganic & Medicinal Chemistry Letters. 2000;**10**: 1687-1690

[22] McCluskey A, Walkom C, Bowyer MC, et al. Cantharidimides, a new class of modified cantharidin analogues inhibiting protein phosphatases 1 and 2A. Bioorganic & Medicinal Chemistry Letters. 2001;**11**:2941-2946

[23] Sakoff JA, Ackland SP, Baldwin ML, et al. Anticancer activity and protein phosphatase 1 and 2A inhibition of a new generation of cantharidin analogues. Investigational New Drugs. 2002;**21**:1-11

[24] McCluskey A, Ackland SP, Bowyer MC, et al. Cantharidin analogues: Synthesis and evaluation of growth inhibition in a panel of selected tumour cell lines. Bioorganic Chemistry. 2003;**31**: 68-79

[25] Hart ME, Chamberlin AR, Walkom C, et al. Modified norcantharidins synthesis, protein phosphatases 1 and 2A inhibition, and anticancer activity. Bioorganic & Medicinal Chemistry Letters. 2004;**14**:1969-1973

[26] Enz A, Zenke G, Pombo-Villar E. 7-Oxa [2.2.1]bicycloheptane-2,3-dicarboxylic acid derivatives as phosphatase inhibitors. Bioorganic & Medicinal Chemistry Letters. 1997;**7**: 2513-2518

[27] Soseoka M, Baba Y, Kobayashi S, et al. Structure-activity relationship of cantharidin derivatives to protein phosphatases 1, 2A1, and 2B. Bioorganic & Medicinal Chemistry Letters. 1997;**7**:1833-1836

[28] Tatlock JH, Linton MA, Hou XJ, et al. Structure-based design of novel calcineurin (PP2B) inhibitors. Bioorganic & Medicinal Chemistry Letters. 1997;**7**:1007-1012

[29] Wang GS. Pharmacology and clinical trials of mylabris and camptotheca acuminata. (Chinese). Beijing Research Information on Prevention and Treatment of Tumors. 1976;**1**:60-62

[30] Chen RT, Hua Z, Yang JL, et al. Studies on antitumor actions of cantharidin. (Chinese). Chinese Journal of Natural Medicines. 1977;**57**:475-448

[31] Dong C, Cheng FH, Jin TY, Li BM. Observations on the antitumor effect with norcantharidin in vitro. (Chinese). Yixue Ziliao (Health Bureau of Fuzhou Military Region). 1982;**1**:52-53

[32] Wang GS, Huang JK, Lu FX, et al. Results of clinical trials in 244 cases of primary hepatoma with norcantharidin. (Chinese). Chinese Pharmaceutical Bulletin. 1986;**21**:90-93

[33] Lın L-H, Huang H-S, Lın C-C, Lee L-W, Lın E-Y. Effects of Cantharidinimides on human carcinoma cells. Chemical & Pharmaceutical Bulletin. 2004;**52**:855-857

[34] Deng Y-Y, Zhang W, Li N-P, Lei X-P, Gong X-Y, Zhang D-M, et al. Cantharidin derivatives from the medicinal insect Mylabris phalerata. Tetrahedron. 2017;**73**:5932-5939

[35] Nakatani T, Jinpo K, Noda N. Cantharimide dimers from the Chinese blister beetle, Mylabris phalerate PALLAS. Chemical & Pharmaceutical Bulletin. 2007;**55**:92-94

[36] Lin PY, Shi SJ, Lin LH, Chen HF, Hsu FL. Synthesis of novel N-pyridylcantharidinimides by using high pressure. Journal of the Chinese Chemical Society. 2001;**48**:49-53

[37] Deng L, Tang S. Norcantharidin analogues: A patent review (2006 – 2010). Expert Opinion on Therapeutic Patents. 2011;**21**:1743-1753

[38] Cheng S-S, Shi Y, Ma X-N, Xing D-X, Liu L-D, Liu Y, et al. Synthesis, crystal structure, spectroscopic properties and potential anti-cancerous activities of four unsaturated bis-norcantharimides. Journal of Molecular Structure. 2016; **1115**:228-240

[39] Bajsa J, McCluskey A, Gordon CP, Stewart SG, Hill TA, Sahu R, et al. The antiplasmodial activity of norcantharidin analogs. Bioorganic & Medicinal Chemistry Letters. 2010;**20**:6688-6695

[40] Zhou J, Ren Y, Tan L, Song X, Wang M, Li Y, et al. Norcantharidin: Research advances in pharmaceutical activities and derivatives in recent years. Biomedicine & Pharmacotherapy. 2020; **131**:110755

[41] Campbell BE, Tarleton M, Gordon CP, Sakoff JA, Gilbert J, McCluskey A, et al. Norcantharidin analogues with nematocidal activity in Haemonchus contortus. Bioorganic & Medicinal Chemistry Letters. 2011;**21**:3277-3281

[42] Hill TA, Stewart SG, Ackland SP, Gilbert J, Sauer B, Sakoff JA, et al. Norcantharimides, synthesis and anticancer activity: Synthesis of new norcantharidin analogues and their anticancer evaluation. Bioorganic & Medicinal Chemistry. 2007;**15**:6126-6134

[43] Hill TA, Stewart SG, Sauer B, Gilbert J, Ackland SP, Sakoff JA, et al. Bioorganic & Medicinal Chemistry Letters. 2007;**17**:3392

[44] Lin PY, Shi SJ, Hsu FL, Chen CF. New Cantharidinimides from Cantharidin and 2-Arylethylamines: Efficient Synthesis under High Pressure. Journal of the Chinese Chemical Society. 1998;**45**:323-236

[45] Lin PY, Shi SJ, Shu HL, Chen HF, Lin CC, Liu PC, et al. A simple procedure for preparation of N-Thiazolyl and N-Thiadiazolylcantharidinimides and evaluation of their Cytotoxicities against human hepatocellular carcinoma cells. Bioorganic Chemistry. 2000;**28**: 266-272

[46] Kok SHL, Chui CH, Lam WS, Chen J, Lau FY, Wong RSM, et al. Synthesis and structure evaluation of a novel cantharimide and its cytotoxicity on SK-Hep-1 hepatoma cells. Bioorganic & Medicinal Chemistry Letters. 2007;**17**: 1155-1159

[47] Hill TA, Stewart SG, Gordon CP, Ackland SP, Gilbert J, Sauer B, et al. Norcantharidin analogues: synthesis, anticancer activity and protein phosphatase 1 and 2A inhibition. ChemMedChem: Chemistry Enabling Drug Discovery. 2008;**3**(12):1878-1892

[48] Campbell BE, Hofmann A, McCluskey A, Gasser RB. Serine/threonine phosphatases in socioeconomically important parasitic nematodes—prospects as novel drug

targets?. Biotechnology advances. 2011; **29**(1):28-39

[49] Robertson MJ et al. Norcantharimide analogues possessing terminal phosphate esters and their anti-cancer activity. Bioorganic & Medicinal Chemistry. 2011;**19**(18):5734-5741

[50] Pan Z, Niu Y, Wang Y, Tang Y, Tang X, Cai C. Intravenous lipid microspheres loaded with alkylated norcantharidin derivative norcantharimide: Improved stability and prolonged half-life. European Journal of Lipid Science and Technology. 2015;**117**(1):55-64

[51] Berube G. Natural and synthetic biologically active dimeric molecules: anticancer agents, anti-HIV agents, steroid derivatives and opioid antagonists. Current Medicinal Chemistry. 2006;**13**(2):131-154

[52] Kok SHL, et al. Induction of apoptosis on carcinoma cells by two synthetic cantharidin analogues. International journal of molecular medicine. 2006;**17**(1):151-157

Section 3

Anti-Inflammatory and Antiviral

Chapter 9

Medicinal Plants as Sources for Drugs and Vaccines

Siham A. Salim

Abstract

In general, vaccines are important biological factors that stimulate human immunity to resist various diseases or their pathogens that invade him. The vaccine includes protein material of the pathogen itself, which is either killed or weakened form, or is made from corresponding artificial protein subunits to help human's immune system for recognizing antigens. However, it has been observed that there are some side effects appeared from using of traditional vaccines, which made trending toward finding alternative solutions is an important goal. In recent years, with the progress in medicinal sciences, genetics and plant biotechnology, the concept of edible vaccines has emerged by biotechnologists in an attempt to use edible plants in the production of alternative vaccines for commercial vaccines that are useful in treating diseases that affect humans without needing for injection or refrigerated storage, which is done through genetically engineering plants to carry antigens through several methods, like bacterial vectors, shot gun or microinjection through plant tissue culture techniques to produce vaccine-bearing plants like banana, maize, potato, rice, tobacco, tomato, legumes and others which makes these plants have two tasks, their suitability for food and to stimulate the body's immune response against many pathogens at once.

Keywords: medicinal plants, immunization, natural drugs, edible vaccines, transgenic plants, vaccination

1. Introduction

Vaccination is a simple, safe, and effective way to protect people from harmful diseases before they are exposed to them via induction of the body's natural defenses to build resistance to specific diseases, as well as strengthen the immune system. Vaccines train the immune system to make antibodies (proteins that the immune system naturally produces to fight disease), just as it does when exposed to a disease, as it recognizes an invading pathogenic organism, such as a virus, bacteria, or others, and fights it. However, because vaccines contain only dead or weakened forms of germs such as viruses or bacteria, they do not cause disease and do not expose the human body to the risk of complications, so most vaccines are given by injection, while others are given orally or sprayed into the nose to treat many diseases such as hepatitis B, measles, tuberculosis, tetanus, diarrhea, diphtheria, etc. [1].

Our immune systems have the ability to remember. Once exposed to one or several doses of a vaccine, we usually remain protected from the disease for years, decades, or even for life. This makes vaccines so effective, as they aim first to protect us from the disease before resorting to treatment after infection [2]. Although traditional vaccination is the safe method used around the world to confront the risk of diseases when exposed to them, it faces some limitations, represented by the cost of production, storage, distribution and the lack of sufficient scientific research on it. Therefore, scientists and researchers have turned to finding safe therapeutic alternatives under the progress made in various biotechnologies and genetic engineering, which made it possible to produce genetically modified plants, which prompted researchers to introduce anti-pathogenic genes into these plants in order to produce plants that are eaten and carry the anti-gene at the same time, which in turn are easy to transport, distribute and store so that they are available to humans when administered as edible vaccines [1, 3].

The concept of edible vaccines, plant-based edible vaccines, or plant-based vaccination appeared in the twentieth century. This is called a GreenVax (a concept developed in the 1990s, which means the consumption of edible tissues of transgenic plants), which refers to food, typically plants, that produce proteins, vitamins, or other nourishments that act as a vaccine against a certain disease. Once the plant, fruit, or plant-derived product is ingested orally, it stimulates the immune system. Specifically, it stimulates both the mucosal and humoral immune systems [4–6]. Edible vaccines offer many benefits over traditional vaccines due to their lower manufacturing cost and lack of negative side effects. However, there are limitations as edible vaccines are still new and developing. Further research will need to be done before they are ready for widespread human consumption.

The plant-based vaccine method works by isolating a specific antigen protein, which triggers a human immune response from the target virus. A protein gene is transferred to the bacteria, which is then used to "infect" plant cells. The plants then begin producing the exact protein that will be used for the vaccine [7]. The possibility of introducing a set of genes of human pathogens (whether viruses or bacteria) into plant cells, thus re-cultivating the plant again so that it can produce biological primary vaccines containing pathogen genes, and by feeding the tissues of these plants to humans or animals, an immune response to vaccines is elicited. The new process will only take 4–6 weeks. Depending on this, if the project succeeds, it will be one of the largest and most powerful vaccine facilities in the world. However, the development and widespread use of new vaccines to improve health conditions at the global level face many challenges. The cost of the new vaccine must be low, the vaccine must be administered orally without injection, and it must remain stable in high temperatures. It should also contain a combination of vaccines to prevent diseases prevalent in developing countries [8, 9].

2. Why the plant-based edible drugs and vaccines are important?

Different drugs and vaccines were used in all countries of the world, as they caused a clear decrease in the death rates among humans, which are caused by various microbial infections with a large percentage, whereas in some cases vaccination leads to death of the vaccinated person [8]. The use of plants that are eaten to act as edible vaccines is an effective, safe alternative to traditional vaccines in controlling various types of diseases and illnesses [3, 10]. To obtain an edible vaccine, the required gene that encodes that active compound as a vaccine is selected and inserted into the desired plant, where this plant manufactures the proteins encoded for this vaccine to perform a systemic immune function that gives the required immunity to the body of the organism when the plant is eaten [11, 12].

Regardless of the way that edible vaccines are consumed, they all share an important common goal of immunizing the human body against different pathogens or before they multiply in quantities that are sufficient to cause disease and the appearance of disease symptoms in the patient. It is well known that the traditional methods of immunization against diseases are done by exposing the person's immune system to killed or very weakened bacteria or viruses [13]. Therefore, when the immune system becomes sensitive to any foreign organism in the vaccine, it will act as if the body is under attack and mobilize all of its forces to eradicate and destroy the attacker after targeting the antibody gene, which the immune system distinguishes as foreign proteins that have entered the body. In fact, there is a rapid suppression of the response, but it leaves behind a guard or a watchdog in the memory of the cells that remain fully prepared when the pathogen enters the body in the future, so some vaccines and serums provide the body with lifelong protection, and the other section fades after time, such as the cholera vaccine and the tetanus vaccine, which requires periodic immunization. It is noted that traditional vaccines have few risks, the most important of which is that the organisms with which the body was vaccinated may live and multiply inside the body, causing diseases that were supposed to be eliminated. From this aspect, most vaccine manufacturers today prefer another type of vaccine called subunit preparations, which consist mainly of antigenic proteins that have been discovered from the genes of the pathogens. Thus, there will be no possible chance for infection to occur in the future. Despite the importance of this modern industry of subunit vaccines, they are criticized for their high production costs due to their manufacture from bacterial cultures or animal cells, as well as their high purity and need for freezing.

The edible vaccines, which are the focus of our discussion, are similar to the subunit preparation in terms of being genetically engineered to ensure that they contain the antigen and do not contain the organism that causes the disease, and both are safe to use. Before starting the production of edible vaccines, scientists raised a number of questions, including whether plants that will be genetically modified to contain the antigen able to produce effective copies of the intended protein. Will the antigen transferred to edible plants be destroyed when consumed by humans or animals? Will the gene decompose in the stomach before it performs its role compared to subunit preparations that are given as injections to avoid their damage? Will the antigen that was produced alert the human or animal immune system, and will the immune system's response be sufficient to the extent of protecting humans or animals from infection with the disease against which they have been vaccinated? Besides that, researchers must know whether the edible vaccine appears in the mucosal immune system because many pathogens enter the body via the mouth, nose, reproductive organs, and others [14–16].

It is noted that when the response of the mucosal immune system is effective, molecules known as secretory antibodies are generated, which are released into the vacuoles of the orifices to resist the attack of pathogenic organisms that they find. An effective interaction may occur that activates the immune system in the body cells and thus kills the attacking pathogen. As it is known, vaccines injected into muscles avoid the mucous membranes, so the immune response to these membranes is weak, while edible vaccines come into contact with the internal walls of the digestive system, so it is assumed that they activate both the response of the mucous membranes and the systemic immunity in the body. It is assumed that this dual effect provides protection against dangerous microorganisms, especially those that cause diarrheal diseases, which prompted researchers to prioritize their research in combating diarrhea causes first, then other pathogens, especially Norwalk Rotavirus [17, 18], against the *Escherichia coli* bacteria [19] that secrete internal toxins causing what is known as traveler's diarrhea, which leads to the death of nearly three million children annually in third world countries, and *Vibrio cholerae* (the bacteria that cause cholera) [20].

In fact, ideas began to circulate among many researchers from different countries of the world since 1995. For example, a gene encoding a protein was isolated from the virus that causes hepatitis B virus (causing liver damage and liver cancer) and transferred to the tobacco plant, which stimulated this plant to protein manufacturing. After injecting the antigen into the mice, it led to the activation of the components of the immune system of the mice to the same extent as what happens when they are infected with the hepatitis virus [21].

3. Methods of preparing the edible vaccine

Several methods exist for genetically modifying plants to obtain edible vaccines, such as gen gun, vector system (bacteria), chimeric viruses, and electroporation (**Figure 1**).

The most important method adopted in the production of an edible vaccine is the vector carrier method, which depends on the *Agrobacterium tumefaciens* bacteria as an intermediate vector in the transfer of the genetic material (antigen proteins) from a virus or other bacteria to the target plant, which is embodied in the immune response of the organism after consuming calculated amounts of the antigen-bearing plant. This process is performed using plant tissue culture technology [22, 23]. It is also

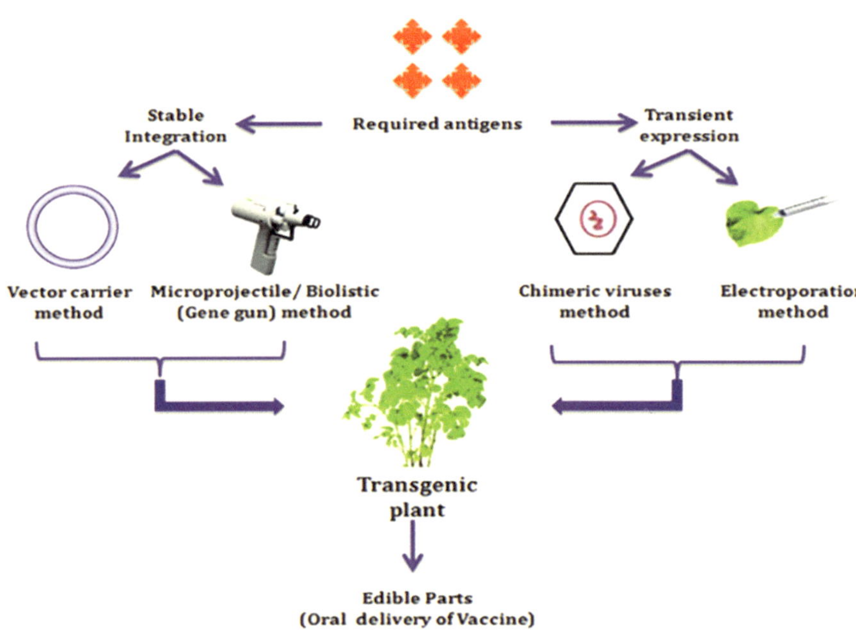

Figure 1.
Schematic representation of various methods for developing an edible plant vaccine.

possible to insert the desired DNA into the plant genome by direct methods in plant genetic engineering. Perhaps it is easier to use the bombardment method by means of an gene gun after shooting it in the cultures of plant embryonic cell suspensions, because they are specialized into embryos, and it is easy to grow into a complete plant carrying the desired gene.

Regardless of how the plant is modified, the desired DNA randomly pairs with the plant's genome, producing different levels of antigen expression that differ from one plant to another. Thus, it is preferable to modify 50–100 plants at the same time, and each plant is considered a separate line, and through this number, the line (plant) that is more expressive of the antigen with less negative effects on the body is elected. The method of producing the edible vaccine using *A. tumefaciens* as an intermediate vector can be summarized in the following steps:

1. Vegetable leaves (leaves from the potato plant, for example, as shown in (**Figure 2**) are separated in good health condition, sterilized superficially, and cut into explants. When the cutting areas of the explant's edges increase, there is a greater chance of infecting them with bacteria carrying the required gene, and thus, the success of the genetic transformation process.

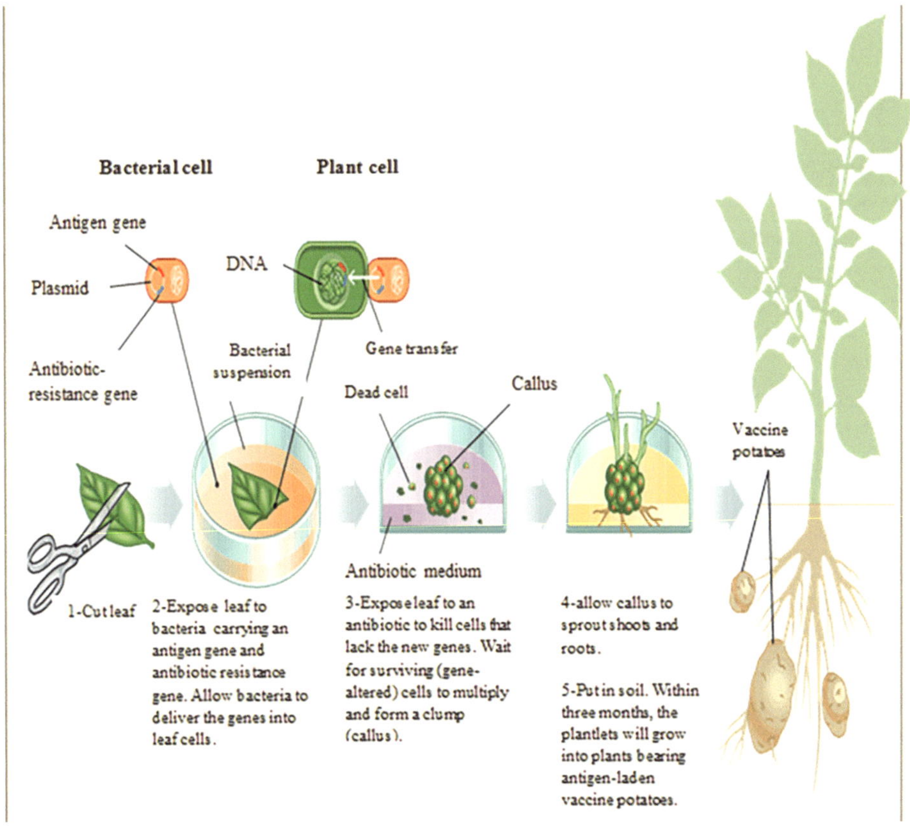

Figure 2.
Schematic showing the method of potatoes-edible vaccine production using Agrobacterium tumefaciens *as intermediate vector through plant tissue culture technique.*

2. Exposing the explants to bacteria (*A. tumefaciens*) carrying the antigen gene and the antibiotic resistance gene in a suitable culture medium which allows the bacteria carrying the two genes to deliver them to the genetic material (DNA) of the plant cell.

3. Exposing the plant cells to the antibiotic to kill the cells that do not carry the new genes and transferring the plant cells that contain the new genes to a suitable nutrient medium to stimulate the formation of callus in an appropriate size.

4. The induced callus mass is transferred to the regeneration medium to form shoots and roots, and the formed plantlets are separated.

5. The plants are acclimatized and transferred to the soil. After 3 months, plants bearing the antigen vaccine are produced. Its gene expression appears in potato tubers, which can be consumed as an edible vaccine.

Some of transgenic plants were invested to be consumed as edible vaccines, as shown in **Table 1**.

With the progress made in medical, agricultural, and pharmaceutical sciences, companies from different countries of the world produced vaccines from potatoes, tomatoes, lettuce, spinach, white clover, and Arabidopsis, where these plants were used as hosts for the production of vaccines [30, 31]. There has also been

Crop	Disease to be treated	Gene expression	Product	Reference
Viral vectors in tobacco	Non-Hodgkin's lymphoma	Antibody	Parts of antibody in single and various chains	[24]
Genetically modified tobacco	Tooth decay	Antibody	CaroRx	[25]
Genetically modified yellow corn and potato tubers	Diarrhea	Vaccine	Thermally stable toxins of *E. coli*	[2, 5]
Genetically modified yellow corn	Cystic fibrosis, pancreatitis	Therapeutic enzyme	Gastric lipase	[16]
Genetically modified potatoes and lettuce	Viral hepatitis type B	Vaccine	Hepatitis B virus surface antigen	[26]
Transgenic Arabidopsis	Vitamin B12 deficiency	Food	Human intrinsic factor	[27]
Genetically modified yellow corn	Intestinal inflammations	Food	Lactoferrin	[28]
Genetically modified potato	Norwalk virus	Vaccine	Norwalk virus capsid protein	[18]
Transgenic Arabidopsis	Diabetes	Hormone	Insulin	[29]
Transgenic rice	Diarrhea	Food	Lactoferrin, lysozyme, human serum albumin	[16]

Table 1.
Some pharmaceutical substances that were derived from plants to treat some human diseases.

progress in the use of plant species that are not edible plants but are medicinal plants and easy to handle in the laboratories for this purpose, such as Aloe vera plant [32], Neem plant [33], and *Chlamydomonas reinhardtii* green algae [34]. In general, each plant has its advantages and disadvantages if it is used as an edible vaccine.

4. Advantages of edible plant vaccines

The most important advantages of plants that make them candidates for the production of edible vaccines:

1. Must they have a long shelf life, as large-scale storage methods such as refrigerated storage are not required, which makes it possible to preserve the plant or the edible part of it without spoiling for a long time.

2. They are characterized by their rapid grow ability for producing them in large quantities, and they are cheap for the purpose of purchasing them by the consumer, because some fruits that grow on trees take a longer time to grow, which makes them expensive accordingly.

3. Easy for genetic transformation of most crops that grow as native or local crops in their regions, which facilitates the possibility of producing edible vaccines and the ease of transportation and distribution to the consumer.

4. The edible vaccines can be taken by eating the plant or part of it without the urgent need to purify or process it.

5. Stimulating the immune response on the mucous surfaces lining the mouth (mucosal immunity), which is the first line of defense in the body.

5. Disadvantages of edible plant vaccines

1. Will the antigen be able to survive the acidic conditions of the host stomach? and if it succeeds in that, will it be able to stimulate the immune system in the right way? Although initial trials have shown promising results in humans, it is not clear what will happen when a person who has eaten an edible vaccine comes into contact with the actual virus or pathogen.

2. The most difficult task remains in how to adjust the dose of the edible vaccine, as there may be a risk that a dose that is too high can evoke oral tolerance to invading bacteria or viruses rather than an immune response against them. In addition, the dosage requirements for children and adults are different, so research is continuing to find solutions to these problems.

3. The availability of limited knowledge regarding plant biotechnology leads to negative public opinion and strict regulations, thus discouraging future investment in the pharmaceutical business to produce edible vaccines.

Plant species	Merits	Demerits	References
Tobacco (*Nicotiana tabacum*)	a. A good model for the evaluation of recombinant proteins. b. The plant can be preserved at a low cost, as it gives large quantities of seeds with the possibility of storing them for a long time. c. Ease of purification of antibodies from seeds. d. It gives a large biomass and is harvested several times a year.	It produces toxic compounds	[2, 15]
Potato (*Solanum tuberosum*)	a. One of the most used crops in laboratory experiments in terms of ease of handling and genetic modification. b. Easy to propagate with buds and tubers. c. Easy to store for long periods with little refrigeration needs.	It needs to be cooked, and this may lead to denature the antigen and reduce its effectiveness for immunization.	[12, 35]
Banana (*Musa sapeintum*)	a. It does not need cooking, and its proteins do not deteriorate even when cooking. b. Relatively cheap and grown in poor countries.	a. The tree needs 2–3 years to reach maturity. b. Transgenic trees need 12 months to bear fruit. c. The crop deteriorates quickly after harvesting, so the need for cold is an expensive process. d. It contains small amounts of protein, so it is difficult to produce large quantities of recombinant proteins.	[11, 36, 37]
Tomato (*Solanum lycopersicum*)	a. It is characterized by rapid growth, so it is grown in a wide range of environmental and geographical conditions. b. Containing vitamins C and A, which increase the immune response. c. The rapid spoilage of fruits can be avoided by freeze-drying techniques. d. The antigen is thermally stable. Tomato powder containing the antigen can be prepared in the form of powder after being freeze-dried, and it can also be made in the form of capsules. e. The possibility of mixing a group of antigens to be given as immunization doses from different diseases.	The fruits spoil quickly.	[1, 11]

Plant species	Merits	Demerits	References
Rice (*Oryza sativa*)	a. It is often used with baby food because it is non-allergic. b. Antigen with high expression of proteins.	Its growth is slow and it consumes large amounts of water.	[11, 36]
Lettuce (*Lactuca sativa*)	a. It grows fast and consumes quickly. b. It has a large biomass and can be grown in more than one season and in different geographical areas.	It cannot be transported over long distances after being genetically modified due to the possibility of its spoilage.	[5]
Carrot (*Daucus carota*)	a. Ease of cultivation with genetic modification and it consumed raw. b. Rich in antioxidants and vitamin A, which strengthens people with weak immunity in resisting the pathogen.	It needs good storage to ensure not denaturation of antigens.	[5, 12]
Algae (*Chlamydomonas reinhardtii*)	a. Having high rates of growth. b. The entire body of algae can be transformed genetically.	The production process is very expensive due to the use of bioreactors for fast growth of algae.	[34]
Pea (*Pisum sativum*)	a. It is characterized as a delicious food consumed by children and adults. b. The plant is seasonal, which helps to genetically engineer it and produce fruits in a short period of time. c. High protein content in seeds.	It needs a cooking process before being consumed as food, which reduces its immunogenicity.	[12]

Table 2.
Merits and demerits of plant species used as edible vaccines.

4. Plants are living organisms that change in terms of growth and response to different environmental conditions, so there may not be a guaranteed continuity here for the production of a required edible vaccine.

6. Most important plant species used as edible vaccines

As a result of the successes achieved from the expression of genes introduced into plants or their parts that are eaten, many plants that have been genetically modified have been produced and tested. **Table 2** shows the most important of these plants, their merits and demerits.

7. Conclusion

It is clear from the above that for many decades, traditional vaccines were the important factors in stimulating the human immune system to resist many diseases that affect him or to become immune to any vectors of diseases, whether viruses or bacteria. However, on the other side, the manufacture of these vaccines required a lot of effort, research, and the high cost of production, storage, and distribution around the world, as well as the occurrence of some side complications for these vaccines.

This led many researchers in the past two decades to find alternative solutions by creating the idea of edible vaccines and heading toward achieving this through the production of edible plant vaccines, which include edible plant parts from fruits, seeds, or plant products to be on hand for consumption by people as food. On the one hand, to stimulate human immunity to resist diseases or pathogens by genetically modifying these plants to contain antigens that fight diseases, in addition to making them easy to cultivate, store, and distribute worldwide as plant products. However, the limitations on this issue are that the concept is new and is not widely accepted at present in developing countries, and the opposition to transgenic plants by injecting them with special genes to make them edible vaccines. This needs to be conducted in many studies in the future.

Author details

Siham A. Salim
Department of Biology, College of Education, Al-Iraqia University, Baghdad, Iraq

*Address all correspondence to: dr.sihamabdalrazzaq22@yahoo.com

References

[1] Saxena J, Rawat S. Edible vaccines. Chapter 12. In: Ravi et al., editors. Advances in Biotechnology. 2014. pp. 207-226

[2] Kurup VM, Thomas J. Edible vaccines: Promises and challenges. Molecular Biotechnology. 2020;**62**(2):79-90. DOI: 10.1007/s12033-019-00222-1

[3] Arntzen CJ. Edible vaccines. Public Health Reports. 1997;**112**(3):190-197

[4] Mishra N, Gupta PN, Khatri K, Goyal AK, Vyas SP. Edible vaccines: A new approach to oral immunization. Indian Journal of Biotechnology. 2008;**7**(3):283-294

[5] Concha C, Canas R, Macuer J, Torres MJ, Herrada AA, Jamett F, et al. Disease prevention: An opportunity to expand edible plant-based vaccines. Vaccine. 2017;**5**(14):1-23. DOI: 10.3390/vaccines5020014

[6] De Silva GO, Aponso MM, Abeysundara AT. A review on edible vaccines: A novel approach to oral immunization as a replacement of conventional vaccines. International Journal of Food Sciences and Nutritio. 2017;**2**(1):19-22

[7] Lee JH, Ko K. Production of recombinant anti-cancer vaccines in plants. Biomolecules & Therapeutics. 2017;**25**(4):345-353. DOI: 10.4062/biomolther.2016.126

[8] Miller E, Moro PL, Cano M, Shimabukuro T. Deaths following vaccination: What does the evidence show. Vaccine. 2015;**33**(29):3288-3292. DOI: 10.1016/j.vaccine.2015.05.023

[9] Laere E, Ling APK, Wong YP, Koh RY, Lila AM, Hussein S. Plant-based vaccines: Production and challenges. Journal of Botany. 2016;**2016**(10):1-11. DOI: 10.1155/2016/4928637

[10] Chaitanya VK, Kumar JU. Edible vaccines: Short review. Sri Ramachandra Journal of Medicine. 2006;**1**(1):33-34

[11] Razzak KS, Jain D, Hossain MN, Bushra A. Plant edible vaccines: A natural way of vaccination. Vigyan Varta. 2020;**1**(5):21-24

[12] Karakas I, Tonk FA. Plants that can be used as plant-based edible vaccines, current situation and recent developments. Virology & Immunology Journal. 2022;**6**(3):1-10. DOI: 10.23880/vij-16000302

[13] Kaya H, Özdemir M. Chapter 2. Vaccine Technologies and Domestic Vaccines. In: Preventive Medicine. 2021. p. 18

[14] Ma JK, Drake PM, Christou P. The production of recombinant pharmaceutical proteins in plants. Nature Reviews Genetics. 2003;**4**(10):794-805. DOI: 10.1038/nrg1177

[15] Bhatia S, Dahiya R. Edible vaccines. In: Modern Applications of Plant Biotechnology in Pharmaceutical Sciences. Chapter 9. London, UK: Academic Press; 2015. pp. 333-343

[16] Gunasekaran B, Gothandam KM. A review on edible vaccines and their prospects. Brazilian Journal of Medical and Biological Research. 2020;**53**(2):e8749. DOI: 10.1590/1414-431X20198749

[17] Tacket CO, Mason HS, Losonsky G, Clements JD, Levine MM, Arntzen CJ. Immunogenicity in humans

of a recombinant bacterial antigen delivered in a transgenic potato. Nature Medicine. 1998;**4**(5):607-609. DOI: 10.1038/nm0598-607

[18] Jiang X, Wang M, Graham DY, Ester MK. Expression, self-assembly, and antigenicity of the Norwalk virus capsid protein. Journal of Virology. 1992;**66**(11):6527-6532. DOI: 10.1128/jvi.66.11.6527-6532.1992

[19] Lal P, Ramachandran VG, Goyal R, Sharma R. Edible vaccine: Current status and future. Indian Journal of Medical Microbiology. 2007;**25**(2):93-102. DOI: 10.4103/0255-0857.32713

[20] Suleiman ABJ. Transformation of salt tolerant *Lactuca sativa* with cholera toxin B gene for production of edible vaccine [PhD dissertation]. College of Science, Al-Nahrain University, Iraq: CRC Press, Woodhead Publishing Ltd; 2012

[21] Bhat SR. Plant transformation for direct genetic change. In: Chopra VL, editor. Plant Breeding: Theory and Practice. 2nd ed. New Delhi, India: Oxford and IBH Publishing Co. Pvt. Ltd.; 2000. pp. 439-459

[22] Langridge WHR. Edible vaccines. Scientific American. 2000;**283**(3):66-71. DOI: 10.1038/scientificamerican0900-66

[23] Hirlekar R, Bhairy S. Edible vaccines: An advancement of oral immunization. Asian Journal of Pharmaceutical and Clinical Research. 2017;**10**(2):71-77

[24] Mccormick A. Tobacco derived cancer vaccines for non-Hodgkin's lymphoma perspectives and progress. Human Vaccines. 2011;7(3):305-312. DOI: 10.4161/hv.7.3.14163

[25] Weintraub JA, Hilton JF, White JM, Hoover CI, Wycoff KL, Yu L, et al. Clinical trial of plant-derived antibody on recolonization of mutant streptococci. Caries Research. 2005;**39**(3):241-250. DOI: 10.1159/000084805

[26] Sobrino F, Saiz M, Jimenez-Clavero MA, Nuaez JI, Rosas MF, Baranowski E, et al. Foot and mouth disease virus: A long known virus, but a current threat. Veterinary Research. 2001;**32**(1):1-30. DOI: 10.1051/vetres:2001106

[27] Fedosov SN, Laursen NB, Nexo E, Moestrup SK, Petersen TE, Jensen EO, et al. Human intrinsic factor expressed in the plant *Arabidopsis thaliana*. European Journal of Biochemistry. 2003;**270**(16):3362-3367. DOI: 10.1046/j.1432-1033.2003.03716.x

[28] Samyn-Petit B, Gruber V, Flahaut C, Wajda-Dubos JP, Farrer S, Pons A, et al. N-glycosylation potential of maize: The human lactoferrin used as a model. Glycoconjugate Journal. 2001;**18**(7):519-527. DOI: 10.1023/a:1019640312730

[29] Nykiforuk CL, Boothe JG, Murray EW, Keon RG, Goren HJ, Markley NA, et al. Transgenic expression and recovery of biologically active recombinant human insulin from *Arabidopsis thaliana* seeds. Plant Biotechnology Journal. 2005;**4**(1):77-85. DOI: 10.1111/j.1467-7652.2005.00159.x

[30] Rigano MM, Alvarez ML, Pinkhasov J, Jin Y, Sala F, Arntzen CJ. Production of a fusion protein consisting of the enterotoxigenic *Escherichia coli* heat-labile toxin B subunit and a tuberculosis antigen in Arabidopsis *thaliana*. Plant Cell Reports. 2004;**22**(7):502-508. DOI: 10.1007/s00299-003-0718-2

[31] Merlin M, Pezzotti M, Avesani L. Edible plants for oral delivery of biopharmaceuticals. British Journal of Clinical Pharmacology. 2017;**83**(1):71-81. DOI: 10.1111/bcp.12949

[32] Kumar S, Tiku AB. Immunomodulatory potential of acemannan (polysaccharide from Aloe vera) against radiation induced mortality in Swiss albino mice. Food and Agricultural Immunology. 2016;**27**(1):72-86

[33] Roy S, Bhattacharyya P. Possible role of traditional medicinal plant neem (*Azadirachta indica*) for the management of COVID-19 infection. International Journal of Pharmaceutical Sciences and Research. 2020;**1**(11):122-125

[34] Patel P, Patel R, Patel S, Patel Y, Patel M, Trivedi R. A nutritional substitute for traditional immunization. Pharmacognosy Reviews. 2022;**16**(32):62-69. DOI: 10.5530/phrev.2022.16.9

[35] Evers D, Deusser H. In: Bouayed J, editor. Potato Antioxidant Compounds: Impact of Cultivation Methods and Relevance for Diet and Health. London, UK: IntechOpen; 2012. ISBN:978-953-51-0125-3. Available from: http://www.intecopen.com/books/

[36] Aryamvally A, Gunasekaran V, Narenthiran KR, Pasupathi R. New strategies toward edible vaccines: An overview. Journal of Dietary Supplements. 2017;**14**(1):101-116. DOI: 10.3109/19390211.2016.1168904

[37] Qureshi MB, Arif S, Rathore SS. Edible plant vaccines: A step towards revolution in the field of immunology. International Journal of Agriculture and Biology. 2023;**29**(5):361-369. DOI: 10.17957/IJAB/15.2041

Chapter 10

Boswellia Carries Hope for Patients with Inflammatory Bowel Disease (IBD)

Sally Elnawasany

Abstract

Boswellia serrata is an ancient and valuable herb that was widely used throughout the centuries. *Boswellia* trees grow in India, Northern Africa, and the Middle East from which Frankincense or olibanum resin is taken. The beneficial effects of *Boswellia* and its active ingredients (Boswellic acids) were thoroughly investigated in many diseases. Where the non-redox and 5-lipoxygenase inhibitory actions were reported. Inflammatory bowel disease (IBD) mainly ulcerative colitis (UC) and Crohn's disease (CD) are chronic inflammatory disorders of the gastrointestinal system. Although the cause is still unclear, the immune system is claimed to have the upper hand in the pathogenesis of IBD. Several studies have demonstrated the ameliorating effect of Boswellic acids on the severity of IBD and the potential role of *Boswellia* in the induction or maintenance of remission. The aim of this chapter is to explore the the possible effect of *Boswellia* in IBD management as a complementary and alternative strategy.

Keywords: *Boswellia*, complementary and alternative medicine, inflammatory bowel disease, ulcerative colitis, Crohn's disease

1. Introduction

Inflammatory bowel disease (IBD) is a chronic inflammatory state classified mainly into ulcerative colitis (UC) and Crohn's disease (CD). IBD is associated with abdominal pain, diarrhea and rectal bleeding [1]. The choice of treatment for IBD varies according to the type and severity. A variety of medications such as aminosalicylates, corticosteroids, immunosuppressive and biologic drugs are involved [2]. Despite their crucial therapeutic role, some medications carry the risk of infection and cancer [3]. In addition, these drugs are taken throughout life, which leads to patient incompliance and hence treatment failure [4]. These obstacles in treatment insist on the search for other treatments, safe and effective as alternative and complementary modalities.

2. Pathogenesis of IBD

The exact mechanism of IBD is not completely clear [5]. The interaction between genetic factors, changes in intestinal flora homeostasis, environmental variabilities, and intestinal hyperimmune response results in chronic intestinal inflammation [6]. Multiple inflammatory mediators are involved in IBD pathogenesis including leukotrienes, cytokines, chemokines, and prostaglandins. Alteration in reactive oxygen and nitrogen species production adds to the pathogenesis [7]. Based on this, attenuation of hyper-stimulated immune response is the target of IBD therapy. Where the treatment passes in 2 ways, the first is to induce remission and the second is to maintain the remission and to ameliorate intestinal chronic inflammation [8]. In normal conditions, there is a balance among many cells in intestinal lamina propria such as macrophages mast cells, neutrophils, dendritic cells (DCs), eosinophils, natural killer (NK), NKT cells, T and B cells. This provides intestinal protection and tolerance. As a response to bacterial infection, the innate immunity cells (macrophages and dendritic cells) upregulate chemokines and cytokines and act as Antigen-presenting cells (APCs) where they feature the microorganism's molecular patterns via toll-like receptors (TLR) [9, 10]. Dendritic cells generate native T cell differentiation in mesenteric lymph nodes [11]. With subsequent generation of T helper (Th) subtypes according to the cytokines produced by APCs [12–15]. Besides proinflammatory cytokines induction, neutrophils stimulate oxidative reactions in the intestinal mucosa [16]. In IBD activity, there is over-expression of many chemokines including macrophage inflammatory proteins (MIP), and Interleukin-8 (IL-8). Under the control of these chemokines, leukocytes are recruited to the inflamed intestine with subsequent oxidative stress [17]. Recruitment of granulocytes and lymphocytes is mediated by adhesion molecules in IBD, such as the intercellular adhesion molecule-1 (ICAM-1), the vascular cell adhesion molecule- (VCAM-) 1, P and E-selectins [18]. In addition, T-cell differentiation and regulation are mediated by cytokines. Th1 differentiation is controlled by IL-12, IL-18, and IL-23 while TNF-α, IL-1β, and IL-6, magnify the immune response by releasing more chemokines and attracting more inflammatory cells [19]. On the other hand, the under-production of IL-10 and Transforming growth factor beta (TGF-β), which are inflammatory attenuating cytokines, contributes to IBD pathogenesis [20, 21].

3. Complementary and alternative (CAM) treatment of IBD

There is a growing worldwide interest in complementary and alternative remedies in IBD treatment [22, 23]. The use of CAM is common for IBD children and adult patients [24, 25]. CAM is found to be commonly used among young, females, at a high educational level or with medication adverse effects [26, 27]. Patients who received massive corticosteroid therapy [28] or suffering from extraintestinal manifestations [24] are more inclined to CAM remedies, as well. Many alternative modalities have been tried in IBD patients including herbs, probiotics, acupuncture and hypnotherapy [8]. Phytochemicals are popularly utilized because of their safety and effectiveness on IBD patients [4, 29]. The variable active herbal ingredients which act on multiple inflammatory pathways and mediators support this preference [25]. *Aloe vera*, *Artemisia absinthium*, *Boswellia serrata* and *Curcuma longa* were widely studied for their effect on IBD [30].

4. *Boswellia serrata*

4.1 Structure of *Boswellia serrata*

Boswellia serrata oleo-gum resin, Indian frankincense was widely used for centuries in traditional medicine. Antioxidant and anti-inflammatory actions have been extensively investigated in several studies on different diseases like colitis, bronchial asthma, arthritis and malignancies [31–35]. *Boswellia serrata* resin is composed of monoterpenes, diterpenes, triterpenes, pentacyclic triterpenic acids (boswellic acids) and tetracyclic triterpenic acids [36–39]. Boswellic acids (BA) compose up 30% of the resin of *Boswellia serrata*. They are organic acids, formed of a pentacyclic triterpene, a carboxyl group and at least one other functional group [40]. Among boswellic acids, 11-keto-β-boswellic acid (KBA) and acetyl-11-keto-β-boswellic acid (AKBA) are the most active [41].

4.2 Pharmacological activities of *Boswellia serrata*

4.2.1 Anti-inflammatory action

Boswellia has variable pharmacological activities, anti-inflammatory properties were widely investigated in many studies. Acetyl-boswellic acids block leukotriene production through the downregulation of enzyme 5-lipoxygenase (5-LOX) mediated by a non-redox reaction [42, 43]. In a double-blind placebo control clinical study alcohol extract was given in a 300 mg thrice daily dose for 6 weeks. 70% of asthmatic patients gained clinical improvement [44]. Similarly, gradual control of asthma, regarding the frequency of attacks, pulmonary function tests improvement, and lowering levels of leukotrienes were obtained by another study [45]. In addition, *Boswellia* ameliorated the inflammation in arthritis [46–48]. Anti-anaphylactic and mast cell stabilizing effects were also reported where *Boswellia* suppressed mast cell degranulation [49]. Moreover, boswellic acids were found to possess anti-complement activity [50]. Roy et al. studied the genetic basis of anti-inflammatory effect of BA. Tumor necrosis factor alpha (TNFa) is one of the most crucial mediators of inflammation. TNF alpha induces inflammation by multiple mechanisms, one of them is by upregulation of the expression of adhesion molecules such as microvascular cellular adhesive molecul-1, VCAM-1 in a system of TNF alpha-induced gene expression in human microvascular endothelial cells (HMEC). Of 522 genes that were induced by TNFa 113 genes were highly sensitive to BE treatment both in vivo and in vitro. The function of these genes is linked to inflammation, and cell adhesion [51].

4.2.2 Anti-microbial action

Boswellia oils showed anti-microbial activity against five organisms. Minimum inhibition concentration ranged from 4 to 16 mg/ml against *Staphylococcus aureus*, 1.5–8.3 mg/ml against *Bacillus cereus*, 4.0–12.0 mg/ml against *Escherichia coli*, 2.0–12.8 mg/ml against *Proteus vulgaris* and 5.3–12.0 mg/ml against *Candida albicans* [52]. This effect may help in controlling intestinal infection in IBD management.

4.2.3 Anti-tumor action

Boswellic acids induced apoptosis through the upregulation of caspase-8 in colon cancer HT-29 cells [53]. In another study, *Boswellia* extract altered DNA methylation in colon cancer cells [54]. 4-Amino analogues prepared from β-boswellic acid and 11-keto-β-boswellic acid showed an apoptotic activity mediated by DNA fragmentation [55]. There are multiple pathways by which *Boswellia* exerts its anti-tumor action such as suppression of topoisomerases I and II [56]. It was also revealed that acetyl keto beta boswellic acid (AKBA) inhibits phosphorylation of ERK pathways and consequently, impairs signal transduction and tumorigenesis [57]. Oxidative stress and nitric oxide production were mediated by *Boswellia* with subsequent apoptosis in human leukemia HL-60 cells [58]. This anti-tumor potential of *Boswellia* protects against colorectal cancer which is a common sequala of IBD [59].

4.2.4 Hepato-protective action

In the models of liver injury, hexane extract of oleo-gum-resin of *Boswellia* in lower doses (87.5 mg/kg p.o.) reduced marker enzymes and prevented the increase in liver weight with histological evidence of hepatoprotection while a mild effect was obtained by higher dose (175 mg/kg p.o.) [60].

4.2.5 Anti-lipidemic action

The aqueous extract of *Boswellia carterii* with other herbs improved the lipid profile of alloxan-induced diabetic rats [61]. *Boswellia* showed therapeutic potential for metabolic syndrome. Where *Boswellia* succeeded in lowering the lipid profile by decreasing the level of TNF-α, IL-1β and increasing the adiponectin level. This action is based on its antioxidant activity [62].

4.2.6 Hypoglycemic action

Herbal formulation containing *B. serrata* oleo-gum-resin induced a significant anti-diabetic activity on non-insulin-dependent diabetes mellitus [63]. Furthermore, a significant reduction in blood glucose levels and HbA1c was observed when *Boswellia serrata* was introduced for 17 days to diabetic rats in another study [64].

Based on the mentioned pharmacological actions, *Boswellia* can help in IBD in many ways, 5-lipoxygenase (5-LO) suppression, downregulation of Tumor necrosis factor alpha (TNF-α) and interleukins, P-selectin-mediated recruitment of inflammatory cells, decreasing reactive oxygen species (ROS), and by modulation intestinal motility [65–68].

4.3 Preclinical studies of *Boswellia* in IBD

In vitro study demonstrated the role of *Boswellia* in the suppression of leukotriene synthesis through interfering with 5-lipoxygenase pathway [69]. The anti-inflammatory effects of boswellic acids extended to inhibit the nuclear transcription factor kappa B (NF-κB) activation, this factor expresses and potentiates the proinflammatory cytokines including TNFα, IL-1β, and IL-6 [70, 71]. This effect was confirmed in an in vitro experimental model of intestinal inflammation where pretreatment of Caco-2 cells monolayers by *Boswellia serrata* oleo-gum extract (BSE) and AKBA abolished nuclear factor kappa B (NF-κB) activation, protected against cellular changes and inhibited reactive oxygen species (ROS) [72]. Furthermore, the anti-inflammatory and antioxidant properties of *Boswellia serrata* were explored in acetic acid (AA) induced UC rat model. Daily administration of 34.2 mg/kg of *Boswellia serrata* extract pre- and post-induction of colitis significantly improved tissue lesions, decreased lipid peroxidation and nitric oxide [73]. A semisynthetic form of acetyl-11-keto-β-boswellic acid ameliorated the disease activity and histology in dextran sodium sulfate (DSS) induced murine colitis. This action was mediated by attenuation of adherent leukocytes and platelets into inflamed tissue by blocking P-selectin stimulation [66]. In an in vitro model of intestinal inflammation, the anti-inflammatory action of *Boswellia serrata* and *C. longa* were tested. *Boswellia serrata* at 1 μg/mL protected the intestinal epithelium with a 25% reduction of ROS generation [74]. The immunosuppressive potential of *Boswellia carteri* gum resin extract was illuminated in an in vitro study, where the extract attenuated human primary T lymphocyte proliferation in a concentration-dependent manner via nuclear factor of activated T-cells (NFAT) dependent mechanism [75]. Metabolism of sphingomyelin induces lipid signals that impact cell proliferation, and inflammation. 3-acetyl-11-keto-β-boswellic acids (AKBA) attenuated the expression of sphingomyelinase in intestinal cells. This provides other anti-inflammatory mechanism of *Boswellia* [76]. A new herbal formulation that includes Punica granatum L, *Boswellia* serrata, and Curcuma longa L extracts, inhibited TNF-alfa-induced release of IL-8, IL-6 and Monocyte chemoattractant protein-1 (MCP-1) in Caco-2 cells [77]. Although many studies support the anti-inflammatory action of *Boswellia*, on the contrary, Kiela et al. claimed that *Boswellia* showed no effect on improving colitis in dextran sulfate sodium (DSS)- or trinitrobenzene sulfonic acid- (TNBS-) induced experimental models [78]. Antifibrotic effect of *Boswellia* was declared when oral *Boswellia* in combination with significantly improved the inflammation of trinitrobenzene sulphonic acid (TNBS)-induced chronic colitis. There was an improvement in the histological features of colonic fibrosis. Together with a significant reduction in the expression of alpha-smooth muscle actin (α-SMA), collagen I–III, connective tissue growth factor (CTGF) and transforming growth factor-beta1 (TGF-β1) [79].

4.4 Clinical studies of *Boswellia* in IBD

Boswellia serrata in a dose of 350 mg thrice daily for 6 weeks achieved remission in 82% of treated UC patients compared to 75% of patients who were treated with sulfasalazine (1 g thrice daily). Inflammatory parameters were better in *Boswellia* treated group as well [80]. Another trial was conducted on 30 patients with chronic colitis for 6 weeks. Of 20 patients who received daily 900 mg *Boswellia serrata,* improvement of inflammatory parameters and remission were noticed in (18 and 14 patients, respectively) while after treatment of 10 patients with 3 gm sulfasalazine daily 6 participants showed improvement in inflammatory parameters and remission was achieved in 4 patients [81]. *Boswellia* tolerability and ability to maintain remission were demonstrated in a 52-week multicenter double-blind, placebo-controlled, randomized Germain study. 82 CD patients were randomly divided, 42 patients received daily *Boswellia* in 400 mg capsules and 40 patients received a placebo. There was no difference in both groups in parameters of inflammation or disease activity or maintaining remission and *Boswellia* was well tolerated [82]. Another 4-week trial was conducted on UC patients who were in remission. *Boswellia serrata* extract (BSE) was introduced orally, in a novel delivery form to 22 patients compared to 21 patients with no treatment. Improvement in clinical parameters, a decrease in medication needs and a lowering in fecal calprotectin levels were observed in *Boswellia* treated group [83]. The effect of *Boswellia* was tried in patients with collagenous colitis as well, where 400 mg oral BSE was given thrice daily for six weeks compared to a placebo. The remission rate was higher in BSE-treated patients but without any changes in histology or quality of life [34]. We conducted a 6-week clinical trial on 60 patients with active ulcerative colitis to investigate the effect of *Boswellia* extract on disease activity. 20 patients received Mesalamine 3, 20 patients received *Boswellia* extract in the form of oral tablets in a dose of 2 gm/day, and 20 patients were given *Boswellia* extract plus Mesalamine in the mentioned doses. Clinical and laboratory improvement was noticed in the three groups without a significant difference. There were no recorded side effects of *Boswellia* during the 6 weeks of the study [84].

5. Conclusion

Based on preclinical and clinical studies, the anti-inflammatory, immune modulation, and anti-cancer activities of *Boswellia* as well as its safety and tolerability recommend its therapeutic use in IBD patients.

Acknowledgements

I thank God Almighty for his countless blessings, including *Boswellia*, and I thank my patients for their patience and trust. I extend my gratitude to Professor Dr. Farid Badria for his continuous support.

Conflict of interest

I confirm that there are no conflicts of interest.

Author details

Sally Elnawasany[1,2]

1 Tanta University, Egypt

2 Al-Rayan College, KSA

*Address all correspondence to: elnawasany_s@hotmail.com

References

[1] Baumgart DC, Sandborn WJ. Inflammatory bowel disease: Clinical aspects and established and evolving therapies. The Lancet. 2007;**369**(9573):1641-1657

[2] Seyedian SS, Nokhostin F, Malamir MD. A review of the diagnosis, prevention, and treatment methods of inflammatory bowel disease. Journal of Medicine and Life. 2019;**12**(2):113-122

[3] Toruner M, Loftus EV Jr, Harmsen WS, Zinsmeister AR, Orenstein R, Sandborn WJ, et al. Risk factors for opportunistic infections in patients with inflammatory bowel disease. Gastroenterology. 2008;**134**(4):929-936

[4] Ng S, Lam Y, Tsoi K, Chan F, Sung J, Wu J. Systematic review: The efficacy of herbal therapy in inflammatory bowel disease. Alimentary Pharmacology & Therapeutics. 2013;**38**(8):854-863

[5] Haag L-M, Siegmund B. Exploring & exploiting our 'other self'–does the microbiota hold the key to the future therapy in Crohn's? Best Practice & Research Clinical Gastroenterology. 2014;**28**(3):399-409

[6] Strober W, Fuss IJ. Proinflammatory cytokines in the pathogenesis of inflammatory bowel diseases. Gastroenterology. 2011;**140**(6):1756-67. e1

[7] Xavier RJ, Podolsky DK. Unravelling the pathogenesis of inflammatory bowel disease. Nature. 2007;**448**(7152):427-434

[8] Bernstein CN. Treatment of IBD: Where we are and where we are going. Official Journal of the American College of Gastroenterology| ACG. 2015;**110**(1):114-126

[9] Mogensen TH. Pathogen recognition and inflammatory signaling in innate immune defenses. Clinical Microbiology Reviews. 2009;**22**(2):240-273

[10] Smith P, Smythies L, Shen R, Greenwell-Wild T, Gliozzi M, Wahl S. Intestinal macrophages and response to microbial encroachment. Mucosal Immunology. 2011;**4**(1):31-42

[11] Cerovic V, Houston S, Scott C, Aumeunier A, Yrlid U, Mowat A, et al. Intestinal CD103– dendritic cells migrate in lymph and prime effector T cells. Mucosal Immunology. 2013;**6**(1):104-113

[12] Usui T, Preiss JC, Kanno Y, Yao ZJ, Bream JH, O'Shea JJ, et al. T-bet regulates Th1 responses through essential effects on GATA-3 function rather than on IFNG gene acetylation and transcription. The Journal of Experimental Medicine. 2006;**203**(3):755-766

[13] Yang XO, Panopoulos AD, Nurieva R, Chang SH, Wang D, Watowich SS, et al. STAT3 regulates cytokine-mediated generation of inflammatory helper T cells. Journal of Biological Chemistry. 2007;**282**(13):9358-9363

[14] Acosta-Rodriguez EV, Napolitani G, Lanzavecchia A, Sallusto F. Interleukins 1β and 6 but not transforming growth factor-β are essential for the differentiation of interleukin 17–producing human T helper cells. Nature Immunology. 2007;**8**(9):942-949

[15] Wilson NJ, Boniface K, Chan JR, McKenzie BS, Blumenschein WM, Mattson JD, et al. Development, cytokine profile and function of human interleukin 17–producing helper T cells. Nature Immunology. 2007;**8**(9):950-957

[16] Seguí J, Gironella M, Sans M, Granell S, Gil F, Gimeno M, et al. Superoxide dismutase ameliorates TNBS-induced colitis by reducing oxidative stress, adhesion molecule expression, and leukocyte recruitment into the inflamed intestine. Journal of Leukocyte Biology. 2004;**76**(3):537-544

[17] Laing KJ, Secombes CJ. Chemokines. Developmental & Comparative Immunology. 2004;**28**(5):443-460

[18] Vainer B. Intercellular adhesion molecule-1 (ICAM-1) in ulcerative colitis: Presence, visualization, and significance. APMIS. 2010;**118**:1-46

[19] Sanchez-MuñozF,Dominguez-LopezA, Yamamoto-Furusho JK. Role of cytokines in inflammatory bowel disease. World Journal of Gastroenterology: WJG. 2008;**14**(27):4280

[20] Izcue A, Coombes JL, Powrie F. Regulatory lymphocytes and intestinal inflammation. Annual Review of Immunology. 2009;**27**:313-338

[21] Li MO, Flavell RA. Contextual regulation of inflammation: A duet by transforming growth factor-β and interleukin-10. Immunity. 2008;**28**(4):468-476

[22] Joos S, Rosemann T, Szecsenyi J, Hahn EG, Willich SN, Brinkhaus B. Use of complementary and alternative medicine in Germany–a survey of patients with inflammatory bowel disease. BMC Complementary and Alternative Medicine. 2006;**6**:1-7

[23] Hilsden RJ, Verhoef MJ, Best A, Pocobelli G. Complementary and alternative medicine use by Canadian patients with inflammatory bowel disease: Results from a national survey. The American Journal of Gastroenterology. 2003;**98**(7):1563-1568

[24] Fernández A, Barreiro-de Acosta M, Vallejo N, Iglesias M, Carmona A, González-Portela C, et al. Complementary and alternative medicine in inflammatory bowel disease patients: Frequency and risk factors. Digestive and Liver Disease. 2012;**44**(11):904-908

[25] Hilsden RJ, Verhoef MJ, Rasmussen H, Porcino A, DeBruyn JC. Use of complementary and alternative medicine by patients with inflammatory bowel disease. Inflammatory Bowel Diseases. 2011;**17**(2):655-662

[26] Opheim R, Hoivik ML, Solberg IC, Moum B, Group IS. Complementary and alternative medicine in patients with inflammatory bowel disease: The results of a population-based inception cohort study (IBSEN). Journal of Crohn's and Colitis. 2012;**6**(3):345-353

[27] Opheim R, Bernklev T, Fagermoen MS, Cvancarova M, Moum B. Use of complementary and alternative medicine in patients with inflammatory bowel disease: Results of a cross-sectional study in Norway. Scandinavian Journal of Gastroenterology. 2012;**47**(12):1436-1447

[28] Langhorst J, Anthonisen IB, Steder-Neukamm U, Lüdtke R, Spahn G, Michalsen A, et al. Amount of systemic steroid medication is a strong predictor for the use of complementary and alternative medicine in patients with inflammatory bowel disease. Results from a German national survey. Inflammatory Bowel Diseases. 2005;**11**(3):287-295

[29] Ke F, Yadav PK, Ju LZ. Herbal medicine in the treatment of ulcerative colitis. Saudi Journal of Gastroenterology: Official Journal of the Saudi Gastroenterology Association. 2012;**18**(1):3

[30] Algieri F, Rodriguez-Nogales A, Rodriguez-Cabezas ME, Risco S,

Ocete MA, Galvez J. Botanical drugs as an emerging strategy in inflammatory bowel disease: A review. Mediators of Inflammation. 2015;**2015**:179616

[31] Moussaieff A, Mechoulam R. Boswellia resin: From religious ceremonies to medical uses; a review of in-vitro, in-vivo and clinical trials. Journal of Pharmacy and Pharmacology. 2009;**61**(10):1281-1293

[32] Siddiqui MZ. Boswellia serrata, a potential antiinflammatory agent: An overview. Indian Journal of Pharmaceutical Sciences. 2011;**73**(3):255

[33] Ernst E. Frankincense: systematic review. British Medical Journal. 2008;**337**:a2813

[34] Madisch A, Miehlke S, Eichele O, Mrwa J, Bethke B, Kuhlisch E, et al. Boswellia serrata extract for the treatment of collagenous colitis. A double-blind, randomized, placebo-controlled, multicenter trial. International journal of colorectal disease. 2007;**22**:1445-1451

[35] Huang MT, Badmaev V, Ding Y, Liu Y, Xie JG, Ho CT. Anti-tumor and anti-carcinogenic activities of triterpenoid, β-boswellic acid. BioFactors. 2000;**13**(1-4):225-230

[36] El Khadem H, El-Shafei Z, El Sekeily M, Rahman MA. Derivatives of boswellic acids. Planta Medica. 1972;**22**(06):157-159

[37] Hamm S, Bleton J, Connan J, Tchapla A. A chemical investigation by headspace SPME and GC–MS of volatile and semi-volatile terpenes in various olibanum samples. Phytochemistry. 2005;**66**(12):1499-1514

[38] Safayhi H, Mack T, Sabieraj J, Anazodo MI, Subramanian LR, Ammon H. Boswellic acids: novel, specific, nonredox inhibitors of 5-lipoxygenase. Journal of Pharmacology and Experimental Therapeutics. 1992;**261**(3):1143-1146

[39] Mahajan B, Taneja S, Sethi V, Dhar K. Two triterpenoids from Boswellia serrata gum resin. Phytochemistry. 1995;**39**(2):453-455

[40] Ammon H. Boswellic acids (components of frankincense) as the active principle in treatment of chronic inflammatory diseases. Wiener Medizinische Wochenschrift (1946). 2002;**152**(15-16):373-378

[41] Abdel-Tawab M, Werz O, Schubert-Zsilavecz M. Boswellia serrata: An overall assessment of in vitro, preclinical, pharmacokinetic and clinical data. Clinical Pharmacokinetics. 2011;**50**:349-369

[42] Krieglstein CF, Anthoni C, Rijcken EJ, Laukötter M, Spiegel H-U, Boden SE, et al. Acetyl-11-keto-β-boswellic acid, a constituent of a herbal medicine from Boswellia serrata resin, attenuates experimental ileitis. International Journal of Colorectal Disease. 2001;**16**:88-95

[43] Ammon H, Safayhi H, Mack T, Sabieraj J. Mechanism of antiinflammatory actions of curcumine and boswellic acids. Journal of Ethnopharmacology. 1993;**38**(2-3):105-112

[44] Gupta I, Gupta V, Parihar A, Gupta S, Lüdtke R, Safayhi H, et al. Effects of Boswellia serrata gum resin in patients with bronchial asthma: Results of a double-blind, placebo-controlled, 6-week clinical study. European Journal of Medical Research. 1998;**3**(11):511-514

[45] Badria FA, Mohammed EA, El-Badrawy MK, El-Desouky M.

Natural leukotriene inhibitor from Boswellia: A potential new alternative for treating bronchial asthma. Alternative & Complementary Therapies. 2004;**10**(5):257-265

[46] Badria FA, El-Farahaty T, Shabana AA, Hawas SA, El-Batoty MF. Boswellia–curcumin preparation for treating knee osteoarthritis: A clinical evaluation. Alternative & Complementary Therapies. 2002;**8**(6):341-348

[47] Bannuru RR, Osani MC, Al-Eid F, Wang C. Efficacy of curcumin and Boswellia for knee osteoarthritis: Systematic review and meta-analysis. Seminars in Arthritis and Rheumatism. 2018;**48**(3):416-429

[48] Singh S, Khajuria A, Taneja SC, Khajuria RK, Singh J, Qazi GN. Boswellic acids and glucosamine show synergistic effect in preclinical anti-inflammatory study in rats. Bioorganic & Medicinal Chemistry Letters. 2007;**17**(13):3706-3711

[49] Pungle P, Banavalikar M, Suthar A, Biyani M, Mengi S. Immunomodulatory activity of boswellic acids of Boswellia serrata Roxb. Indian journal of experimental biology. 2003;**41**(12):1460-1462

[50] Kapil A, Moza N. Anticomplementary activity of boswellic acids--an inhibitor of C3-convertase of the classical complement pathway. International Journal of Immunopharmacology. 1992;**14**(7):1139-1143

[51] Roy S, Khanna S, Shah H, Rink C, Phillips C, Preuss H, et al. Human genome screen to identify the genetic basis of the anti-inflammatory effects of Boswellia in microvascular endothelial cells. DNA and Cell Biology. 2005;**24**(4):244-255

[52] Van Vuuren SF, Kamatou GP, Viljoen AM. Volatile composition and antimicrobial activity of twenty commercial frankincense essential oil samples. South African Journal of Botany. 2010;**76**(4):686-691

[53] Liu J-J, Nilsson Å, Oredsson S, Badmaev V, Zhao W-Z, Duan R-D. Boswellic acids trigger apoptosis via a pathway dependent on caspase-8 activation but independent on Fas/Fas ligand interaction in colon cancer HT-29 cells. Carcinogenesis. 2002;**23**(12):2087-2093

[54] Shen Y, Link A, Takahashi M, Balaguer F, Hur K, Boland CR, et al. Boswellia extracts induce DNA methylation changes in colon cancer cells. Gastroenterology. 2011;**5**(140):S-400

[55] Shah BA, Kumar A, Gupta P, Sharma M, Sethi VK, Saxena AK, et al. Cytotoxic and apoptotic activities of novel amino analogues of boswellic acids. Bioorganic & Medicinal Chemistry Letters. 2007;**17**(23):6411-6416

[56] Syrovets T, Büchele B, Gedig E, Slupsky JR, Simmet T. Acetyl-boswellic acids are novel catalytic inhibitors of human topoisomerases I and IIα. Molecular Pharmacology. 2000;**58**(1):71-81

[57] Park YS, Lee JH, Bondar J, Harwalkar JA, Safayhi H, Golubic M. Cytotoxic action of acetyl-11-keto-β-boswellic acid (AKBA) on meningioma cells. Planta Medica. 2002;**68**(05):397-401

[58] Bhushan S, Kumar A, Malik F, Andotra SS, Sethi VK, Kaur IP, et al. A triterpenediol from Boswellia serrata induces apoptosis through both the intrinsic and extrinsic apoptotic pathways in human leukemia HL-60 cells. Apoptosis. 2007;**12**:1911-1926

[59] Lucafò M, Curci D, Franzin M, Decorti G, Stocco G. Inflammatory bowel disease and risk of colorectal cancer: An overview from pathophysiology to pharmacological prevention. Frontiers in Pharmacology. 2021;**12**:772101

[60] Jyothi Y, Kamath JV, Asad M. Effect of hexane extract of Boswellia serrata oleo-gum resin on chemically induced liver damage. Pakistan Journal of Pharmaceutical Sciences. 2006;**19**(2):125-129

[61] Helal EG, Shahat M. Hypolipidimic effect of some medicinal plants on diabetic rats. The Egyptian Journal of Hospital Medicine. 2006;**23**(1):200-211

[62] Mahdian D, Abbaszadeh-Goudarzi K, Raoofi A, Dadashizadeh G, Abroudi M, Zarepour E, et al. Effect of Boswellia species on the metabolic syndrome: A review. Iranian Journal of Basic Medical Sciences. 2020;**23**(11):1374-1381

[63] Al-Awadi F, Fatania H, Shamte U. The effect of a plants mixture extract on liver gluconeogenesis in streptozotocin induced diabetic rats. Diabetes Research (Edinburgh, Scotland). 1991;**18**(4):163-168

[64] Azemi ME, Namjoyan F, Khodayar MJ, Ahmadpour F, Padok AD, Panahi M. The antioxidant capacity and anti-diabetic effect of Boswellia serrata Triana and planch aqueous extract in fertile female diabetic rats and the possible effects on reproduction and histological changes in the liver and kidneys. Jundishapur Journal of Natural Pharmaceutical Products. 2012;**7**(4):168-175

[65] Poeckel D, Werz O. Boswellic acids: Biological actions and molecular targets. Current Medicinal Chemistry. 2006;**13**(28):3359-3369

[66] Anthoni C, Laukoetter MG, Rijcken E, Vowinkel T, Mennigen R, Muller S, et al. Mechanisms underlying the anti-inflammatory actions of boswellic acid derivatives in experimental colitis. American journal of physiology-gastrointestinal and liver. Physiology. 2006;**290**(6):G1131-G11G7

[67] Ammon H. Modulation of the immune system by Boswellia serrata extracts and boswellic acids. Phytomedicine. 2010;**17**(11):862-867

[68] Borrelli F, Capasso F, Capasso R, Ascione V, Aviello G, Longo R, et al. Effect of Boswellia serrata on intestinal motility in rodents: Inhibition of diarrhoea without constipation. British Journal of Pharmacology. 2006;**148**(4):553

[69] Stanke-Labesque F, Pofelski J, Moreau-Gaudry A, Bessard G, Bonaz B. Urinary leukotriene E4 excretion: A biomarker of inflammatory bowel disease activity. Inflammatory Bowel Diseases. 2008;**14**(6):769-774

[70] Takada Y, Ichikawa H, Badmaev V, Aggarwal BB. Acetyl-11-keto-β-boswellic acid potentiates apoptosis, inhibits invasion, and abolishes osteoclastogenesis by suppressing NF-κB and NF-κB-regulated gene expression. The Journal of Immunology. 2006;**176**(5):3127-3140

[71] Wang H, Syrovets T, Kess D, Büchele B, Hainzl H, Lunov O, et al. Targeting NF-κB with a natural triterpenoid alleviates skin inflammation in a mouse model of psoriasis. The Journal of Immunology. 2009;**183**(7):4755-4763

[72] Catanzaro D, Rancan S, Orso G, Dall'Acqua S, Brun P, Giron MC, et al. Boswellia serrata preserves intestinal epithelial barrier from oxidative and inflammatory damage. PLoS One. 2015;**10**(5):e0125375

[73] Hartmann RM, Fillmann HS, Morgan Martins MI, Meurer L, Marroni NP.

Boswellia serrata has beneficial anti-inflammatory and antioxidant properties in a model of experimental colitis. Phytotherapy Research. 2014;**28**(9):1392-1398

[74] Governa P, Marchi M, Cocetta V, De Leo B, Saunders PT, Catanzaro D, et al. Effects of Boswellia Serrata Roxb. And Curcuma longa L. in an in vitro intestinal inflammation model using immune cells and Caco-2. Pharmaceuticals. 2018;**11**(4):126

[75] Zimmermann-Klemd AM, Reinhardt JK, Nilsu T, Morath A, Falanga CM, Schamel WW, et al. Boswellia carteri extract and 3-O-acetyl-alpha-boswellic acid suppress T cell function. Fitoterapia. 2020;**146**:104694

[76] Zhang Y, Duan R-D. Boswellic acid inhibits expression of acid sphingomyelinase in intestinal cells. Lipids in Health and Disease. 2009;**8**:1-8

[77] Petti L, De Santis F, Vetrano S. In vitro anti-inflammatory and protective effects of ibidì® on intestinal epithelial cells. European Journal of Medicinal Plants. 2014;**4**(9):1022

[78] Kiela PR, Midura AJ, Kuscuoglu N, Jolad SD, Sólyom AM, Besselsen DG, et al. Effects of Boswellia serrata in mouse models of chemically induced colitis. American journal of physiology-gastrointestinal and liver. Physiology. 2005;**288**(4):G798-G808

[79] Latella G, Sferra R, Vetuschi A, Zanninelli G, D'angelo A, Catitti V, et al. Prevention of colonic fibrosis by Boswellia and Scutellaria extracts in rats with colitis induced by 2, 4, 5-trinitrobenzene sulphonic acid. European Journal of Clinical Investigation. 2008;**38**(6):410-420

[80] Gupta I, Parihar A, Malhotra P, Singh GB, Lüdtke R, Safayhi H, et al. Effects of Boswellia serrata gum resin in patients with ulcerative colitis. European Journal of Medical Research. 1997;**2**(1):37-43

[81] Gupta I, Parihar A, Malhotra P, Gupta S, Lüdtke R, Safayhi H, et al. Effects of gum resin of Boswellia serrata in patients with chronic colitis. Planta Medica. 2001;**67**(5):391-395

[82] Holtmeier W, Zeuzem S, Preiβ J, Kruis W, Böhm S, Maaser C, et al. Randomized, placebo-controlled, double-blind trial of Boswellia serrata in maintaining remission of Crohn's disease: Good safety profile but lack of efficacy. Inflammatory Bowel Diseases. 2011;**17**(2):573-582

[83] Pellegrini L, Milano E, Franceschi F, Belcaro G, Gizzi G, Feragalli B, et al. Managing ulcerative colitis in remission phase: Usefulness of Casperome®, an innovative lecithin-based delivery system of Boswellia serrata extract. European Review for Medical and Pharmacological Sciences. 2016;**20**(12):2695-2700

[84] Elnawasany S, Assal F, Awad A, Eldesoky K, Badria F. Study of the effect of Boswellia serrata on patients with ulcerative colitis [dissertation for the doctorate]: Tanta University; 2012. Available from: https://tanta.edu.eg

Chapter 11

A Review of South African Traditional Medicinal Plants Used for Treating Fungal Coinfections in COVID-19 Patients with Respiratory Diseases

Moleboheng Emily Binyane, Sitheni Samson Mashele and Polo-Ma-Abiele Hildah Mfengwana

Abstract

Fungal infections are still most prevalent in the South African population. Fungal respiratory infections and diseases are the cause of severe clinical challenges and mortality in patients with compromised immune systems. Clinical signs of coronavirus disease of 2019 (COVID-19) such as lung injury, hyperglycemia due to diabetes, host iron and zinc depletion, hypoxia, immunosuppression, steroid therapy, and long-term hospitalization predispose patients to opportunistic fungal infections. Fungal pathogens, including *Cryptococcus*, *Aspergillus*, and *Candida* species, cause coinfections in patients infected with (COVID-19), and this has a negative impact on the patients' pharmacological management goals. *Cryptococcus*, *Aspergillus*, and *Candida* species cause respiratory infections and illnesses including pneumonia, pulmonary aspergillosis, pulmonary candidiasis, and pulmonary cryptococcosis. South African traditional medicinal plants have been used in the treatment of respiratory symptoms and diseases caused by these fungal pathogens. Medicinal plants contain secondary metabolites possessing antifungal activity against *Cryptococcus*, *Aspergillus*, and *Candida* species. Moreover, medicinal plants are cheaper and easily accessible and are believed to be safe. This review documents the use of South African traditional medicinal plants including *Artemisia absinthium*, *Artemisia afra*, *Dicoma anomala*, *Felicia* species, *Mentha* species, *Ruta graveolens*, and *Seasia erosa* in the treatment of fungal infections and diseases caused by these pathogens.

Keywords: fungal coinfections, traditional medicinal plants, COVID-19, cryptococcosis, aspergillosis

1. Introduction

Coronavirus disease of 2019 (COVID-19) patients with asymptomatic, mild, moderate, severe, and critical disease states are at risk of developing coinfection with pathogenic fungal species including *Aspergillus*, *Candida*, and *Cryptococcus* [1, 2]. Research reports suggest that COVID-19 predisposes patients to fungal, and other viral coinfections, and superinfections [3]. Concurrently occurring coinfections pose a massive challenge because it complicates diagnoses and COVID-19 management [3]. COVID-19 by severe acute respiratory coronavirus 2 (SARS-CoV-2) [1–4] causes respiratory symptoms such as shortness of breath, fever, fatigue, runny nose, headache, chest pain, congestion, anosmia, ageusia, sore throat, confusion, and vomiting [3, 5, 6], similar to those caused by *Aspergillus*, *Candida*, and *Cryptococcus* species infections [3]. An estimated 15% of COVID-19 patients admitted to the hospital's intensive care units (ICU) become coinfected by *Aspergillus* [7]. *Aspergillus* causes pulmonary aspergillosis including allergic bronchopulmonary aspergillosis (ABPA), chronic pulmonary aspergillosis (CPA), and invasive pulmonary aspergillosis (IPA) [8]. COVID-19-associated pulmonary aspergillosis (CAPA) is reported to have a 52% death rate [9]. *Aspergillus fumigatus*/*A. fumigatus* and *A. flavus* are the most common *Aspergillus* species causing coinfection in COVID-19 patients [4]. Conducted cohort studies on COVID-19-associated pulmonary aspergillosis have described its incidence to be between 2 and 33% [2, 10]. Aspergillosis is treated by the antifungal drug class, triazoles [1, 11], voriconazole, and isavuconazole being the first-line therapies [7, 9]. However, there are challenges associated with treatment therapy including the occurrence of azole-resistant *A. fumigatus* [11] and drug-drug interactions associated with the use of voriconazole, which lead to increased cardiotoxic effects of anti-SARS-CoV-2 agents [1]. The study conducted on COVID-19 patients who were severely and critically ill has revealed that dexamethasone is associated with increased pulmonary aspergillosis risk and death [12]. COVID-19-associated candidiasis (CAC) has occurred in various hospitals across countries [3]. CAC is an opportunistic infection caused by fungal species of *Candida* genus [3, 13]. Studies conducted in various countries, including the UK, Italy, Egypt, China, Iran, India, Gharbia, and Cairo, have revealed that *Candida* species including *C. albicans*, *C. tropicalis*, *C. glabrata*, *C. auris*, and *C. parapsilosis* are implicated in CAC [4, 13, 14]. Treatment of *Candida* infections includes azoles, echinocandin, Amphotericin B, and its liposomes [15]. However, there is an emergence of multidrug-resistant *Candida* species, including *C. glabrata*, *C. auris*, inherently resistant *C. krusei*, *C auris*-resistant fluconazole, and Amphotericin B, and fluconazole-resistant *C. parapsilosis* and *C. tropicalis* [4, 15]. Moreover, COVID-19 patients receiving treatment therapy, including tocilizumab, interferon type 1β, and lopinavir-ritonavir, are at an elevated risk of developing coinfections with *Candida* spp. [16]. Chloroquine, hydroxychloroquine, azithromycin, and protease inhibitors can cause direct myocardial toxicity, arrhythmias, and death [1]. COVID-19 patients coinfected with human immunodeficiency virus (HIV) or those with compromised immune systems are at risk of developing cryptococcosis [15]. The literature reveals a growing number of cryptococcosis cases in COVID-19 patients who were receiving corticosteroids and immunomodulators [17–19]. Pulmonary cryptococcosis is caused by two cryptococcal pathogenic species, namely *C. neoformans* and *C. gattii* [20, 21]. The recommended treatment therapy for cryptococcosis includes initial treatment with Amphotericin B in combination with flucytosine, followed by maintenance therapy with fluconazole [15, 22]. However, fluconazole-resistant *Cryptococcus* has been reported, and there is also an increased risk of antifungal toxicity [19]. Phytotherapy is an important solution for treating respiratory infections and diseases in adults and children [23]. Research reports that medicinal plants contain a variety of active secondary metabolites including alkaloids, saponins, and terpenoids with antifungal activity [24]. In South Africa

(SA), the majority of people utilize traditional medicinal plants (TMPs) more than Western medicines because TMPs are cheaper, widely available, and considered to be more effective [25]. South African TMPs such as *Artemisia absinthium*, *Artemisia afra*, *Dicoma anomala, Felicia* species, *Mentha* species, *Ruta graveolens*, and *Searsia erosa* have been shown to possess antifungal activity against fungal pathogens, including *Cryptococcus*, *Aspergillus*, and *Candida* species [19, 26–29].

2. South African traditional medicinal plants used in the treatment of respiratory diseases caused by fungal pathogens

2.1 *Artemisia* species

Artemisia is the most widely distributed genus belonging to the Asteraceae family [26, 27]. It consists of over 500 plant species of small herbs and shrubs, which are classified as annual, biennial, and perennial natural plants [27, 30]. These plants are used as traditional medicines [26]. Among all 500 *Artemisia* species, two species, *Artemisia afra* and *Artemisia absinthium* are the most used in SA [30]. *Artemisia afra* Jacq. ex Willd (**Figure 1**), also known as Wilde als in Afrikaans, African wormwood in English, Lengana in Sesotho, Umhlonyane in isiXhosa, and Mhlonyane in isiZulu, is a South African medicinal plant commonly used to treat respiratory symptoms and conditions such as bronchitis, asthma, colds, coughs, fever, pneumonia, sore throat, chills, whooping cough and headache [6, 19, 28, 30, 31]. *A. afra* is also used in combination with other TMPs such as *E. globulus* and *Lippia asperifolia* as prophylaxis for lung inflammation and to treat influenza [28]. The crude extract of *A. afra* has shown antifungal activity against *Candida albicans*, *Cryptococcus neoformans*, and *Aspergillus* species including *Aspergillus ochraceus*, *Aspergillus niger*, and *Aspergillus parasiticus* (**Table 1**) [19, 28, 32]. The leaves of *A. afra* contain numerous phenolic compounds with antimicrobial activity [33]. *A. afra* methanolic crude extract contains scopoletin, betulinic acid, and acacetin with good antimicrobial activity [34]. Other secondary metabolites including alkaloids, tannins, saponins, steroids, cardiac glycosides, and anthraquinones, are found in the crude extract and essential oil of *A. afra* [35]. Toxicity testing results of *A. afra* extract on McCoy fibroblast cell lines indicated moderate toxicity [19].

Figure 1.
Artemisia afra.

South African TMPs	Venicular names	Traditional uses in respiratory conditions	Inhibited fungal pathogens implicated in coinfections in COVID-19 patients	Secondary metabolites responsible for the antifungal activity
Artemisia afra	Wild als, African wormwood, Lengana, Umhlonyane, Mhlonyane [6, 19, 28, 30, 31]	Asthma, bronchitis, colds, coughs, sore throat, chills, fever headaches, lung, inflammation, influenza, whooping cough, pneumonia [6, 19, 28, 30, 31]	*C. albicans*, *C. neoformans* [19, 28, 32]	Phenolic compounds, scopoletin, betulinic acid, acacetin, alkaloids, tannins, saponins, steroids, cardiac glycosides, anthraquinones [33–35]
Artemisa absinthium	Wormwood, Green ginger, Absinthium, Absinthe [27, 29]	Fever [29]	*C. albicans*, *A. niger*, *A. flavus* [27, 29]	Lactones, terpenoids, flavonoids, flavonoid glycosides, organic acids, tannins, phenols [27]
Dicoma anomala	Fever bush, Hloenya, Maagbitterwortel, Inyongana, Isihlabamakho-ndlwane [36, 37]	Cold, cough, fever, sore throat [36–38]	*C. albicans*, *A. niger* [36, 39]	Phenolic acids, flavonoids, tannins, saponins, triterpene, phytosterols, acetylenic compounds, sesquiterpene, lactones, diterpene [40]
Felicia muricata	White Felicia, Ihbosisi [41–43]	Headaches, fever [41, 43–45]	*A. niger*, *A. flavus* [41]	Phenols, proanthocyanidins, flavonols, sesquiterpene, lactones, triterpenoids flavonoids [41, 46]
Mentha spicata	Spearmint, brown mint, Garden mint, Lady's mint, Imboza [47, 48]	Asthma, cold, fever, flu [48–50]	*A. niger*, *C. neoformans*, *C. albicans* [48, 51–53]	Biopeptides, flavonoids, tannins, sterols, polyphenols, sterols, triterpenes, glycosides [53, 54]
Mentha longifolia	Wild mint, Horsemint, Silver mint, Koena, Inxina, Inzinziniba [49, 55–59]	Common cold, cough, sore throat, fever, headache, flu [60, 61]	*C. albicans* *C. glabrata*, *A. flavus*, *A. fumigatus*, *A. niger* [55, 57, 62]	Flavonoids, ceramides, cinnamates, ester, ketones, monoterpenes, phenols, polyene, sesquiterpenes [60]
Ruta graveolens	Ruta, rue, Garden rue, Herb of grace, Wynruit [63–66]	Fever, headache, colds, influenza [64, 66]	*C. albicans*, *C. tropicalis* *C. parapsilopsis*, *C. glabrata* *A. flavus*, *A. fumigatus*, *A. niger* *C. neoformans* [67–70]	Coumarins, coumarin dimers, dihydrofuranocoumarins, quinolone, furoquinoline, dihydrofuroquinoline, phenolic acids, alkaloids, flavonoids [71]

South African TMPs	Venicular names	Traditional uses in respiratory conditions	Inhibited fungal pathogens implicated in coinfections in COVID-19 patients	Secondary metabolites responsible for the antifungal activity
Searsia erosa	Broom karee, Besembos, Tšilabele [68, 72, 73]	Colds [72, 73]	*C. neoformans* [74]	Alkaloids, flavonoids, terpenoids, saponins, tannins [74]
Searsia lancea	African sumuc, Willow rhus, [68]	Colds, influenza [68]	*A. flavus* [75]	Flavonoids, tannins, phenols [76]
Searcia natalensis	Natal rhus [76]	Influenza [76]	*C. albicans*, *A. Niger* [75]	epicatechin, 3β-sitosterol, 3β-sitosterol glucoside stigmasterol, lupeol [75]

Table 1.
Traditional medicinal plants used in respiratory diseases caused by fungal pathogens causing coinfections in COVID-19 patients.

Figure 2.
Artemisia absinthium.

Figure 3.
Dicoma anomala.

A. absinthium (**Figure 2**), also known as Wormwood, Green ginger, Absinthium, or Absinthe in English, is used traditionally to treat fever [27, 29]. However, when used for a long period, *A. absinthium* is reported to be responsible for the central nervous system associated-adverse effects in patients such as convulsions, hallucination, and insomnia [27]. It contains secondary metabolites including lactones, terpenoids, flavonoid glycosides, organic acids, tannins, and phenols [27]. Moreover, *A. absinthium* has antifungal activity against *C. albicans*, *A. niger*, and *A. flavus* (**Table 1**) [29]. *A. absinthium* is reported to be nontoxic when tested on Wistar Hannover rats for 13 weeks [77].

2.2 *Dicoma anomala*

Dicoma anomala (**Figure 3**) is a herbaceous plant belonging to the Asteraceae family of plants [36, 37]. It is known as Maagbitterwortel in Afrikaans, Fever bush in English, Hloenya in Sesotho, Inyongana in isiXhosa, and Isihlabamakhondlwane in isiZulu [36, 37]. In SA, *Dicoma anomala* is distributed in various provinces including the Free State, Limpopo, Gauteng, Northwest, Northern Cape, and Kwazulu natal [36, 38, 78]. Two subspecies, *Dicoma anomala* and *Dicoma gerrardi* are found in SA [37]. *Dicoma anomala* is used traditionally to treat respiratory symptoms and diseases including coughs, colds, and fever [36–38]. It has antifungal activity against *C. albicans*, and *A. niger* (**Table 1**) [36, 39]. *Dicoma anomala* produces bioactive compounds including phenolic acids, flavonoids, tannins, saponins, triterpenes, phytosterols, acetylenic compounds, sesquiterpene, lactones, and diterpene [40]. Results of acute and subchronic oral toxicity assessment of aqueous root extract of *Dicoma anomala* in rats for 14-day acute and 90-day subchronic toxicity testing have revealed that *Dicoma anomala* is not toxic at 0.5 to 2000 mg/kg [39]. *Dicoma anomala* dichloromethane: Methanol extract was found to be nontoxic at concentrations below 200 μg/ml when tested on Chang liver cells [79].

2.3 *Felicia muricata*

The genus *Felicia* consists of small shrubs of 85 known species of annual and perennial herbaceous plants [80]. *Felicia muricata* (**Figure 4**) is an aromatic herb belonging to the Asteraceae family [41, 42]. It is known as white *Felicia* in English and Ihbosisi or Ubosisi in isiXhosa [41–43]. *Felicia muricata* is widely distributed in SA, and in the Eastern Cape province, it is used traditionally to treat respiratory symptoms including headaches and fever [41, 43–45]. It has antifungal activity against *Aspergillus* species including *A. niger* and *A. flavus* (**Table 1**) [41]. *Felicia muricata* contains secondary metabolites including phenols, proanthocyanidins, flavonols, sesquiterpene lactones, triterpenoids, and flavonoids [41, 46]. The study conducted in Wistar rats using *Felicia muricata* aqueous leaf extract at 50, 100, and 200 mg/kg body weight for 14 days revealed that the plant is mildly toxic and safe for oral use, and requires further investigation [81].

Figure 4.
Felicia muricata.

2.4 *Mentha* species

Genus *Mentha* is a perennial and annual plant belonging to the Lamiaceae family [82, 83]. *Mentha spicata* (**Figure 5**) is also known as Spearmint, Brown mint, Garden mint, Lady's mint in English, and Imboza in isiXhosa [47]. It is a creeping rhizomatous and perennial herb cultivated in various tropical to temperate regions including SA [47]. *Mentha spicata* is used traditionally to treat respiratory symptoms and conditions such as asthma, flu, cold, and fever [48–50]. It has antifungal activity against *A. niger*, *Cryptococcus neoformans*, and *Candida albicans* (**Table 1**) [48, 51–53]. *Mentha* extracts and oils contain biopeptides responsible for their antifungal activity [54, 84]. *Mentha spicata* contains secondary metabolites including flavonoids, tannins, sterols, polyphenols, sterols, triterpenes, and glycosides [53]. Toxicity investigational study of *Mentha spicata* methanolic extract in mice using 500, 1000, 2000, and 5000 mg/kg for 24 hours to 14 days revealed that the plant is safe for oral administration [53]. *Mentha longifolia* (**Figure 6**), also known as Wild mint, Silver mint, and Horsemint in English, Koena in Sesotho, Inxina, and Inzinziniba in isiXhosa, is naturally present

Figure 5.
Mentha spicata.

Figure 6.
Mentha longifolia.

in SA [49, 55–59]. It is traditionally used to treat respiratory conditions including the common cold, cough, sore throat, headache, flu, and fever [60, 61]. *Mentha longifolia* has antifungal activity against *Candida albicans*, *Candida glabrata*, *A. flavus*, *A. fumigatus*, and *A. niger* (**Table 1**) [55, 57, 62]. The essential oil of *Mentha longifolia* contains a terpenoid and methanol, that has fungistatic and fungicidal activities [85]. *Mentha longifolia* possesses other secondary metabolites such as flavonoids, ceramides, cinnamates, ester, ketones, monoterpenes, phenols, polyene, and sesquiterpenes [60]. A toxicity testing study of *Mentha longifolia* methanolic extract in rats revealed that the acute oral dose was nontoxic [86].

2.5 *Ruta graveolens*

Ruta graveolens (**Figure 7**) belongs to the *Rutaceae* family [63]. It is commonly known as Ruta, Rue, Garden rue, and Herb of grace in English, and Wynruit in Afrikaans [63–66]. *Ruta graveolens* is distributed worldwide including in SA [64, 66]. It is used traditionally to treat respiratory symptoms and diseases including fever, headache, colds, and influenza [64, 66]. *Ruta graveolens* has antifungal activity against *Candida albicans*, *Candida tropicalis*, *Candida parapsilosis*, *Candida glabrata*, *Aspergillus flavus*, *Aspergillus fumigatus*, and *Cryptococcus neoformans* (**Table 1**) [67–70]. The essential oil of *Ruta graveolens* contains ketones responsible for antimicrobial activity [67]. *Ruta graveolens* is rich in bioactive compounds including coumarins, coumarin dimers, dihydrofuranocoumarins, quinolone, furoquinoline, dihydrofuroquinoline, phenolic acids, alkaloids, and flavonoids [71]. Toxicity investigation of *Ruta graveolens* in Wistar rats has shown that the plant's seeds extract at 50 mg/kg/day was not toxic after oral administration for 4 weeks [87].

2.6 *Seasia* species

The genus *Searsia* (previously known as *Rhus*) belongs to the family *Anacardiaceae*. It is widely distributed in tropics and subtropics areas globally mostly in the African continent, especially southern Africa [88, 89]. Most *Searsia* species such as *Searsia*

Figure 7.
Ruta graveolens.

erosa, *Searsia divaricate*, *Searsia lancea*, *Searcia natalensis*, and *Searsia undulata* are traditionally used to treat respiratory illnesses including colds, influenza, and microbial infections [68, 90]. *Searsia* species have pharmacological activities including anti-inflammatory, anticancer, antiviral, antimalarial, antidiarrheal, and antioxidant activities [91]. *Searsia erosa* (**Figure 8**), also known as Broom karee, Besembos in English, and Tśilabele in Sesotho [68, 72, 73], is used traditionally to treat respiratory diseases including colds [72, 73]. It has antifungal activity against *Cryptococcus*

Figure 8.
Searsia erosa.

Figure 9.
Searsia lancea.

Figure 10.
Searsia natalensis.

neoformans (**Table 1**) [74]. Aqueous extracts of *Searsia erosa* were found to be non-toxic when tested using the brine shrimp lethality assay [74]. *Searsia lancea* (**Figure 9**) also known as African sumuc, and Willow rhus in English is used to treat colds and influenza [68]. It contains bioactive compounds including flavonoids, tannins, and phenols [75]. *Searsia lancea* has antifungal activity against *A. flavus* (**Table 1**) [76]. *Searsia natalensis* (**Figure 10**), also known as Natal rhus in English is used to treat influenza [76], possesses secondary metabolites including epicatechin, 3β-sitosterol, 3β-sitosterol, glucoside stigmasterol, lupeol [75]. *Searsia natalensis* has antifungal activity against *C. albicans* and *A. Niger* [75]. There are no studies documenting the toxicity analysis of reported *Searsia* species, and further studies are warranted to determine the safety of these medicinal plants.

Table 1 shows the traditional use of South African TMPs in respiratory conditions including, asthma, bronchitis, colds, coughs, sore throat, headaches, lung inflammation, influenza, chills, whooping cough, pneumonia, and fever [6, 19, 28–31, 63, 70, 85]. These TMPs are also reported to possess antifungal activity against *Aspergillus*, *Candida*, and *Cryptococcus* species, which are implicated in coinfections with COVID-19.

3. Conclusions

This review has summarized TMPs commonly used in the treatment of respiratory diseases caused by fungal pathogens such as *Aspergillus*, *Candida,* and *Cryptococcus* species implicated in coinfection in COVID-19 patients. *Artemisia absinthium*, *Artemisia afra*, *Dicoma anomala*, *Felicia* species, *Mentha* species, *Ruta graveolens*, and *Searsia erosa* have been used in SA for the treatment of respiratory symptoms and diseases including asthma, bronchitis, colds, coughs, sore throat, headaches, lung inflammation, influenza, chills, whooping cough, pneumonia, fever, and flu. These TMPs contain secondary metabolites responsible for their antifungal activities. *In vitro* and *in vivo* toxicity studies have confirmed that these TMPs are nontoxic for oral administration. However, further testing using animal models and clinical studies are required to profile the pharmacokinetics and pharmacodynamics of these TMPs before recommendations to use in coinfections in COVID-19 patients.

Acknowledgements

We acknowledge the Central University of Technology, Department of Health Sciences and Walter Sisulu University, Department of Internal Medicine and Pharmacology, and National Research Foundation (Black Academics Advancement Programme).

Conflict of interest

The authors declare no conflict of interest.

Author details

Moleboheng Emily Binyane[1], Sitheni Samson Mashele[2]
and Polo-Ma-Abiele Hildah Mfengwana[2]*

1 Faculty of Health Sciences, Department of Internal Medicine and Pharmacology, Walter Sisulu University, Mthatha, South Africa

2 Faculty of Health and Environmental Sciences, Department of Health Sciences, Central University of Technology, Bloemfontein, South Africa

*Address all correspondence to: pntsoeli@cut.ac.za

References

[1] Lai CC, Yu WL. COVID-19 associated with pulmonary aspergillosis: A literature review. Journal of Microbiology, Immunology, and Infection. 2021;**54**:46-53. DOI: 10.1016/j.jmii.2020.09.004

[2] Baddley JW, Thompson GR, Chen SC, White L, Johnson MD, Nguyen H, et al. Coronavirus disease 2019–associated invasive fungal infection. Open Forum Infectious Diseases. 2021;**8**(12):1-11. DOI: 10.1093/ofid/ofab510

[3] Amin A, Vartanian A, Poladian A, Voloshko A, Yegiazaryan A, Al-Kassir AL, et al. Root causes of fungal coinfections in COVID-19 infected patients. Infectious Disease Reports. 2021;**13**(4):1018-1035. DOI: 10.3390/idr13040093

[4] Kundu R, Singla N. COVID-19 and plethora of fungal infections. Current Fungal Infection Reports. 2022;**16**:47-54. DOI: 10.1007/S12281-022-00432-2

[5] Larsen JR, Martin MR, Martin JD, Kuhn P, Hicks JB. Modeling the onset of symptoms of COVID-19. Frontiers in Public Health. 2020;**8**(473):1-14. DOI: 10.3389/fpubh.2020.00473

[6] Binyane ME, Mfengwana PH. Traditional medicinal plants as the potential adjuvant, prophylactic and treatment therapy for COVID-19 disease: A review. Medicinal Plants. 2022;**1**:51-70. DOI: 10.5772/intechopen.104491

[7] Kuehn BM. Aspergillosis is common among COVID-19 patients in the ICU. Journal of American Medical Association. 2021;**326**(16):1573. DOI: 10.1001/jama.2021.17973

[8] Kanj A, Abdallah N, Soubani AO. Review article the spectrum of pulmonary aspergillosis. Respiratory Medicine. 2018;**141**:121-131. DOI: 10.1016/j.rmed.2018.06.029

[9] Verweij PE, Brüggemann RJM, Azoulay E, Bassetti M, Blot S, Buil JB, et al. Martin-Loeches I. taskforce report on the diagnosis and clinical management of COVID-19 associated pulmonary aspergillosis. Intensive Care Medicine. 2021;**47**:819-834. DOI: 10.1007/s00134-021-06449-4

[10] Borman AM, Palmer MD, Fraser M, Patterson Z, Mann C, Oliver D, et al. COVID-19-associated invasive aspergillosis: Data from the UK national mycology reference laboratory. Journal of Clinical Microbiology. 2021;**59**(1):1-12. 20. DOI: 10.1128/JCM.02136-20

[11] Sabino R, Veríssimo C. Perspective novel clinical and laboratorial challenges in aspergillosis. Microorganisms. 2022;**10**(259):1-7. DOI: 10.3390/microorganisms 10020259

[12] Leistner R, Schroeter L, Adam T, Poddubnyy D, Stegemann M, Siegmund B, et al. Corticosteroids as risk factor for COVID-19-associated pulmonary aspergillosis in intensive care patients. Critical Care. 2022;**26**:30-41. DOI: 10.1186/s13054-022-03902-8

[13] Ahmed N, Mahmood MS, Ullah A, Araf Y, Rahaman TI, Moin AT, et al. COVID-19-associated candidiasis: Possible patho-mechanism, predisposing factors, and prevention strategies. Current Microbiology. 2022;**79**(5):1-15. DOI: 10.1007/s00284-022-02824-6

[14] Arastehfar A, Carvalho A, Nguyen MH, Hedayati MT, Netea MG, Perlin DS, et al. COVID-19-associated candidiasis (CAC): An underestimated

complication in the absence of immunological predispositions? Journal of Fungi. 2020;**6**(211):1-13. DOI: 10.3390/jof6040211

[15] Song G, Liang G, Liu W. Fungal coinfections associated with global COVID-19 pandemic: A clinical and diagnostic perspective from China. Mycopathologia. 2020;**185**:599-606. DOI: 10.1007/s11046-020-00462-9

[16] Segrelles-Calvo G, de Araújo GRD, Llopis-Pastor E, Carrillo J, Hernandez-Hernandez M, Rey L, et al. *Candida spp.* co-infection in COVID-19 patients with severe pneumonia: Prevalence study and associated risk factors. Respiratory Medicine. 2021;**188**(106619):1-6. DOI: 10.1016/j.rmed.2021.106619

[17] Gil Y, Gil YD, Markou T. The emergence of cryptococcemia in COVID-19 infection: A case report. Cureus. 2021;**13**(11):1-4. DOI: 10.7759/cureus.19761

[18] Isaac S, Pasha MA, Isaac S, Lal A, Kyei-nimako E. Pulmonary cryptococcosis complicating post-COVID-19 pulmonary fibrosis. Chest. 2021;**160**(4):A467. DOI: 10.10 16/j.chest.2021.07.457

[19] More G, Lall N, Hussein A, Tshikalange TE. Antimicrobial constituents of *Artemisia afra* Jacq. Ex Willd. Against periodontal pathogens. Evidence-Based Complementary and Alternative Medicine. 2012;**252758**:1-7. DOI: 10.1155/2012/252758

[20] Sharma S, Agrawal G, Das S. COVID-19-associated pulmonary cryptococcosis: A rare case presentation. Indian Journal of Critical Care Medicine. 2022;**26**(1):129-132. DOI: 10.5005/jp-journals-10071-24084

[21] Wang Y, Pawar S, Dutta O, Wang K, Rivera A, Xue C. Macrophage mediated immunomodulation during cryptococcus pulmonary infection. Frontiers in Cellular and Infection Microbiology. 2022;**12**(859049):1-15. DOI: 10.3389/fcimb.2022.859049

[22] Bhatt K, Agolli A, Patel MH, Garimella R, Devi M, Garcia E, et al. High mortality co-infections of COVID-19 patients: Mucormycosis and other fungal infections. Discover. 2021;**9**(1):1-12. DOI: 10.15190/d.2021.5

[23] Dutu LE, Popescu ML, Purdel CN, Ilie EI, Lută EA, Costea L, et al. Review traditional medicinal plants-a possible source of antibacterial activity on respiratory diseases induced by *chlamydia pneumoniae*, *haemophilus influenzae, Klebsiella pneumoniae* and *Moraxella catarrhalis*. Diversity. 2022;**14**(145):1-34. DOI: 10.3390/d14020145

[24] Mishra KK, Kaur CD, Sahu AK, Panik R, Kashyap P, Mishra SP, et al. Medicinal Plants Having Antifungal Properties. London, UK: IntechOpen; 2020. pp. 1-15. DOI: 10.5772/intechopen.90 674

[25] Bhuda MT, Marumo P. African traditional medicine and healing in South Africa: Challenges and prospects before and during COVID 19. Gender and Behaviour. 2020;**18**(4):16718-16732

[26] Obistioiu D, Cristina RT, Schmerold I, Chizzola R, Stolze K, Nichita I, et al. Chemical characterization by GC-MS and *in vitro* activity against *Candida albicans* of volatile fractions prepared from *Artemisia dracunculus*, *Artemisia abrotanum*, *Artemisia absinthium* and *Artemisia vulgaris*. Chemistry Central Journal. 2014;**8**:6-16. DOI: 10.1186/1752-153X-8-6

[27] Batiha GE, Olatunde A, El-Mleeh A, Hetta HF, Al-Rejaie S, Alghamdi S,

et al. Review bioactive compounds, pharmacological actions, and pharmacokinetics of wormwood (*Artemisia absinthium*). Antibiotics. 2020;**9**(353):1-25. DOI: 10.3390/antibiotics9060353

[28] Liu NQ, der Kooy F, Verpoorte R. Review *Artemisia afra*: A potential flagship for African medicinal plants. South African Journal of Botany. 2009;**75**:185-195. DOI: 10.1016/j.sajb.2008.11.001

[29] Szopa A, Pajor J, Klin P, Rzepiela A, Elansary HO, Al-Mana FA, et al. *Artemisia absinthium* L.-importance in the history of medicine, the latest advances in phytochemistry and therapeutical, cosmetological and culinary uses. Plants. 2020;**9**(1063):1-33. DOI: 10.3390/plants9091063

[30] Nigam M, Atanassova M, Mishra AP, Pezzani R, Devkota HP, Plygun S, et al. Review bioactive compounds and health benefits of *Artemisia* species. Natural Product Communications. 2019;14(7):1-17. DOI: 10.1177/1934578X19850354

[31] Shirinda H, Leonard C, Candy G, van Vuuren S. Antimicrobial activity and toxicity profile of selected southern African medicinal plants against neglected gut pathogens. South African Journal of Science. 2019;**11**(12):1-10. DOI: 10.17159/sajs.2019/6199

[32] Setianingrum F, Rautemaa-Richardson R, Denning DW. Review article pulmonary cryptococcosis: A review of pathobiology and clinical aspects. Medical Mycology. 2019;**57**:133-150. DOI: 10.1093/mmy/myy086

[33] Gundidza M. Antifungal activity of essential oil from *Artemisia afra* Jacq. Central African Journal of Medicine. 1993;**7**(39):140-142. DOI: DOI/pdf/10.10520/AJA00089176_71

[34] Haile AB, Jiru TM. Antibacterial effects of *Artemisia afra* leaf crude extract against some selected multi-antibiotic resistant clinical pathogens. Ethiopian Journal of Health Science. 2022;**32**(3):351-660. DOI: 10.4314/ejhs.v32i3.22

[35] Yimam BB, Desalew A. Phytochemical screening, antibacterial effect, and essential oil extract from the leaf of *Artemisia afra* against on selected pathogens. Advances in Microbiology. 2022;**12**(7):386-397. DOI: 10.4236/aim.2022.127028

[36] Balogun FO, Tshabalala NT, Ashafa AOT. Antidiabetic medicinal plants used by the Basotho tribe of eastern free state: A review. Journal of Diabetes Research. 2016;**46028201**:1-13. DOI: 10.1155/2016/4602820

[37] Munodawafa T, Chagonda LS, Moyo SR. Antimicrobial and phytochemical screening of some Zimbabwean medicinal plants. Journal of Biologically Active Products from Nature. 2013;**3**(5-6):323-330. DOI: 10.1080/22311866.2013.782759

[38] Muto T, Watanabe T, Okamura M, Moto M, Kashida Y, Mitsumori K. Thirteen-week repeated dose toxicity study of wormwood (*Artemisia absinthium*) extracts in rats. The Journal of Toxicological Sciences. 2003;**28**(5):471-478. DOI: 10.2131/jts.28.471

[39] Tripathy S, Rademan S, Matsabisa MG. Effects of silver nanoparticle from *Dicoma anomala* Sond. Root extract on MCF-7 cancer cell line and NF54 parasite strain: An *in vitro* study. Biological Trace Element Research. 2020;**195**:82-94. DOI: 10.1007/s12011-019-01822-3

[40] Balogun FO, Omotayo A, Ashafa T. Acute and subchronic oral

toxicity evaluation of aqueous root extract of *Dicoma anomala* Sond. in wistar rats. Evidence-based Complementary and Alternative Medicine. 2016;**3509323**:1-11. DOI: 10.1155/2016/35 09323

[41] Ashafa AOT, Grierson DS, Afolayan AJ. *In vitro* antioxidant activity of extracts from the leaves of *Felicia muricata* Thunb. An underutilized medicinal plant in the eastern cape province, South Africa. African Journal of Traditional, Complementary and Alternative Medicines. 2010;**7**(4):296-302. DOI: 10.4314/ajtcam.v7i4.56695

[42] Elshorbagy AM, Fayed MAA, Sallam A, Badria FA. Phytochemical, ethnopharmacological, and potential therapeutic uses of the genus *Felicia*. Asian Journal of Phytomedicine and Clinical Research. 2019;**7**(4):163-171

[43] Ashafa AOT, Grierson DS, Afolayan AJ. Antimicrobial activity of extracts from *Felicia muricata* Thunb. Journal of Biological Sciences. 2008;**8**(6):1062-1066. DOI: 10.3923/jbs.2008.1062.1066

[44] Ashafa AOT, Yakubu MT, Grierson DS, Afolayan AJ. Evaluation of aqueous extract of *Felicia muricata* leaves for anti-inflammatory, antinociceptive, and antipyretic activities. Pharmaceutical Biology. 2010;**48**(9):994-1001. DOI: 10.3109/138802009033 73664

[45] Ashafa AOT, Grierson DS, Afolayan AJ. Effects of drying methods on the chemical composition of essential oil from *Felicia muricata* leaves. Asian Journal of Plant Sciences. 2008;**7**(6):603-606. DOI: 10.3923/ajps.2008.603.606

[46] Ashafa AOT, Yakubu MT, Grierson DS, Afolayan AJ. Toxicological evaluation of the aqueous extract of *Felicia muricata* Thunb. Leaves in Wistar rats. African Journal of Biotechnology. 2009;**8**(6):949-954

[47] Božović M, Pirolli A, Ragno R. Mentha suaveolens Ehrh. (*Lamiaceae*) essential oil and its main constituent piperitenone oxide: Biological activities and chemistry. Molecules. 2015;**20**:8605-8633. DOI: 10.3390/molecules20058605

[48] Sevindik M. Pharmacological properties of *Mentha* species. Journal of Traditional Medicine and Clinical Naturopathy. 2018;**7**(1):1-7. DOI: 10.4172/2573-4555.10 00259

[49] Mikaili P, Mojaverrostami S, Moloudizargari M, Aghajanshakeri S. Pharmacological and therapeutic effects of *Mentha Longifolia* L. and its main constituent, menthol. Ancient Science of Life. 2013;**33**(2):131-138. DOI: 10.4103/g

[50] Aziz Eftekhari Khusro A, Hasanzadeh A, Dizaj SM, Hasanzadeh A, Cucchiarini M. Phytochemical and nutra-pharmaceutical attributes of *Mentha* spp. A comprehensive review. Arabian Journal of Chemistry. 2021;**14**(103106):1-13. DOI:10.1016/j.arabjc.2021.103106

[51] Kee LA, Shori AB, Baba AS. Bioactivity and health effects of *Mentha spicata*. Integrative Food, Nutrition and Metabolism. 2017;**5**(1):1-2. DOI: 10.15761/IFNM.1000 203

[52] Asowata-Ayodele AM, Afolayan AJ, Otunola GA. Ethnobotanical survey of culinary herbs and spices used in the traditional medicinal system of Nkonkobe municipality, eastern cape, South Africa. South African Journal of Botany. 2016;**104**:69-75. DOI: 10.1016/j.sajb.2016.01.001

[53] Ahmad RS, Imran A, Arshad MS, Hussain MB, Waheed M, Safdar S, et al. Introductory Chapter: *Mentha Piperita*

(a Valuable Herb): Brief Overview. London, UK: IntechOpen; 2020. pp. 1-11. DOI: 10.5772/intechopen.93627

[54] Ojewumi ME, Adedokun SO, Ayoola AA, Taiwo OS. Evaluation of the Oil Extract from *Mentha spicata* and its Chemical Constituents. Available from: https://core.ac.uk/download/pdf/162043603.pdf [Accessed: July 30, 2022]

[55] Brahmi M, Adli DEH, Boufadi MY, Arabi W, Kahloula K, Slimani M. AntimicrobialandAntiochratoxicactivities of *Mentha spicata* essential oil. Phytothérapie. 2021;19:397-403. DOI: 10.3166/phyto-2021-0278

[56] Menyiy NE, Mrabti HN, Omari NE, Bakili AE, Bakrim S, Mekkaoui M, et al. Review article medicinal uses, phytochemistry, pharmacology, and toxicology of *Mentha spicata*. Hindawi Evidence-Based Complementary and Alternative Medicine. 2022;**7990508**:1-32. DOI: 10.1155/2022/7990508

[57] Yassin MT, Mostafa AA, Al-Askar AA. Anticandidal and anti-carcinogenic activities of *Mentha longifolia* (wild mint) extracts *in vitro*. Journal of King Saud University-Science. 2020;**32**:2046-2052. DOI: 10.1016/j.jksus.2020.02.008

[58] Ali HM, Abo Elgat WAA, EL-Hefny M, Salem MZM, Taha AS, Al Farraj DA, et al. New approach for using of *Mentha longifolia* L. and *Citrus reticulata* L. essential oils as wood-biofungicides: GC-MS, SEM, and MNDO quantum chemical studies. Materials. 2021;**14**(1361):1-8. DOI: 10.3390/ma14061361

[59] Vining KJ, Zhang Q, Tucker AO, Smith C, Davis TM. *Mentha longifolia* (L.) L.: A model species for mint genetic research. HortScience. 2005;**40**(5):1225-1229. DOI: 10.21273/HORTSCI.40.5.1225

[60] Farzaei MH, Bahramsoltani R, Ghobadi A, Farzaei F, Najafi F. Pharmacological activity of *Mentha longifolia* and its phytoconstituents. Journal of Traditional Chinese Medicine. 2017;**37**(5):710-720. DOI: 10.1016/S0254-6272(17)30327-8

[61] Tafrihi M, Imran M, Tufail T, Gondal TA, Caruso G, Sharma S, et al. Review the wonderful activities of the genus *Mentha*: Not only antioxidant properties. Molecules. 2021;**26**(1118):1-22. DOI: 10.3390/molecules26041118

[62] Patti F, Palmioli A, Vitalini S, Bertazza L, Redaelli M, Zorzan M, et al. Anticancer effects of wild mountain *Mentha longifolia* extract in adrenocortical tumor cell models. Frontiers in Pharmacology. 2020;**10**(1647):1-11. DOI: 10.3389/fphar.20 19.01647

[63] Abbood SM, Al-Rawi KF, Qaddoori HT, Mohammed MT, Kadhim SM. Antioxidant activity and acute oral toxicity of the methanol extract from *Mentha Longifolia* L. ssp. in Iraq. Systemic Review in Pharmacy. 2020;**11**(12):743-746. DOI: 10.31838/srp.2020.12.118

[64] Águila L, Ruedlinger J, Mansilla K, Ordenes J, Salvatici R, Ribeiro de Campos R, et al. Relaxant effects of a hydroalcoholic extract of *Ruta graveolens* on isolated rat tracheal rings. Biological Research. 2015;**48**(1):22-28. DOI: 10.1186/s40659-015-0017-8

[65] Tobyn G, Whitelegg M. *Ruta graveolens*, rue. Research Gate. 2011;**27**:283-295. DOI: 10.1016/B978-0-443-10344-5.00032-X

[66] Parray SA, Bhat JU, Ahmad G, Jahan N, Sofi G, Iqbal SM. *Ruta*

graveolens: From traditional system of medicine to modern pharmacology: An overview. American Journal of Pharm Tech Research. 2012;**2**(2):239-252 ISSN:2249-3387

[67] Szewczyk A, Marino A, Molinari J, Ekiert H, Miceli N. Phytochemical characterization, and antioxidant and antimicrobial properties of agitated cultures of three rue species: *Ruta chalepensis*, *Ruta corsica*, and *Ruta graveolens*. Antioxidants. 2022;**11**(3):592. DOI: 10.3390/antiox11030592

[68] Cock IE, Van Vuuren SF. The traditional use of southern African medicinal plants in the treatment of viral respiratory diseases: A review of the ethnobotany and scientific evaluations. Journal of Ethnopharmacology. 2020;**262**(113194):1-25. DOI: 10.1016/j.jep.2020.113194

[69] Reddy DN, Al-Rajab JA. Chemical composition, antibacterial and antifungal activities of *Ruta graveolens* L. volatile oils. Cogent Chemistry. 2016;**2**(1220055):1-11. DOI: 10.1080/23312009.2016.1220055

[70] Donadu MG, Peralta-Ruiz Y, Usai D, Maggio F, Molina-Hernandez JB, Rizzo D, et al. Colombian essential oil of *Ruta graveolens* against nosocomial antifungal resistant *Candida* strains. Journal of Fungi. 2021;**7**(383):1-17. DOI: 10.3390/jof7050383

[71] Kengar A, Paratkar G. Antifungal activity of phytoconstituents of *Ruta graveolens* L. Bionano Frontier. 2014;7(1):61-64. ISSN 0974-0678

[72] Mashimbye NN, Moteetee A, Oskolskii A. Ethnobotanical Uses, Anatomical Features, Phytochemical Properties, Antimicrobial Activity, and Cytotoxicity of the Sotho Medicinal plant *Searsia erosa* (*Anacardiaceae*). 2019. pp. 1-102. file:///C:/Users/moleb/Downloads/Mashimbye_Nhlamulo_NN_MSc_2019.pdf.

[73] Seleteng-Kose L, Kobisi K, Pool-Stanvliet, Mohapi K. A rapid biodiversity assessment of Lesotho's first proposed biosphere reserve: A case study of Bokong nature reserve and Tséhlanyane national park. Bothalia. 2021;**51**(2):1-34. DOI: 10.38201/btha.abc.v51.i2.6

[74] Koki M, Yalo M, Makhaba M, Nako N, Rautenbach F, Badmus JA, et al. Phytochemical investigation and biological studies on selected *Searsia* species. Plants. 2022;**11**(20):2793. DOI: 10.3390/plants11202 793

[75] Njoroge PW, Opiyo SA. Some antibacterial and antifungal compounds from root bark of *Rhus natalensis*. American Journal of Chemistry. 2019;**9**(5):150-158. DOI: 10.5923/j.chemistry.20190905.03

[76] Gundidza M, Gweru N, Mmbengwa V, Ramalivhana NJ, Magwa Z, Samie A. Phytoconstituents and biological activities of essential oil from *Rhus lancea* L. F. African Journal of Biotechnology. 2008;**7**(16):2787-2789. DOI: 10.5897/AJB 08.136

[77] Maroyi A. Dicoma anomala Sond.: A review of its botany, ethnomedicine, phytochemistry and pharmacology. Asian Journal of Pharmaceutical and Clinical Research. 2018;**11**(6):70-77. DOI: 10.22159/ajpcr.2018.v11i6.25538

[78] Becker JVW, van der Merwe MM, van Brummelen AC, Pillay P, Crampton BG, Mmutlane EM, et al. *In vitro* anti-plasmodial activity of *Dicoma anomala* subsp. *gerrardii* (Asteraceae): Identification of its main active constituent, structure-activity relationship studies and gene expression profiling. Malaria

Journal. 2011;**10**(295):1-12. DOI: 10.1186/1475-2875-10-295

[79] Makuwa SC, Serepa-Dlamini MH. The antibacterial activity of crude extracts of secondary metabolites from bacterial endophytes associated with *Dicoma anomala*. International Journal of Microbiology. 2021;**8812043**:1-12. DOI: 10.1155/2021/88 12043

[80] Matsabisa M G, Chukwuma CI, Chaudhary SK, Kumar CS, Baleni R, Javu M, et al. *Dicoma anomala* (Sond.) abates glycation and DPP-IV activity and modulates glucose utilization in Chang liver cells and 3T3-L1 adipocytes. South African Journal of Botany. 2020;**128**:182-188. DOI: 10.1016/j.sajb.2019.09.013

[81] Hyde MA, Wursten BT, Ballings P. Coates Palgrave *M. Flora* of Zimbabwe: Species Information: Individual Images: *Felicia muricata*. https://www.Zimbabweflora.co.zw/speciesdata/imagedisplay.php?species_id=158820&image_id=1 [Accessed: July 12, 2022]

[82] Jaiswal R, Verma NK, Singh AK, Vikas Y, Srivastava A. Pharmacological properties of *Felicia Muricata* Thunb. (NEES): A review. International journal of modern. Pharmaceutical Research. 2021;**5**(3):15-18. Available from: www.ijmpronline.com [Accessed: May 22, 2023]

[83] Ashafa AOT, Grierson DS, Afolayan AJ. Foliar micromorphology of *Felicia muricata* thumb., a south African medicinal plant. Pakistan Journal of Biological Sciences. 2008;**11**(13):1713-1717. DOI: 10.3923/pjbs.2008.1713.1717

[84] Piras A, Porcedda S, Falconieri D, Maxia A, Gonçalves M, Cavaleiro C, et al. Antifungal activity of essential oil from *Mentha spicata* L. and *Mentha pulegium* L. growing wild in Sardinia island (Italy). Natural Product Research. 2019;**35**(6):993-999. DOI: 10.1080/14786419.2019.1610755

[85] Moteetee A, Van Wyk BE. Sesotho names for exotic and indigenous edible plants in southern Africa. Bothalia. 2006;**36**(1):25-32. DOI: 10.4102/abc.v36i1.328

[86] Dold AP, Cocks ML. Preliminary list of Xhosa plant names from eastern cape, South Africa. Bothalia. 1999;**29**(2):267-292. DOI: 10.4102/abc.v29i2.601

[87] Nahar L, El-Seedi HR, Khalifa SAM, Mohammadhosseini M, Sarker SD. Ruta essential oils: Composition and bioactivities. Molecules. 2021;**26**(4766):1-13. DOI: 10.33 90/molecules26164766

[88] Adam SIY, Ahmed NNA, Eltayeb AM, Saad H, Taha KA. Toxicity of *Ruta graveolens* seeds' extracts on male Wistar rats. International Journal of Animal and Veterinary Advances. 2014;**6**(3):92-96. DOI: 10.19026/ijava.6.5624

[89] Yang Y, Meng Y, Wen J, Sun H, Nie Z. Phylogenetic analyses of *Searsia* (*Anacardiaceae*) from eastern Asia and its biogeographic disjunction with its African relatives. South African Journal of Botany. 2016;**106**:129-136. DOI: 10.1016/j.sajb.20 16.05.021

[90] Moteetee A, Moffet RO, Seleteng-K L. A review of the ethnobotany of the basotho of Lesotho and the Free State province of South Africa (south Sotho). South African Journal of Botany. 2019;**122**:21-56. DOI: 10.1016/j.sajb.2017.12.012

[91] Nhlamulo N, Mashimbye NN, Moteetee AN, Oskolski AA. Stem and leaf structure of *Searsia erosa* (Thunb.) Moffett (*Anacardiaceae*) with systematic, ecological and ethnobotanical implications. Botanica Pacifica. A journal of plant science and conservation. 2020;**9**(2):103-112. DOI: 10.17581/bp.2020.09214

Chapter 12

Immunomodulatory Plant Based Foods, It's Chemical, Biochemical and Pharmacological Approaches

Bamidele Sekinat Olayem, Origbemisoye Babawande Olaitan and Akinbode Badiu Akinola

Abstract

There has been a growing interest in research focused on enhancing immune function, given its crucial role in maintaining human health and preventing illnesses. While antibiotics are commonly employed in clinical settings to treat and prevent various diseases, their synthetic nature often leads to undesirable side effects. Since the beginning of time, medicinal plants have been employed in healthcare. Global research has been done to confirm their efficacy, and some of the results have sparked the development of plant-based medications; also, plant-based diets have emerged as leading contenders in the field of chronic disease prevention. They offer affordability, natural origins, and easy accessibility. One key reason for their effectiveness is their Immunomodulatory effect, whereby they stimulate immune cells and influence the development of immune molecules. This comprehensive review aims to explore the potential of medicinal plant as well as plant-based foods while examining their medicinal properties and their utilization in preventing and managing disease through their chemicals, biochemical components, and pharmacological approaches.

Keywords: medicinal plant, plant-based foods, bioactive components, immune system, diseases

1. Introduction

The use of plants as a primary source of medicines can be traced back to early civilizations of the world. They are natural and less expensive products, which are becoming more and more popular for both preventative and therapeutic purposes as a result of the negative side effects of continuous use of conventional medications [1]. Several plants are known to have natural healing capabilities for a variety of diseases due to their high contain of bioactive substance. These properties have contributed significantly to the development of modern medicine. Researchers have successfully assisted in identifying the potencies of these plants to cure diseases, thanks to generations' worth of knowledge [2–4]. These insights have helped to understand the various uses of different medicinal plants in different cultures around the world. Also, functional meals derived from plants have been demonstrated to

IntechOpen

have immunomodulatory effects due to the presence of bioactive components that have been utilized to elucidate the biological and chemical activities in the human body system, and these have developed as a new trend. According to Origbemisoye and Bamidele [5] reports, utilizing these immunostimulatory foods and herbs can strengthen the immune system and safeguard the body against COVD-19 and any other ailments.

Immune dysfunction has been exacerbated by stress and unhealthy lifestyle choices, which has increased demand for functional and nutraceutical foods. Depending on how they perform, the bioactive elements in foods made from plants and herbs are divided into primary and secondary metabolites. Protein, carbohydrates, lipids, and nucleic acids are primary metabolites that the body uses to support, grow, and maintain daily functions. Secondary metabolites, on the other hand, have biological properties like antioxidant activity, antimicrobial activity, enzyme detoxification regulation, immune system modulation, reduced platelet aggregation, hormone metabolism, and anticancer property [6], which are frequently attributed to their high concentration of phenolic chemicals, flavonoids, curcumin, saponin, glucosides, lignans, phenolic acids, alkaloids, terpenes, and steroid [6] that are typically found in medicinal plants, foods, and ingredients eaten every day, such as legumes, cereals, fruits and vegetables, herbs, spices, and essential oils [7]. This study reviews the bioactive compound in medicinal plants and functional plant-based foods that exhibit immunomodulatory effects as well as their chemical, biochemical, and pharmacological approaches.

2. Medicinal plants, their bioactive components, and pharmacology approach

2.1 Plant and herbs

There are several phytochemicals with important qualities found in all kinds of plants. Several antioxidant compounds that are present in naturally occurring plant sources and function as active oxygen or free radical scavengers are included in the comprehensive antioxidative defense mechanism that plants have developed, according to Youwei et al. [8]. Plants are a potential source of new compounds with antioxidant activity since they produce a lot of antioxidants to counteract oxidative stress. As a result, dietary antioxidants have lately generated more research interest. A lower frequency of illnesses brought on by oxidative stress from free radicals has been associated with dietary antioxidant intake from plant materials [9].

2.1.1 Phyllanthus niruri (stone breaker)

It is a native of the Amazon rainforest and other tropical nations like India, China, the Bahamas, and the Philippines [10, 11]. It is a widely used plant that is said to have anticancer, anticarcinogenic, hypolipidaemic, hepatoprotective, antiviral, antihypertension, and antidiabetic qualities. *P. niruri* contains a number of bioactive compounds, including lignin, phyllanthin, hypophyllanthin, flavonoids, glycosides, and tannins [12]. All of the plant's components, including the fruits and leaves, are employed in the medicinal formulations.

Phyllanthin, a bitter component of lignans, and hypophyllanthin, a non-bitter component, were isolated from *P. niruri* [13]; these lignans are significant because of their wide range of therapeutic properties, including hepatoprotection, antitumor, antimitotic, and antiviral properties [14–17] as well as antioxidant activities. According to reports, leaves contain the highest concentrations of phyllanthin (0.7% w/w) and hypophyllanthin (0.3% w/w), whereas the stem contains only trace amounts of both [18]. Other lignans with significant therapeutic potentials are reported in *Phyllanthus amarus* including niranthin, phyltetralin, nirtetralin, isonirtetralin, hinokinin, lintetralin, isolintetralin, demethylenedioxy-niranthin, 5-demethoxy-niranthin, and so on.

Flavonoids are polyphenolic substances that belong to the class of secondary metabolites found in plants. The many categories include catechins, chalcones, favanone, favones, favonols, isofavones, and their derivatives. The many bioactivities of *P. niruri* are also owed to this family of chemicals. The main flavonoids found in the plantincludes rutin, astragalin, kaempferol, quercetin, and so on, which contribute to the herb's antioxidant properties. With a wide variety of structural types, biosynthesis processes, and pharmacological effects, alkaloids are among the most diverse categories of secondary metabolites; a wide variety of structural types, biosynthesis processes, and pharmacological actions were discovered. Alkaloids are cyclic nitrogenous chemicals with a low molecular weight. The *Angiospermae*, or flowering plants, are the main source of alkaloids; they contain roughly 20% of them. Since ancient times, their broad variety of pharmacological properties, notably in mammals like humans, have drawn attention, in addition to their role in plant defense against herbivores and pathogens. Among their broad class of secondary metabolites, *P. niruri* is also known to contain a number of alkaloids, such as securinine, epibubbialine, and isobubbialine, which are also accountable for the herb's many supported therapeutic effects.

2.1.2 Garcinia kola (bitter Kola)

A species of flowering plant known as *G. kola* is a member of the *Clusiaceae* or *Guttiferae* family of tropical plants. It is a domesticated giant forest tree that is highly prized for its palatable nuts throughout most of West and Central Africa. It is a plant that has long been valued for both its medicinal and nutritive properties. The seeds offer potential therapeutic effects due to the concentration of the flavonoid and other bioactive components, and they are also employed in folk medicine in many herbal preparations [19, 20]. All of this plant's parts, including the nut, leaf, stem, bark, and root, have been discussed in several ethnobotanical and pharmacological studies, albeit the nut is still the one that is most frequently employed.

Flavonoids predominate among the phytochemical components of *G. kola* seeds, which also include proteins, glycosides, reducing sugar, starch, sterols, and triterpenoids, according to Esimone et al. [21]. Other chemical analyses of the seeds have revealed that they contain a complex mixture of phenolic compounds, including GB-type biflavonoids; xanthones; benzophenones; cycloartenol; triterpenes [22, 23]; kolaviron [23, 24]; the chromanols, garcioic and garcinal [25]; biflavonoids; xanthones; kolanone; ameakoflavone; 2,4,3-methylenecyclartenol; coumarine; prenylate benzophenones [26]; and oleoresin [27]. Traditional African medicine makes considerable use of extracts from *G. kola* [28, 29] particularly when creating treatments for laryngitis, coughing, and liver conditions [30]. As a purgative, antiparasitic, antimicrobial, antiviral, and anti-inflammatory; antidote to the effects of *Strophanthus gratus*; and remedy for guinea-worm infection and for gastroenteritis, rheumatism, asthma, menstrual cramps, throat infections, headache relief, colic relief, chest colds, cough, and liver disorders are just a few of its additional medical uses [31, 32].

2.1.3 Aloe vera

Aloe vera, often referred to as the "miraculous plant" or "wonder plant," has been utilized for medicinal purposes by various cultures for over 3000 years [33]. In the Democratic Republic of the Congo, aloe vera has been traditionally employed as a botanical medicine to treat illnesses and has shown significant potential against COVID-19. It is also known for its soothing properties and has been traditionally used for various health purposes. It has been reported to exhibit antiviral activity against certain viruses, including herpes simplex virus and influenza virus. Additionally, aloe vera has been shown to possess immunomodulatory effects by enhancing immune responses and stimulating the production of cytokines. Experimental studies have revealed that aloe vera exhibits potent virucidal properties with a broad spectrum of action. Notably, the toxicity of these plant extracts has been demonstrated to be benign both in vitro and in vivo. Aloe vera contains various antiviral compounds, including anthraquinones, which function independently or in conjunction with pharmaceutical targets such as the SARSCov-2 protease 3CLPro. These antiviral properties complement the plant's inherent anti-inflammatory and immunomodulatory capabilities [34]. It is plausible that a phytodrug based on aloe vera extracts could attenuate the expression of pro-inflammatory factors and receptors associated with acute respiratory distress, the primary cause of COVID-19 mortality, while simultaneously weakening the immune system. Consequently, aloe vera and its key secondary metabolites may play a crucial role in the treatment of COVID-19 and cardiovascular diseases, especially when combined with viral protease inhibitors, which represent the optimal therapeutic choice [5].

2.1.4 Stinging nettle (Urtica dioica)

Stinging nettle (*U. dioica*) has a long history of use in traditional medicine across various nations. It is believed to offer therapeutic benefits for the nervous, immune, cardiovascular, and digestive systems [35]. Previous reports have indicated that specific lectins, such as the agglutinin lectin from *U. dioica* (UDA), the agglutinin lectin from leeks (Allium porrum Agglutinin or APA), and the NICTABA lectin from tobacco (*Nicotiana tabacum*), isolated from the rhizomes of *U. dioica*, have shown the strongest inhibition against the proliferation of the Covid-19 virus, with an EC50 (50% effective concentration) of 1.3 g/ml in in vitro studies. These pure extracts exhibited low toxicity [36]. Furthermore, UDA has demonstrated inhibitory effects on the SARS-CoV virus in in vitro studies [37].

2.1.5 Torreya nucifera (nutmeg-yew from Japan)

The Taxaceae tree *T. nucifera* has a history of use in traditional Asian medicine for the treatment of stomachaches, hemorrhoids, and rheumatoid arthritis [38]. This tree is found in snowy regions near the sea of Jeju Island in Korea. It has been investigated as a potential inhibitor of SARS-CoV 3CLpro. Ethanol extracts of *T. nucifera* leaves were obtained and evaluated for their inhibitory activity against SARS-CoV 3CLpro using a fluorescence resonance energy transfer (FRET) technique. The leaves' ethanol extracts yielded 12 phytochemicals with inhibitory activity against SARS-CoV 3CLpro, including eight diterpenoids and four biflavonoids. The findings suggest the potential of *T. nucifera* as a source of natural compounds with inhibitory effects on SARS-CoV 3CLpro.

2.1.6 Alder bark

Alder bark is known to contain salicin, an anti-inflammatory compound that is converted into salicylic acid in the body. It also contains diarylheptanoids, a type of secondary metabolite [39]. Red alder bark (*Alnus rubra*) has been traditionally used in several Native American cultures to treat conditions such as poison ivy, bug bites, and skin irritations. The Blackfeet tribe, in particular, has utilized an infusion made from red alder bark to treat tuberculosis and lymphatic ailments. In a study by Park et al. [40], the inhibitory potential of nine diarylheptanoid derivatives (platyphyllenone, hirsutenone, platyphyllone, platyphyllonol-5-xylopyranoside, hirsutanonol, oregonin, rubranol, rubranoside B, and rubranoside A) isolated from *Alnus japonica* Steud (Betulaceae) of Korean origin was investigated. The study evaluated the inhibitory effects of these compounds against both SARS-CoV 3CLpro and PLpro using a continuous fluorometric assay.

2.1.7 Licorice root

Licorice (*Glycyrrhiza glabra*) root contains a major component called glycyrrhizin [41, 42]. This compound has a long history of traditional use for the treatment of gastritis, bronchitis, and jaundice. It is known to possess antioxidant and anti-inflammatory properties and has been reported to stimulate the production of interferons in the body [43]. Glycyrrhizin has been shown to inhibit the attachment of SARS-CoV to cells, particularly during the initial phase of the virus infection cycle [44]. Licorice root also contains flavonoids, glycyrrhetinic acid, β-sitosterol, and hydroxyl coumarins [43]. Cinatl et al. [45] demonstrated the anti-SARS-CoV activity of glycyrrhizin, and later, Pilcher [44] suggested licorice and glycyrrhizin as potential candidates for the development of drugs against SARS-CoV. However, it should be noted that the development of a commercial drug for SARS-CoV is still a long process. Further research by Chen et al. [46] confirmed the anti-SARS-CoV properties of glycyrrhizin, and numerous review articles have been published highlighting its positive antiviral activity [47–49].

2.2 Plant-based functional foods and their pharmacology uses

2.2.1 Legumes

Legumes, a member of the *Fabaceae* family, are dried seeds that have been consumed by humans for over 10,000 years. They are primarily cultivated for human consumption and include popular varieties such as soybeans, peanuts, lentils, lupins, alfalfa, beans, tamarind, and clovers. Legumes have been utilized to prevent and manage diet-related diseases such as metabolic disorders, inflammatory bowel disease, diabetes, and cardiovascular disease, primarily due to the presence of bioactive components with immunomodulatory properties [50]. Flavonoids are abundant in legumes, and various phenolic acids, including p-hydroxybenzoic, protocatechuic, syringic, gallic, vanillic, caffeic, and sinapic acids, have been identified [51, 52]. Phenolic acids commonly found in legumes include trans-ferulic acid, trans-p-coumaric acid, and syringic acid, with navy bean, lima bean, and cowpea exhibiting the highest concentrations, respectively [51]. Flavonoids in legumes exist in the form of glycosides or aglycones. Chemically, flavonoids are a class of phenolic compounds with a C6-C3-C6 skeleton [53]. Their structure provides hydroxyl groups in the

B-ring, enabling the stabilization of radicals by supplying hydrogen and electrons to hydroxyl, peroxyl, and peroxynitrile radicals [54].

As a result, stable flavonoid radicals are generated, and flavonoids can be classified into various classes based on the position of the B ring and the replacement pattern of the C ring. Flavonols, flavanones, isoflavones, anthocyanidins, and flavones are among the flavonoids found in legumes [55–57]. The consumption of pulses has been suggested to reduce the risk of cancer according to the World Cancer Research Fund/American Institute for Cancer Research (WCRF/AICR) 2010 report [58]. Several research studies have investigated the effects of commonly consumed legumes on cancer cell proliferation [59–62]. The high fiber content of pulses has been associated with a reduced risk of colon cancer. Fiber from common beans has been shown to have an antiproliferative effect and induce apoptosis in colon cancer cells [63, 64]. The mineral composition of pulses, including zinc and selenium, which reduce oxidative stress and inhibit tumor cell growth, may also contribute to their anticancer properties [65, 66]. The anticancer effects of pulses are attributed to biologically active components such as tannins, phytic acid, saponins, and protease inhibitors [67, 68]. Saponins, for example, have been found to inhibit the growth of leukemia, colon, lung, and other cancer cells [69, 70]. Protease inhibitors exert an anticancer effect by reducing the rate at which cancer cells divide and by blocking tumors from secreting proteases that could otherwise kill nearby cells [71].

Both pinto bean trypsin inhibitor [65] and pea protease inhibitor [55] have demonstrated in vitro anti proliferative effects. The phytochemicals' anticancer effects could be attributed to their phytoestrogenic characteristics [66]. For instance, daidzin and genistin found in soybeans exhibit this effect by binding to the human estrogen receptor [67]. Hormonal imbalance is implicated in several hormone-dependent malignancies such as breast and prostate cancer. Soybean isoflavones, which resemble mammalian estrogen, have been shown to have protective effects against hormone-dependent cancers by binding to the human estrogen receptor in both agonistic and antagonistic manners, thus exhibiting beneficial effects on hormone-dependent tumors [66].

Evidence from numerous epidemiological studies has demonstrated a link between pulse consumption and a decrease in cardiovascular diseases (CVDs) [67]. Pulses have an impact on various factors related to CVDs, including lipid profiles, blood pressure, and inflammation [64]. Consuming legumes has been shown to reduce LDL-C and total cholesterol levels [68, 69]. The mono- and polyunsaturated fatty acids and sterol content in pulses contribute to increased levels of HDL cholesterol while decreasing total and LDL cholesterol [70–72]. The association between pulse consumption and lower levels of total and LDL cholesterol has been demonstrated in clinical trials and meta-analyses [73–75]. Isoflavones in pulses also play a crucial role in the treatment of CVDs through their antihypertensive, anti-atherosclerotic, and antiplatelet activities [76]. When consumed in high amounts, the isoflavones found in kidney and black beans stimulate adipocytes to produce adiponectin, a cardioprotective hormone that exhibits anti-inflammatory effects on blood vessel cells and is associated with a lower risk of heart attack [64, 77, 78]. There is an inverse relationship between pulse consumption and blood pressure [79]. Randomized clinical studies and meta-analyses have shown that diets high in pulses can lower systolic and mean arterial blood pressure [80–82].

Diabetic patients have also shown a decrease in blood pressure and heart rate [80, 83, 84]. Pulses are often chosen as a meal by diabetics due to their high fiber content and potential benefits for glycemic management [85–87]. Increased consumption of beans in the diet may help with glycemic management and lower the risk of diabetes [88]. Randomized controlled trials (RCTs) and epidemiological research have demonstrated that pulses reduce fasting insulin and blood glucose levels, as well as fructosamine and glycosylated hemoglobin when combined with low glycemic index (GI) and high fiber diets [89–91]. The polyphenols present in legumes are also associated with regulating the absorption of carbohydrates in the intestine. Polyphenols such as diadzein, caffeic acid, ferulic acid, syringic acid, naringenin, and kaempferol inhibit the pancreatic enzymes alpha-amylase and alpha-glucosidase, which are responsible for breaking down starch in the intestine, thereby lowering blood sugar levels [92]. Furthermore, the antioxidant properties of polyphenols protect against oxidative stress generated by hyperglycemia-related free radical production. Mungbean flour, for example, has a low glycemic index and is rich in fiber [93].

The low glycemic index of mungbeans indicates that consuming them can help prevent diabetes. Soybeans, especially, have a low glycemic index and are an important component in the diet of diabetics. Legumes, particularly soybeans, are the main source of isoflavones, which have estrogenic activity. Adding isoflavones like genistein, daidzein, and glycosides to the diet may help prevent osteoporosis. Experimental investigations have shown that isoflavones work in various ways, including promoting osteoblast activity and proliferation to preserve bone mass against the action of osteoclast cells, which break down bone [94]. Research by Frassetto et al. [95] found an inverse relationship between plant protein intake and hip fracture occurrence, indicating that plant protein intake, including from legumes, is beneficial for bone health. Meals consisting of lentils, chickpeas, and yellow peas have been found to be more satisfying and effective at reducing hunger compared to a breakfast of cereals, vegetables, and cheese [90, 96]. Pulses may help manage appetite and limit calorie intake in diets. Epidemiological data also suggests a favorable relationship between bean consumption and body weight [79]. Greater weight loss was observed with a diet including four servings of pulses per week for 8 weeks within a 30% energy-restricted diet compared to an energy-restricted diet that excluded pulses [73]. Reduction in body weight has been reported after 3 months with a high-pulse-low-GI diet [97]. Although significant weight reduction was not observed in other trials of pulse-rich diets over a period of 6–18 months, a reduction in waist circumference was observed [98]. Consuming pulses has been associated with potential benefits for longevity and overall health [64]. A study by Darmadi-Blackberry et al. [99] found that elderly people who consume a lot of legumes exhibit persistent and significant protective effects. Additionally, frequent bean consumption in older adults has been associated with reduced stress, anxiety, and depression [100].

2.2.2 Cereals

Cereals, such as wheat and rice, have been recognized as functional foods and nutraceuticals due to their high protein, dietary fiber, and bioactive component content, which contribute to antioxidant and anti-inflammatory activities. These properties make cereals beneficial in preventing diseases associated with metabolic syndromes, including obesity, cardiovascular disease, and type 2 diabetes [101, 102]. Cereals like wheat, sorghum, barley, millet, and rice have been found to possess antihypertensive and antioxidant activities, which help regulate hormonal control processes, lower blood pressure, and mitigate other noncommunicable disorders [103–106].

2.2.2.1 Rice (Zizania spp.)

Rice (*Zizania* spp.) contains phenolic acids, flavonoids, and phytochemicals with antioxidant capabilities that contribute to the prevention of chronic diseases [107, 108]. Rice consumption has been associated with lower rates of chronic diseases, likely due to the presence of phytochemical antioxidants [109]. Wild rice (*Zizania latifolia*) has been shown to have antihypertensive effects, attributed to its polyphenol content, particularly quercetin, which has demonstrated blood pressure-lowering and protective effects against cardiovascular diseases [110]. Research on virgin rice bran oil has also demonstrated its potential in preventing hypertension induced by L-N-G nitroarginine methyl ester (L-name) in rats, improving hemodynamic changes and reducing oxidative stress and vascular inflammation [111]. Rice bran is rich in biologically active substances that benefit human health, supporting immune, antioxidant, anticancer, and antidiabetic functions [112]. Rice's anti-inflammatory, anti-allergy, anti-atherosclerosis, anti-influenza, anti-obesity, and antitumor properties contribute to defense against various chronic and degenerative diseases, including hypertension, and may play a role in preventing COVID-19 infection.

2.2.2.2 Wheat (Triticum spp.)

Wheat (*Triticum* spp.) is a crucial staple food, providing a significant portion of starch and calories in meals. Studies have shown that regular consumption of wheat, particularly dietary fiber and other bioactive substances, are associated with a decrease in chronic diseases [113]. Wheat contains various phytochemicals, including phenolic acids, terpenoids, tocopherols, and sterols, with whole wheat containing significant amounts of phenolic acids [114, 115]. The processing and utilization of wheat significantly impact the composition of bioactive compounds and, consequently, the health benefits, such as improvements in colon functions, cancer prevention, protection against obesity, weight loss promotion, and other positive effects [116–125].

Research on wheat's pharmacology has identified ACE-inhibitory peptides derived from wheat gluten protein hydrolysates, which have potential cardiovascular benefits [126]. Structure plays a role in the inhibitory activity of these peptides, and certain sequences with specific amino acids exhibit greater inhibitory activity toward ACE [127]. The isolation of phenols from protein fractions in wheat flour has been found to enhance antihypertensive activity and antioxidant characteristics while reducing allergenicity [128]. Polysaccharides found in wheat have also shown the ability to boost the immune response to infectious diseases by activating protein pathways in cells like macrophages, thereby stimulating immune response control processes [129].

2.2.2.3 Millet

Millet is made up of several species that are not genetically linked. It does, however, include a variety of phytochemicals, phenolic compounds, phytosterols, policosanols, and bioactive peptides [130]. Foxtail millet (*Setaria italica* Beauv) has antioxidant and anticancer properties and lowers cholesterol levels [131, 132]. Foxtail millet also has antihypertensive properties. Studies have also documented this cereal's hydrolyzed proteins' ability to suppress ACE [133]. Eating whole grains helps lower blood pressure. Forty-five middle-aged hypertensive patients who consumed 50 g of whole grains of pulverized foxtail millet extruded as bread or millet pancakes for 12 weeks showed a significant decrease in SBP of 133.61 and 129.48 mmHg as well as a decrease in the mass index and body fat [134]. Cereals have antihypertensive benefits because they improve endothelial function by blocking the effects of vasoconstrictors like Ang II itric oxide-based vasodilatation induced by substances like Ang II, which also affects the vasorelaxation tracts. In addition to the cereal already described, millet ingestion can support immune function modulation, which helps to guard against the COVID-19 illness [135].

2.2.2.4 Sorghum

Tannins, phenolic acids, anthocyanins, and phytosterols are all present in sorghum (*Sorghum* spp.). These phytochemicals may have a major impact on human health by boosting cardiovascular health by lowering plasma levels of hepatic cholesterol and low-density lipoproteins [136]. Between 16 and 131 mg/g and 41 and 444 mg/g, respectively, of benzoic and cinnamic acids are present in sorghum [137]. The most researched sorghum flavonoids are anthocyanins. According to Awika et al. [138], black sorghum bran has anthocyanin levels at least twice as high as those of red sorghum (10.1 mg/g) and three to four times higher than those of whole grain. Though approximate concentrations of 44 to 72 mg/100 g have been reported [138, 139], there is a lack of quantitative information regarding the phytosterols contained in sorghum. The human immunodeficiency virus (HIV), herpes simplex 1 (VHS-1), hepatitis B and C (VHB/VHC), influenza A (H1N1), and, more recently, the virus that caused the COVID-19 disease (SARS-CoV-2) have all been shown to be susceptible to the antiviral effects of polyphenols [140]. In addition to having antiviral properties, phenolic chemicals also have antihypertensive properties. Irondi et al. [141] investigated the inhibitory activity of several enzymes, including ACE, in raw and toasted red sorghum grain flour (150 and 180°C). They discovered that the presence of phenolic acids (gallic, chlorogenic, caffeic, ellagic, and p-coumaric) and flavonoids (quercetin, luteolin, and apigenin) in high concentrations in raw grains led to their high inhibitory activities (19.64 g/mL), as increasing the temperature during toasting reduces the presence of phenolic compounds and, consequently, results in a decrease in inhibitory activity, having an IC50 of 20.99 g/mL in grains roasted at 150°C and 22.81 g/mL in grains roasted at 180°C. As a result, the corresponding decrease in the inhibitory activity of the enzymes and the phenolic composition of the grains with increasing toasting temperature shows that phenolic acids and flavonoids may be the primary inhibitors of the grain enzymes.

2.2.3 Root and tubers

Plants that produce starchy roots, tubers, rhizomes, corms, and stems are essential for human nutrition and health. In tropical places around the world, roots and tuber crops are essential agricultural staple energy sources, second only to cereals. They consist of potatoes, cassava, sweet potatoes, yams, and aroids, which are members of distinct botanical families but are grouped together. Several different carcinoma cell lines and animal models have shown that root and tuber phytochemicals have numerous pharmacology activities.

2.2.3.1 Yams (Dioscorea sp.)

Yam are root crops that contain various bioactive substances with potential health benefits. They are rich in mucin, dioscin, dioscorin, allantoin, choline, polyphenols, diosgenin, carotenoids, tocopherols, and vitamins [142, 143]. Yam extracts have demonstrated hypoglycemic, antibacterial, and antioxidant properties [144, 145]. They can increase the activity of digestive enzymes in the small intestine and promote the growth of stomach epithelial cells [146]. Additionally, yams have shown potential in bone marrow cell regeneration and splenocyte proliferation [147]. Dioscorin, a component of fresh yam (*Dioscorea batatas*), has exhibited DPPH radical scavenging activity and has been found to have beneficial effects in lowering blood pressure [148–150].

Yam cultivars also possess phenolic compounds that have antibacterial activity. Methanolic extracts from *Dioscorea* yams, such as *Dioscorea dumetorum* and *Dioscorea hirtiflora*, have demonstrated antioxidant and antibacterial activity [151]. In vitro studies using agar diffusion tests showed strong antibacterial activity against *Proteus mirabilis* by the nonedible *D. dumetorum*. Methanolic extracts from *D. hirtiflora* exhibited susceptibility against various species, including *Staphylococcus aureus*, *E. coli*, *Bacillus subtilis*, *P. mirabilis*, *Salmonella typhi*, *Candida albicans*, *Aspergillus niger*, and *Penicillium chrysogenum*.

Furthermore, dioscorin from yams has shown inhibitory and antihypertensive effects on the angiotensin-converting enzyme (ACE) in rats with spontaneous hypertension [143, 151]. Yam dioscorin has also displayed actions such as dehydroascorbate reductase (DHA), monodehydroascorbate reductase (MDA), trypsin inhibitor, and immunomodulatory effects [152, 153]. Chinese yam, in particular, contains diosgenin, an immunoactive steroidal saponin with prebiotic properties that promote the development of enteric lactic acid bacteria [154]. In animal studies, an ethanol extract of *Dioscorea alata* tubers exhibited antidiabetic action in rats with alloxan-induced diabetes [155]. Yam extract administered to diabetic rats resulted in lower creatinine levels, potentially improving renal function by reducing plasma glucose levels and subsequent glycosylation of renal basement membranes. Bioactive components of yam have also shown potential in reducing the risk of cancer and cardiovascular disorders in postmenopausal women [156]. Chronic administration of *Dioscorea* has been associated with changes in hormonal activities, bone remodeling, and improved bone strength [157].

2.2.3.2 Ipomoea batatas L. (sweet potato)

Sweet potatoes (*I. batatas*) have their origins in Central America and are now widely cultivated in tropical and subtropical regions across the world. They are ranked as the seventh-largest food crop and can be grown throughout the year under ideal climatic conditions, making them a reliable "insurance crop" due to their resistance to complete crop loss from unfavorable weather conditions. The roots of sweet potatoes contain a soluble protein called sporamin, which constitutes 60–80% of the total proteins in the roots and serves as their primary protein store. Sporamin has been shown to possess antioxidant properties related to stress tolerance, including activities such as DHA and MDA reductase activities.

Alkaloids are another class of compounds found in sweet potatoes. Alkaloids are primarily used by plants as phytotoxins, antibacterials, insecticides, and fungicides to protect against feeding by insects, herbivorous animals, and mollusks. Commercial sweet potatoes mostly contain two glycoalkaloids, chaconine and solanine, which are glycosylated solonidine aglycone derivatives. Solasonine, a glycoalkaloid found in eggplants and wild potatoes (*Solanum chacoense*), belongs to the *Solanum* species.

Solanum species have been found to possess various biological properties, including antitumor, antifungal, teratogenic, antiviral, and antiestrogenic activities. Some glycoalkaloids are even used as anticancer agents.

The peels of sweet potatoes have been reported by several authors to have potent wound healing effects, likely due to the free radical scavenging activity of the phytoconstituents and their ability to inhibit lipid oxidation. Studies conducted in rat models have shown that sweet potato fiber can have a positive impact on burn or decubital wounds, leading to changes in wound quantity and quality. In these studies, rats treated with sweet potato fiber coating showed smaller wound areas compared to the control group. Furthermore, a sweet potato petroleum ether extract was found to significantly reduce the amount of scarring necessary for complete epithelialization compared to the control. Human studies have explored the effects of consuming potatoes with different flesh colors on oxidative stress and inflammatory damage. In a randomized trial, participants were given white-, yellow-, or purple-fleshed potatoes once daily. The results indicated that the color of the potatoes affected oxidative stress and inflammatory damage, with purple-fleshed potatoes, high in anthocyanin and phenolic acids, showing positive effects. According to Hwang et al. [158], purple sweet potatoes have the potential to be effective in fighting obesity. Anthocyanin fractions from purple sweet potatoes were found to reduce the accumulation of hepatic lipids by activating the adenosine monophosphate-activated protein kinase (AMPK) signaling pathways. AMPK is a critical regulator of lipid production in metabolic tissues. In a study involving mice fed with purple sweet potatoes for 4 weeks, an intake of 200 mg/kg of body weight per day of anthocyanins led to decreased weight gain, reduced hepatic triacylglycerol buildup, and improved serum lipid parameters. The study also found that AMPK and acetyl coenzyme A carboxylase (ACC) were more frequently phosphorylated in the liver and HepG2 hepatocytes after anthocyanin treatment.

2.2.3.3 Cassava (Manihot esculenta)

Cassava (*M. esculenta*) is the most extensively grown root crop in the tropics and can only be cultivated in tropical and subtropical regions due to its lengthy growth season, which typically ranges from 8 to 24 months. It belongs to the Euphorbiaceae family and is a perennial shrub. Among the 98 species that make up the genus *Manihot*, *M. esculenta* is the most widely cultivated species. The antioxidant properties of cassava roots have been the focus of several investigations. A recent study has highlighted that cassava tubers grown organically exhibited stronger antioxidant activities compared to roots treated with mineral-based fertilizers [159]. The researchers found that the total phenolic content (TPC) and flavonoid content (FC) of cassava cultivated with organic fertilizers were significantly higher than those of cassava treated with inorganic fertilizers. Antioxidants play a crucial role in protecting the body against oxidative stress and combating the damaging effects of free radicals. It is worth noting that cassava is primarily consumed as a staple food in many tropical regions, providing a significant source of carbohydrates. However, it is important to note that cassava contains cyanogenic glycosides, which are potentially toxic compounds. These compounds can release cyanide when consumed in large quantities or when cassava is prepared improperly. Therefore, proper processing methods such as soaking, fermenting, and cooking are essential to remove the cyanide content and make cassava safe for consumption.

2.3 Spices

In addition to fruits and vegetables, spices and herbs are further sources of natural antioxidants [160]. Spices and herbs can be categorized using both taxonomical characteristics and flavor or taste. Depending on the taxonomic categorization, these spices and herbs are classified as either flowering plants or members of the Angiospermae class [161]. These herbs and spices are an abundant supply of phenolic compounds, and because of their useful characteristics, they are frequently used as food additives and in the pharmaceutical industry [162]. Spices' chemical constituents, which in turn depend on the polyphenolic content and bioactive chemicals, are responsible for their antioxidant activity [160].

2.3.1 Basil (Tulsi)

In addition to preventing the growth of a variety of bacteria, yeasts, and molds, holy basil has been shown in a small research to improve immune system performance by raising certain immune cells in the blood. In addition to lowering blood sugar levels before and after meals and reducing anxiety and anxiety-related depression, holy basil has been linked to other health benefits. Reduces inflammation, reduces lipid peroxidation, and acts as an antioxidant [163].

2.3.2 Dill

Dill and other greens are used to produce a variety of regional meals that are served with rotis or chapatis in India. Dill was a common ingredient in traditional treatments for a wide range of ailments, including jaundice, headaches, boils, a lack of appetite, stomach issues, nausea, liver issues, and many others. Additionally, dill seeds can be used to make herbal tea. Which contains antioxidative and antimicrobial activities [164].

2.3.3 Marjoram

The tops of marjoram plants are trimmed as they start to flower and are then slowly dried in the shade. Marjoram is grown for its aromatic leaves, either green or dry, which are used in cooking. It is frequently combined with other herbs to make dishes like za'atar and herbes de Provence. Marjoram's flowering leaves and tops are steam-distilled to create an essential oil with a yellowish hue that eventually turns brown. It has a variety of chemical constituents, including pinene, borneol, and camphor. Antibacterial and antioxidant properties [165].

2.3.4 Thyme (Ajwain ke Phool)

Common thyme (*Thymus vulgaris*) essential oil, known as oil of thyme, contains 20–54% thymol. Along with these extra substances, thyme essential oil also includes p-cymene, myrcene, borneol, and linalool. Listerine and other commercially available mouthwashes contain the antimicrobial thymol as an active component. Oil of thyme was employed to treat bandages before to the development of current antibiotics. Additionally, it has been demonstrated to work well against a number of fungi that frequently infect toenails. Some all-natural, alcohol-free hand sanitizers use thymol as their main active ingredient [166]. The plant can be infused in water to create a tisane that can treat bronchitis and coughing. The antioxidant presence in thyme helps to prevent the loss of bone.

2.3.5 Rosmarinic acid

Acid is the name of the rosemary's active component. It has been demonstrated that this chemical reduces nasal congestion and allergic reactions. In a study, rosmarinic acid doses of 50 and 200 mg were found to reduce allergy symptoms. Along with less congestion, nasal mucus' immune cell count fell. Anti-inflammatory, anti-carcinogen, decreased bone resorption, and antioxidant [167]. Herbs and spices are a great method to enhance the flavor of food without adding extra calories because they are naturally low in calories, fat, saturated fat, carbohydrates, and sodium. In reality, adding herbs and spices to food can help consumers' diets contain less harmful elements. Studies on the ethnopharmacology of spices' antioxidant and anti-inflammatory effects on food and drinks [168] show that some of the most widely utilized natural antibacterial agents in food are in addition to these spices. Various spices contain natural substances that have antibacterial properties [169]. Because of this, steps must be done to control the issue by employing plant extracts that include photochemicals with antibacterial effects [170].

2.3.6 Garlic (Lehsun)

Allium sativum, a member of the Alliance family, is thought to have its origins in Central Asia. To improve physical and mental health, it is used internationally as a flavoring ingredient, a traditional remedy, and a functional food. Antioxidant inhibits cerebral aging, decreases blood pressure, elevates HDL cholesterol, reduces inflammation, and strengthens immunity [170].

2.3.7 Ginger (Adrakh)

Since ancient times, ginger, also known as *Zingiber officinale* Roscoe or Zingiberacae, has been used extensively in Chinese, Ayurvedic, and Tibb-Unani herbal medicines for a variety of unrelated ailments, such as arthritis, rheumatism, sprains, muscular aches and pains, sore throats, cramps, constipation, indigestion, vomiting, hypertension, dementia, fever, infectious diseases, and Antioxidant, alleviates knee osteoarthritis, antiemetic, anti-inflammatory, strengthens immunity, and antimicrobial (**Table 1**) [170].

3. Future perspective of medicinal plant and functional foods

Given that there are around 500,000 plants in the world, most of which have not yet been investigated in medical practice, and given that both current and future research on medical activities may be successful in treating diseases, the future of medicinal herbs is bright. There is a long history of the use of medicinal plants; however, using the entire plant or raw materials for treatment or experimentation has many disadvantages, including changes in the plant's compounds in different climates, the concurrent development of synergistic compounds that lead to

Diseases	Spices	Reference
Cardiovascular diseases including heart attack	Ginger, Turmeric, Garlic	[171–178]
Neurodegenerative diseases	Mint, Onions	[179]
Antidiabetic action	Cinnamon, bay leaf, wormwood, fenugreek, mustard, pomegranate	[180–182]
Hypertension	Cardamom, cinnamon	[183, 184]
Hepatic diseases	Caraway, Cardamom	[185–187]
Endocrine diseases	Ginger, Turmeric	[188]
Against DNA Oxidation	Basil	[189]
Obesity	Saffron, Turmeric	[190]
Bone diseases	Cloves	[191]
Immunomodulatory action	Turmeric	[192, 193]
Renal diseases	Garlic, Fennelflower, ginger	[194, 195]
Gastrointestinal diseases	Fenugreek, garlic	[196, 197]
Antiulcer action	Ginger	[198]
Pigment cell growth inhibition	Turmeric	[199]
Reduction of cortisol level in saliva	Lavender, rosemary	[197]
Against alcohol abuse	Thyme, ginger	[199]
Against gum disease	Licorice	[197]

Table 1.
Therapeutic effects of spices in different diseases.

antagonistic effects or other unexpected changes in bioactivity, and changes or loss of bioactivity due to variability and accumulation, storage, and preparation of raw materials.

However, novel approaches are therefore required to identify bioactive ingredients from medicinal plants and plant-based functional foods, evaluate their efficacy in human and animal models, and develop a sustainable and natural means of treating or preventing disease.

4. Conclusion

Medicinal plant and plant-based functional foods have garnered significant interest due to their immune-enhancing properties and potential to address immune dysfunction. Extensive research has been conducted to understand the cellular and molecular mechanisms through which the bioactive ingredients in these foods exert their immune-enhancing effects. Compared to conventional drugs, plant-based functional foods offer several advantages, including lower risk of side effects, stability, and sustained efficacy.

By incorporating a variety of fruits and vegetables into our diet, such as mangoes, tomatoes, carrots, beetroot, bananas, grapes, apples, pomegranates, and oranges, we can benefit from their rich content of dietary fiber, vitamins, minerals, and phytochemicals with potent antioxidant properties. These components have been associated with a reduced risk of chronic diseases, including heart disease, stroke, cancer, and age-related macular degeneration, as well as improved immune function.

Furthermore, exploring the potential of byproducts and waste materials from fruit and vegetable processing, such as peels and seeds, can lead to the discovery of additional bioactive compounds and contribute to reducing waste in the food industry. While medicinal plants and plant-based functional foods offer promising immune-enhancing properties, it is important to continue conducting research to fully understand their mechanisms of action and optimize their use in promoting overall health and well-being.

Author contributions

Conceptualization, Bamidele S.O and Origbemisoye B.A Writing—original draft preparation, Bamidele S.O Writing—review and editing, Origbemisoye B.A and Akinbode B.A; authors have read and agreed to the published version of the manuscript.

Funding

This research was not funded by any organization.

Conflict of interest

The authors declare that the research was conducted in the absence of any commercial or financial relationships that could be construed as a potential conflict of interest.

Author details

Bamidele Sekinat Olayem[1*], Origbemisoye Babawande Olaitan[1]
and Akinbode Badiu Akinola[2]

1 Federal University of Technology, Akure, Ondo State, Nigeria

2 Gdansk University of Technology, Gdansk, Poland

*Address all correspondence to: sekinatolayemi@gmail.com

References

[1] Bolen S, Feldman L, Vassy J. Systematic review: Comparative effectiveness and safety of oral medications for type 2 diabetes mellitus. Annals of Internal Medicine. 2007;**147**(6):386-399

[2] Petrovska BB. Historical review of medicinal plants' usage. Pharmacognosy Reviews. 2012;**6**(47):80-82

[3] Di Fabio G, Romanucci V, Zarrelli M, Giordano M, Zarrelli A. C-4 gem-dimethylated oleanes of Gymnema sylvestre and their pharmacological activities. Molecules. 2013;**18**(12): 14892-14919

[4] Ekor M. The growing use of herbal medicines: Issues relating to adverse reactions and challenges in monitoring safety. Frontiers in Neurology. 2014;**4**:Article No. 177

[5] Origbemisoye WA, Bamidele SO. Immunomodulatory foods and functional plants for COVID-19 prevention: A review. Asian Journal of Medical Principles and Clinical Practice. 2020;**3**(4):15-26

[6] Saxena M, Saxena J, Nema R, Singh D, Gupta A. Phytochemistry of medicinal plants. Journal Pharmacognotic Phytochemistry. 2013;**1**(6):168-182

[7] Charalampopoulos D, Wang R, Pandiella SS, Webb C. Application of cereals and cereal components infunctional foods: Are view. International Journal Food Microbiology. 2002;**79**:131-141. DOI: 10.1016/S0168

[8] Youwei Z, Junlian Z, Peng Y. A comparative study on the free radical scavenging activities of some fresh flowers in southern China. Journal of Pharmacology. 2008;**41**(9):1586-1591

[9] Wojcikowski K, Stevenson L, Leach D, Wohlmuth H, Gobe G. Antioxidant capacity of 55 medicinal herbs traditionally used to treat the urinary system: A comparison using a sequential three-solvent extraction process. The Journal of Alternative & Complementary Medicine. 2007;**13**(1):103-109

[10] Morton JF. Atlas of medicinal plants of middle America. In: Library of Congress Cataloging in Publication Data. Bahama to Yucatan. Middle, Collectanea, University of Miami, Coral Gables. Florida, USA: Thomas Books; 1981. p. 1420. Ref.563

[11] Chevallier A. Encyclopedia of Herbal Medicine: Natural Health. 2nd ed. USA: Dorling Kindersley Book; 2000. p. 336

[12] Sabir SM, Rocha JBT. Waterextractable phytochemical from Phyllanthus Niruri exhibit distinct in vitro antioxidant and in vivo hepatoprotective activity against paracetamol-induced liver damage in mice. Journal Food Chemistry. 2008;**4**:60

[13] Row LR, Satyanarayana P, Subba Rao GSR. Crystalline constituents of Euphorbiaceae—The synthesis and absolute confguration of phyllanthin. Tetrahedron. 1967;**23**:1915

[14] MacRae WD, Towers GHN. Biological activities of lignans. Phytochemistry. Journal of Biochemistry. 1984;**23**: 1207-1220

[15] Calixto JB, Santos AR, Cechinel Filho V, Yunes RA. A review of the plants of the Phyllanthus: Their chemistry, pharmacology, and therapeutic potential. MedResRev. 1998;**18**:225-258

[16] Ayres DC, Loike JD. Lignans. In: Chemical, biological and clinical properties. Cambridge: Cambridge University Press; 1990

[17] Negi AS, Kumar JK, Luqman S, Shanker K, Gupta MM, Khanuja SP. Recent advances in plant hepatoprotectives: A chemical and biological profle of some important leads. Med Reserve Review. 2008;**28**:746-772

[18] Sharma A, Singh RT, Anand S. Estimation of phyllanthin and hypophyllanthin by high performance liquid chromatography in Phyllanthus amarus. Photochemical Analysis. 1993;**4**:226-229

[19] Akintonwa A, Essien AR. Protective effects of Garcinia kola seed extract against paracetamol-induced hepatotoxicityin rats. Journal of Ethnopharmacology. 1990;**29**:207-219

[20] Okunji CO, Ware TA, Hicks RP, Iwu MM, Skanchy DJ. Capillary electrophoresis determination of biflavanones from Garcinia kola in three traditional African medicinal formulations. Planta Medica. 2002;**68**:440-444

[21] Esimone CO, Adikwu MU, Nworu CS, Okoye FBC, Odimegwu DC. Adaptogenic potentials of Camellia sinensis leaves, Garcinia kola and Kola nitida seeds. Science Research Essays. 2007;**2**:232-237

[22] Antia BS, Pansanit A, Ekpa OD, Ekpe UJ, Mahidol C, Kittakoop P. Alpha-glucosidase inhibitory, aromatase inhibitory and antiplasmodial activities of a biflavonoid GB1 from Garcinia kola stem bark. Planta Medica. 2010;**76**(3):276-277

[23] Adaramoye OA, Farombi EO, Adeyemi EO, Emerole GO. Inhibition of human low density lipoprotein oxidation by flavonoids of Garcinia kola seeds. Pakistan Journal Medicine Science. 2005a;**21**(3):331-339

[24] Lacmata ST, Kuete V, Dzoyem JP, Tankeo SB, Teke GN, Kuiate JR, et al. Antibacterial activities of selected cameroonian plants and their synergistic effects with antibiotics against bacteria expressing MDR phenotypes based complementary. Journal of Alternative medicine. 2012;**6**:23-72

[25] Terashima K, Takaya NM. Powerful antioxidative agents based on garcinoic acid from Garcinia kola. Bioorganic. Medicinal Chemistry. 2002;**10**(5):1619-1625

[26] Narcisi EM, Sacor NE. In vitro effect of tinidazole and furazolidone on metronidazole resistant trichomonas vaginalis. Antimicrobial Agents and Chemotherapy. 1996;**40**:1121-1126

[27] Odebunmi EO, Oluwanili OO, Awolola GV, Adediji OD. Proximate and nutritional composition of Kola nut (cola nitida), bitter kola (Garcinia kola), and Alligator pepper (Afromomum melegueta). African Journal of Biotechnology. 2009;**8**(2):308-310

[28] Seanego CT, Ndip RN. Identification and antibacterial evaluation of bioactive compounds from Garcinia kola (Heckel) seeds. Molecules. 2012;**17**(6):6569-6584

[29] Seanego CT, Ndip RN. Identification and antibacterial evaluation of bioactive compounds from Garcinia kola (Heckel) seeds. Molecules. 2012, 1996;**17**(6):6569-6584

[30] Xu XY, Li F, Zhang X, Li PC, Zhang X, Wu ZX. In Vitro Synergistic Antioxidant Activity and Identification of Antioxidant Components from

Astragalus membranceus and paeonia lectiflora. Plos ONE. 2014;**9**:e96780

[31] Farombi EO, Owoeye O. Antioxidative and chemopreventive properties of Vernonia amygdalina and Garcinia biflavonoid. International Journal of Environment Research and Public Health. 2011;**8**(6):2533-2555

[32] Iwu MM. Anti-hepatotoxicity of Garcinia kola seeds. Experimentia. 1985;**41**:679-700

[33] Iwu MM, Igboko OA, Onwuchekwa U, Okunji CO. Evaluation of the anti-hepatotoxicity of the biflavonoids of Garcinia kola seeds. Journal of Ethnopharmacology. 1987;**21**:127-142

[34] Iwu MM, Igboko OA, Okunji CO, Tempesta MS. Anti-diabetic and aldose reductase activities of biflavanones of Garcinia Kola. Journal of Pharmacy and Pharmacoogy. 1990;**42**:2903-2922

[35] Mukherjee PK, Nema NK, Maity N, Mukherjee K, Harwansh RK. Phytochemical and therapeutic profile of Aloe vera. Journal of Natural Remedies. 2014;**14**(1):1-26

[36] Kahlon JB, Kemp MC, Carpenter RH, McAnalley BH, McDaniel HR, Shannon WM. Inhibition of AIDS virus replication byacemannan invitro. Molecular Biotherapy. 1991;**3**:127-135

[37] Bernard SG, Hughes BG, Sidwell RW. Evaluation of the antiviral activity of anthraquinones, anthrones and anthraquinone derivatives against human cytomegalovirus. Antiviral Research. 1992;**35**:2463-2466

[38] Semple SJ, Pyke SM, Reynolds GD, Flower RL. Invitro antiviral activity of the anthraquin one chrysophanic acid against poliovirus. Journal of Antiviral Reserve. 2001;**49**(3):169-178

[39] Bhalsinge RR, Rajbhoj SR, Limaye MV, Vaidya MU, Rane PS, Tilak AV. Anti inflammatory and immunomodulatory activity of ethanol extract of aloe veragel. IJPSR. 2017;**9**(2):832-835

[40] Dhouibi R, Affes H, Ben Salem M, Hammami S, Sahnoun Z, Zeghal KM, et al. Screening of pharmacological uses of Urtica dioica and others benefits. Progress in Biophysics and MolecularBiology. 2020;**150**:67-77

[41] Vander Meer FJUM, Haan CAM, Schuurman NMP, Haijema BJ, Verheije MH, Bosch BJ, et al. Thecarbohydrate-binding plant lectins and the nonpeptidic antibiotic pradimicin a target the glycans of the coronavirus envelop eglycoproteins. Journal of Antimicrobial Chemotherapy. 2007;**4**:741-749

[42] Katsuki T, Luscombe D. Torreyanucifera. The IUCN red list of threatened species. IUCN. 2013;**4**:298-7599

[43] Ewing S. The Great Alaska Nature Factbook: A Guide to the state's Remarkable Animals, Plants and Natural Alaska, USA. 2nd ed. Vol. 106. Graphic Arts Books; 2012. pp. 142-180

[44] Keivan Z, Moloud AZ, Kohzad S, Zahra R. Anti viral activity of Aloe vera against herpes simplex virus type 2: An in vitro study. African Journal of Biotechnology. 2007;**6**(15): 1770-1773

[45] Sosulski FW, Dabrowski KJ. Composition of free and hydrolysable phenolic acids in flours and hulls of ten legume species. Journal Agric Food Chemistry. 1984;**32**:131-133

[46] Madhujith T, Amarowicz R, Shahidi F. Phenolic antioxidants in beans and their effects on inhibition of radical induced DNA damage. Journal of the American Oil Chemists' Society. 2004;**81**:691-696

[47] Madhavi DL, Singhal RS, Kulkarni PR. Technological aspects of food antioxidants. In: Madhavi DL, Deshpande SS, Salunkhe DK, editors. Food Antioxidants: Technological, Toxicological, and Health Perspectives. New York: Marcel Dekker; 1996. pp. 159-265

[48] Cao G, Sofic E, Prior RL. Antioxidant capacity of tea and common vegetables. Journal Agric Food Chemistry. 1996;**44**:3426-3431

[49] Duenas M, Estrela I, Hernandez T. Occurrence of phenolic compounds in the seed coat and the cotyledon of peas (Pisum sativum L.). European Food Reserve Technology. 2004;**219**:116-123

[50] Diaz-Batalla L, Widholm JM, Fahey GC, Castano Tostado E, Paredes-Lopez O. Chemical components with health implications in wild and cultivated Mexican common bean seeds (Phaseolus vulgaris L.). Journal of Agricultural and Food Chemistry. 2006;**54**:2045-2052

[51] Duenas M, Fernandez D, Hernandez T, Estrella I, Munoz R. Bioactive phenolic compounds of cowpeas. (Vigna sinensis L.). modifications by fermentation with natural microflora and with lactobacillus plantarum ATCC 14977. Journal Science Food Agriculture. 2005;**85**:297-304

[52] Amarowicz R, Troszynska A. Antioxidant activity of extract of pea and its fractions of low molecular phenolics and tannins. Poland Journal Food Nutrition Science. 2003;**53**:10-15

[53] Lopez-Amoros ML, Hernandez T, Estrella I. Effect of germination on legume phenolic compounds and their antioxidant activity. Journal Food Composition Analyses. 2006;**19**:277-283

[54] Amarowicz R, Estrella I, Hernandez T, Troszynska A. Antioxidant activity of extract of adzuki bean and its fractions. Journal Food Lipids. 2008;**15**:119-136

[55] Duenas M, Hernandez T, Estrella I. Assessment of in vitro antioxidant capacity of the seed coat and the cotyledon of legumes in relation to their phenolic contents. Food Chemistry. 2006;**98**:95-103

[56] Clinton SK, Giovannicei EL, Hurshing SD. WCRF/AICR. Food, Nutrition, Physical Activity, and the Prevention of Cancer impact and future direction. Journal Nutrition. 2020;**150**(4):663-671

[57] Xu B, Chang SKC. Comparative study on antiproliferation properties and cellular antioxidant activities of commonly consumed food legumes against nine human cancer cell lines. Food Chemistry. 2012;**134**:1287-1296

[58] Kim DK, Jeong SC, Gorinstein S, Chon SU. Total polyphenols, antioxidant and antiproliferative activities of different extracts in mungbean seeds and sprouts. Plant Foods Human Nutrition. 2012;**67**:71-75

[59] Clemente A, Carmen M, Jiménez E, Carmen AM, Domoney C. The anti-proliferative effect of TI1B, a major Bowman-birk isoinhibitor from pea (Pisum sativum L.), on HT29 colon cancer cells is mediated through protease inhibition. British Journal of Nutrition. 2012;**108**(Suppl. 1):S135-S144

[60] Campos-Vega R, Oomah BD, Loarca-Pina G, Vergara-Castaneda HA. Common beans and their non-digestible fraction: cancer inhibitory activity—An overview. Food. 2013, 2013;**2**(3):374-392. DOI: 10.3390/foods2030374

[61] Hayde VC, Ramon GG, Lorenzo GO, Dave OB, Rosalia RC, Paul W, et al. 2012. Non-digestible fraction of beans (Phaseolus vulgaris L.) modulates signalling pathway genes at an early stage of colon cancer in Sprague–Dawley rats. British Journal of Nutrition. 2012;**108**:S145-S154. DOI: 10.1017/S0007114512000785

[62] Eide DJ. The oxidative stress of zinc deficiency. Metallomics. 2011;**3**(11):1124-1129. DOI: 10.1039/C1MT00064K

[63] Greeder GA, Milner JA. Factors influencing the inhibitory effect of selenium on mice inoculated with Ehrlich ascites tumor cells. Science. 1980;**209**(4458):825-827. DOI: 10.1126/science.7406957

[64] Dai J, Mumper RJ. Plant phenolics: Extraction, analysis and their antioxidant and anticancer properties. Molecules. 2010;**15**(10):7313-7352. DOI: 10.3390/molecules15107313

[65] Kerem Z, German-Shashoua H, Yarden O. Microwave-assisted extraction of bioactive saponins from chickpea (Cicer arietinum L). Journal Science Food Agriculture. 2005;**85**(3):406-412

[66] Fan Y, Guo DY, Song Q, Li T. Effect of total saponin of aralia taibaiensis on proliferation of leukemia cells. Journal of Chinese Medicinal Materials. 2013;**36**(4):604-607

[67] Shi J, Arunasalam K, Yeung D, Kakuda Y, Mittal G, Jiang Y. Saponins from edible legumes: Chemistry, processing, and health benefits. Journal Medicinal Food. 2004;**7**(1):67-78. DOI: 10.1089/109662004322984734

[68] Mudryj AN, Yu N, Aukema HM. Nutritional and health benefits of pulses. Applied Physiology, Nutrition, and Metabolism. 2014;**39**(11):1197-1204

[69] Chan YS, Zhang Y, Sze SCW, Ng TB. A thermostable trypsin inhibitor with antiproliferative activity from small pinto beans. Journal of Enzyme Inhibition and Medicinal Chemistry. 2013;**29**(4):485-490. DOI: 10.3109/14756366.2013.805756

[70] Wang S, Meckling KA, Marcone MF, Kakuda Y, Tsao R. Can phytochemical antioxidant rich foods act as anti-cancer agents? Food Reserve International. 2011;**44**:2545-2554

[71] Hwang CS, Kwak HS, Lim HJ, Lee SH, Kang YS, Choe TB. Isoflavone metabolites and there in vitro dual functions: They can act as an estrogenic agonist or antagonist depending on the estrogen concentration. Journal Steroid Biochemistry Molecule Biology. 2006;**101**:246-265

[72] Bazzano LA, He J, Ogden LG, Loria C, Vupputuri S, Myers L, et al. Legume consumption and risk of coronary heart disease in US men and women: NHANES I epidemiologic follow-up study. Archives of Internal Medicine. 2001;**161**:2573-2578

[73] Messina MJ. Legumes and soybeans: Overview of their nutritional profiles and health effects. American Journal Clinical Nutrition. 1999;**70**:439S-450S

[74] Winham DM, Hutchins AM. Baked bean consumption reduces serum cholesterol in hypercholesterolemic adults. Nutritional Reserve. 2007;**27**:380-386

[75] Iqbal A, Khalil IA, Ateeq N, Sayyar KM. Nutritional quality of important food legumes. Food Chemistry. 2006;**97**(2):331-335

[76] Lovejoy JC. Fat: The good, the bad, and the ugly. In: Wilson T, Bray GA, Temple NL, Struble MB, editors. Nutrition Guide for Physicians. New York: Humana Press; 2010. pp. 1-11

[77] Patterson CA, Maskus H, Dupasquier C. Pulse Crops for Health. Cereals Foods World. Canada: AACC International Inc. 2009;**54**(3):108-112

[78] Jenkins DJ, Kendall CW, Augustin LS, Mitchell S, Sahye-Pudaruth S, Blanco Mejia S, et al. Effect of legumes as part of a low glycemic index diet on glycemic control and cardiovascular risk factors in type 2 diabetes mellitus: A randomized controlled trial. Archives of Internal Medicine. 2012;**172**:1653-1660

[79] Bazzano LA, Thompson AM, Tees MT, Nguyen CH, Winham DM. Non-soy legume consumption lowers cholesterol levels: A meta-analysis of randomized controlled trials. Nutrition, Metabolism, and Cardiovascular Diseases: NMCD. 2011;**21**:94-103

[80] Anderson JW, Major AW. Pulses and lipaemia, short- and long-term effect: Potential in the prevention of cardiovascular disease. British Journal of Nutrition. 2002;**88**(Suppl. 3):S263-S271

[81] Ha V, Sievenpiper JL, de Souza RJ, Jayalath VH, Mirrahimi A, Agarwal A, et al. Effect of dietary pulse intake on established therapeutic lipid targets for cardiovascular risk reduction: A systematic review and meta analysis of controlled feeding trials. CMAJ. 13, 2014;**186**(8):52-62. DOI: 10.1503/cmaj.131727

[82] Hertog MG, Feskens EJ, Hollman PC, Katan MB, Kromhout D. Dietary antioxidant flavonoids and risk of coronary heart disease: The Zutphen elderly study. The Lancet. 1993;**342**(8878):1007-1011

[83] Pischon T, Girman CJ, Hotamisligil GS, Rifai N, Hu FB, Rimm EB. Plasma adiponectin levels and risk of myocardial infarction in men. JAMA. 2004;**291**(14):1730-1737. DOI: 10.1001/jama.291.14.1730

[84] Teede HJ, McGrath BP, DeSilva L, Cehun M, Fassoulakis A, Nestel PJ. Isoflavones reduce arterial stiffness: A placebo-controlled study in men and postmenopausal women. Arteriosclerosis Thrombosis and Vascular Biology. 2003;**23**(6):1066-1071. DOI: 10.1161/01.ATV.0000072967.97296.4A

[85] Papanikolaou Y, Fulgoni VL 3rd. Bean consumption is associated with greater nutrient intake, reduced systolic blood pressure, lower body weight, and a smaller waist circumference in adults: Results from the National Health and nutrition examination survey 1999-2002. Journal of the American College of Nutrition. 2008;**27**(5):569-576. DOI: 10.1080/07315724.2008.10719740

[86] Hermsdorff HH, Zulet MA, Abete I, Martinez JA. A legume-based hypocaloric diet reduces proinflammatory status and improves metabolic features in overweight/obese subjects. European Journal Nutrition. 2011;**50**:61-69

[87] Jayalath VH, de Souza RJ, Sievenpiper JL, Ha V, Chiavaroli L, Mirrahimi A, et al. Effect of dietary pulses on blood pressure: A systematic review and meta-analysis of controlled feeding trials. American Journal of Hypertension. 2014;**27**(1):56-64. DOI: 10.1093/ajh/hpt155. Epub 2013 Sep 7

[88] Belski R, Mori TA, Puddey IB, Sipsas S, Woodman RJ, Ackland TR, et al. Effects of lupin-enriched foods on body composition and cardiovascular disease risk factors: A 12-month randomized controlled weight loss trial. International Journal Obesity. 2011;**35**:810-819

[89] Lee YP, Mori TA, Puddey IB, Sipsas S, Ackland TR, Beilin LJ, et al. Effects of lupin kernel flour-enriched bread on blood pressure: A controlled intervention study. American Journal of Clinical Nutrition. 2009;**89**:766-772

[90] Jenkins DJ, Wolever TM, Taylor RH, Barker H, Fielden H, Baldwin JM, et al. Glycemic index of foods: A physiological basis for carbohydrate exchange. American Journal of Clinical Nutrition. 1981;**34**(3):362-366

[91] Rizkalla SW, Bellisle F, Slama G. Health benefits of low glycaemic index foods, such as pulses, in diabetic patients and healthy individuals. British Journal of Nutrition. 2002;**88**(S3):255-262. DOI: 10.1079/BJN2002715

[92] Jenkins DJ, Kendall CW, Augustin LS, Mitchell S, Sahye-Pudaruth S, Mejia SB, et al. Effect of legumes as part of a low glycemic index diet on glycemic control and cardiovascular risk factors in type 2 diabetes mellitus: A randomized controlled trial effect of legumes on glycemic control. Arch International Medicine. 2012;**172**(21):1653-1660. DOI: 10.1001/2013.jamainternmed.70

[93] Jenkins DJ, Wolever TM, Taylor RH, Barker HM, Fielden H. Exceptionally low blood glucose response to dried beans: Comparison with other carbohydrate foods. Bristish Medicine Journal. 1980;**281**:578-580

[94] Mollard RC, Zykus A, Luhovyy BL, Nunez MF, Wong CL, Anderson GH. The acute effects of a pulse-containing meal on glycaemic responses and measures of satiety and satiation within and at a later meal. British Journal of Nutrition. 2011;**108**(3):509-517. DOI: 10.1017/S0007114511005836 Epub 2011 Nov 7

[95] Nestel P, Cehun M, Chronopoulos A. Effects of long-term consumption and single meals of chickpeas on plasma glucose, insulin, and triacylglycerol concentrations. Am Journal Clinical Nutrition. 2004;**79**:390-395

[96] Sievenpiper JL, Kendall CW, Esfahani A, Wong JM, Carleton AJ, Jiang HY, et al. Effect of non-oil-seed pulses on glycaemic control: A systematic review and meta-analysis of randomised controlled experimental trials in people with and without diabetes. Diabetologia. 2009;**52**:1479-1495

[97] Shobana S, Sreerama YN, Malleshi NG. Composition and enzyme inhibitory properties of finger millet (Eleusine coracana L.) seed coat phenolics: Mode of inhibition of α-glucosidase and pancreatic amylase. Food Chemistry. 2009;**115**:1268-1273

[98] Evelyn M, Mendoza T. Development of functional foods in the Philippines. Food Science and Technology Research. 2007;**13**(3):179-186

[99] Frassetto LA, Todd KM, Morris RC Jr, Sebastian A. Worldwide incidence of hip fracture in elderly women: Relation to consumption of animal and vegetable foods. Journal Gerontol Biology Science Medicine Science. 2000;**55**:M585-M592

[100] Mollard RC, Wong CL, Luhovyy BL, Anderson GH. First and second meal effects of pulses on blood glucose, appetite, and food intake at a later meal. Applied Physiological Nutrition Metabolic. 2011;**36**:634-642

[101] Murty CM, Pittaway JK, Ball MJ. Chickpea supplementation in an Australian diet affects food choice, satiety and bowel health. Appetite. 2010;**54**:282-288

[102] Venn BJ, Perry T, Green TJ, Skeaff CM, Aitken W, Moore NJ, et al. The effect of increasing consumption of pulses and whole grains in obese people: A randomized controlled trial. Journal of the American College of Nutrition. 2010;**29**:365-372

[103] Björck I, Östman E, Kristensen M, Mateo Anson N, Price RK, Haenen GRMM, et al. Cereal grains for nutrition and health benefits: Overview of results from in vitro, animal and human studies in the HEALTHGRAIN project. Trends Food Science Technology. 2012;**25**:87-100

[104] Zamaratskaia G, Mhd Omar NA, Brunius C, Hallmans G, Johansson J-E, Andersson S-O, et al. Consumption of whole grain/bran rye instead of refined wheat decrease concentrations of TNF-R2, e-selectin, and endostatin in an exploratory study in men with prostate cancer. Clinical Nutrition. 2020;**39**:159-165

[105] Alu'Datt MH, Ereifej K, Abu-Zaiton A, Alrababah M, Almajwal A, Rababah T, et al. Anti-oxidant, anti-diabetic, and anti-hypertensive effects of extracted phenolics and hydrolyzed peptides from barley protein fractions. International Journal Food Proportion. 2012;**15**:781-795

[106] Gupta R, Meghwal M, Prabhakar PK. Bioactive compounds of pigmented wheat (Triticum aestivum): Potential benefits in human health. Trends Food Science Technology. 2021;**110**:240-252

[107] Ho HVT, Sievenpiper JL, Zurbau A, Blanco Mejia S, Jovanovski E, Au-Yeung F, et al. The effect of oat β-glucan on LDL-cholesterol, non-HDL-cholesterol and apoB for CVD risk reduction: A systematic review and meta-analysis of randomised-controlled trials. Britian journal. Nutrition. 2016;**116**:1369-1382

[108] Kim B, Woo S, Kim M-J, Kwon S-W, Lee J, Sung SH, et al. Identification and quantification of flavonoids in yellow grain mutant of rice (Oryza sativa L.). Food Chemistry. 2018;**241**:154-162

[109] Gong ES, Liu C, Li B, Zhou W, Chen H, Li T, et al. Phytochemical profiles of rice and their cellular antioxidant activity against ABAP induced oxidative stress in human hepatocellular carcinoma HepG2 cells. Food Chemistry. 2020;**318**:126484

[110] Yu X, Chu M, Chu C, Du Y, Shi J, Liu X, et al. Wild rice (Zizania spp.): A review of its nutritional constituents, phytochemicals, antioxidant activities, and health-promoting effects. Food Chemistry. 2020;**331**:127293

[111] Okarter N, Liu RH. Health benefits of whole grain phytochemicals. Critical Reverse Food Science Nutrition. 2010;**50**:193-208

[112] Deng Y, Luo Y, Qian B, Liu Z, Zheng Y, Song X, et al. Antihypertensive effect of few-flower wild rice (Zizania latifolia Turcz.) in spontaneously hypertensive rats. Food science. Biotechnology. 2014;**23**:439-444

[113] Gammoh S, Alu'datt MH, Alhamad MN, Rababah T, Al-Mahasneh M, Qasaimeh A, et al. The effects of protein-phenolic interactions in wheat protein fractions on allergenicity, antioxidant activity and the inhibitory activity of angiotensin I-converting

enzyme (ACE). Food Bioscience. 2018;**24**:50-55

[114] Chen J, Duan W, Ren X, Wang C, Pan Z, Diao X, et al. Effect of foxtail millet protein hydrolysates on lowering blood pressure in spontaneously hypertensive rats. European Journal Nutrition. 2017;**56**:2129-2138

[115] Jan-on G, Sangartit W, Pakdeechote P, Kukongviriyapan V, Sattayasai J, Senaphan K, et al. Virgin rice bran oil alleviates hypertension through the upregulation of eNOS and reduction of oxidative stress and in flammation in L-NAME À induced hypertensive rats. Nutrition. 2020;**69**:110575

[116] Asoodeh A, Haghighi L, Chamani J, Ansari-Ogholbeyk MA, Mojallal-Tabatabaei Z, Lagzian M. Potential angiotensin converting enzyme inhibitory peptides from gluten hydrolysate: Biochemical characterization and molecular docking study. Journal Cereal Science. 2014;**60**:92-98

[117] Luthria DL, Lu Y, John KMM. Bioactive phytochemicals in wheat: Extraction, analysis, processing, and functional properties. Journal Functional Foods. 2015;**18**:910-925

[118] Andersson AAM, Dimberg L, Åman P, Landberg R. Recent findings on certain bioactive components in whole grain wheat and rye. Journal Cereal Science. 2014;**59**:294-311

[119] Wieser H, Koehler P, Scherf KA. (eds.) chapter 6—Nutritional value of wheat. In: Wheat—An Exceptional Crop. Sawston, UK: Woodhead Publishing; 2020. pp. 133-148. ISBN 978-0-12-821715-3

[120] Kim SK, Ngo DH, Vo TS. Chapter 16—Marine fish-derived bioactive peptides as potential antihypertensive agents. In Marine Medicinal Foods; Kim, S.-K., Ed.; Vol. 65. Cambridge, MA, USA: Academic Press; 2012. pp. 249-260. ISBN 1043-4526

[121] Barbosa JR, de Carvalho Junior RN. Occurrence and possible roles of polysaccharides in fungi and their influence on the development of new technologies. Carbohydrate Polymerase. 2020;**246**:116613

[122] Deng C, Fu H, Shang J, Chen J, Xu X. Dectin-1 mediates the immunoenhancement effect of the polysaccharide from Dictyophora indusiata. International Journal Biology. Macromolecules. 2018;**109**:369-374

[123] Shen T, Wang G, You L, Zhang L, Ren H, Hu W, et al. Polysaccharide from wheat bran induces cytokine expression via the toll-like receptor 4-mediated p38 MAPK signaling pathway and prevents cyclophosphamideinduced immunosuppression in mice. Food Nutrition Reserve. 2017;**61**:1344523

[124] Duodu KG, Awika JM. Chapter 8—Phytochemical-related health-promoting attributes of Sorghum and millets. In Taylor JRN, Duodu KG, editors. Sorghum and Millets. 2nd ed. Washington, DC, USA: AACC International Press; 2019. pp. 225-258. ISBN 978-0-12-811527-5

[125] Choi Y-Y, Osada K, Ito Y, Nagasawa T, Choi M-R, Nishizawa N. Effects of dietary protein of Korean foxtail millet on plasma adiponectin, HDL-cholesterol, and insulin levels in genetically type 2 diabetic mice. Bioscience Biotechnology Biochemistry. 2005;**69**:31-37

[126] Shan S, Li Z, Newton IP, Zhao C, Li Z, Guo M. A novel protein extracted from foxtail millet bran displays anti-carcinogenic effects in human

colon cancer cells. Toxicology Letters. 2014;**227**:129-138

[127] Baksi AJ, Treibel TA, Davies JE, Hadjiloizou N, Foale RA, Parker KH, et al. A meta-analysis of the mechanism of blood pressure change with aging. Journal of the American College of Cardiology. 2009;**54**:2087-2092

[128] Althwab S, Carr TP, Weller CL, Dweikat IM, Schlegel V. Advances in grain sorghum and its co-products as a human health promoting dietary system. Food Reserve International. 2015;**77**:349-359

[129] Chiremba C, Taylor JRN, Rooney LW, Beta T. Phenolic acid content of sorghum and maize cultivars varying in hardness. Food Chemistry. 2012;**134**:81-88

[130] Awika JM, Rooney LW, Waniska RD. Anthoycanins from black sorghum and their antioxidant properties. Food Chemistry. 2004;**90**:293-301

[131] Bean SR, Wilson JD, Moreau RA, Galant A, Awika JM, Kaufman RC, et al. Structure and composition of the Sorghum grain. Sorghum. 2019;**58**:173-214

[132] Bhandari S, Lee Y-S. The contents of phytosterols, squalene, and vitamin E and the composition of fatty acids of Korean landrace Setaria italica and Sorghum bicolar seeds. Korean Journal Plant Resources. 2013;**26**:663-672

[133] Paraiso IL, Revel JS, Stevens JF. Potential use of polyphenols in the battle against COVID-19. Curriculum Opinion Food Science. 2020;**32**:149-155

[134] Irondi EA, Adegoke BM, Effion ES, Oyewo SO, Alamu EO, Boligon AA. Enzymes inhibitory property, antioxidant activity and phenolics profile of raw and roasted red sorghum grains in vitro. Food Science Human. Wellness. 2019;**8**:142-148

[135] Liu Y-W, Shang HF, Wang CK, Hsu FL, Hou WC. Immunomodulatory activity of dioscorin, the storage protein of yam (Dioscorea alata cv. Tainong No. 1) tuber. Food and Chemical Toxicology. 2007;**45**(11):2312-2318

[136] Iwu MM, Okunji CO, Ohiaeri ZGO, Akah GOP, Corley D, Tempesta MS. Hypoglycaemic activity of dioscoretine from tubers of Dioscorea dumetorum in normal and alloxan diabetic rabbits. Planta Medica. 1990;**56**(3):264-267

[137] Bhandari MR, Kasai T, Kawabata J. Nutritional evaluation of wild yam (Dioscorea spp.) tubers of Nepal. Food Chemistry. 2003;**82**(4):619-623

[138] Scott GJ. Transforming traditional food crops: Product development for roots and tubers. Product Development for Root and Tuber Crops. 1992;**1**:3-20

[139] Nassar NMA, Hashimoto DYC, Fernandes SDC. Wild Manihot species: Botanical aspects, geographic distribution and economic value. Genetics and Molecular Research. 2008;**7**(1):16-28

[140] Chan YC, Hsu CK, Wang MF, Su TY. A diet containing yam reduces the cognitive deterioration and brain lipid peroxidation in mice with senescence accelerated. International Journal of Food Science and Technology. 2004;**39**(1):99-107

[141] Chen HL, Wang CH, Chang CT, Wang TC. Effects of Taiwanese yam (Dioscorea japonica Thunb var. pseudojaponica Yamamoto) on upper gut function and lipid

metabolism in Balb/c mice. Nutrition. 2003;**19**(7-8):646-651

[142] Huang CH, Cheng JY, Deng MC, Chou CH, Jan TR. Prebiotic effect of diosgenin, an immunoactive steroidal sapogenin of the Chinese yam. Food Chemistry. 2012;**132**(1):428-432

[143] Hou WC, Chen HJ, Lin YH. Dioscorins, the major tuber storage proteins of yam (Dioscorea batatas Decne), with dehydroascorbate reductase and monodehydroascorbate reductase activities. Plant Science. 1999;**149**(2):151-156

[144] Hou WC, Chen HJ, Lin YH. Dioscorins from different Dioscorea species all exhibit both carbonic anhydrase and trypsin inhibitor activities. Botanical Bulletin of Academia Sinica. 2000;**41**(3):191-196

[145] Hou WC, Lee MH, Chen HJ, et al. Antioxidant activities of dioscorin, the storage protein of yam (Dioscorea batatas Decne) tuber. Journal of Agricultural and Food Chemistry. 2001;**49**(10):4956-4960

[146] Hsu FL, Lin YH, Lee MH, Lin CL, Hou WC. Both dioscorin, the tuber storage protein of yam (Dioscorea alata cv. Tainong No. 1), and its peptic hydrolysates exhibited angiotensin converting enzyme inhibitory activities. Journal of Agricultural and Food Chemistry. 2002;**50**(21):6109-6113

[147] Lin JY, Lu S, Liou YL, Liou HL. Antioxidant and hypolipidaemic effects of a novel yam-boxthorn noodle in an in vivo murine model. Food Chemistry. 2006;**94**(3):377-384

[148] Shewry PR. Tuber storage proteins. Annals of Botany. 2003;**91**(7):755-769

[149] Hou WC, Lin YH. Dehydroascorbate reductase and monodehydroascorbate reductase activities of trypsin inhibitors, the major sweet potato (Ipomoea batatas [L.] lam) root storage protein. Plant Science. 1997;**128**(2):151-158

[150] Friedman M. Potato glycoalkaloids and metabolites: Roles in the plant and in the diet. Journal of Agricultural and Food Chemistry. 2006;**54**(23):8655-8681

[151] Kuo KW, Hsu SH, Li YP, et al. Anticancer activity evaluation of the Solanum glycoalkaloid solamargine: Triggering apoptosis in human hepatoma cells. Journal Biochemical Pharmacology. 2000;**60**(12):1865-1873

[152] Liu LF, Liang CH, Shiu LY, Lin WL, Lin CC, Kuo KW. Action of solamargine on human lung cancer cellsenhancement of the susceptibility of cancer cells to TNFs. FEBS Letters. 2004;**577**(1-2):67-74

[153] Chimkode R, Patil MB, Jalalpure SS. Wound healing activity of tuberous root extracts of Ipomoea batatas. Advances in Pharmacology and Toxicology. 2009;**10**:69-72

[154] Panda V, Sonkamble M. Anti-ulcer activity of Ipomoea batatas tubers (sweet potato). Functional Foods in Health and Disease. 2011;**2**(3):48-61

[155] Yashin A, Yashin Y, Xia X. Antioxidant activity of spices and their impact on human health: A review. Antioxidants. 2017;**6**(3):70

[156] Embuscado ME. Bioactives from culinary spices and herbs: A review. Journal of Food Bioactives. 2019;**6**:68-99

[157] Embuscado ME. Spices and herbs: Natural sources of antioxidants – A mini review. Journal of Functional Foods. 2015;**18**:811-819. DOI: 10.1016/j.jff.2015.03.005

[158] Islam SMD. Transient Receptor Potential Channels. Springer. 2011;**3**:50. ISBN 978-94-007-0265-3

[159] Davidson A. Seafood of South East Asia. 2nd ed. Ten Speed Press. 2003;**4**:216. ISBN 1-58008-452-4

[160] Pimple BP, Patel AN, Kadam PV, Patil MG. Marjoram: Global warming. 2012. www.time.com

[161] Grieve, Maud. Thyme. A Modern Herbal. botanical.com (Hypertext version of the 1931 ed.). 2008

[162] Pandey B, Khan S, singh S. A study of antimicrobial activity of some spices. International Journal Curriculum Microbiology Applied Science. 2014;**3**(3):643-650

[163] Megha S, Alka G, Ranu P. A review on herbs, spices and functional food used in diseases. International Journal of Research & Review. 2017;**4**:4-1

[164] Suzuki T, Tada H, Sato E, Sagae Y. Application of sweet potato fiber to skin wound in rat. Biological and Pharmaceutical Bulletin. 1996;**19**(7):977-983

[165] Su PF, Li CJ, Hsu CC, et al. Dioscorea phytocompounds enhance murine splenocyte proliferation ex vivo and improve regeneration of bone marrow cells in vivo. Evidence-based Complementary and Alternative Medicine. 2011;**2011**, Article ID 731308:11

[166] Kaspar KL, Park JS, Brown CR, MAthison BD, Navarre DA, Chew BP. Pigmented potato consumption alters oxidative stress and inflammatory damage in men. Journal of Nutrition. 2011;**141**(1):108-111

[167] Sonibare MA, Abegunde RB. In vitro antimicrobial and antioxidant analysis of Dioscorea dumetorum (Kunth) Pax and Dioscorea hirtiflora (Linn.) and their bioactive metabolites from Nigeria. Journal of Applied Biosciences. 2012;**51**:3583-3590

[168] Maithili V, Dhanabal SP, Mahendran S, Vadivelan R. Antidiabetic activity of ethanolic extract of tubers of Dioscorea alata in alloxan induced diabetic rats. Indian Journal of Pharmacology. 2011;**43**(4):455-459

[169] Wu WH, Liu LY, Chung CJ, Jou HJ, Wang TA. Estrogenic effect of yam ingestion in healthy postmenopausal women. Journal of the American College of Nutrition. 2005;**24**(4):235-243

[170] Chen JH, Wu JSS, Lin HC, et al. Dioscorea improves the morphometric and mechanical properties of bone in ovariectomised rats. Journal of the Science of Food and Agriculture. 2000;**88**(15):2700-2706

[171] Hwang YP, Choi JH, Han EH, et al. Purple sweet potato anthocyanins attenuate hepatic lipid accumulation through activating adenosine monophosphate–activated protein kinase in human HepG2 cells and obese mice. Nutrition Research. 2011;**31**(12):896-906

[172] Panda V, Sonkamble M. Phytochemical constituents and pharmacological activities of Ipomoea batatas l. (lam)—A review. International Journal of Research in Phytochemistry & Pharmacology. 2012;**2**(1):25-34

[173] Kim DE, Min JS, Jang MS, Lee JY, Shin YS, Park CM. Natural bis benzylisoquinoline alkaloids-tetrandrine, fangchinoline, and cepharanthine, inhibit human coronavirus OC43 infection of MRC-5 human lung cells. Biomolecules. 2019;**9**(11):696

[174] Gorinstein S, Leontowicz H, Leontowicz M, Namiesnik J, Najman K, Drzewiecki J, et al. Comparison of the mainbioactivecompoundsandantioxidant activities in garlic and white and red onions after treatment protocols. Journal Agricultural Food Chemistry. 2008;**56**:4418-4426

[175] Gorinstein S, Leontowicz H, Leontowicz M, Jastrzebski Z, Najman K, Tashma Z, et al. The influence of raw and processed garlic and onions on plasma classical and non-classical atherosclerosis indices: Investigations in vitro and in vivo. Phytotherapy Reserve. 2010;**24**:706-714

[176] Rastogi S, Mohan Pandey M, Kumar Singh Rawat A. Spices: Therapeutic potential in cardiovascular health. Curriculum Pharmacology Dessertation. 2017;**23**:989-998

[177] Hwang IK, Lee CH, Yoo KY, Choi JH, Park OK, Lim SS, et al. Neuroprotective effects of onion extract and quercetin against ischemic neuronal damage in the gerbil hippocampus. Journal of Medicinal Food. 2009;**12**:990-995

[178] Mills E, Koren G. From type 2 diabetes to antioxidant activity: A systematic review of the safety and efficacy of common and cassia cinnamon bark. Cancer Journal Physiological Pharmacology. 2007;**85**:837-847

[179] Khan A, Zaman G, Anderson RA. Bay leaves improve glucose and lipid profile of people with type 2 diabetes. Journal Clinical Biochemical Nutrition. 2009;**44**:52-56

[180] Mehmood MH, Gilani A. Pharmacological basis for the medicinal use of black pepper and piperine in gastrointestinal disorders. Journal Medicine Food. 2010;**13**:1086-1096

[181] Speroni, E, Cervellati R, Dall'Acqua S, Guerra MC, Greco E, Govoni P, Innocenti G. Gastroprotective effect and antioxidant properties of different Laurus nobilis L. leaf extracts. Journal Medicine Food. 2011;**14**:499-504

[182] Edwards RL, Lyon T, Litwin SE, Rabovsky A, Symons JD, Jalili T. Quercetin reduces blood pressure in hypertensive subjects. Journal Nutrition. 2007;**137**:2405-2411

[183] Davis PA, Yokoyama W. Cinnamon intake lowers fasting blood glucose: Meta-analysis. Journal Medicine Food. 2011;**14**:884-889

[184] Mallikarjuna K, Sahitya Chetan P, Sathyavelu Reddy K, Rajendra W. Ethanol toxicity: Rehabilitation of hepatic antioxidant defense system with dietary ginger. Fitoterapia. 2008;**79**:174-178

[185] Beric T, Nikolic B, Stanojevic J, Vukovic-Gacic B, Knezevic-Vukcevic J. Protective effect of basil (Ocimum basilicum L.) against oxidative DNA damage and mutagenesis. Food Chemistry Toxicology. 2008;**46**: 724-732

[186] Alappat L, Awad AB. Curcumin and obesity: Evidence and mechanisms. Nutrition Reviews. 2010;**68**:729-738

[187] Karmakar S, Choudhury M, Das AS, Maiti A, Majumdar S, Mitra C. Clove (Syzygium aromaticum Linn) extract rich in eugenol and eugenol derivatives shows bone-preserving efficacy. National Production Reserve. 2012;**26**:500-509

[188] Kaviarasan S, Vijagalakshmi K, Anuradha CV. Polyphenol-rich extract of fenugreek seeds protect erythrocytes from oxidative damage. Plant Foods Human Nutrition. 2004;**59**:143-147

[189] Majdalawieh AF, Carr RI. In vitro investigation of the potential immunomodulatory and anti-cancer activities of black pepper (Piper nigrum) and cardamom (Elettaria cardamomum). Journal Medicine Food. 2010;**13**:371-381

[190] Salem ML. Immunomodulatory and therapeutic properties of nigella sativa L. seed. International Immunopharmacology. 2005;**5**:1749-1770

[191] Elbarbry F, Gazarin S, Shoker A. The protective effect of thymoquinone, an anti-oxidant and anti-inflammatory agent, against renal injury: A review. Saudi Journal Kidney Disease Transplant. 2009;**20**:741-752

[192] Mahmoud MF, Diaai AA, Ahmed F. Evaluation of the efficacy of ginger, Arabic gum and Boswellia in acute and chronic renal failure. Renal Failures. 2012;**34**:73-82

[193] Hlavackova Singh PK, Kaur IP. Synbiotic (probiotic and ginger extract) loaded floating beads: A novel therapeutic option in an experimental paradigm of gastric ulcer. Journal Pharmacology. 2012;**64**:207-217

[194] Alex AF, Spitznas M, Tittel AP, Kurts C, Eter N. Inhibitory effect of epigallocatechin gallate (EGCG), resveratrol, and curcumin on proliferation of human retinal pigment epithelial cells in vitro. Curriculum Eye Reserve. 2010;**35**:1021-1033

[195] Butt MS, Sultan MT. Nigella sativa-reduces the risk of various maladies. Critical Reviews in Food Science and Nutrition. 2010;**50**:654-655

[196] Samojlik I, Laki N, Mimica-Duki N, Dakovi-Svajcer K, Bozin B. Antioxidant and hepatoprotective potential of essential oils of coriander (Coriandrum sativum L.) and caraway (Carum carvi L.) (Apiaceae). Journal Agriculture Food Chemistry. 2010;**58**:8848-8853

[197] Atsumi T, Tonosaki K. Smelling lavender and rosemary increases free radical scavenging activity and decreases cortisol level in saliva. Psychiatry Research. 2007;**150**:89-96

[198] Shati AA, Elsaid FG. Effects of water extracts of thyme (Thymus vulgaris) and ginger (Zingiber officinale roscoe) on alcohol abuse. Food and Chemical Toxicology. 2009;**47**:1945-1949

[199] Geoghegan F, Wong RW, Rabie AB. Inhibitory effect of quercetin on periodontal pathogens in vitro. Phytotherapy Research. 2010;**24**:817-820

Chapter 13

Medicinal Plants Used for the Treatment and Management of Bilharziasis and Other Parasitic Infections Affecting Humans in Zimbabwe: A Systematic Review

Elliot Nyagumbo, Trust Nyirenda, Cephas Mawere, Ian Mutasa, Emmanuel Kademeteme, Alfred M. Mutaramutswa, Donald Kapanga, Godwins Ngorima, Leroy Nhari, Fabian Maunganidze, Michael Bhebhe, William Pote and Lucy Mabaya

Abstract

The World Health Organization (WHO) estimated that at least 251.4 million people from 78 countries were in need of preventative care for bilharziasis in 2021. Globally, soil-transmitted helminth infections are present in at least 24% of the world's population. Tropical and subtropical areas have a wide distribution of infections with a high prevalence in the sub-Saharan Africa. The aim of this study was to document plants that have been traditionally used in Zimbabwe to manage bilharziasis and other parasitic infections. The literature review was based on published papers and abstracts retrieved from the online databases. Books, book chapters, scientific reports and theses from universities in Zimbabwe that were available online were also used in this review. Plants with the reported traditional usage against bilharziasis and other parasitic infections were recorded from the data retrieved. In total, 68 species were used to treat and manage bilharzia and other parasitic infections. Most of these medicinal plants were used to treat and manage schistosomes (fluke or worm). A total of 76.5% of the medicinal plants reported have been scientifically validated and documented to exhibit anthelmintic activity. In conclusion, Zimbabwe has a plethora of medicinal plants that can be used to manage bilharziasis and other parasitic infections.

Keywords: ethnobotanical, bilharzia, schistosomiasis, worms, pharmacological, toxicology, traditional plants, anthelmintic, Zimbabwe

1. Introduction

Schistosomiasis is a neglected parasitic tropical disease caused by blood flukes (trematode worms) of the genus Schistosoma. According to WHO [1] estimates, at least 251.4 million people reported from 78 countries were in need of preventative care for bilharziasis in 2021. Schistosomiasis, a disease caused by *Schistosoma mansoni* (intestinal) and *S. haematobium* (urogenital) species, is mainly concentrated in sub-Saharan Africa, where about 90% of the disease burden exists. The transmission of these species occurs through feces and urine, respectively [2]. Endemic areas where the infection is prevalent are inhabited by over 700 million people, particularly in tropical and subtropical regions. Schistosomiasis is more prevalent in impoverished rural communities, particularly in regions where fishing and agricultural activities are prevalent. These areas are often characterized by poor communities lacking access to potable water and adequate sanitation [3]. Morbidity reduction can therefore be accomplished with the use of preventative therapy, which should be repeated over a number of years in endemic areas with moderate to high transmission [4]. According to WHO [5] the most frequent illnesses worldwide are soil-transmitted helminth (STH) infections, which mostly afflict the poorest and most destitute populations. Where sanitation is inadequate, the soil is polluted with egg-infected human feces and fecal waste [5]. The primary causative species that infect humans are whipworms (*Trichuris trichiura*), hookworms (*Ancylostoma duodenale, Necator americanus* and *Ancylostoma duodenale*) and roundworms (*Ascaris lumbricoides*) [6]. STH infections afflict at least 24% of the world's population, which exceeds 1.5 billion individuals. Infections are widely distributed in tropical and subtropical locations, with the Americas, China, East Asia and sub-Saharan Africa having the highest frequency [5].

Pharmacotherapy is the most effective approach for decreasing the incidence of schistosomiasis infections. The WHO recommends preventive chemotherapy using praziquantel as the strategy for managing schistosomiasis. School-age children (5 to 15 years old) are the target population for this therapy due to their high infection burden and ability to be effectively targeted through schools [7]. Zimbabwe aims to eradicate bilharzia and intestinal worms, by 2030. During a mass treatment campaign carried out from April 3–9, 2022, over 1.8 million children received free oral treatment for schistosomiasis (bilharzia) and soil-transmitted helminthiases (intestinal worms) [8].

Despite not achieving complete elimination of bilharzia, Zimbabwe has significantly reduced the burden of the disease through the annual national treatment campaigns. In 2014 a study revealed that the district of Chiredzi in Masvingo province had the highest prevalence of *S. mansoni* at 43.7%, followed by the Hwedza district in Mashonaland East and Nyanga in Manicaland province, with prevalence rates of 32.3 and 31.5%, respectively [9]. Despite the great prevalence of parasitic diseases worldwide and the substantial amount of suffering caused by these parasites, the majority are considered neglected diseases. Only malaria treatment and prevention receive significant financing, but there is an urgent need for more action to be done to alleviate the suffering of the large populations of people who are infected with other parasitic diseases [10]. Similarly, despite those parasitic infections accounting for more than 10% of the world's disease burden, drug discovery efforts for parasitic diseases are limited, with only 1% of new medications addressing parasitic diseases in the last 40 years [6].

The purpose of the study is to find medicinal plant species that are used as an effective treatment against schistosomiasis in Zimbabwe. The study involved the compilation of a list of medicinal plants used in Zimbabwe to treat parasitic infections in humans. Bilharzia, gastrointestinal worms and helminths, ectoparasites, trichomoniasis, leishmaniasis and trypanosomiasis are among the diseases covered. Due to the broad scope of these topics, veterinary usage and malaria were excluded. This systematic review will create a comprehensive digital database of medicinal plants used in traditional practice which holds the potential to expand treatment options, improve access to healthcare, preserve traditional knowledge and promote sustainable practices in disease management. Moreover, the potential development of drug resistance has sparked an ongoing debate about the future efficacy of praziquantel. The emergence of drug-resistant strains of schistosomes will pose a significant challenge in controlling the disease. Thus, exploring medicinal plants may help to identify novel compounds that can overcome drug resistance and provide alternative treatment options.

2. Objectives

This systematic review was therefore undertaken to:

1. Determine the medicinal plants traditionally used in Zimbabwe for the management of parasitic infections.

2. Describe the common names, scientific names, plant family, plant parts used, modes of preparation, traditional uses, distribution and conservation status.

3. Compile a comprehensive document that describes the ethnobotany of medicinal plants in Zimbabwe that are traditionally used to treat and manage parasitic infections in humans.

4. Generate integrated and sufficient traditional evidence to support its medicinal use.

5. Determine the endangered medicinal plant species to prioritize for conservation amidst the growing destruction of natural resources for settlement, industrialization, construction and energy production [11–13].

2.1 Inclusion criteria

Plants used to treat the following parasitic infectious diseases were included in this study

1. Bilharzia

2. Gastrointestinal worms

3. Helminths

4. Ectoparasites

5. Trichomoniasis

6. Leishmaniasis

7. Typanosomiasis

2.2 Exclusion criteria

Plants used to treat the following parasitic infectious diseases were excluded from this study

1. Malaria

2. Veterinary parasites

3. Materials and methods

3.1 Research protocol and reporting

The Preferred Reporting Items for the Systematic Reviews and Meta-Analyses (PRISMA) guidelines were used in the reporting of this study (**Figure 1**). The protocol used in this systematic review was as previously reported [14].

3.2 Literature search

Electronic data on the ethnobotany of medicinal plants used in Zimbabwe were retrieved from electronic databases such as Google, Google Scholar, Springer Link, Researchgate, PubMed, Science Direct and JSTOR. The keywords set "medicinal plants AND (antiparasitic OR antihelmintic) AND Zimbabwe" were used. The retrieved articles were downloaded and stored in EndNote X9 (Thomson Reuters, San Francisco, CA and USA). Duplicate articles were then removed from the file. Further, a manual search from the reference lists of screened eligible articles and deposited electronic copies of dissertations and theses in online Universities' repositories and National Herbarium and Botanic Gardens (SRGH) libraries up to 31 December 2020 were done. Other sources utilized in this study included books [15–18], book chapters, scientific reports and theses available at universities [19, 20] and National Herbarium and Botanic Gardens (SRGH) libraries. The authors continuously received notifications of any new "similar reports" meeting the search criteria from Science Direct, Scopus and Google Scholar.

The plant names were verified with http://www.theplantlist.org and https://www.zimbabweflora.co.zw. Plants with the reported traditional usage against bilharziasis and other parasitic infections were identified and compiled from the information collected and gathered. A master list was prepared including all the medicinal plants used in Zimbabwe for the treatment and management of bilharziasis and other parasitic infections (**Table 1**). The above-mentioned databases were also searched for pharmacological and toxicological properties providing scientific evidence of medicinal usage comparable to their ethnomedicinal usage. All the information was summarized in three tables (**Tables 1–3**) and five figures (**Figures 2–6**). The review excluded medicinal plants for veterinary use and those against malaria to limit the plants to those used in the treatment and management of bilharziasis and other parasitic infections in humans.

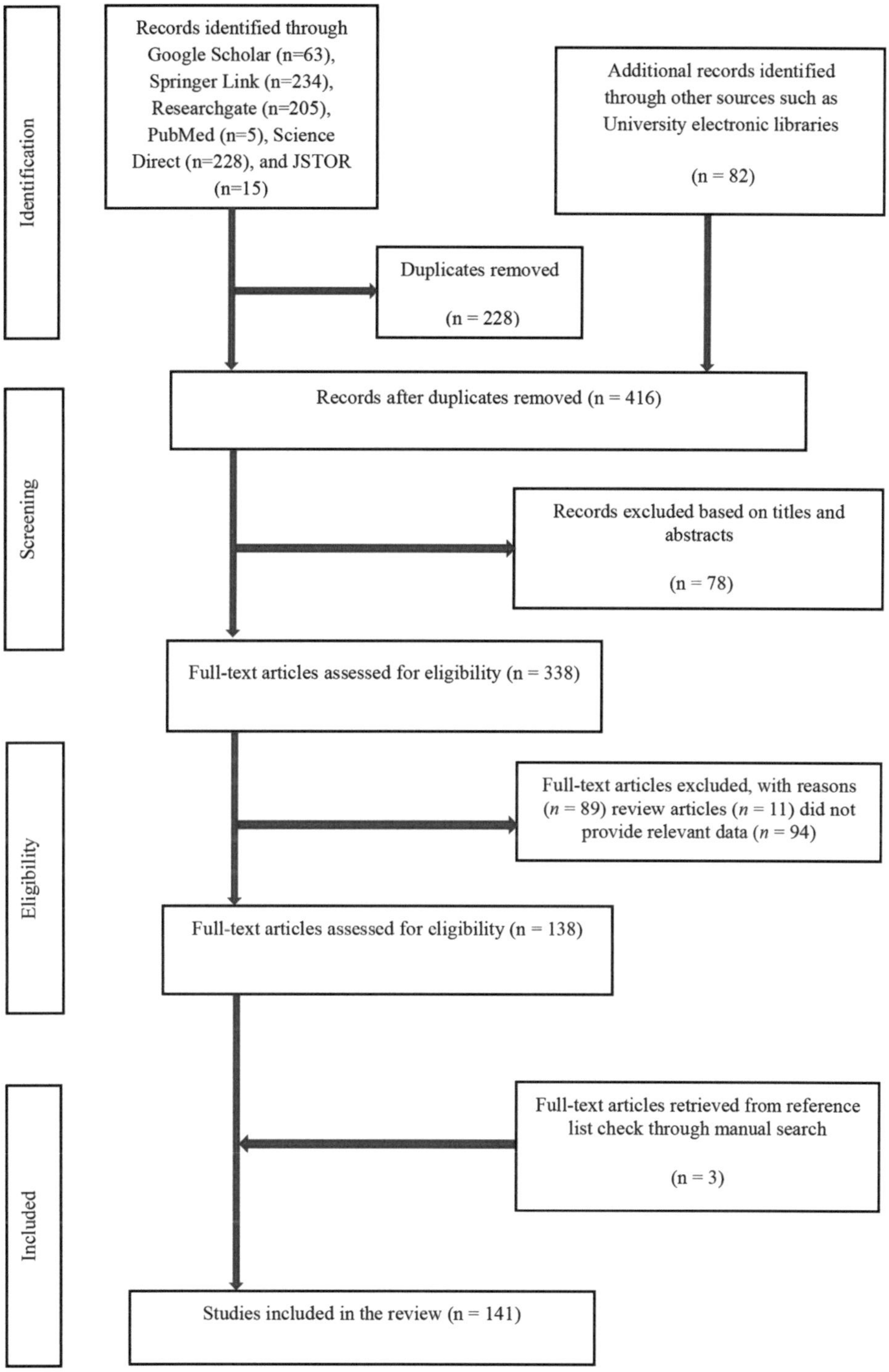

Figure 1.
PRISMA flow diagram showing the search and retrieval steps of the study (adopted from Moher et al. [14]).

Scientific name [Family]	[Growth habit] Distribution	Vernacular names and other names	Parts used and mode of preparation	Parasitic infection ethnomedicinal uses [References]
Abrus precatorius L. subsp. *africanus* Verde. [Fabaceae]	Woody, deciduous climber. [Climber] N, W, C, E, S	Chonjo (Shona) Lucky bean creeper (English) Minimini (Shona) Munhutuwaro (Shona)	Roots (fresh) Leaves and stem	Bilharziasis (Urinary schistosomiasis) [21–23]
Acacia karoo Hayne [Fabaceae]	Small to medium-sized tree. [Tree] N, W, C, E, S	Isinga (Ndebele) Mubayamhondoro (Shona) Muhunga (Shona) Muzunga (Shona) Sweet thorn (English)	Leaves and roots Roots decoction	Bilharziasis (Urinary schistosomiasis) [21, 23, 24]
Albizia amunesiana Harms [Fabaceae]	Small to medium-sized tree. [Tree] N, W, C, E, S	Muriranyenze (Shona) Purple-leaved albizia (English) Umnonjwana (Ndebele)	Roots, bark, leaves and stem.	Bilharziasis (Urinary schistosomiasis) [21, 23]
Asparagus spp except *A. asparagoides L* [Asparagaceae]	Perennial herbs or shrubs [Herb or shrub] N, W, C, E, S		Roots – mixed with seeds of *Vigna unguiculata* in soup and taken orally.	Bilharziasis [16]
Burkea africanus Hook. [Fabaceae]	Deciduous tree, to 8 m. [Tree] N, W, C, E, S	Burkea (English) False ash (English) Mukarati (Shona) Umnondo (Ndebele) Wild syringa (English)		Bilharziasis (Urinary schistosomiasis) [21]
Carissa spinarum L. *Carrisa edulis* Vahl [Apocynaceae]	Scrambling shrub or small tree. [Tree] N, W, C, E, S	Mudyabveni (Shona) Mudzambara (Shona) Muhlababzunzi (Shona) Mumbingwa (Shona) Muruguru (Shona) Mutsamviringa (Shona) Simple-spined num-num (English) Umlugulu (Ndebele)		Bilharziasis (Urinary schistosomiasis) [21]
Cassia abbreviata Oliv. [Fabaceae]	Shrub or small rounded tree. [Tree] N, W, C, E, S	Isihaqa (Ndebele) Long-tail cassia (English) Muremberembe (Shona) Muvheneka (Shona)	Leaves, roots and bark.	Bilharziasis (Urinary schistosomiasis) [21, 23]
Catunaregam swynnertonii (S. Moore) Bridson *Catunaregum spinosa sensu* Verdcourt subsp. *spinosa* [Rubiaceae]	Spiny, deciduous shrub or small tree growing up to 7 metres. [Tree or shrub] E, S	Sand bone-apple (English)	Leaves	Intestinal worms [25]
Celtis africana Burm. f. [Ulmaceae]	Deciduous tree 5–15(−35) m [Tree] [Cultivated]	Common celtis (English) Kamutuna (Shona) Mugara (Shona) Muguru (Shona) Mukonachando (Shona) Musvutaderere (Shona) Umdlawuthu (Ndebele) White stinkwood (English)	Leaves and roots.	Bilharziasis (Urinary schistosomiasis) [21, 23]

Scientific name [Family]	[Growth habit] Distribution	Vernacular names and other names	Parts used and mode of preparation	Parasitic infection ethnomedicinal uses [References]
Cissampelos mucronata A. Rich. [Menispermaceae]	Liane with a woody root stock [Liane] N, W, C, E, S	Hairy heartleaf (English) Heart-leaved vine (English) Nyakuta (Shona) Ruzambu (Shona)	Roots – infusion taken orally three times a day for three consecutive days.	Bilharziasis [16]
Cissus quadrangularis L [Vitaceae]	Succulent climbing herb with tendrils. [Climber] N, W, C, E, S	Murunjurunju (Shona) Muvengahonye (Shona) Renja (Shona)	Whole plant – crushed and applied on wounds.	To treat wounds infested with maggots [16]
Clerodendrum ternatum Schinz [Verbenaceae]	Suffrutex growing from an extensive rhizomatous woody roots stock [Herb] N, W, C, E, S	Dwarf cat's whiskers (English) Umalanjana (Ndebele) Umqotshanja (Ndebele)	Roots – are ground into powder and taken orally.	Tapeworm and hookworm. [16]
Combretum hereroense Schinz subsp. *heteroense* [Combretaceae]	Shrub or small tree. [Tree or shrub] N, W, C, E, S	Ithetshane (Ndebele) Mouse-eared combretum (English) Murovamhuru (Shona) Mutechani (Shona) Russet bushwillow (English)		Bilharziasis [26]
Combretum imberbe Wawra [Combretaceae]	Medium to large deciduous tree. [Tree] N, W, C, E, S	Leadwood (English) Monzo (Shona) Muchenarota (Shona) Mutsviri (Shona) Umtshenalotha (Ndebele) Umtshwili (Ndebele)	Roots – infusion taken orally.	Bilharziasis [16, 27]
Combretum zeyheri Sond [Combretaceae]	Small to medium-sized tree. [Tree] N, W, C, E, S	Large-fruited bushwillow (English) Muchenja (Shona) Mupembere-kono (Shona) Muruka (Shona) Umbhondo (Ndebele) Umbula (Ndebele)		Bilharziasis and hookworm [28]
Crossopteryx febrifuga (Afzel. ex G. Don) Benth [Rubiaceae]	Shrub or small deciduous tree. [Tree] N, W, C, E, S	Common crown-berry (English) Crystal-bark (English) Mubakatirwa (Shona) Mugoko (Shona) Mukombegwa (Shona) Mukombigo (Shona) Muteyo (Shona) Umphokophokwana (Ndebele)		Bilharziasis (Urinary schistosomiasis) [21]
Croton gratissimus Burch [Euphorbiaceae]	Shrub or small tree. [Tree or shrub] N, W, C, E, S	Gunukira (Shona) Lavender croton (English) Mubangwa (Shona) Mufandemengwe (Shona) Mufarata (Shona)		Internal worms and ascariasis [25]
Cynanchum viminale (L.) L. subsp. *viminale Sarcostemma*	Succulent vine. [Climber] N, W, C, E, S	Caustic vine (English) Chifure (Shona) Imvubu (Ndebele) Ingotsha (Ndebele) Ingotshayeganga	Latex – infusion taken orally.	Worm scours [16]

Scientific name [Family]	[Growth habit] Distribution	Vernacular names and other names	Parts used and mode of preparation	Parasitic infection ethnomedicinal uses [References]
viminale (L) R.Br [Apocynaceae]		(Ndebele) Milk rope (English) Runyakadomdo (Shona) Rusungwe (Shona)		
Dicoma anomala Sond. [Asteraceae]	Prostrate, decumbent or erect perennial herb [Herb] N, W, C, E, S	Fever bush (English) Stomach bush (English) Chifumuro (Shona)	Roots (fresh) decoction	Bilharziasis (Urinary schistosomiasis) Intestinal worm parasites [21–23, 29]
Diplorhynchus condylocarpon (Müll. Arg.) Pichon [Apocynaceae]	Shrub or as a small, graceful tree [Tree or shrub] N, W, C, E, S	Rhodesian rubber tree, Horn-pod tree, Wild rubber (English) Mutowa (Shona)		Bilharziasis (Urinary schistosomiasis) [21]
Elephantorrhiza goetzei (Harms) [Fabaceae]	[Shrub] N, W, C, E, S	Intolwane (Ndebele) Mugudzuru (Shona) Narrow-pod elephant roots (English) Ndorani (Shona)	Roots – is mixed with *Bauhinia thonningii* and a decoction is prepared which is taken orally. Roots decoction and/or infusion. Roots (fresh), fruits, stem bark and stem	Bilharziasis (Urinary schistosomiasis) [16, 21, 23, 30–32]
Eriosema englerianum Harms. [Fabaceae]	A many-stemmed perennial, growing from a woody root stock. [Herb] N, W, C	Blue bush (English) Mashona fire bean (English)	Roots – is mixed with *Vigna unguiculata* seeds and soup drunk.	Bilharziasis [16]
Erythrina abyssinica Lam. ex DC. [Fabaceae]	Small to medium-sized tree of wooded grassland. [Tree] N, C, E, S	Lucky-bean tree (English) Munhimbiti (Shona) Mutete (Shona) Mutiti (Shona) Mutsiti (Shona) Red-hot-poker tree (English) Umgqogqogqo (Ndebele)	Roots – is mixed with *Vigna unguiculata* seeds and soup drunk.	Bilharziasis (Urinary schistosomiasis) [16, 21]
Euclea divorum Hiern [Ebenaceae]	Evergreen shrub or small tree. [Tree or shrub] N, W, C, E, S	Diamond-leaved euclea (English) Magic guarri (English) Mubhununu (Shona) Mudziviriratsuro (Shona) Mugarazvuru (Shona) Muguurameno (Shona) Munyenya (Shona) Mushangura (Shona) Umtshekesane (Ndebele)	Roots – mixed with *Vigna unguiculata* seeds and cooked in soup.	Bilharziasis (Urinary schistosomiasis) [16, 21]
Euphorbia schinzii Pax [Euphorbiaceae]	Shrub under 2 m. [Shrub] N, W, C, E, S			Bilharziasis (Urinary schistosomiasis) [21]
Faurea saligna Harv [Proteaceae]	Small to medium-sized tree.	Mutsatstsi (Shona) African beech (English) Isidwadwa (Ndebele) Kapfutsana (Shona)	Leaves	Bilharziasis and helminthiasis. [33]

Scientific name [Family]	[Growth habit] Distribution	Vernacular names and other names	Parts used and mode of preparation	Parasitic infection ethnomedicinal uses [References]
	[Tree] N, W, C, E, S	Mugarahungwe (Shona) Munyaganza (Shona) Mushangwa (Shona) Muzazati (Shona) Umpembele (Ndebele) Willow beechwood (English)		
Flacourtia indica (Burm.f) Merr [Salicaceae]	Tree, shrub over 2 m. [Tree or shrub] N, W, C, E, S	Batoka plum (English) Governor's plum (English) Mududwe (Shona) Munhunguru (Shona) Mutombototo (Shona) Mutudza (Shona) Mutunguru (Shona)	Roots – infusion taken orally.	Bilharziasis and intestinal worms [16, 19, 20]
Gymnosporia senegalensis (Lam.) Loes. *Maytenus senegalensis* (Lam) Exell [Celastraceae]	Tree, shrub over 2 m. [Tree or shrub] N, W, C, E, S	Chivhunabadza (Shona) Chizhuzhu (Shona) Confetti tree (English) Isihlangu (Ndebele) Mugaranjiva (Shona) Mukokoba (Shona) Musosaguva (Shona) Musosawafa (Shona) Musukameno (Shona) Mutotova (Shona) Mutsotsova (Shona) Red spike-thorn (English)	Leaves and stem. Roots Roots and bark	Bilharziasis (Urinary schistosomiasis) [21, 23]
Hydnora abyssinica A. Braun ex Schweinf. *Hydnora solmsiana* Dinter [Hydnoraceae]	A subterranean roots parasite, lacking chlorophyll [Roots parasite] W, C	Emerging flower (English)		Bilharziasis (Urinary schistosomiasis) [21]
Khaya anthotheca (Welw.) C. DC. *Khaya nyasica* Bak. f. [Meliaceae]	Large to very large evergreen tree with a long straight stem. [Tree] N, E, S	Mubarwa (Shona) Mururu (Shona) Muwawa (Shona) Red mahogany (English)		Bilharziasis (Urinary schistosomiasis) [21]
Kigellia africana (Lam.) Benth. *Kigelia pinnata* (Jacq.) DC. [Bignoniaceae]	Medium to large tree. [Tree] N, W, C, E, S	Mubveve (Shona) Musonya (Shona) Muvhumati (Shona) Sausage tree (English) Umvebe (Ndebele)	Fruit, bark and roots.	Tapeworm [19]
Landolphia kirkii Dyer ex Hook. f. [Apocynaceae]	[Climber, liane.] E	Mukanga (Shona) Muungu (Shona) Rubber vine (English) Runyangarwapene (Shona) Sand apricot-vine (English)		Bilharziasis (Urinary schistosomiasis) [21]
Lannea discolor (Sond.) Engl. [Anacardiaceae]	Medium-sized deciduous tree. [Tree] N, W, C, E, S	Chizhenje (Shona) Live-long (English) Mugan'acha (Shona) Muhumbukumbu (Shona) Mumbumbu (Shona) Mupuri (Shona)	Roots – decoction taken orally. Leaves and stem. Roots and bark.	Bilharziasis (Urinary schistosomiasis) [21, 23, 34, 35]

Scientific name [Family]	[Growth habit] Distribution	Vernacular names and other names	Parts used and mode of preparation	Parasitic infection ethnomedicinal uses [References]
		Mushamba (Shona) Tree grape (English)		
Lannea edulis (Sond.) Engl. [Anacardiaceae]	Shrub under 2 m [Shrub] N, W, C, E, S	Intakubomvu (Ndebele) Mutsambatsi (Shona) Tsombori (Shona) Wild grape (English)	Roots – infusion or decoction taken orally. Roots can be ground into powder and mixed with porridge and taken orally. Roots (fresh) Leaves and stem.	Bilharziasis (Urinary schistosomiasis) [16, 21–23, 31, 36]
Lecaniodiscus fraxinifolias Bak. [Sapindaceae]	Tree, shrub over 2 m. [Tree or shrub] N, E, S	Chikuhlule (Hlengwe) Musando (Shona) Mutarara (Tonga: Zimbabwe) River-litchi (English)	Leaves and stem. Roots.	Bilharziasis (Urinary schistosomiasis) [21, 23]
Loranthus on *Dichrostachys cinerea,* [Loranthaceae]	Parasitic shrub with long spreading stems. [Shrub] N, W, C, E, S	Sicklebush (English)	Whole plant – infusion taken orally.	Bilharziasis [16]
Mondia whitei (Hook.f.) Skeels [Apocynaceae]	[Climber, liane] N, W, C, E, S	Mungurauwe (Shona) Tonic roots (English) White's ginger (English)	Roots – ground into powder and mixed in porridge.	Bilharziasis [16]
Mucuna coriacea Baker subsp. irritans (Burtt Davy) Verdc. [Fabaceae]	Climber, shrub over 2 m. [Climber or shrub] N, C, E, S	Buffalo bean (English) Huriri (Shona) Uriri (Shona)		Bilharziasis (Urinary schistosomiasis) [21]
Musa sp. [Musaceae]	Tall herbs, perennial [Herb] Cultivated	Mubhanana (Shona)		Bilharziasis (Urinary schistosomiasis) [21]
Ozoroa reticulata (Baker f.) R. Fern. & A. Fern. *Ozoroa insignis* Del. subsp. *reticulata* (Bak.f.) Gillett *Heeria reticulata* Engl. [Anacardiaceae]	A small, much-branched deciduous tree. [Tree] N, W, C, E, S	Isafice (Ndebele) Muacha (Shona) Mubedu (Shona) Mudyamombe (Shona) Mugaragunguwo (Shona) Mulilila (Tonga: Zimbabwe) Murungu (Shona) Raisin bush (English) Tar berry (English)	Roots – infusion mixed with porridge and taken orally. Roots – infusion taken orally. Roots (fresh) Leaves Stem and bark. Roots bark.	Tapeworm and hookworm. Bilharziasis (Urinary schistosomiasis) [16, 21–23]
Peltophorum africanum Sond. [Fabaceae]	Tree, shrub over 2 m. [Tree or shrub] N, W, C, E, S	African wattle (English) Dzedze (Shona) Mudjiza (Shona) Mupumhamauva (Shona) Musambanyoka (Shona) Mutandarombo (Shona) Muzeze (Shona) Nyakambariro (Shona) Nyamanyoka (Shona) Umkahla (Ndebele) Umsehla (Ndebele) Zeze (Shona)	Leaves and stem. Roots and roots bark. Roots	Bilharziasis (Urinary schistosomiasis) Intestinal parasites [21, 23, 29]
Phaseolus vulgaris L [Fabaceae]	Herbs or subshrubs, erect, prostrate	Common bean (English) French bean (English)	Seeds – dried	Bilharziasis (Urinary schistosomiasis) [21]

Scientific name [Family]	[Growth habit] Distribution	Vernacular names and other names	Parts used and mode of preparation	Parasitic infection ethnomedicinal uses [References]
	or climbing. [Herb or shrub] Cultivated			
Piliostigma thonningii (Schumach.) Milne-Redh. [Fabaceae]	[Tree] N, W, C, E, S	Camel-foot (English) Ihabahaba (Ndebele) Monkey bread (English) Mubaba (Shona) Muhuku (Shona) Musakasa (Shona) Musekesa (Shona) Mutukutu (Shona)	Roots – are mixed with *Elephantorrhiza goetzei* roots and an infusion are prepared which is taken orally. Leaves and stem. Roots and roots bark. Roots	Bilharziasis (Urinary schistosomiasis) [16, 21, 23, 31]
Pogonarthria squarrosa (Roem. & Schult.) Pilg. [Poaceae]	Erect perennial tufted grass, up to 1 m. [Grass] N, W, C, E, S	Cross grass (English) Herringbone grass (English) Meerjarige denneboomgras (Afrikaans) Sekelgras (Afrikaans)	Roots – decoction taken orally.	Bilharziasis [16]
Pterocarpus angolensis DC. [Fabaceae]	Tree, shrub over 2 m. [Tree or shrub] N, W, C, E, S	Bloodwood (English) Mubvamakovo (Shona) Mubvamaropa (Shona) Mubvinziropa (Shona) Mukambira (Shona) Mukonambiti (Shona) Mukula (Tonga: Zimbabwe) Mukurambira (Shona) Mukwa (English) Mukwa (Shona) Mukwirambira (Shona) Mushambaropa (Shona) Muzwamulowa (Tonga: Zimbabwe) Umvagazi (Ndebele)	Bark – infusion taken orally. Flowers – applied to incision on area affected. Roots (fresh) Leaves Stem Bark	Bilharziasis (Urinary schistosomiasis) [16, 21–23]
Ricinus communis L [Euphorbiaceae]	Tree, annual, perennial, shrub over 2 m, shrub under 2 m. [Tree or shrub] Introduced	Castor-oil plant (English)	Roots – infusion taken orally. Leaves and roots – infusion taken orally.	Bilharziasis (Urinary schistosomiasis) [16, 21, 37]
Sansevieria hyacinthoides (L.) Druce [Dracaenaceae]	Evergreen, perennial herb [Herb] N, C, E, S	Mother-in-law's tongue (English) Piles roots (English) Bowstring hemp (English)	Leaves, rhizome and roots.	Intestinal parasites and worms. [38]
Sclerocarya birrea (A. Rich.) Hochst. subsp. *caffra* (Sond.) Kokwaro [Anacardiaceae]	Medium-sized deciduous tree. [Tree] N, W, E, S	Marula (English) Mufuna (Shona) Mupfura (Shona) Mushomo (Shona) Umganu (Ndebele)	Roots – infusion taken orally. Bark	Bilharziasis [16, 19]
Securidaca longipedunculata Fresen [Polygalaceae]	Tree, shrub over 2 m. [Tree or shrub] N, W, C, E, S	Chipvufanana (Shona) Mufufu (Shona) Munyapunyapu (Shona) Munyazvirombo (Shona) Mutangeni (Shona) Umfufu (Ndebele) Violet tree (English)	Roots – infusion taken orally.	Tapeworm and hookworm. Bilharziasis (Urinary schistosomiasis) [16, 21, 37]
Senna italica Mill. subsp. *micrantha*	Prostrate or ascending	The Port Royal senna (English)		Bilharziasis (Urinary

Scientific name [Family]	[Growth habit] Distribution	Vernacular names and other names	Parts used and mode of preparation	Parasitic infection ethnomedicinal uses [References]
(Brenan) Lock *Cassia italica* (Mill.) F.W. Andr. [Fabaceae]	perennial herb or small shrub [Herb or shrub] W, C, S			schistosomiasis) [21]
Senna petersiana (Bolle) Lock *Cassia petersiana* Bolle [Fabaceae]	Tree, shrub over 2 m. [Tree or shrub] N, C, E, S	Eared senna (English) Monkey pod (English)		Bilharziasis (Urinary schistosomiasis) [21]
Senna singueana (Delile) Lock *Cassia singueana* Delile [Fabaceae]	Tree, shrub over 2 m. [Tree or shrub] N, W, C, E, S	Isihaqa esincinyane (Ndebele) Mudyamhungu (Shona) Munzungu (Shona) Mushayanyoka (Shona) Scrambled egg (English) Sticky pod (English) Winter cassia (English) Winter-flowering senna (English)	Leaves, stem, roots and bark.	Bilharziasis (Urinary schistosomiasis) [23]
Solanum campylacanthum *Solanum delegoense* Dunal *S. incanum* [Solanaceae]	[Shrub] Introduced	Nhundurwa (Shona) Bitter apple (English) Intume (Ndebele) Munhomboro (Shona) Munhundurwa (Shona) Poison apple (English) Snake apple (English) Sodom apple (English) Thorn apple (English) Umdulukwa (Ndebele)	Leaves Roots	Bilharziasis (Urinary schistosomiasis) [16, 21, 37]
Steganotaenia araliacea Hochst. [Apiaceae]	[Tree] N, C, E, S	Carrot Tree (English)	Leaves and stem. Roots and stem.	Bilharziasis (Urinary schistosomiasis) [38]
Strychnos occuloides Bak. [Loganiaceae]	Small deciduous tree. [Tree] N, W, E, S	Corky monkey-orange (English) Muhumi (Shona) Mushumwi (Shona) Mutamba muzhinyu (Shona)		Bilharziasis (Urinary schistosomiasis) [16, 19]
Terminalia brachystemma Welw. ex Hiern [Combretaceae]	Shrub or small semi-deciduous tree. [Tree or shrub] N, W, C, E, S	Kalahari cluster-leaf (English)	Leaves Roots Fruit	Bilharziasis (Urinary schistosomiasis) [16, 21, 37]
Terminalia sericea Burch ex. DC [Combretaceae]	Small to medium-sized deciduous tree. [Tree] Introduced	Mangwe (Shona) Mukonono (Shona) Mususu (Shona) Mutabvu (Shona) Silver cluster-leaf (English) Silver terminalia (English) Umangwe (Ndebele)	Roots – mixed with *Vigna unguiculata* seeds and cooked in soup. Roots – decoction taken orally.	Bilharziasis Worms in anus and arms. [16, 21, 37]
Toddalia asiatica (L.) Lam. *Toddalia aculeata* Pers. [Rutaceae]	[Climber, liane.] N, C, E, S	Chikafusi (Shona) Climbing orange (English) Cockspur orange (English) Gato (Shona) Mubhatakhamba (Ndau) Rukato (Shona)		Bilharziasis (Urinary schistosomiasis) [38]

Scientific name [Family]	[Growth habit] Distribution	Vernacular names and other names	Parts used and mode of preparation	Parasitic infection ethnomedicinal uses [References]
Trichilia emetica Vahl subsp. *emetica* *Trichilia roka* Chiov. [Meliaceae]	Medium-sized to large evergreen tree. [Tree] N, W, E, S	Banket mahogany (English) Muchichiri (Shona) Mutsikiri (Shona) Natal mahogany (English)	Leaves Roots and roots bark.	Bilharziasis (Urinary schistosomiasis) [16, 19]
Trichodesma ambacense Welw. subsp. hockii (De. Wild) Brummitt [Boraginanceae]	Perennial herb. [Herb] N, W, C, E, S	Blue Bells of St. Mary's (English) Gwiramwaka (Shona)	Tuber – is ground into powder and taken orally.	Bilharziasis [16, 21, 37]
Vangueria infausta Burch.subsp. *infausta* [Rubiaceae]	Small deciduous tree. [Herb] W, C, E, S	Munjiro (Shona) Munzviro (Shona) Munzvirwa (Shona) Umthofu (Ndebele) Umviyo (Ndebele) Velvet wild medlar (English)	Fruit, leaves and roots – decoction taken orally. Roots	Parasitic worms [29, 39] Roundworm
Vernonia amydalina Del [Asteraceae]	Tree, shrub over 2 m. [Tree or shrub] N, W, C, E, S	Bitter-tea vernonia (English) Dembezeko (Shona) Inyathelo (Ndebele) Musikavakadzi (Shona) Muzhozho (Shona) Nyareru (Shona) Tree vernonia (English)	Roots – are mixed with *Vigna unguiculata* seeds and a soup is prepared and taken orally. Leaves and stem. Roots. Roots bark	Bilharziasis (Urinary schistosomiasis) [16, 23]
Vernonia musofensis S. Moore var. *miamensis* (S. Moore) G.V. Pope *Vernonia philipsoniana* Lawaltree [Asteraceae]	[Herb] N, C, E		Roots – are mixed with *Vigna unguiculata* seeds and a soup is prepared and taken orally.	Bilharziasis
Warburgia sulcate [Canellaceae]	Evergreen tree. [Tree] No information			Bilharziasis [40]
Ximenia caffra Sond. [Olacaceae]	Tree, shrub over 2 m. [Tree or shrub] N, W, C, E, S	Munhengeni (Shona) Mutengeni (Shona) Mutsvanzva (Shona) Sourplum (English) Umthunduluka (Ndebele)	Roots – infusion taken orally. Roots (fresh) Leaves and stem. Roots bark.	Bilharziasis (Urinary schistosomiasis) [16, 21–23, 40]
Zanthoxylum chalybeum Engl. [Rutaceae]	Tree, shrub over 2 m. Tree or shrub N, W, C	Kundanyoka knobwood (English) Mukundanyoka (Shona)		Bilharziasis (Urinary schistosomiasis) [21]
Ziziphus mucronata Willd [Rhamnaceae]	Small to medium-sized tree. [Tree] N, W, C, E, S	Buffalo-thorn (English) Chinanga (Shona) Muchecheni (Shona) Umpasamala (Ndebele) Umphafa (Ndebele)	Roots – infusion taken orally. Leaves and stem Roots bark Roots	Bilharziasis (Urinary schistosomiasis) [16, 21, 23, 40]

Table 1.
Medicinal plants used to treat and manage bilharziasis and other parasitic infections in Zimbabwe: Family and botanical name, local name, part used, mode of preparation, growth form, distribution and ethnomedicinal uses.

Scientific name	Pharmacological properties	Biological target	Toxicological evaluation	Reference
Abrus precatorius L. subsp. *africanus* Verde.	*Anthelmintic, analgesic, antimicrobial, antimigraine, anti-bacterial, anti-fungal, anti-tumor, anti-spasmodic, anti-diabetic, anti-serotonergic and anti-inflammatory activities.	Cestodes Schistosomes	Safe $LD_{50} > 5000$ mg/kg	[23, 41–46]
Acacia karoo Hayne	Analgesic, HIV1 reverse transcriptase, antilisterial, anti-gonococcal, anti-diabetic, anti-inflammatory, antioxidant, antibacterial, antifungal, antimalarial, antimicrobial, *anthelmintic and anti-mycobacterial activities.	Cestodes	Weak or low toxicity or mildly toxic $LD_{50} < 1600$ mg/kg	[23, 47–49]
Albizia antunesiana Harms	*Anthelmintic and oxidant activities.	Cestodes	no records found	[23, 31, 50]
Asparagus spp except *A. asparagoides L*	Analgesic, diuretic, *anthelmintic, anti-inflammatory and antimicrobial activities.	Nematodes: *Haemonchus contortus*	Safe $LD_{50} > 5000$ mg/kg	[51–54]
Burkea africanus Hook.	Antioxidant, anti-diarrhoeal, antibacterial, analgesic, anti-inflammatory, anti-cholinesterase and *anthelmintic activities.	Nematodes: *Haemonchus contortus*	Safe $LD_{50} > 5000$ mg /kg	[55–57]
Carissa spinarum L. *Carrisa edulis* Vahl	Anti-plasmodial, diuretic, antioxidant, *antihelmintic, antiherpetic, anti-inflammatory and antiviral activities.	Earthworms: *Pheretima posthuma.*	Safe $LD_{50} > 2000$ mg in rats	[58–62]
Cassia abbreviata Oliv.	*Anthelmintic, antiviral, antioxidant, antimicrobial, abortifacient, anti-diabetic, anti-inflammatory, hepatoprotective and antimicrobial activities.	Cestodes	Safe LC_{50}– 1319.37 ± 356.63 μg/ml	[23, 63–68]
Catunaregam swynnertonii (S. Moore) Bridson *Catunaregum spinosa sensu* Verdcourt subsp. *spinosa*	*Anthelmintic, antioxidant, emetic, nauseant, anti-allergic, antipyretic, anti-inflammatory, expectorant, abortifacient, antibacterial, human cyclooxygenase (COX)-2 inhibitory effects, analgesic, immunomodulatory and a prominent protection of DNA activities.	Nematodes Earthworm: *Eisinia Fetida*	Safe LD_{50} up to 2000 mg/kg.	[56, 69, 70]

Scientific name	Pharmacological properties	Biological target	Toxicological evaluation	Reference
Celtis africana Burm. f.	*Anthelmintic, prokinetic, laxative, antidiarrheal, spasmolytic, antioxidant, anti-inflammatory andweak to moderate acetylcholinestrease enzyme inhibition activities.	Cestodes	no records found	[23, 71, 72]
Cissampelos mucronata A. Rich.	Hypoglycemic, antivenin, anti-diabetic, anti-ulcer, antispasmodic, anti-diarrhoeal and possess significant effects on male fertility.	no records found	Safe $LD_{50} > 5000$ mg/kg	[73–75]
Cissus quadrangularis L.	Bone healing, *anthelmintic, antiulcer, anti-inflammatory, anti-tumor, molluscicidal, gastro-protective, anti-osteoporotic, antioxidant and antimicrobial activities.	Earthworms: *Pheretima posthuma*	Safe LD_{50} - 3000 mg/kg	[76–79]
Cleridendrum ternatum Schinz	no records found	no records found	no records found	
Combretum heteroense Schinz *subsp. heteroense*	*Anthelmintic (antischistosomal), antifungal, anti-inflammatory and cytotoxicity activities.	Nematode: Worms of *C. elegans* var. Bristol Schistosomes: Worms of *S. haematobium*	no records found	[80–82]
Combretum imberbe Wawra	Antibacterial, anthelmintic, antioxidant, antifungal, *anthelmintic (antischistosomal), anti-hyperglycemic, anti-malarial, anti-snake and anti-inflammatory activities.	Nematode: Worms of *C. elegans* var. Bristol Schistosomes: Worms of *S. haematobium*	Highly toxic LC_{50}–168.6 μg/mL	[80–85]
Combretum zeyheri Sond	Antibacterial, anti-inflammatory, cytotoxicity against human cancer cell line, *anthelmintic (antischistosomal), antioxidant, antifungal and anti-proliferative activities.	Schistosomes: Worms of *S. haematobium*	Highly toxic LC_{50}–16 μg/ml to 159 μg/ml	[28, 56, 80, 81]
Crossopteryx febrifuga (G. Don) Benth.	Anti-inflammatory, *anthelmintic, anticonvulsant, analgesic, anti-plasmodial, antipyretic, antihyperglycemic, anti-proliferative and hypolipidemic activities.	Nematodes: *Haemonchus contortus*	Safe LD_{50}–5000 mg/kg	[51, 86–91]

Scientific name	Pharmacological properties	Biological target	Toxicological evaluation	Reference
Croton gratissimus Burch	Good antioxidant, anti-diabetic, anti-inflammatory, antibiotic, antiviral, analgesic, anticonvulsant, *anthelmintic (antiprotozoal and antileishmanial), anticancer, antiulcer, immunomodulatory, anti-pyretic, anti-plasmodial, hypolipidemic, antiarthritic, anti-eczemic, antihistimic and anti-coronary activities.	Protozoa: *T. b. rhodesiense* Leishmania: *Leishmania donovani*	Highly toxic LC_{50} Hexane fraction - 25.3 ± 0.87 μg/ml. DCM fraction - 67.3 ± 0.32 μg/ml.	[92–98]
Cynanchum viminale (L.) L. subsp. *viminale Sarcostemma viminale* (L) R. Br	Antipyretic, analgesic, and anti-inflammatory activities.	no records found	no records found	[99]
Dicoma anomala Sond.	*Anthelmintic, anticancer, antioxidant, antihyperglycemic, anti-inflammatory and antimicrobial activities.	Cestodes	Safe LC_{50} value of 3040 ± 1060 μg/ml	[34, 35, 100]
Diplorhynchus condylocarpon (Muell Arg.) Pich.	Sympatholytic, anti-amoebic, anti-plasmodial, analgesic, antibacterial, antimalarial, anti-inflammatory and antioxidant activities	no records found	Safe LD_{50} > 2000 mg/kg	[56, 101–103]
Elephantorrhiza goetzei (Harms)	*Anthelmintic, antifungal, antioxidant, antibacterial, antiviral, and cytotoxicity activities.	Cestodes Schistosomes	Moderately toxic LC_{50}_356.55 ± 24.55 μg/ml.	[23, 32]
Eriosema englerianum Harms.	Antimicrobial, antibacterial, antifungal and antioxidant activities.	no records found	no records found	[104, 105]
Erythrina abyssinica DC.	Antimycobacterial, antifungal, hypoglycemi, antiplasmodial, hepatoprotective, *antihelminthic and antimicrobial activities.	Nematodes: *Ascaridia galli*	Safe LC_{50}–5440 ± 0 μg/ml.	[100, 106–109]
Euclea divorum Hiern	Antimicrobial, diuretic cytotoxic, antibacterial, oxytocic, antifungal, diuretic, antioxidant, *antihelminthic and anti-plasmodial activities.	Nematodes: *Caenorhabditis elegans*	Safe LD_{50} - 2000 mg/kg	[110–113]

Scientific name	Pharmacological properties	Biological target	Toxicological evaluation	Reference
Euphorbia schinzii Pax	no records found	no records found	no records found	
Faurea saligna Harv	Antifungal activity.	no records found	no records found	[114]
Flacourtia indica (Burm.f) Merr	Antimicrobial, anti-diabetic, anthelmintic hepatoprotective, antiviral, *anthelmintic (antitrypanosomal and antileishmanial), anti-inflammatory, antimalarial, anti-plasmodial, antioxidant and anti-asthmatic activities.	Trypanosome: *Trypanosoma brucei brucei, Trypanosoma brucei rhodesiense, Trypanosoma cruzi* Leishmania: *Leishmania donovani*	Moderately toxic LC_{50} - 467.31 ± 39.01 µg/ml	[19, 66, 115–119]
Gymnosporia senegalensis Loes *Maytenus senegalensis* (Lam) Exell	*Anthelmintic (antileishmanial), antioxidant, antiviral, antibacterial and antifungal activities.	Cestodes	Safe LC_{50}–2185.61 ± 872. 25 µg/ml LD_{50} > 1600 mg/kg	[19, 23, 66, 120–122]
Khaya anthotheca (Welw.) C. DC. *Khaya nyasica* Bak.f.	Antimicrobial, antioxidant, *anthelmintic (antitrypanosomal and antileishmanial), antiplatelet, antiviral and anti-plasmodial activities.	Trypanosome: Trypanosoma brucei rhodesiense and Trypanosoma cruzi Leishmania: Leishmania donovani	Moderately toxic LC50 - 482.19 ± 43.49 µg/ml.	[19, 66, 116, 123, 124]
Hydnora abyssinica A. Braun ex Schweinf. *Hydnora solmsiana* Dinter	Antibacterial, antioxidant, anti-diarrhoeal, antiglycation and antifungal activities.	no records found	Weak or low toxicity or mildly toxic LD_{50} > 1600 mg/kg	[125–128]
Kigellia africana (Lam.) Benth. *Kigelia pinnata* (Jacq.) DC.	Antiplasmodial, antiviral, antiulcer, anticancer, anti-diarrhoeal, antimicrobial, antioxidant, anti-diabetic, *anthelmintic (antitrypanosomal), effects on reproductive system and anti-inflammatory activities.	Trypanosomes: *Trypanosoma brucei* and *Trypanosoma bruceirhodesiense*	Safe LD_{50} > 5000 mg/kg	[66, 129–131]
Landolphia kirkii Dyer ex Hook. f.	no records found	no records found	no records found	
Lannea discolor (Sond.) Engl.	*Anthelmintic (nematicidal), antibacterial, antimycobacterial, antifungal, antioxidant and antiplasmodial activities.	Cestodes	Weak or low toxicity or mildly toxic LC_{50} values ranging 0.408 mg/mL to >1.0 mg/mL	[23, 34, 35, 132]

Scientific name	Pharmacological properties	Biological target	Toxicological evaluation	Reference
Lannea edulis (Sond.) Engl.	*Anthelmintic, anti-human immunodeficiency virus, antihyperglycemic, antihyperlipidemic, antimalarial, antimicrobial, antioxidant and cytotoxicity activities.	Cestodes	Safe $LD_{50} > 6000$ mg/kg	[23, 36, 100]
Lecaniodiscus fraxinifolias Bak.	*Anthelmintic activity.	Cestodes	no records found	[23]
Loranthus on *Dichrostachys cinerea,*	no records found	no records found	no records found	
Mondia whitei (Hook.f.) Skeels	Antidepressant, anti-diarrheal, antiepileptic, antibacterial, aphrodisiac, anti-convulsant, pro-erectile, antimicrobial, tyrosinase-inhibitory, anti-inflammatory, androgenic *anthelmintic, anti-tyrosinase, antioxidant, anticancer anti-spermatogenic and antifertility activities.	Schistosomes	Safe $LD_{50} > 5000$ mg/kg	[133–136]
Mucuna coriacea Baker subsp. irritans (Burtt Davy) Verdc.	no records found	no records found	no records found	
Musa sp.	Antioxidant, antibacterial, antiviral, anti-ulcerogenic, antithrombotic, anti-allergic, anti-inflammatory, antiallergenic, anti-diabetic, diuretic, mutagenecity, wound healing, antidiarrhoeal, *anthelmintic, antimalarial, anti-snake venom and vasodilatory activities.	Nematodes: *Haemonchus contortus* and *Trichostrongylus colubriformis*	Safe $LD_{50} > 5000$ mg/kg	[137–141]
Ozoroa reticulata (Baker f.) R. Fern. & A. Fern. *Ozoroa insignis* Del. subsp. *reticulata (*Bak. f.) Gillett *Heeria reticulata* Engl.	Antimicrobial, cytotoxic, antibacterial and *anthelmintic activities.	Cestodes Schistosomes	Highly toxic LC_{50} ranging 2.21–10.63 µg/ml	[23, 31, 142–145]
Peltophorum africanum Sond.	Antibacterial, antifungal, antiviral, antioxidant and *anthelmintic activities.	Schistosomes	Weak or low toxicity LC_{50} ranging 882–1060 ± 106 µg/ml	[23, 65, 100]

Scientific name	Pharmacological properties	Biological target	Toxicological evaluation	Reference
Phaseolus vulgaris L	Hypocholesterolemic, nephroprotective, anticancer, anti-hypertensive, diuretic, neuroprotective, antifertility, analgesic, antibacterial, hepatoprotective, antiobesity, osteoprotective, anti-inflammatory, antioxidant, antidiabetic, *anthelmintic, immunostimulatory, cardio-protective, litholytic, trypsin and α-amylase inhibitor activities.	Nematodes: *Trichostrongylus colubriformis* and *Teladorsagia circumcincta*	Safe LD $_{50}$ up to 2000 mg/kg	[146–151]
Piliostigma thonningii (Schumach.) Milne-Redh.	*Anthelmintic (antileishmanial), anti-oxidative, antiviral, antipyretic, antibacterial and anti-inflammatory activities.	Cestodes Nematodes: *Haemonchus contortus*	Safe LD_{50} > 5000 mg/kg in rats.	[23, 51, 152–154]
Pogonarthria squarrosa (Licht.) Pilg.	no records found	no records found	no records found	
Pterocarpus angolensis DC.	*Anthelmintic, antibacterial, anti-plasmodial, anti-inflammatory and antifungal activities.	Cestodes	Weak or low toxicity or mildly toxic LC_{50} ranging 478–1320 ± 266 μg/ml.	[23, 100, 155–157]
Ricinus communis L	Anticonceptive, antioxidant, antidiabetic, antifertility, anti-inflammatory, antioxidant, *anthelmintic, hepatoprotective, insecticidal, wound-healing, anti-asthmatic, lipolytic, immunomodulatory and antimicrobial activities.	Nematodes	Safe LD_{50}–8000 mg/kg	[158–161]
Sansevieria hyacinthoides (L.) Druce	*Anthelmintic, antibacterial, antifungal and antioxidant activities.	Nematode: *Caenorhabditis elegans*	no records found	[31, 38, 110, 162–164]
Sclerocarya birrea (A. Rich.) Hochst. subsp. *caffra* (Sond.) Kokwaro	Anti-diarrhoeal, anti-diabetic, anti-inflammatory, anti-plasmodial, *anthelmintic, antimicrobial, antioxidant, antihypertensive, anti-convulsant and antinociceptive activities.	Nematode: *Haemonchus contortus*	Safe LC_{50}–1112.37 ± 210.04 μg/ml.	[19, 51, 66, 165]

Scientific name	Pharmacological properties	Biological target	Toxicological evaluation	Reference
Securidaca longipedunculata Fresen	Antibacterial, anti-plasmodial, *anthelmintic and antifungal activities.	Nematode: *Haemonchus contortus*	Moderately toxic LD_{50} value of 771 mg/kg	[51, 100, 155, 157, 166]
Senna italica Mill. subsp. *micrantha* (Brenan) Lock *Cassia italica* (Mill.) F.W. Andr.	Antibacterial, hypoglycemic effect anti-inflammatory, antipyretic, uteruscontractions, *anthelminthic, antioxidant, antiproliferative, analgesic, prostaglandin (PG) release, antineoplastic and antiviral activities.	Nematodes: *Haemonchus contortus*	Safe LD_{50} > 5000 mg/kg	[167–175]
Senna petersiana (Bolle) Lock *Cassia petersiana* Bolle	*Anthelmintic, antibacterial, antifungal, cyclooxygenase (COX) inhibitory, antiviral and antimicrobial activities.	Nematode: *Caenorhabditis elegans*	no records found	[176, 177]
Senna singueana (Delile) Lock *Cassia singueana* Del	*Anthelmintic, antioxidant, antiplasmodial, antiulcer, antipyretic, anti-inflammatory and analgesic activities.	Cestodes	Safe LD_{50}–2150 mg/kg	[23, 178, 179]
Solanum campylacanthum Solanum delegoense Dunal *S. incanum*	*Anthelmintic, antinociceptive effect, antipyretic, analgesic, antimicrobial, anti-inflammatory and anti-cytotoxic activities.	Cestodes	Safe LD_{50} > 2000 mg/kg	[23, 180–184]
Steganotaenia araliacea Hochst.	*Anthelmintic (antileishmanial and larvicidal), antimitotic, antitubulin, uterotonic, antioxidant, antibacterial, diuretic and antiplasmodial activities.	Cestodes	no records found	[23, 185–187]
Strychnos cocculoides Bak.	Antimalarial and antioxidant activities.	no records found	no records found	[31, 188, 189]
Termilia sericea Burch ex. DC	Antibacterial, antifungal, anti-neurodegenerative, anticancer, antioxidant, antiviral, anti-HIV, anti-fungal, antibacterial, anticancer, lipolytic, wound healing, *anthelmintic (antiprotozoal), anti-inflammatory and anti-oxidant activities.	Protozoa: *Trichomonas vaginalis*	Toxic LC_{50} < 300 μg/ml.	[19, 66, 190–196]
Terminalia brachystemma Welw.	*Anthelmintic, antifungal and antioxidant activities.	Cestodes	no records found	[23, 197, 198]

Scientific name	Pharmacological properties	Biological target	Toxicological evaluation	Reference
Toddalia asiatica (L.) Lam. *Toddalia aculeata* Pers.	Anti-inflammatory, anti-bacterial, anti-tumor, antifeedant analgesic, anti-HIV, anti-plasmodial, antiviral, *anthelmintic, analgesic, antiplatelet aggregation, wound healing, anticancer, spasmolytic activity and skin whitening activities.	Protozoa: *Ichthyophthirius multifiliis*	Weak or low toxicity or mildly toxic $LD_{50} > 1000$ mg/kg	[199–206]
Trichilia emetica Vahl subsp. *emetic* *Trichilia roka* Chiov.	*Anthelmintic (antischistosomal and antitrypanosomal), anti-diarrhoeal, antifungal, anti-oxidant, anti-infective, anti-inflammatory, antiviral, anticonvulsant, antifeedant, anti-plasmodial, antitussive, antimutagenic, bactericidal hepatoprotective and growth regulating activities.	Cestodes Schistosomes Trypanosomes: *Trypanosoma brucei brucei* and *Trypanosoma brucei rhodesiense*	Safe $2000 < LD_{50} < 5000$ mg/kg	[23, 207–212]
Trichodesma ambacense Welw. subsp. hockii (De. Wild) Brummitt	no records found	no records found	no records found	
Vangueria infausta Burch. subsp. *infausta*	Antibacterial, antifungal, antimycobacterial, anti-inflammatory, antioxidant, *anthelmintic (antileishmanial), anti-plasmodial, antifeedant and prostaglandin synthesis inhibitory activities.	Leishmania: *Leishmania donovani*	Moderately toxic LC_{50} values ranging 338–416 ± 28.3 μg/mL	[39, 100, 101, 213–217]
Vernonia amydalina Del	Antimicrobial, antimalarial, antithrombotic, antioxidant, antipyretic, analgesic, anti-diabetic, laxative, hypoglycemic, anticancer, *antihelmintic, antifertility, anti-inflammatory, cathartic, antifungal and antibacterial activities.	Cestodes	Safe LD_{50}–5152.3 mg/kg. LD_{50}–3721 mg/kg.	[23, 218–220]
Vernonia philipsoniana Lawaltree	no records found	no records found	no records found	
Vigna unguiculata (L.) Walp.	Antioxidant, *anthelmintic, antibacterial, anti-diabetic, anti-depressant, anti-sickling, antifungal, antiviral, antimicrobial, antinociceptive, hypocholesterolemic, thrombolytic and hypolipidaemic activities.	Earthworms: *Edrilus euginiae*	Safe $LD_{50} > 2000$ mg/kg	[221–225]

Scientific name	Pharmacological properties	Biological target	Toxicological evaluation	Reference
Warburgia sulcata	no records found	no records found	no records found	
Ximenia caffra Sond.	*Anthelmintic, anti-amoebic, antibacterial, antigonococcal agent, antifungal, anti-inflammatory, antioxidant, anti-parasitic, anti-proliferative, insecticidal, HIV-1 reverse transcriptase (RT) inhibitory and non-mutagenic activities.	Cestodes	Highly Toxic LC_{50}–11.25 μg/ml	[23,142, 226–229]
Zanthoxylum chalybeum Engl.	Anti-plasmodial, antibacterial, *anthelmintic (antitrypanosomal), antifungal antiviral, anti-hyperglycemic, anti-hyperlidemic and anti-measles virus activities.	Nematodes: *Ascaris suum*	Safe LD_{50} > 5000 mg/kg	[230–235]
Ziziphus mucronata Willd	Antimicrobial, antiviral, anti-diabetic, anti-inflammatory, anti-oxidant, anti-plasmodial, *anthelmintic and anti-anemic activities.	Schistosomes Nematode: *Caenorhabditis elegans*	Safe LC_{50} > 1000 μg/ml LD_{50} > 5000 g/kg	[23, 236, 237]

In vitro *experiments were carried-out and reported by Mølgaard et al. [23] demonstrating dose-dependent anthelmintic activity. Anthelmintic activities reported were conducted on a number of biological targets (**Figure 6**): Cestodes; Earthworms:* Pheretima posthuma, Edrilus euginiae, Eisinia fetida*; Leishmania:* Leishmania donovani*; Nematode:* Ascaris suum, Ascaridia galli, Caenorhabditis elegans, Chabertia ovina, Cooperia spp., Haemonchus contortus, Trichostrongylus spp., Trichostrongylus colubriformis, Teladorsagia circumcincta, Teladorsagia spp*; Protozoa:* Ichthyophthirius multifiliis, Trichomonas vaginalis, T. b. rhodesiense*; Schistosomes:* Schistosoma haematobium*; Trypanosome:* Trypanosoma brucei brucei, Trypanosoma brucei rhodesiense, Trypanosoma cruzi. *More studies should be carried out on more prevalent biological targets such as Schistosomes:* Schistosoma haematobium.
** signifies the major pharmacological activity attributed by the different plant species of plants which is the anthelmintic activity.*

Table 2.
Pharmacological and toxicological evaluation of medicinal plants used to treat and manage bilharziasis and other parasitic infections in Zimbabwe.

Scientific name	Anthelmintic activity Lethal concentrations after 1 h
Abrus precatorius	[a]Cestodes: Stem 3.0 Root 1.2 mg/ml [a]Schistosomes: Stem 1.5 Root 0.6 mg/ml
Acacia karoo	[a]Cestodes: Leaves 3.1 mg/ml Root 17 mg/ml [c]Schistosomes: Leaves 103 mg/ml
Albizia antunesiana	[a]Cestodes: Leaves and stem 16.8 mg/ml Root bark 6.3 [c]Schistosomes: Root bark 100 mg/ml
Cassia abbreviata	[a]Cestodes: Root and root bark 17.1 mg/ml
Celtis africana	[b]Cestodes: Leaves 63.0 mg/ml
Dicoma anomala	[a]Cestodes: Root 31.0 mg/ml

Scientific name	Anthelmintic activity Lethal concentrations after 1 h
Elephantorrhiza goetzei	[a]Cestodes: Stem bark 4.2 mg/ml fruits 12.1 mg/ml Root 17.4 Leaves and stem 15.9 mg/ml [a]Schistosomes: Stem bark 0.8 mg/ml
Gymnosporia senegalensis *Maytenus senegalensis*	[a]Cestodes: Leaves and stem 25.0 mg/ml Root 25.0 mg/ml Root bark 2.5 mg/ml [b]Schistosomes: Root bark 100 mg/ml
Lannea discolor	[b]Cestodes: Leaves and stem 10 mg/ml Root and root bark 12.9 mg/ml
Lannea edulis	[b]Cestodes: Leaves and stem 4.0 mg/ml
Lecaniodiscus fraxinifolias	[a]Cestodes: Leaves and stem 6.3 mg/ml Root 6.4 mg/ml
Ozoroa reticulata *Ozoroa insignis*	[a]Cestodes: Leaves 2.5 mg/ml Schistosomes: 34.0 mg/ml [a]Cestodes: Stem bark 51.0 mg/ml [a]Cestodes: Root bark 0.8 mg/ml [a]Schistosomes: Root bark 26.3 mg/ml Leaves + stem 25.3 mg/ml
Peltophorum africanum	[c]Schistosomes: Leaves and stem 100 mg/ml
Piliostigma thonningii	[a]Cestodes: Root and root bark 30.8 mg/ml
Pterocarpus angolensis	[a]Cestodes: Leaves 102.0 mg/ml Bark 51.3 mg/ml [c]Schistosomes: Leaves 102, stem 117 mg/ml
Senna singueana *Cassia singueana*	[a]Cestodes: Leaves ad stem 15.9 mg/ml Root bark 17.2 mg/ml
Solanum campylacanthum *Solanum delegoense*	[a]Cestodes: Leaves 50.2 mg/ml
Steganotaenia araliacea	[a]Cestodes; Leaves and stem 62.7 mg/ml
Terminalia brachystemma	[a]Cestodes: Leaves 13.4 mg/ml, Root 33.3 mg/ml, Fruit 1 66.5 mg/ml, Fruit 2 15.4 mg/ml
Trichilia emetica *Trichilia roka*	[b]Cestodes: Root and root bark 84.7 mg/ml [b]Schistosomes: Root 6.25 mg/ml [b]Trypanosomes: Leaves *Trypanosoma brucei brucei* - 14.9 mg/ml and *Trypanosoma brucei rhodesiense* - 8.6 mg/ml
Vernonia amydalina	[b]Cestodes: Root 1.7 mg/ml Root bark 59.5 mg/ml
Ziziphus mucronata	[c]Schistosomes: Root bark 101 mg/ml

[a]*All cestodes: schistosomules died.*
[b]*Some of the cestodes: schistosomules died.*
[c]*No cestodes: schistosomules died.*
Schistosomules of Schistosoma mansoni, *Cestodes of* Hymenolepis dimin.

Table 3.
Anthelmintic screening of Zimbabwean plants traditionally used against schistosomiasis [23].

3.3 Screening

Retrieved articles were first screened based on the titles and abstracts for relevance to the study excluding articles that reported on malaria and on veterinary use of medicinal plants. For example, we excluded articles on bovine mastitis and Oriental medicines, although they appeared in the search results. However, articles that

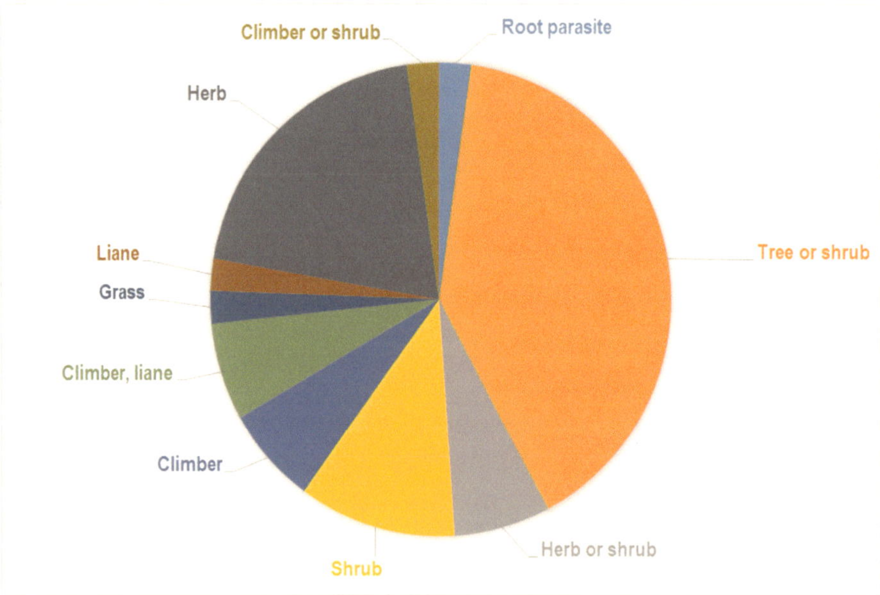

Figure 2.
Growth habit of medicinal plant species used to treat and manage bilharziasis and other parasitic infections in Zimbabwe.

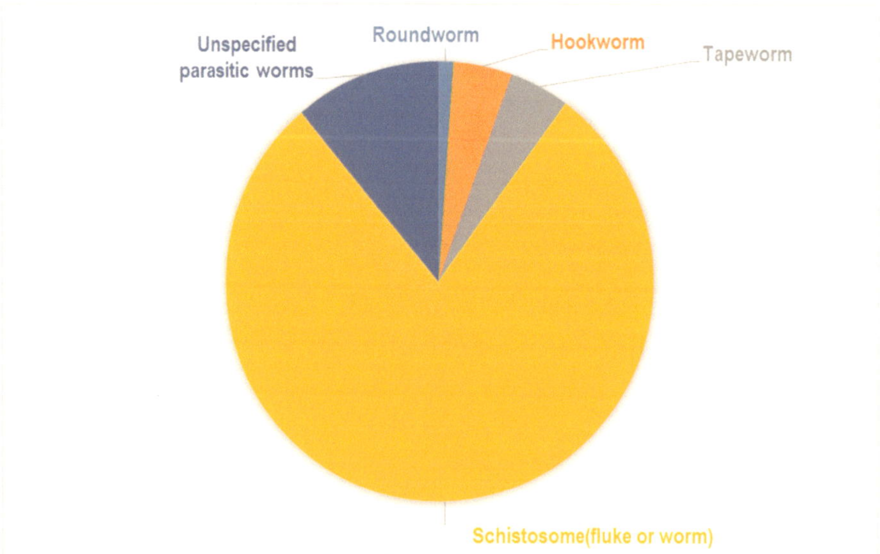

Figure 3.
Parasites managed or treated using medicinal plants in Zimbabwe.

included both malaria and schistosomiasis were considered. The eligible full articles were then assessed further for inclusion in the study using the inclusion/exclusion criteria.

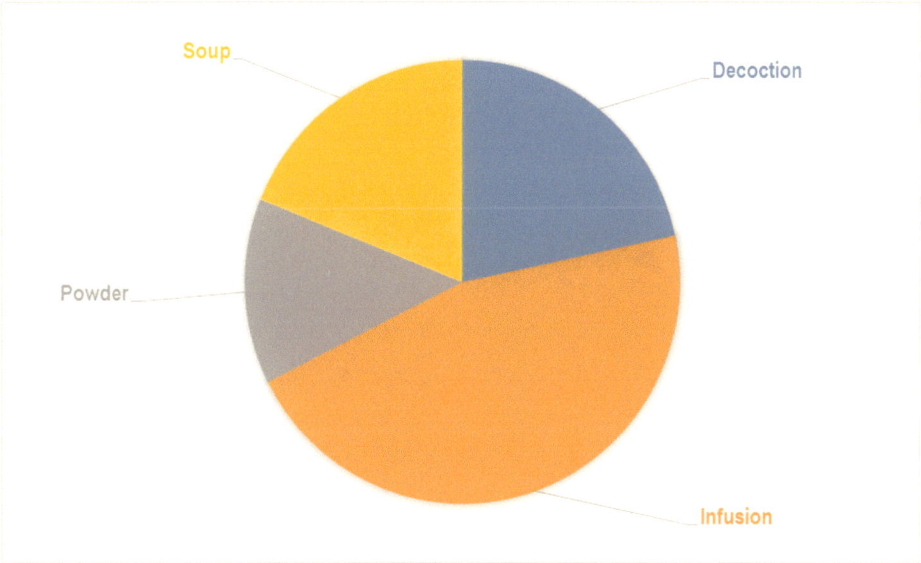

Figure 4.
Mode of preparation of medicinal plant species used to treat and manage bilharziasis and other parasitic infections in Zimbabwe.

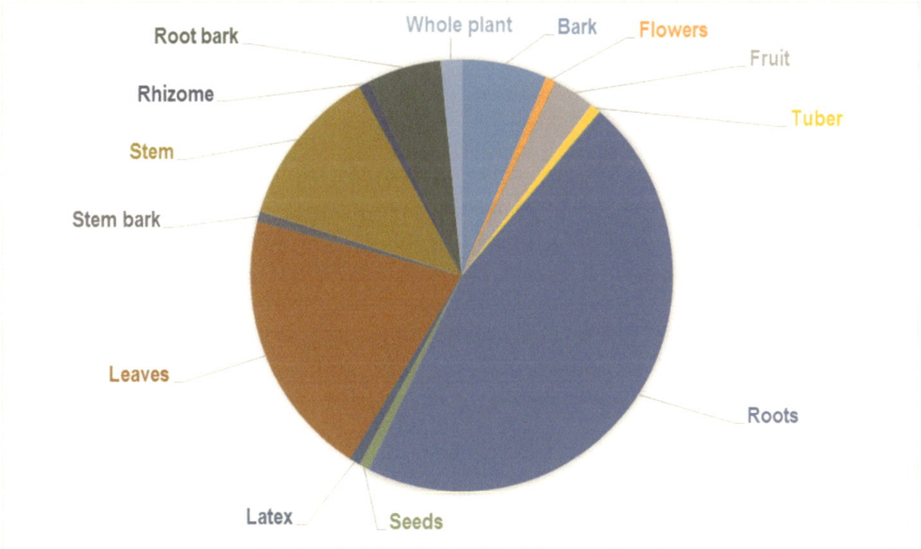

Figure 5.
Plant parts used for medicinal preparations used for the management of bilharziasis and other parasitic infections in Zimbabwe.

3.4 Inclusion and exclusion criteria

Full-text articles that at least reported on ethnobotany of Zimbabwean medicinal plants written in English and published in peer-reviewed journals, reports, books,

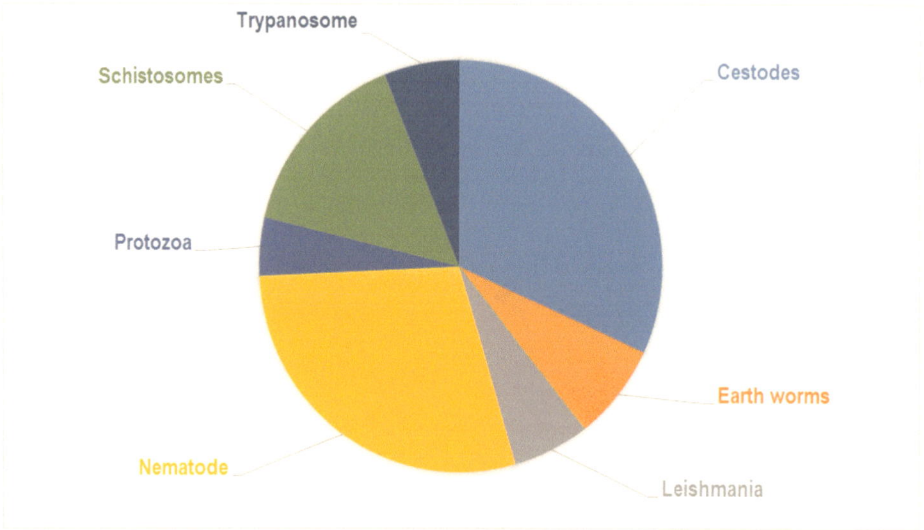

Figure 6.
Biological targets of tested parasites of medicinal plants reported.

theses and dissertations dated 31 December 2020 were considered. All publishing years were included without any geographical restrictions. Articles that reported data not relevant to the study or reviews or those not written in English were excluded from the study.

3.5 Data extraction

A data collection tool was designed in Microsoft Excel (Microsoft Corporation, USA) to capture data on different aspects of Zimbabwean medicinal plants. Three reviewers independently extracted relevant data from the included articles regarding the ethnobotany of Zimbabwean medicinal plants. For ethnobotanical data, the diseases or ailments managed, parts used and mode of preparation and administration were captured. The collected data were checked for completeness and processed independently by two other reviewers.

4. Results and discussion

From the several scientific papers reviewed based on ethnobotanical surveys of different areas of Zimbabwe, the results are presented in the following sections.

4.1 Literature search and publications

A total of 750 reports were retrieved out of which 138 met the inclusion criteria and were reviewed. Most of the articles were published in the 2010–2019 decade, indicating a lot of research is being done as compared to the preceding decades. This could be due to: (1) the growing need for more effective and less toxic medicinal products of plant origin, (2) emerging antimicrobial resistance that has rendered most chemotherapeutic agents less effective, (3) new disease outbreaks like COVID-19 and (4) increase in noncommunicable diseases such as cancers, hypertension, diabetes mellitus and sexual dysfunction that require readily available, affordable, effective and safe therapies.

4.2 Ethnobotanical surveys and distribution of medicinal plants traditionally used to treat and manage bilharziasis and other parasitic infections in Zimbabwe

Based on soil, rainfall regime and several other factors, Zimbabwe is divided into 5 agro-ecological regions. A total of 43 of the medicinal plants reported in this review are widely distributed throughout the Northern (N), Eastern (E), Central (C), Western (W) and Southern (S) regions of Zimbabwe as represented in **Figure 7**. The remaining plant species were distributed in several regions across the country with n = 9 plant species distributed in 4 regions, n = 8 in 3 regions, n = 2 in 2 regions and n = 1 in 1 region. A total of n = 3 plant species are being cultivated [*Celtis africana, Musa sp., Phaseolus vulgaris*] and n = 1 has been recently introduced *Ricinus communis*. *Warburgia sulcata* had no information on distribution in Zimbabwe (**Table 1**).

The current review indicates that there are at least 68 species of plants belonging to 63 genera in 33 families used to treat and manage bilharziasis and other parasitic infections in Zimbabwe (**Table 1**).

Generally, the family with the highest number of medicinal plants in Zimbabwe was the *Fabaceae* family represented with a total of 17 plants followed by *Combretaceae* (n = 5), *Apocynaceae* (n = 5), *Anacardiaceae* (n = 4), *Rubiaceae* (n = 3), *Euphorbiaceae* (n = 3) *Asteraceae* (n = 3), *Rutaceae* (n = 2) and *Meliaceae* (n = 2). A further 24 more plant families which only had one plant represented were also

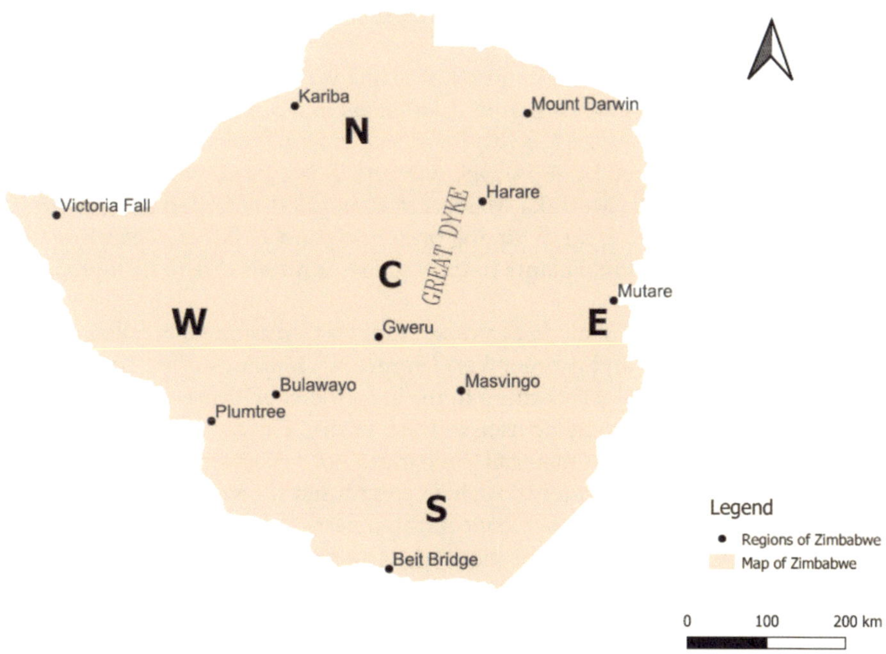

Figure 7.
General distribution of medicinal plants in different floristic regions of Zimbabwe.

recorded, giving a total of 33 families. These included *Apiaceae, Asparagaceae, Bignoniaceae, Boraginanceae, Canellaceae, Celastraceae, Dracaenaceae, Ebenaceae, Hydroraceae, Loganiaceae, Lorantaceae, Menispermaceae, Musaceae, Olacaceae, Poaceae, Polygalaceae, Proteaceae, Rhamnaceae, Salicaceae, Sapindaceae, Solanaceae, Ulmaceae, Verbenaceae* and *Vitaceae*.

Hutchings et al. [17] reported similar use of some medicinal plants reported in this study to treat and manage bilharziasis: *Abrus precatorius, Cassia abbreviata, Cissampelos mucronata, Euclea divorum, Faurea saligna, Gymnosporia senegalensis (Maytenus senegalensis), Mondia whitei, Pterocarpus angolensis, Sclerocarya birrea and Ximenia caffra*. Other studies reported similar anthelmintic medicinal plants; *Dicoma anomala* - Intestinal worms [15]; *Pterocarpus angolensis* - General use against intestinal worms, *Sclerocarya birrea* - Intestinal worms [18]; *Securidaca longipedunculata* – Tapeworm, *Vangueria infausta* - Roundworm [238]; *Ximenia caffra* - Intestinal worms [226]. These medicinal plants have been compiled by Cock et al. [10] review of Southern Africa.

4.3 Growth habit, parts used and mode of preparation of medicinal plants used to treat and manage bilharziasis and other parasitic infections in Zimbabwe

According to **Figure 2**, the frequency and type of plants used to treat and manage bilharziasis and other parasitic infections is as follows; tree (n = 23), tree, tree or shrub (n = 18), herb (n = 9), shrub (n = 5), climber, liane (n = 3), herb or shrub (n = 3), climber (n = 3), grass (n = 1), liane (n = 1), root parasite (n = 1) and shrub or climber (n = 1).

According to **Figure 3**, the parasites managed or treated are schistosomes (fluke or worm) 79%, unspecified parasitic worms 11%, hookworm 5%, tapeworm 4% and roundworm 1%. Midzi et al. [9] carried out a nationwide survey in Zimbabwe in 2010 and 2011 to map schistosomiasis and STH. The survey was conducted among primary school children. The study reported a high national prevalence of schistosomiasis (22.7%) and STH (5.5%). The common schistosome was *Schistosoma haematobium* with a prevalence of 18.0% while that of *Schistosoma mansoni* was 7.2%. The most common STH were hookworms (*Ascaris lumbricoides* and *Trichuris trichiura*) with a prevalence of 3.2% followed by *A. lumbricoides* and *T. trichiura* with prevalence of 2.5 and 0.1%, respectively [9]. Mutsaka-Makuvaza et al. [239] recorded a 13.3% prevalence in Madziwa, Shamva District among preschool-aged children. Therefore, there has been high use of medicinal plants to treat schistosomiasis due to its high prevalence in Zimbabwe.

The most frequently used mode of preparation was infusion 46% followed by decoction 22%, soup 19% and powder 13% (**Figure 4**). Methods of preparation of plant medicines seem to vary according to the area and subculture of the people in that region. Plant materials may be used as fresh or dry. However, the review observed a high usage of fresh material. Preparation of decoctions is carried out by boiling the plant material in water to such an extent that the volume of water is reduced to half. An infusion is a less concentrated version of a decoction and usually prepared by adding the plant material to water. There is a predominant use of decoctions and infusions which when both combined contribute to 68% of the gross mode of preparation. This may be attributed to the quick, low cost and easy to administer properties of these methods. Unfortunately, some of the ethomedicinal papers did not highlight the mode of preparation of the medicinal plants used [21, 22, 25, 26, 28].

The plant parts that are frequently used to treat and manage bilharziasis and other parasitic infections are shown in **Figure 5**. It appears the roots (46%) are the main target plant parts used. The use of the roots, bark and / or stem are the least environmentally sustainable part of the plant as its collection may lead to death of the plant however, they are the most preferred source of medicine. A number of papers did not highlight the plant parts used: [21, 22, 25, 26, 28].

4.4 Pharmacological properties of medicinal plants traditionally used to treat and manage bilharziasis and other parasitic infections in Zimbabwe

Some of the plant species have demonstrated a wide range of medicinal uses across different clinical conditions and therefore utilizing scientific methods to fully understand their pharmacological consequences could be vital. We have summarized the results of the pharmacological properties of 61 (89.7%) of the plant species (**Table 2**). The activities that were reported to be key in the treatment of bilharzia and parasitic infections were mainly dominated by the anthelmintic/antiparasitic properties. A medicinal plant with anthelmintic activity is responsible for treating and managing infections caused by a broad range of parasites (trematodes, worms, cestodes and nematodes) [240] (**Table 3**). Other complementary pharmacological properties include antioxidant, antibacterial and antifungal activities responsible for managing and treating parasitic infections (**Table 2**).

4.5 Toxicological evaluation of medicinal plants used to treat and manage bilharziasis and other parasitic infections in Zimbabwe

Out of the medicinal plants listed in **Table 1**, a total of 47 species (69.1%) have been subjected to toxicological evaluation studies, while the remaining 21 species (30.9%) lacked documented studies in this regard (**Table 2**). According to Kumari and Kotecha [241] ensuring the safety of herbal medicines is crucial in herbal research due to the potential for adverse effects and interactions. Of the 47 plants with toxicological profiles, the toxicological activities of the extracts were evaluated in several ways, including their effects on liver chang cells, cytotoxic activities on human monocyte cells, genotoxicity and anticancer properties among others. According to Kumari and Kotecha [241] toxicity assessment of herbal medicines involves various techniques, including *in vivo, in vitro* and cell line studies, as well as modern methods like microarray analysis. The BSLT and rodent acute toxicity experiments were the most common methods used to assess the toxicity of the 47 plants with available toxicological profiles (**Table 2**). Munodawafa et al. [100] reported that the BSLT and rodent acute toxicity tests were the most common methods used to assess the toxicity of herbal extracts. This is probably because the tests are relatively reliable, accurate, simple and cost-effective.

Munodawafa et al. [100] and Erhabor et al. [242] classified BSLT toxicity by determining the lethal concentration [LC50] of medicinal plant extracts that resulted in 50% mortality in brine shrimps, and the lethal dose [LD50] causing 50% mortality in mice/rats for rodent acute toxicity studies. In the classification of BSLT toxicity, high toxicity was assigned to [LC50] values below 249 μg/mL, moderate toxicity encompassed the range of 250–499 μg/mL, concentrations between 500 and 999 μg/mL were regarded as weak or low in toxicity and values exceeding 1000 μg/mL were considered safe Bussmann et al. [243] and Erhabor et al. [242]. In the rodent acute toxicity tests conducted by Malebo et al. [121], substances with [LD50] values below 50 mg/kg body weight were classified as highly toxic, those within the range of 50–300 mg/kg body weight were considered toxic, 300–1000 mg/kg body weight fell under the category of moderately toxic, 1000–2000 mg/kg body weight were mildly toxic and 2000 up to 5000 mg/kg body weight were classified as non-toxic. Among the 47 plants used for the treatment and management of bilharziasis and other parasitic infections in Zimbabwe, 30 plants (63.8%) were deemed safe/non-toxic, 6 plants (12.8%) exhibited weak or low toxicity or mild toxicity, 5 plants (10.6%) showed moderate toxicity, 1 plant (2.1%) was classified as toxic and 5 plants (10.6%) were highly toxic (**Table 4**).

In vitro investigations play a crucial role in the initial screening of compounds; however, these studies do not yield insights regarding the bioavailability, toxicity and *in vivo* efficacy of the tested extract/compound. Consequently, it is imperative to conduct future *in vivo* studies utilizing appropriate animal models to comprehensively

Toxicological profile	No of plants	Names of the plant species
Safe or nontoxic $LC_{50} \geq 1000$ µg/ml $2000 \leq LD_{50} \leq 5000$ mg/kg body weight	30	*Abrus precatorius, Asparagus spp, Burkea africanus, Carissa spinarum, Cassia abbreviata, Senna italica, Catunaregam swynnertonii, Cissampelos mucronata, Cissus quadrangularis, Crossopteryx febrifuga, Dicoma anomala, Diplorhynchus condylocarpon, Erythrina abyssinica, Euclea divorum, Gymnosporia senegalensis, Kigellia africana, Lannea edulis, Mondia whitei, Musa sp., Phaseolus vulgaris, Piliostigma thonningii, Ricinus communis, Sclerocarya birrea, Senna singueana, Solanum campylacanthum, Trichilia emetica, Vernonia amydalina, Vigna unguiculata, Zanthoxylum chalybeum and Ziziphus mucronata*
Weak or low toxicity or mildly toxic $500 \leq LC_{50} \leq 999$ µg/ml $1000 \leq LD_{50} \leq 2000$ mg/kg body weight	6	*Acacia karoo, Hydnora abyssinica, Lannea discolor, Peltophorum africanum, Pterocarpus angolensis and Toddalia asiatica*
Moderately toxic $250 \leq LC_{50} \leq 499$ µg/ml $300 \leq LD_{50} \leq 1000$ mg/kg body weight	5	*Elephantorrhiza goetzei, Flacourtia indica, Khaya anthotheca, Securidaca longipedunculata and Vangueria infausta*
Toxic $50 \leq LD_{50} \leq 300$ mg/kg body weight	1	*Termilia sericea*
Highly toxic $LC_{50} \leq 249$ µg/ml $0 \leq LD_{50} \leq 50$ mg/kg body weight	5	*Combretum imberbe, Combretum zeyheri, Croton gratissimus, Ozoroa reticulata and Ximenia caffra*
No records found	21	*Albizia antunesiana, Celtis africana, Cleridendrum ternatum, Combretum heteroense, Cynanchum viminale, Eriosema englerianum, Euphorbia schinzii, Faurea saligna, Landolphia kirkii, Lecaniodiscus fraxinifolias, Loranthus on Dichrostachys cinerea, Mucuna coriacea, Pogonarthria squarrosa, Sansevieria hyacinthoides, Senna petersiana, Steganotaenia araliacea, Strychnos cocculoides, Terminalia brachystemma, Trichodesma ambacense, Vernonia philipsoniana and Warburgia sulcata*

Table 4.
Toxicological evaluation of medicinal plants used to treat and manage bilharziasis and other parasitic infections in Zimbabwe.

comprehend the pharmacokinetics and pharmacodynamics of the tested extract/compound. The majority of *in vivo* studies fail to provide evidence concerning the toxicity and mechanism of action of medicinal plants/compounds, thereby highlighting the neglected nature of this aspect. Researchers are strongly encouraged to assess the toxicity levels and pharmacological actions of the tested plant/compound.

Conflict of interest

The authors declare no conflict of interest.

Author details

Elliot Nyagumbo[1*], Trust Nyirenda[2], Cephas Mawere[3], Ian Mutasa[4], Emmanuel Kademeteme[4], Alfred M. Mutaramutswa[1], Donald Kapanga[1], Godwins Ngorima[1], Leroy Nhari[1], Fabian Maunganidze[1], Michael Bhebhe[1], William Pote[4] and Lucy Mabaya[1]

1 Midlands State University, Gweru, Zimbabwe

2 Khalifa University, Abu Dhabi, United Arab Emirates

3 Harare Institute of Technology, Zimbabwe

4 Great Zimbabwe University, Masvingo, Zimbabwe

*Address all correspondence to: nyagumboelliot@gmail.com

References

[1] World Health Organization. Schistosomiasis: Key Facts. Geneva: World Health Organization; April 17, 2019. Available from: https://www.who.int/news-room/fact-sheets/detail/schistosomiasis; 2023 [Accessed: July 28, 2023]

[2] Deol AK, Fleming FM, Calvo-Urbano B, Walker M, Bucumi V, Gnandou I, et al. Schistosomiasis—Assessing progress toward the 2020 and 2025 global goals. New England Journal of Medicine. 2019;**381**(26):2519-2528

[3] World Health Organization. Schistosomiasis (Bilharzia). 2018 [Accessed: July 31, 2023]. Available from: https://www.who.int/health-topics/schistosomiasis#tab=tab_1

[4] World Health Organization. Schistosomiasis. 2021 [Accessed: May 31, 2021]. Available from: https://www.who.int/news-room/fact-sheets/detail/schistosomiasis

[5] World Health Organization. Soil-transmitted helminth infections. 2020 [Accessed: May 31, 2021]. Available from: https://www.who.int/news-room/fact-sheets/detail/soil-transmitted-helminth-infections

[6] Loukas A, Hotez PJ, Diemert D, Yazdanbakhsh M, McCarthy JS, Correa-Oliveira R, et al. Hookworm infection. Nature Reviews Disease Primers. 2016; **2**(1):1-8

[7] Montresor A, Engels D, Ramsan M, Foum A, Savioli L. Field test of the 'dose pole'for praziquantel in Zanzibar. Transactions of the Royal Society of Tropical Medicine and Hygiene. 2002; **96**(3):323-324

[8] World Health Organization. Over 1.8 million children receive treatment for bilharzia and intestinal worms. 2022. [Accessed: August 2, 2023]. Available from: https://www.afro.who.int/countries/zimbabwe/news/over-18-million-children-receive-treatment-bilharzia-and-intestinal-worms-0

[9] Midzi N, Mduluza T, Chimbari MJ, Tshuma C, Charimari L, Mhlanga G, et al. Distribution of schistosomiasis and soil transmitted helminthiasis in Zimbabwe: Towards a national plan of action for control and elimination. PLOS Neglected Tropical Diseases. 2014;**8**(8):e3014

[10] Cock IE, Selesho MI, Van Vuuren SF. A review of the traditional use of southern African medicinal plants for the treatment of selected parasite infections affecting humans. Journal of Ethnopharmacology. 2018;**220**:250-264

[11] Moyo M, Aremu AO, Van Staden J. Medicinal plants: An invaluable, dwindling resource in sub-Saharan Africa. Journal of Ethnopharmacology. 2015;**174**:595-606

[12] Chandra LD. Bio-diversity and conservation of medicinal and aromatic plants. Advances in Plants & Agriculture Research. 2016;**5**(4):00186

[13] Gafna DJ, Obando JA, Kalwij JM, Dolos K, Schmidtlein S. Climate change impacts on the availability of anti-malarial plants in Kenya. Climate Change Ecology. 2023;**5**:100070

[14] Moher D, Liberati A, Tetzlaff J, Altman DG, PRISMA Group*. Preferred reporting items for systematic reviews and meta-analyses: The PRISMA statement. Annals of Internal Medicine. 2009;**151**(4):264-269

[15] Watt JM, Breyer-Brandwijk MG. The medicinal and poisonous plants of

southern and eastern Africa being an account of their medicinal and other uses, chemical composition, pharmacological effects and toxicology in man and animal. In: The Medicinal and Poisonous Plants of Southern and Eastern Africa Being an Account of their Medicinal and Other Uses, Chemical Composition, Pharmacological Effects and Toxicology in Man and Animal. 2nd ed. South Afica: E & S. Livingstone; 1962

[16] Gelfland M, Mavi S, Drummond RB, Ndemera B. The Traditional Medical Practitioner in Zimbabwe: His Principles of Practice and Pharmacopoeia. Zimbabwe: Mambo Press; 1985

[17] Hutchings A. Zulu Medicinal Plants: An Inventory. South Africa: University of Natal Press; 1996

[18] Van Wyk BE. Oudtshoorn BV. Briza: Gericke N. Medicinal Plants of South Africa; 1997

[19] Viol DI. Screening of Traditional Medicinal Plants from Zimbabwe for Photochemistry, Antioxidant, Antimicrobial, Antiviral and Toxicological Activities [doctoral dissertation] University of Zimbabwe

[20] Marekerah L. A Survey on the Biological Activities of Selected Plants Used to Manage Diarrhoea and Cancer in Vumba, Zimbabwe. Online: Afribary; 2015

[21] Ndamba J, Nyazema N, Makaza N, Anderson C, Kaondera KC. Traditional herbal remedies used for the treatment of urinary schistosomiasis in Zimbabwe. Journal of Ethnopharmacology. 1994; **42**(2):125-132

[22] Nyazema NZ, Ndamba J, Anderson C, Makaza N, Kaondera KC. The doctrine of signatures or similitudes: A comparison of the efficacy of praziquantel and traditional herbal remedies used for the treatment of urinary schistosomiasis in Zimbabwe. International Journal of Pharmacognosy. 1994;**32**(2):142-148

[23] Mølgaard P, Nielsen SB, Rasmussen DE, Drummond RB, Makaza N, Andreassen J. Anthelmintic screening of Zimbabwean plants traditionally used against schistosomiasis. Journal of Ethnopharmacology. 2001; **74**(3):257-264

[24] Maroyi A. Acacia karroo Hayne: Ethnomedicinal uses, phytochemistry and pharmacology of an important medicinal plant in southern Africa. Asian Pacific Journal of Tropical Medicine. 2017;**10**(4):351-360

[25] Mangoyi R, Chitemerere T, Chimponda T, Chirisa E, Mukanganyama S. Multiple anti-infective properties of selected plant species from Zimbabwe. Novel Plant Bioresources: Applications in Food, Medicine and Cosmetics. 2014;**3**:179-190

[26] Magwenzi R, Nyakunu C, Mukanganyama S. The effect of selected Combretum species from Zimbabwe on the growth and drug efflux systems of Mycobacterium aurum and mycobacterium smegmatis. Journal of Microbial & Biochemical Technology. 2014;**3**(003):1-7

[27] Rodgers CB, Verotta L. Chemistry and biological properties of the African Combretaceae. In: Hostettmann K, Chinyanganya M, Maillard M, Wolffender JL, editors. Chemistry, Biological and Pharmacological Properties of African Medicinal plants. Zimbabwe: University of Zimbabwe Publications; 1996. p. 121-141

[28] Mapfunde S, Sithole S, Mukanganyama S. In vitro toxicity

determination of antifungal constituents from Combretum zeyheri. BMC Complementary and Alternative Medicine. 2016;**16**:1-1

[29] Munodawafa T. Screening of some traditional medicinal plants from Zimbabwe for biological and anti-microbial activity. [Master's thesis]. University of Zimbabwe. 2012

[30] Maroyi A. An ethnobotanical survey of medicinal plants used by the people in Nhema communal area, Zimbabwe. Journal of Ethnopharmacology. 2011; **136**(2):347-354

[31] Maroyi A. Traditional use of medicinal plants in south-Central Zimbabwe: Review and perspectives. Journal of Ethnobiology and Ethnomedicine. 2013;**9**(1):1-8

[32] Maroyi A. Phytochemical and ethnopharmacological review of Elephantorrhiza goetzei (harms) harms. Asian Pacific Journal of Tropical Medicine. 2017;**10**(2):107-113

[33] Chimponda T, Mukanganyama S. Antimycobacterial activities of selected medicinal plants from Zimbabwe against Mycobacterium aurum and Corynebacterium glutamicum. Tropical Biomedicine. 2010;**27**(3):595-610

[34] Maroyi A. Dicoma anomala sond.: A review of its botany, ethnomedicine, phytochemistry and pharmacology. Asian Journal of Pharmaceutical and Clinical Research. 2018;**11**:70-77

[35] Maroyi A. Lannea discolor: Its botany, ethnomedicinal uses, phytochemistry, and pharmacological properties. Asian Journal of Pharmaceutical and Clinical Research. 2018;**11**(10):49

[36] Maroyi A. Medicinal uses, biological and chemical properties of wild grape (Lannea edulis): An indigenous fruit plant of tropical Africa. Asian Journal of Pharmaceutical and Clinical Research. 2019;**12**(9):16-20

[37] Maroyi A. Local plant use and traditional conservation practices in Nhema communal area, Zimbabwe. International Journal of African Renaissance Studies-Multi-, Inter- and Transdisciplinarity. 2012;**7**(1):109-128

[38] Maroyi A. Sansevieria hyacinthoides (L.) Druce: A review of its botany, medicinal uses, phytochemistry, and biological activities. Asian Journal of Pharmaceutical and Clinical Research. 2019;**12**(9):21-26

[39] Maroyi A. Nutraceutical and ethnopharmacological properties of Vangueria infausta subsp. infausta. Molecules. 2018;**23**(5):1089

[40] Tan A. Turkey: Country Report to the FAO International Technical Conference on Plant Genetic Resource. Leipzig, Germany; 1996. p. 46

[41] Rashmi A, Gill NS, Sukhwinder K, Jain AD. Phytopharmacological evaluation of ethanolic extract of the seeds of Abrus precatorius Linn. Journal of Pharmacology and Toxicology. 2011; **6**(6):580-588

[42] Sunday RM, Ilesanmi OR, Obuotor EM. Acute and subacute toxicity of aqueous extract of Abrus precatorius seed in Wister rats. The Internet Journal of Pharmacology. 2013; **11**(1):1-7

[43] Ragasa CY, Lorena GS, Mandia EH, Raga DD, Shen CC. Chemical constituents of Abrus precatorius. American Journal of Essential Oils and Natural Products. 2013;**1**(2):7-10

[44] Sheikh SG, Hedge K. Therapeutic uses of Abrus precatorius: A review.

International Journal of Pharma and Chemical Research. 2017:196-201

[45] Bhakta S, Das SK. The medicinal values of Abrus precatorius: A review study. Journal of Advanced Biotechnology and Experimental Therapeutics. 2020;**3**(2):84-91

[46] Dahikar GK, Rathi B, Kamble SB. Critical review on pharmacological uses of Gunja (Abrus precatorious). Journal of Indian System of Medicine. 2020;**8**(3): 155-161

[47] Adedapo AA, Sofidiya MO, Masika PJ, Afolayan AJ. Anti-inflammatory and analgesic activities of the aqueous extract of acacia Karroo stem bark in experimental animals. Basic & Clinical Pharmacology & Toxicology. 2008;**103**(5):397-400

[48] Nielsen TR, Kuete V, Jäger AK, Meyer JJ, Lall N. Antimicrobial activity of selected south African medicinal plants. BMC Complementary and Alternative Medicine. 2012;**12**:1-6

[49] Njanje I, Bagla VP, Beseni BK, Mbazima V, Lebogo KW, Mampuru L, et al. Defatting of acetone leaf extract of Acacia karroo (Hayne) enhances its hypoglycaemic potential. BMC Complementary and Alternative Medicine. 2017;**17**(1):1-1

[50] Chipiti T, Ibrahim MA, Koorbanally NA, Islam MS. In vitro antioxidant activities of leaf and root extracts of Albizia antunesiana harms. Acta Poloniae Pharmaceutica. 2013; **70**(6):1035-1043

[51] Koné WM, Atindehou KK, Dossahoua T, Betschart B. Anthelmintic activity of medicinal plants used in northern Côte d'Ivoire against intestinal helminthiasis. Pharmaceutical Biology. 2005;**43**(1):72-78

[52] Hassan HS, Ahmadu AA, Hassan AS. Analgesic and anti-inflammatory activities of Asparagus africanus root extract. African Journal of Traditional, Complementary and Alternative Medicines. 2008;**5**(1):27-31

[53] Kebede S, Afework M, Debella A, Ergete W, Makonnen E. Toxicological study of the butanol fractionated root extract of Asparagus Africanus Lam., on some blood parameter and histopathology of liver and kidney in mice. BMC Research Notes. 2016;**9**:1-9

[54] Matowa PR, Gundidza M, Gwanzura L, Nhachi CF. A survey of ethnomedicinal plants used to treat cancer by traditional medicine practitioners in Zimbabwe. BMC Complementary Medicine and Therapies. 2020;**20**(1):1-3

[55] Toua V, Ahmadou A, Dieudonne N. In vitro effect of Burkea Africana Burke, 1840 (Fabaceae-cesalpinoideae) ethanolic bark extract on the nematode Haemonchus contortus rudolphi, 1803. Indo American Journal of Pharmaceutical Sciences. 2017;**4**(12):4733

[56] Moura I, Duvane JA, Ribeiro N, Ribeiro-Barros I. Woody species from the Mozambican Miombo woodlands: A review on their ethnomedicinal uses and pharmacological potential. Journal of Medicinal Plants Research. 2018;**12**(2): 15-31

[57] Namadina MM, Aliyu BS, Haruna H, Sunusi U, Kamal RM, Balarabe S, et al. Pharmacognostic and acute toxicity study of Burkea Africana root. Journal of Applied Sciences and Environmental Management. 2020;**24**(4):565-573

[58] Woode E, Ansah C, Ainooson GK, Abotsi WM, Mensah AY, Duweijua M. Anti-inflammatory and antioxidant properties of the root extract of Carissa

edulis (Forsk.) Vahl (Apocynaceae). Journal of Science and Technology (Ghana). 2007;**27**(3):5-15

[59] Harwansh RK, Garabadu D, Rahman MA, Garabadu PS. In vitro anthelmintic activity of different extracts of root of Carissa spinarum. International Journal of Pharmaceutical Sciences and Research. 2010;**1**(10):84

[60] Ibrahim H, Williams FE, Salawu KM, Usman AM. Phytochemical screening and acute toxicity studies of crude ethanolic extract and flavonoid fraction of Carissa edulis leaves. Biokemistri. 2015;**27**(1):39-43

[61] Osseni R, Akoha S, Adjagba M, Azonbakin S, Lagnika L, Awede B, et al. In vivo toxicological assessment of the aqueous extracts of the leaves of Carissa edulis (Apocynaceae) in Wistar rats. European Journal of Medicinal Plants. 2016;**15**(1):1

[62] Kaunda JS, Zhang YJ. The genus Carissa: An ethnopharmacological, phytochemical and pharmacological review. Natural Products and Bioprospecting. 2017;**7**:181-199

[63] Parry O, Matambo C. Some pharmacological actions of aloe extracts and Cassia abbreviata on rats and mice. Central African Journal of Medicine. 1992;**38**(10):409-414

[64] Okeleye BI, Mkwetshana NT, Ndip RN. Evaluation of the antibacterial and antifungal potential of Peltophorum africanum: Toxicological effect on human chang liver cell line. The Scientific World Journal. 2013;**2013**:1-9

[65] Mongalo NI. Peltophorum africanum Sond [Mosetlha]: A review of its ethnomedicinal uses, toxicology, phytochemistry and pharmacological activities. Journal of Medicinal Plants Research. 2013;**7**(48):3484-3491

[66] Viol DI, Chagonda LS, Moyo SR, Mericli AH. Toxicity and antiviral activities of some medicinal plants used by traditional medical practitioners in Zimbabwe. American Journal of Plant Sciences. 2016;**7**(11):1538

[67] Mujuru S. Flavonoid content, anti-bacterial and anti-inflammatory activity of cassia abbreviata pods [doctoral dissertation] BUSE

[68] Sobeh M, Esmat A, Petruk G, Abdelfattah MA, Dmirieh M, Monti DM, et al. Phenolic compounds from Syzygium jambos (Myrtaceae) exhibit distinct antioxidant and hepatoprotective activities in vivo. Journal of Functional Foods. 2018;**41**:223-231

[69] Conde P, Figueira R, Saraiva S, Catarino L, Romeiras M, Duarte MC. The botanic mission to Mozambique (1942-1948): Contributions to knowledge of the medicinal flora of Mozambique. História, Ciências, Saúde-Manguinhos. 2014;**21**:539-585

[70] Saini H, Dwivedi J, Paliwal H, Kataria U, Sharma M. An ethno-pharmacological evaluation of Catunaregam spinosa (thumb.) tirveng for antioxidant activity. Journal of Drug Delivery and Therapeutics. 2019;**9**(4-s):280-284

[71] Al-Taweel AM, Perveen S, El-Shafae AM, Fawzy GA, Malik A, Afza N, et al. Bioactive phenolic amides from Celtis Africana. Molecules. 2012;**17**(3):2675-2682

[72] Akhlaq A, Mehmood MH, Rehman A, Ashraf Z, Syed S, Bawany SA, et al. The prokinetic,

laxative, and antidiarrheal effects of Morus nigra: Possible muscarinic, Ca2+ channel blocking, and antimuscarinic mechanisms. Phytotherapy Research. 2016;**30**(8):1362-1376

[73] Tanko Y, Yaro AH, Isa AI, Yerima M, Saleh MI, Mohammed A. Toxicological and hypoglycaemic studies on the leaves of Cissampelos mucronata (Menispermaceae) on blood glucose levels of streptozotocin-induced diabetic Wistar rats. Journal of Medicinal Plant Research: Planta Medica. 2007;**1**(5):113-116

[74] Garba SH, Jacks TW, Onyeyili PA, Nggada HA. Testicular and andrological effects of the methanol extract of the root of Cissampelos mucronata (A. Rich) in rats. Journal of Biological Sciences and Bioconservation. 2014;**6**:18-30

[75] Maroyi A. A synthesis and review of medicinal uses, phytochemistry and pharmacological properties of Cissampelos mucronata A. Rich. (Menispermaceae). Journal of Pharmacy and Nutrition Sciences. 2020;**10**: 2013-2022

[76] Pathak AK, Kambhoja S, Dhruv S, Singh HP, Chand H. Anthelimintic activity of Cissus quadraangularis Linn stem. Pharmacology. 2010;**3**:15-18

[77] Buddhadev S, Buddhadev S. A review update on plant Cissus quadrangularis L. Punarnav. 2014;**2**:1

[78] Shukla R, Pathak A, Kambuja S, Sachan S, Mishra A, Kumar S. Pharmacognostical, phytochemical and pharmacological overview: Cissus quadrangularis Linn. Indian Journal of Pharmaceutical and Biological Research. 2015;**3**(3):59

[79] Kavitha A, Babu AN, Nadendla RR. Acute toxicity study of Cissus quadrangularis in wiss albino mice. Panacea Journal of Pharmacy and Pharmaceutical Sciences. 2018;**7**(1): 748-756

[80] McGaw LJ, Rabe T, Sparg SG, Jäger AK, Eloff JN, Van Staden J. An investigation on the biological activity of Combretum species. Journal of Ethnopharmacology. 2001;**75**(1):45-50

[81] de Morais Lima GR, de Sales IR, Caldas Filho MR, de Jesus NZ, de Sousa FH, Barbosa-Filho JM, et al. Bioactivities of the genus Combretum (Combretaceae): A review. Molecules. 2012;**17**(8):9142-9206

[82] Roy S, Gorai D, Acharya R, Roy R. Combretum (combretaceae): Biological activity and phytochemistry. American Journal of Pharm Research. 2014;**4**(11): 5266-5299

[83] Peloewetse E, Thebe MM, Ngila JC, Ekose GE. Inhibition of growth of some phytopathogenic and mycotoxigenic fungi by aqueous extracts of Combretum imberbe (Wawra) wood. African Journal of Biotechnology. 2008;**7**(16):2934-2939

[84] Masoko P, Picard J, Howard RL, Mampuru LJ, Eloff JN. In vivo antifungal effect of Combretum and Terminalia species extracts on cutaneous wound healing in immunosuppressed rats. Pharmaceutical Biology. 2010;**48**(6): 621-632

[85] Mangoyi R, Mafukidze W, Marobela K, Mukanganyama S. Antifungal activities and preliminary phytochemical investigation of Combretum species from Zimbabwe. Microbial and Biochemical Technology. 2012;**4**:037-044

[86] Ramalhet C, Lopes D, Mulhovo S, Rosário VE, Ferreira MJ. Antimalarial activity of some plants traditionally used

in Mozambique. In: Workshop Plantas Medicinais e Fitoterapêuticas Nos Trópicos. IICT/CCCM 2008 Oct 29. Vol. 29. p. 30

[87] Salawu OA, Chindo BA, Tijani AY, Obidike IC, Salawu TA, Akingbasote AJ. Acute and sub-acute toxicological evaluation of the methanolic stem bark extract of Crossopteryx febrifuga in rats. African Journal of Pharmacy and Pharmacology. 2009;**3**(12): 621-626

[88] Nnatuanya IN, Ohadoma SC. Pharmacological evaluation for antitrypanosomal activity of aqueous stem bark extract of Grossopteryx verbrifuga in rats. Journal of Applied Sciences. 2014;**17**(2): 11282-11291

[89] Bassoueka DJ, Taiwe Sotoing G, Nsonde Ntandou G, Ngo BE. Anticonvulsant activity of the decoction of Crossopteryx febrifuga in mice. International Journal of Sciences and Research. 2016;**3**:112-116

[90] Idris MM, Nenge HP. Antihyperglycaemic and antilipidaemic properties of ethanol stem bark extract of Crossopteryx febrifuga in alloxan-induced diabetic rats. ChemSearch Journal. 2019;**10**(2): 130-137

[91] Uchogu AP, Yahaya TA, Salawu OA, Adamu MA, Ameh FS. Anti-proliferative potential of some common vegetables and plants in Nigeria. Journal of Pharmaceutical Development and Industrial Pharmacy. 2020;**2**:2

[92] Grace OM, Prendergast HD, Jäger AK, Van Staden J, Van Wyk AE. Bark medicines used in traditional healthcare in KwaZulu-Natal, South Africa: An inventory. South African Journal of Botany. 2003;**69**(3):301-363

[93] Okokon JE, Ofodum KC, Ajibesin KK, Danladi B, Gamaniel KS. Pharmacological screening and evaluation of antiplasmodial activity of croton zambesicus against plasmodium berghei berghei infection in mice. Indian Journal of Pharmacology. 2005;**37**(4): 243

[94] Okokon JE, Nwafor PA, Noah K. Nephroprotective effect of croton zambesicus root extract against gentimicin-induced kidney injury. Asian Pacific Journal of Tropical Medicine. 2011;**4**(12):969-972

[95] Salatino A, Salatino ML, Negri G. Traditional uses, chemistry and pharmacology of croton species (Euphorbiaceae). Journal of the Brazilian Chemical Society. 2007;**18**:11-33

[96] Okokon JE, Dar A, Choudhary MI. Immunomodulatory, cytotoxic and antileishmanial activity of phytoconstituents of croton zambesicus. Phytopharmacology Journal. 2013;**4**(1): 31-40

[97] Mfotie Njoya E, Eloff JN, McGaw LJ. Croton gratissimus leaf extracts inhibit cancer cell growth by inducing caspase 3/7 activation with additional anti-inflammatory and antioxidant activities. BMC Complementary and Alternative Medicine. 2018;**18**(1):1-1

[98] Mahmoud AB, Danton O, Kaiser M, Khalid S, Hamburger M, Mäser P. HPLC-based activity profiling for antiprotozoal compounds in Croton gratissimus and Cuscuta hyalina. Frontiers in Pharmacology. 2020;**11**:1246

[99] Safari VZ, Ngugi MP, Orinda J, Njagi EM. Antipyretic, anti-inflammatory and analgesic activities of aqueous stem extract of Cynachum viminale (L.) in albino mice. Medicinal

and Aromatic Plants. 2016;**5**(236): 2167-0412

[100] Munodawafa T, Moyo S, Chipurura B, Chagonda L. Brine shrimp lethality bioassay of some selected Zimbabwean traditional medicinal plants. International Journal of Phytopharmacology. 2016; **7**(4):229-232

[101] Clarkson C, Maharaj VJ, Crouch NR, Grace OM, Pillay P, Matsabisa MG, et al. In vitro antiplasmodial activity of medicinal plants native to or naturalised in South Africa. Journal of Ethnopharmacology. 2004;**92**(2-3):177-191

[102] Mulyangote LT. Ethnobotany and bioactivity of medicinal plants used to treat symptoms associated with gastro-intestinal infections in Namibia [doctoral dissertation] University of Namibia

[103] Mokoka TA. The discovery and characterization of antiprotozoal compounds from South African medicinal plants by a HPLC-based activity profiling technique [doctoral dissertation]

[104] Mmbengwa V, Samie A, Gundidza M, Matikiti V, Ramalivhana NJ, Magwa ML. Biological activity and phytoconstituents of essential oil from fresh leaves of Eriosema englerianum. African Journal of Biotechnology. 2009;**8**(3):361-364

[105] Lawal OA, Ogunwande IA. Essential oils from the medicinal plants of Africa. In: Medicinal Plant Research in Africa. Nigeria: Elsevier; 2013. pp. 203-224

[106] Bunalema L, Kirimuhuzya C, Tabuti JR, Waako P, Magadula JJ, Otieno N, et al. The efficacy of the crude root bark extracts of Erythrina abyssinica on rifampicin resistant mycobacterium tuberculosis. African Health Sciences. 2011;**11**(4):587-593

[107] Lagu C, Kayanja FI. The in vitro antihelminthic efficacy of erythrina abyssinica extracts on Ascaridia galli. In: Insights from Veterinary Medicine. London: IntechOpen; 2013. pp. 269-281

[108] Chitopoa W, Muchachaa I, Mangoyi R. Evaluation of the antimicrobial activity of Erythrina abyssinica leaf extract. Journal of Microbial & Biochemical Technology. 2019;**11**(2):43-46

[109] Macharia FK, Mwangi PW, Yenesew A, Bukachi F, Nyaga NM, Wafula DK. Hepatoprotective effects of erythrina abyssinica lam ex dc against non alcoholic fatty liver disease in Sprague dawley rats. BioRxiv. 2019: 577007

[110] McGaw LJ, Jäger AK, Van Staden J. Antibacterial, anthelmintic and anti-amoebic activity in South African medicinal plants. Journal of Ethnopharmacology. 2000;**72**(1-2): 247-263

[111] Kama-Kama F, Midiwo J, Nganga J, Maina N, Schiek E, Omosa LK, et al. Selected ethno-medicinal plants from Kenya with in vitro activity against major African livestock pathogens belonging to the "mycoplasma mycoides cluster". Journal of Ethnopharmacology. 2016;**192**:524-534

[112] Woldemedhin B, Nedi T, Shibeshi W, Sisay M. Evaluation of the diuretic activity of the aqueous and 80% methanol extracts of the root of Euclea divinorum Hiern (Ebenaceae) in Sprague Dawley rats. Journal of

Ethnopharmacology. 2017;**202**: 114-121

[113] Al-Fatimi M. Antifungal activity of Euclea divinorum root and study of its ethnobotany and phytopharmacology. PRO. 2019;**7**(10):680

[114] Mangoyi R, Mukanganyama S. In vitro antifungal activities of selected medicinal plants from Zimbabwe against Candida albicans and Candida krusei. The African Journal of Plant Science and Biotechnology. 2011;**5**(1):1-7

[115] Kota GC, Karthikeyan M, Kannan M. Flacourtia indica (Burm. f.) Merr.-A phytopharmacological review. International Journal of Research in Pharmaceutical and Biomedical Sciences. 2012;**3**(1):78-81

[116] Obbo CJ, Makanga B, Mulholland DA, Coombes PH, Brun R. Antiprotozoal activity of Khaya anthotheca, (Welv.) CDC a plant used by chimpanzees for self-medication. Journal of Ethnopharmacology. 2013; **147**(1):220-223

[117] Sashidhara KV, Singh SP, Singh SV, Srivastava RK, Srivastava K, Saxena JK, et al. Isolation and identification of β-hematin inhibitors from Flacourtia indica as promising antiplasmodial agents. European Journal of Medicinal Chemistry. 2013;**60**:497-502

[118] Hussain SM, Hussain MS, Ahmed A, Arif N. Characterization of isolated bioactive phytoconstituents from Flacourtia indica as potential phytopharmaceuticals-an in silico perspective. Journal of Pharmacognosy and Phytochemistry. 2016;**5**(6):323-331

[119] Taderera T, Chagonda LS, Gomo E, Shai LJ. Inhibitory activity of α-glucosidase and α-amylase by Annona stenophylla root extract as mechanism for hypoglycaemic control of DM. International Journal of Pharmacy, Photon. 2015;**106**:436-444

[120] Khalid SA, Friedrichsen GM, Christensen SB, El Tahir A, Satti GM. Isolation and characterization of pristimerin as the antiplasmodial and antileishmanial agent of Maytenus senegalensis (Lam.) Exell. Archive for Organic Chemistry. 2007;**2007**(9): 129-134

[121] Malebo HM, Wiketye V, Katani SJ, Kitufe NA, Nyigo VA, Imeda CP, et al. In vivo antiplasmodial and toxicological effect of maytenus senegalensis traditionally used in the treatment of malaria in Tanzania. Malaria Journal. 2015;**14**:1-7

[122] Makgatho ME, Nxumalo W, Raphoko LA. Anti-mycobacterial,-oxidative,-proliferative and-inflammatory activities of dichloromethane leaf extracts of Gymnosporia senegalensis (Lam.) Loes. South African Journal of Botany. 2018; **114**:217-222

[123] Lee SE, Kim MR, Kim JH, Takeoka GR, Kim TW, Park BS. Antimalarial activity of anthothecol derived from Khaya anthotheca (Meliaceae). Phytomedicine. 2008; **15**(6–7):533-535

[124] Suleiman MM, Bagla V, Naidoo V, Eloff JN. Evaluation of selected South African plant species for antioxidant, antiplatelet, and cytotoxic activity. Pharmaceutical Biology. 2010;**48**(6): 643-650

[125] Saadabi AM, Ayoub SM. Comparative bioactivity of Hydnora abyssinica A. Braun against different groups of fungi and bacteria. Journal of

Medicinal Plants Research. 2009:**3**(4): 262-265

[126] Osman HM. Anti-diarrhoeal activity of hydnora abyssinica aqueous root extract in rats [doctoral dissertation] Department of Preventive Medicine and Veterinary Public Health, Faculty of Veterinary Medicine, University of Khartoum

[127] Yagi S, Drouart N, Bourgaud F, Henry M, Chapleur Y, Laurain-Mattar D. Antioxidant and antiglycation properties of Hydnora johannis roots. South African Journal of Botany. 2013;**84**:124-127

[128] Al-Fatimi M, Ali NA, Kilian N, Franke K, Arnold N, Kuhnt C, et al. Ethnobotany, chemical constituents and biological activities of the flowers of Hydnora abyssinica A. Br. (Hydnoraceae). Die Pharmazie-An International Journal of Pharmaceutical Sciences. 2016;**71**(4):222-226

[129] Moideen SV, Houghton PJ, Rock P, Croft SL, Aboagye-Nyame F. Activity of extracts and naphthoquinones from Kigelia pinnata against Trypanosoma brucei brucei and Trypanosoma brucei rhodesiense. Planta Medica. 1999; **65**(06):536-540

[130] Sharma UK, Sharma US, Singh A, Agarwal VI. Diuretic activity of Kigelia pinnata bark extract. Journal of Pharmacology Research. 2010;**1**(2):17-20

[131] Atawodi SE, Olowoniyi OD. Pharmacological and therapeutic activities of Kigelia Africana (Lam.) Benth. Annual Research & Review in Biology. 2015;**5**:1-7

[132] Chakuma N, Chipurura B, Muchuweti M, Chitindingu K, Bhebhe M, Chagonda L. Total phenolic content, free radical scavenging and antioxidant potential of Lannea discolor (Sond.) Engl bark and root extracts. Journal of Biologically Active Products from Nature. 2015;**5**(1):71-77

[133] Aremu AO, Cheesman L, Finnie JF, Van Staden J. Mondia whitei (Apocynaceae): A review of its biological activities, conservation strategies and economic potential. South African Journal of Botany. 2011;**77**(4): 960-971

[134] Gakunga NJ, Sembajwe LF, John K, Patrick V. Phytochemical screening and antidiarrheal activity of ethanolic fresh root bark extract of Mondia whitei in albino rats. Journal of Pharmaceutical and Scientific Innovation. 2013;**2**(6):1-6

[135] Oketch-Rabah HA. Mondia whitei, a medicinal plant from Africa with aphrodisiac and antidepressant properties: A review. Journal of Dietary Supplements. 2012;**9**(4):272-284

[136] Joseph O, Kihdze TJ, Katusiime B, Imanirampa L,Waako P, Bajunirwe F, Ganafa AA. Toxicity of four herbs used in erectile dysfunction; Mondia whiteii, Cola acuminata, Urtica massaica, and Tarenna graveolensin male rats. African Journal of Pharmacy and Phamacology. 2015;**9**(30):756-763

[137] Imam MZ, Akter S. Musa paradisiaca L. and Musa sapientum L.: A phytochemical and pharmacological review. Journal of Applied Pharmaceutical Science. 2011;**30**:14-20

[138] Pereira A, Maraschin M. Banana (Musa spp) from peel to pulp: Ethnopharmacology, source of bioactive compounds and its relevance for human health. Journal of Ethnopharmacology. 2015;**160**:149-163

[139] Neuwirt N, Gregory L, Yoshihara E, Gorniak SL. Effect of Musa spp. extract on eggs and larvae of gastrointestinal

nematodes from infected sheep. Semina: Agricultural Sciences. 2015;**36**(6): 3751-3756

[140] Gregory L, Yoshihara E, Ribeiro BL, Silva LK, Marques EC, Meira EB, et al. Dried, ground banana plant leaves (Musa spp.) for the control of Haemonchus contortus and Trichostrongylus colubriformis infections in sheep. Parasitology Research. 2015;**114**:4545-4551

[141] Ugbogu EA, Ude VC, Elekwa I, Arunsi UO, Uche-Ikonne C, Nwakanma C. Toxicological profile of the aqueous-fermented extract of musa paradisiaca in rats. Avicenna Journal of Phytomedicine. 2018;**8**(6):478

[142] Moshi MJ, Cosam JC, Mbwambo ZH, Kapingu M, Nkunya MH. Testing beyond ethnomedical claims: Brine shrimp lethality of some Tanzanian plants. Pharmaceutical Biology. 2004;**42**(7): 547-551

[143] Nyaberi MO, Onyango CA, Mathooko FM, Maina JM, Makobe M, Mwaura F. Bioactive fractions in the stem charcoal of Ozoroa insignis used by the pastoral communities in West Pokot to preserve milk. Journal of Applied Biosciences. 2010;**26**: 1653-1658

[144] Haule EE, Moshi MJ, Nondo RS, Mwangomo DT, Mahunnah RL. A study of antimicrobial activity, acute toxicity and cytoprotective effect of a polyherbal extract in a rat ethanol-HCl gastric ulcer model. BMC Research Notes. 2012;**5**:1-9

[145] Nyaberi MO. Studies on the use of herbs to preserve meat and milk among the pastoral communities of West Pokot in Kenya [doctoral dissertation]

[146] Obiro WC, Zhang T, Jiang B. The nutraceutical role of the Phaseolus vulgaris α-amylase inhibitor. British Journal of Nutrition. 2008;**100**(1):1-2

[147] Ríos-de Álvarez L, Jackson F, Greer AW, Grant G, Jackson E, Morrison AA, et al. Direct anthelmintic and immunostimulatory effects of oral dosing semi-purified phytohaemagglutinin lectin in sheep infected with Teladorsagia circumcincta and Trichostrongylus colubriformis. Veterinary Parasitology. 2012;**187**(1–2): 267-274

[148] Saleem ZM, Ahmed S, Hasan MM. Phaseolus lunatus Linn: Botany, medicinal uses, phytochemistry and pharmacology. World Journal of Pharmacy and Pharmaceutical Sciences. 2016;**5**(11):87-93

[149] Ganesan K, Xu B. Polyphenol-rich dry common beans (Phaseolus vulgaris L.) and their health benefits. International Journal of Molecular Sciences. 2017;**18**(11):2331

[150] Jawaid T, Kamal M, Kumar S. Antihypertensive effect of the alcoholic extract of seeds of Phaseolus vulgaris Linn.(Fabaceae) on high salt diet induced hypertension in male rats. International Journal of Pharmaceutical Sciences and Research. 2017;**8**(7): 3092-3097

[151] Ramadhani UP, Chandra B, Rivai H. Overview of phytochemistry and pharmacology of chickpeas (Phaseolus vulgaris). World Journal of Pharmacy and Pharmaceutical Sciences. 2020;**9**(9): 442-461

[152] Kone WM, Atindehou KK, Kacou-N A, Dosso M. Evaluation of 17 medicinal plants from northern Côte d'Ivoire for their in vitro activity against Streptococcus pneumoniae. African Journal of Traditional, Complementary

and Alternative Medicines. 2007;**4**(1): 17-22

[153] Ukwuani A, Ihebunna O, Samuel RM, Peni IJ. Acute oral toxicity and antiulcer activity of Piliostigma thonningii leaf fraction in albino rats. Bulletin of Environment, Pharmacology and Life Sciences. 2012;**2**: 41-45

[154] Afolayan M, Srivedavyasasri R, Asekun OT, Familoni OB, Orishadipe A, Zulfiqar F, et al. Phytochemical study of Piliostigma thonningii, a medicinal plant grown in Nigeria. Medicinal Chemistry Research. 2018;**27**:2325-2330

[155] Chipinga JV. Efficacy of Pterocarpus angolensis crude extracts against Candida krusei, Staphylococcus aureus, Streptococcus agalactiae and Escherichia coli. Malawi Medical Journal. 2018;**30**(4):219-224

[156] Sigidi MT, Anokwuru CP, Zininga T, Tshisikhawe MP, Shonhai A, Ramaite ID, et al. Comparative in vitro cytotoxic, anti-inflammatory and anti-microbiological activities of two indigenous Venda medicinal plants. Translational Medicine Communications. 2016;**1**:1-7

[157] Zininga T, Anokwuru CP, Sigidi MT, Tshisikhawe MP, Ramaite II, Traoré AN, et al. Extracts obtained from Pterocarpus angolensis DC and Ziziphus mucronata exhibit antiplasmodial activity and inhibit heat shock protein 70 (Hsp70) function. Molecules. 2017; **22**(8):1224

[158] Sadashiv PS. Acute toxicity study for Ricinus communis. Der Pharmacia Lettre. 2011;**3**(5):132-137

[159] Franke H, Scholl R, Aigner A. Ricin and Ricinus communis in pharmacology and toxicology-from ancient use and “papyrus Ebers” to modern perspectives and “poisonous plant of the year 2018”. Naunyn-Schmiedeberg's Archives of Pharmacology. 2019;**392**:1181-1208

[160] Khan, Marwat S, Khan EA, Baloch MS, Sadiq M, Ullah I, Javaria S, et al. Ricinus cmmunis: Ethnomedicinal uses and pharmacological activities. Pakistan Journal of Pharmaceutical Sciences. 2017;**30**(5):1815-1827

[161] Fomum WS, Nsahlai VI. In vitro evaluation of anthelmintic efficacy of some plant species possessing proteinases and/or other nitroge-nous compounds in small ruminants. Journal of Alternative Complementary & Integrative Medicine. 2017;**3**:038

[162] Aliero AA, Jimoh FO, Afolayan AJ. Antioxidant and antibacterial properties of Sansevieria hyacinthoides. International Journal of Pure and Applied Sciences. 2008;**2**(3):103

[163] Sultana N, Rahman MM, Ahmed S, Akter S, Haque MM, Parveen S, et al. Antimicrobial compounds from the Rihzomes of Sansevieria hyacinthoides. Bangladesh Journal of Scientific and Industrial Research. 2011;**46**(3):329-332

[164] Akhalwaya S, Van Vuuren S, Patel M. An in vitro investigation of indigenous south African medicinal plants used to treat oral infections. Journal of Ethnopharmacology. 2018; **210**:359-371

[165] Ojewole JA, Mawoza T, Chiwororo WD, Owira PM. Sclerocarya birrea (A. Rich) Hochst. [‘Marula’] (Anacardiaceae): A review of its phytochemistry, pharmacology and toxicology and its ethnomedicinal uses. Phytotherapy Research: An International Journal Devoted to

Pharmacological and Toxicological Evaluation of Natural Product Derivatives. 2010;**24**(5):633-639

[166] Auwal SM, Atiku MK, Wudil AM, Sule MS. Phytochemical composition and acute toxicity evaluation of aqueous root bark extract of Securidaca longipedunculata (Linn). Bayero Journal of Pure and Applied Sciences. 2012;**5**(2): 67-72

[167] Jain SC, Jain R, Sharma RA, Capasso F. Pharmacological investigation of Cassia italica. Journal of Ethnopharmacology. 1997;**58**(2): 135-142

[168] Ahangarpour A, Oroujan AA. The effects of Cassia italica leaves aqueous extract on non-pregnant uterus contraction in rats. Iranian Journal of Reproductive Medicine. 2010;**8**(4): 179-184

[169] Qamar F, Afroz S, Feroz Z, Siddiqui S, Ara A. Evaluation of hypoglycemic effect of Cassia italica. Journal of Basic & Applied Sciences. 2011;**7**(1)

[170] Dabai YU, Kawo AH, Aliyu RM. Phytochemical screening and antibacterial activity of the leaf and root extracts of Senna italica. African Journal of Pharmacy and Pharmacology. 2012; **6**(12):914-918

[171] Chitra A, Senthilkumar N, Ashraf AM. Antioxidant and antitumor activities on catunaregumspinosa. International Journal of Research in Pharmacology & Pharmacotherapeutics. 2013;**2**:464-470

[172] Nadro MS, Onoagbe IO. Effects of the aqueous and ethanolic extracts of Cassia italica leaf in normal rats. American Journal of Research Communication. 2014;**2**(8):72-80

[173] Bayala B, Zohoncon TM, Djigma FW, Nadembega C, Baron S, Lobaccaro JM, et al. Antioxidant and antiproliferative activities on prostate and cervical cultured cancer cells of five medicinal plant extracts from Burkina Faso. International Journal of Biological and Chemical Sciences. 2020;**14**(3): 652-663

[174] Mahmuda A, Sani M, Adamu T, Sanda A, Gobir LG. In vivo anthelminthic activity of Ethanolic leaf extract of Senna italica on rats with Hymenolepis diminuta infection. Advances in Research. 2020; **21**(8):18-27

[175] Yongwa G, Ngnoda BF, Ndjonka D, Saotoing P. In vitro anthelmintic activity of aqueous and ethanolic extract of Senna italica (Caesalpiniaceae) on three-stages of Haemonchus contortus. Journal of Pharmaceutical Research International. 2020;**32**(3):25-34

[176] Tshikalange TE, Meyer JJ, Hussein AA. Antimicrobial activity, toxicity and the isolation of a bioactive compound from plants used to treat sexually transmitted diseases. Journal of Ethnopharmacology. 2005;**96**(3): 515-519

[177] Aremu AO, Ndhlala AR, Fawole OA, Light ME, Finnie JF, Van Staden J. In vitro pharmacological evaluation and phenolic content of ten south African medicinal plants used as anthelmintics. South African Journal of Botany. 2010;**76**(3):558-566

[178] Umar AB. Phytochemical evaluation, toxicity study and graded dose response of the methanol crude extract of Cassia singueana (del.) on experimental animals. International Journal of Academic Research. 2019; **1**(4):36-50

[179] Adzu B, Abbah J, Vongtau H, Gamaniel K. Studies on the use of Cassia singueana in malaria ethnopharmacy. Journal of Ethnopharmacology. 2003; **88**(2-3):261-267

[180] Mandal K, Dobhal Y, Joshi BC. An updated review on Solanum viarum Dunal. Recent Trends in Pharmaceutical Sciences and Research. 2019;**2**:1-6

[181] Indhumathi T, Mohandass S. Efficacy of ethanolic extract of Solanum incanum fruit extract for its antimicrobial activity. International Journal of Current Microbiology and Applied sciences. 2014;**3**(6):939-949

[182] Mwonjoria JK, Ngeranwa JJ, Githinji CG, Kahiga T, Kariuki HN, Waweru FN. Suppression of nociception by Solanum incanum (Lin.) Diclomethane root extract is associated anti-inflammatory activity. Suppression of nociception by Solanum incanum (Lin.) The Journal of Phytopharmacology. 2014;**3** (3):156-162

[183] Dakone D, Guadie A. A review on ethnomedicinal use, nutritional value, phytochemistry and pharmacological characteristics of Solanum incanum L. An important medicinal plant. International Journal of Scientific and Technology Research. 2016;**5**(6):350-354

[184] Anwar S. Pharmacological investigation of solanum incanum against P. Falciparum, L. infantum, T. Cruzi and T. Brucei: A role of antioxidant effect and clinical overview. Biomedical and Pharmacology Journal. 2018;**11**(2): 653-660

[185] Wang RW, Rebhun LI, Kupchan SM. Antimitotic and antitubulin activity of the tumor inhibitor steganacin. Cancer Research. 1977;**37**(9):3071-3079

[186] Demoz MS, Gachoki KP, Mungai KJ, Negusse BG. GC-MS analysis of the essential oil and methanol extract of the seeds of steganotaenia araliacea hochst. American Journal of Plant Sciences. 2014;**5**(26): 3752

[187] Mailu JK, Nguta JM, Mbaria JM, Okumu MO. Medicinal plants used in managing diseases of the respiratory system among the Luo community: An appraisal of Kisumu East Sub-County, Kenya. Chinese Medicine. 2020; **15**:1-27

[188] Sunghwa F, Koketsu M. Phenolic and bis-iridoid glycosides from Strychnos cocculoides. Natural Product Research. 2009;**23**(15): 1408-1415

[189] Ngadze RT, Verkerk R, Nyanga LK, Fogliano V, Ferracane R, Troise AD, et al. Effect of heat and pectinase maceration on phenolic compounds and physicochemical quality of Strychnos cocculoides juice. PLoS One. 2018;**13**(8): e0202415

[190] Moshi MJ, Mbwambo ZH. Some pharmacological properties of extracts of Terminalia sericea roots. Journal of Ethnopharmacology. 2005;**97**(1):43-47

[191] Lembede BW. Effect of dietary Terminalia sericea aqueous leaf extracts on high-fructose diet fed growing Wistar rats [doctoral dissertation] University of Witwatersrand

[192] Mongalo NI, McGaw LJ, Segapelo TV, Finnie JF, Van Staden J. Ethnobotany, phytochemistry, toxicology and pharmacological properties of Terminalia sericea Burch. Ex DC.(Combretaceae)–A review. Journal of Ethnopharmacology. 2016; **194**:789-802

[193] Parkar H. Wound healing potential of Terminalia sericea [doctoral dissertation] University of Pretoria

[194] Beigi M, Haghani E, Alizadeh A, Samani ZN. The pharmacological properties of several species of Terminalia in the world. International Journal of Pharmaceutical Sciences and Research. 2018;**9**(10):4079-4088

[195] Nair AA, Anjum N, Tripathi YC. A review on ethnomedicinal, phytochemical, and pharmacological significance of Terminalia sericea Burch. Ex DC. Journal of Pharmacy Research. 2018;**12**(3):420

[196] Sobeh M, Mahmoud MF, Hasan RA, Abdelfattah MA, Osman S, Rashid HO, et al. Chemical composition, antioxidant and hepatoprotective activities of methanol extracts from leaves of Terminalia bellirica and Terminalia sericea (Combretaceae). PeerJ. 2019;7: e6322

[197] Masoko P, Eloff JN. Screening of twenty-four south African Combretum and six Terminalia species (Combretaceae) for antioxidant activities. African Journal of Traditional, Complementary and Alternative Medicines. 2007;**4**(2):231-239

[198] Liu M, Katerere DR, Gray AI, Seidel V. Phytochemical and antifungal studies on Terminalia mollis and Terminalia brachystemma. Fitoterapia. 2009;**80**(6):369-373

[199] Rajkumar M, Chandra R, Asres K, Veeresham C. Toddalia asiatica (Linn.) Lam.-A comprehensive review. Pharmacognosy Reviews. 2008;**2**(4):386

[200] Madhava MS, Srivastava S, Sharma S. Ethnomedicinal plants used by the villagers of district Udham Singh Nagar, Uttarakhand, India. International Journal of Medicinal and Aromatic Plants. 2012;**2**(3):417-421

[201] Orwa JA, Ngeny L, Mwikwabe NM, Ondicho J, Jondiko IJ. Antimalarial and safety evaluation of extracts from Toddalia asiatica (L). Lam.(Rutaceae). Journal of Ethnopharmacology. 2013; **145**(2):587-590

[202] Nattudurai G, Gopinath R, Kavimani S, Jayakumararaj R. Antimicrobial activity of Toddalia asiatica against some human pathogens. International Journal of Pharmacy and Pharmaceutical Sciences. 2014;**6**(3): 378-381

[203] Shan XF, Kang YH, Bian Y, Gao YH, Wang WL, Qian AD. Isolation of active compounds from methanol extracts of Toddalia asiatica against Ichthyophthirius multifiliis in goldfish (Carassius auratus). Veterinary Parasitology. 2014;**199**(3-4):250-254

[204] Kimang'a A, Gikunju J, Kariuki D, Ogutu M. Safety and analgesic properties of ethanolic extracts of Toddalia Asiatica (L) Lam.(Rutaceae) used for central and peripheral pain management among the east African ethnic communities. Ethiopian Journal of Health Sciences. 2016;**26**(1):55-66

[205] Zhu Y, Chen Y, Yao X, Zhang X, Yang F, Fang X. Chemical constituents from Toddalia asiatica. Natural Product Research & Development. 2019;**31**(2): 225-230

[206] Omara T. Plants used in antivenom therapy in rural Kenya: Ethnobotany and future perspectives. Journal of Toxicology. 2020;**2020**:1-9

[207] Germanò MP, D'Angelo V, Biasini T, Sanogo R, De Pasquale R, Catania S. Evaluation of the antioxidant

properties and bioavailability of free and bound phenolic acids from Trichilia emetica Vahl. Journal of Ethnopharmacology. 2006;**105**(3): 368-373

[208] Komane BM, Olivier EI, Viljoen AM. Trichilia emetica (Meliaceae)–A review of traditional uses, biological activities and phytochemistry. Phytochemistry Letters. 2011;**4**(1):1-9

[209] Prisca DA, Félix YH, Gnahoué Kouadio AE, David NJ, Joseph DA. Phytochemical and Acute Toxicity Study of Trichilia Emetica (Meliaceaes) bark of trunk Extract in Albinos Rats. American Journal of Bio-pharmacology Biochemistry and Life Sciences. 2015;**4** (1):1-8

[210] Konaté K, Yomalan K, Sytar O, Zerbo P, Brestic M, Patrick VD, et al. Free radicals scavenging capacity, antidiabetic and antihypertensive activities of flavonoid-rich fractions from leaves of Trichilia emetica and Opilia amentacea in an animal model of type 2 diabetes mellitus. Evidence-Based Complementary and Alternative Medicine. 2014;**2014**:1-13

[211] Konaté K, Yomalan K, Sytar O, Brestic M. Antidiarrheal and antimicrobial profiles extracts of the leaves from Trichilia emetica Vahl. (Meliaceae). Asian Pacific Journal of Tropical Biomedicine. 2015;**5**(3):242-248

[212] Rukayyah SS, Jigam AA, Aisha MT. In vivo antiplasmodial and effects of subchronic administration of Trichilia emetica leaves extracts. International Journal of Natural Sciences Research. 2015;**3**(2):1-5

[213] Lindsey K, Jäger AK, Raidoo DM, van Staden J. Screening of plants used by southern African traditional healers in the treatment of dysmenorrhoea for prostaglandin-synthesis inhibitors and uterine relaxing activity. Journal of Ethnopharmacology. 1998;**64**(1):9-14

[214] de Boer HJ, Kool A, Broberg A, Mziray WR, Hedberg I, Levenfors JJ. Anti-fungal and anti-bacterial activity of some herbal remedies from Tanzania. Journal of Ethnopharmacology. 2005; **96**(3):461-469

[215] Mbukwa E, Chacha M, Majinda RR. Phytochemical constituents of Vangueria infausta: Their radical scavenging and antimicrobial activities. ARKIVOC. 2007;**9**:104-112

[216] Bapela MJ, Kaiser M, Meyer JJ. Antileishmanial activity of selected South African plant species. South African Journal of Botany. 2017;**108**: 342-345

[217] Gwatidzo L, Chowe L, Musekiwa C, Mukaratirwa-Muchanyereyi N. In vitro anti-inflammatory activity of Vangueria infausta: An edible wild fruit from Zimbabwe. African Journal of Pharmacy and Pharmacology. 2018;**12**(13):168-175

[218] Alara OR, Abdurahman NH, Mudalip SK, Olalere OA. Phytochemical and pharmacological properties of Vernonia amygdalina: A review. Journal of Chemical Engineering and Industrial Biotechnology. 2017;**2**(1):80-96

[219] Tijjani MA, Mohammed GT, Alkali YT, Adamu TB, Abdurahaman FI. Phytochemical analysis, analgesic and antipyretic properties of ethanolic leaf extract of Vernonia amygdalina Del. Journal of Herbmed Pharmacology. 2017;**6**(3):95-99

[220] Danladi S, Hassan MA, Masa'ud IA, Ibrahim UI. Vernonia amygdalina Del: A mini review. Research Journal of

Pharmacy and Technology. 2018;**11**(9): 4187-4190

[221] Kapravelou G, Martínez R, Andrade AM, Lopez Chaves C, López-Jurado M, Aranda P, et al. Improvement of the antioxidant and hypolipidaemic effects of cowpea flours (Vigna unguiculata) by fermentation: Results of in vitro and in vivo experiments. Journal of the Science of Food and Agriculture. 2015;**95**(6):1207-1216

[222] Sayeed VK, Satish S, Kumar A, Hegde K. Pharmacological activities of Vigna unguiculata (L) Walp: A review. International Journal of Pharma and Chemical Research. 2017; **3**(1):44-49

[223] Akinpelu LA, Adegbuyi TA, Agboola SS, Olaonipekun JK, Olawuni IJ, Adegoke AM, et al. Antidepressant activity and mechanism of aqueous extract of vigna unguiculata ssp. Dekindtiana (L.) walp dried aerial part in mice. International Journal of Neuroscience and Behavioral Science. 2017;**5**(1):7-18

[224] Abdoulaye T, Constant AA, Faustin KA, Claude KA, Etienne EK, Alette ZE, et al. Antibacterial activity and acute toxicity studies of culinary leaves from Corchorus olitorius L., Vigna unguiculata L. Walp and Hibiscus sabdariffa L. used in the north of cote d'Ivoire. Research Journal of Pharmaceutical Biological and Chemical Sciences. 2018;**9**(5):485-494

[225] Zaheer M, Ahmed S, Hassan MM. Vigna unguiculata (L.) Walp. (Papilionaceae): A review of medicinal uses, Phytochemistry and pharmacology. Journal of Pharmacognosy and Phytochemistry. 2020;**9**(1):1149-1152

[226] Maroyi A. Ximenia caffra Sond. (Ximeniaceae) in sub-Saharan Africa: A synthesis and review of its medicinal potential. Journal of Ethnopharmacology. 2016;**184**:81-100

[227] Mulaudzi RB, Ndhlala AR, Kulkarni MG, Finnie JF, Van Staden J. Antimicrobial properties and phenolic contents of medicinal plants used by the Venda people for conditions related to venereal diseases. Journal of Ethnopharmacology. 2011;**135**(2): 330-337

[228] Mboweni HF. Antimicrobial, cytotoxic and prelimenary phytochemical analysis of four medicinal plants and their formulation [doctoral dissertation]

[229] Nair JJ, Mulaudzi RB, Chukwujekwu JC, Van Heerden FR, Van Staden J. Antigonococcal activity of Ximenia caffra Sond.(Olacaceae) and identification of the active principle. South African Journal of Botany. 2013; **86**:111-115

[230] Olila D, Opuda-Asibo J. Antibacterial and antifungal activities of extracts of Zanthoxylum chalybeum and Warburgia ugandensis, Ugandan medicinal plants. African Health Sciences. 2001;**1**(2):66-72

[231] Nalule AS, Mbaria JM, Kimenju JW. In vitro anthelmintic potential and phytochemical composition of ethanolic and aqueous crude extracts of Zanthoxylum chalybeum Engl

[232] Bbosa GS, Mwebaza N, Lubega A, Musisi N, Kyegombe DB, Ntale M. Antiplasmodial activity of leaf extracts of Zanthoxylum chalybeum Engl. British Journal of Pharmaceutical Research. 2014;**4**(6):705

[233] Ngugi DN. Study of antiplasmodial activity, cytotoxicity and acute toxicity of Zanthoxylum chalybeum ENGL, and

Vernonia lasiopus o. Hoffman [doctoral dissertation], University of Nairobi; 2014

[234] Agwaya M, Nandutu A, Vuzi P. Protective effects of Zanthoxylum chalybeum in diabetes-induced myocardial dysfunction in rats. European Journal of Medicinal Plants. 2016;**12**(1):1

[235] Nantongo JS, Odoi JB, Abigaba G, Gwali S. Variability of phenolic and alkaloid content in different plant parts of Carissa edulis Vahl and Zanthoxylum chalybeum Engl. BMC Research Notes. 2018;**11**(1):1-5

[236] Waterman C, Smith RA, Pontiggia L, DerMarderosian A. Anthelmintic screening of sub-Saharan African plants used in traditional medicine. Journal of Ethnopharmacology. 2010;**127**(3): 755-759

[237] Mongalo NI, Mashele SS, Makhafola TJ. Ziziphus mucronata Willd. (Rhamnaceae): It's botany, toxicity, phytochemistry and pharmacological activities. Heliyon. 2020;**6**(4):1-20

[238] Koenen EV. Medicinal, Poisonous and Edible Plants in Namibia. Namibia: Klaus Hess Verlag; 1996

[239] Mutsaka-Makuvaza MJ, Matsena-Zingoni Z, Katsidzira A, Tshuma C, Chin'ombe N, Zhou XN, et al. Urogenital schistosomiasis and risk factors of infection in mothers and preschool children in an endemic district in Zimbabwe. Parasites & Vectors. 2019; **12**:1-5

[240] Silva C, Vareda J, Sousa A, Perestrelo R. Forensic attribution profiling of food using liquid chromatography–Mass spectrometry. In: Food Toxicology and Forensics. Portugal: Elsevier Academic Press; 2021. pp. 97-121

[241] Kumari R, Kotecha M. A review on the standardization of herbal medicines. International Journal of Pharma Sciences and Research. 2016;**7**(2):97-106

[242] Erhabor JO, Komakech R, Kang Y, Tang M, Matsabisa MG. Ethnopharmacological importance and medical applications of Myrothamnus flabellifolius Welw. (Myrothamnaceae)-A review. Journal of Ethnopharmacology. 2020;**252**:112576

[243] Bussmann RW, Malca G, Glenn A, Sharon D, Nilsen B, Parris B, et al. Toxicity of medicinal plants used in traditional medicine in Northern Peru. Journal of Ethnopharmacology. 2011; **137**(1):121-140

Chapter 14

Exploring the Potential of *Calotropis procera* in Pharmacological Approaches

Poonam Bansal, Sunayna Choudhary, Tanvi Taneja, Sonali Sangwan, Bhupesh Gupta, Soniya Goyal, Raman Kumar and Pooja Sharma

Abstract

Medicinal plants have been a source of treatments for many ailments for thousands of years. The WHO estimates that 80% of worldwide population use traditional medicines to treat common health issues. Plant derived bioactive substances constitute 50% of Western medications. The increase in incidents of emerging medical challenges, including post-COVID syndrome, rising multidrug-resistant (MDR), and many more, has raised annual fatalities. To address these issues, novel medications and strategic approaches are urgently required. Designing novel drugs relies on exploring medicinal plants, which have great scope in combating diseases. *Calotropis procera* is a medicinal plant belongs to Apocynaceae family and subfamily Asclepiadoideae that have been exploring for developing novel drugs. *C. procera* consists of numerous phytochemicals including flavonoids, terpenoids, cardenolides, steroids and oxypregnanes. Therefore, its phytoconstituents have been used to treat a variety of conditions including cancer, asthma, epilepsy and snake bite. *C. procera* is reported to have anti-inflammatory, antitumor, anthelmintic, antibacterial, antinociceptive and antimalarial properties. Roots, leaves and flower of *C. procera* have been used in wide range of ethnomedicinal and pharmacological actions including leukoderma, malaria and eczema. Recent ongoing techniques including computational tools using the phytoconstituents of *C. procera* against various diseases will open up avenues for developing novel drugs.

Keywords: *Calotropis procera*, anti-inflammatory, antibacterial, antimalarial, drugs

1. Introduction

For thousands of years, plants have been the only source of treatments to treat both human and animal illnesses [1]. Medicinal plants (MPs) are the primary source of basic healthcare in underdeveloped nations [2, 3]. According to World Health Organization (WHO), approximately 80% of the world's population relies on traditional medicines, primarily on MPs, for their everyday health problems. Also, 50% of Western medications contain bioactive substances derived from plants [4].

A dramatic increase in fungal diseases over the past few decades has caused the dispersion of fungal spores across the soil and the environment. As a result of excessive fungal spore exposure, numerous illnesses, such as sinusitis, lung infections, and skin infections, are reported to be increased in people with impaired immune systems [5]. Similar to fungal diseases, microbial diseases have historically been the leading cause of mortality [6]. Currently, multidrug-resistant (MDR is solely to blame for about 230,000 of the 700,000 annual deaths caused by resistant infections. By 2050, drug-resistant illnesses will result in 10 million annual fatalities [6].

To address antibiotic resistance, new medications and alternative therapies (like traditional plant-based medicines, bacteriophage therapies, and combinational therapies) are urgently required [7, 8]. World Health Organization strongly emphasizes developing novel antibiotics to combat resistant diseases [9]. Since the dawn of civilization, phytochemicals such as alkaloids, terpenoids, tannins, steroids, coumarins, and flavonoids derived from medicinal plants have a great scope to combat diseases. Essential oils and phenolic acids from *Petroselinum crispum*, *Levisticum officinale* Koch, *Ocimum basilicum*, *Thymus vulgaris*, *Syzygium aromaticum* alter the physiology of bacteria such as *Bacillus cereus*, *Staphylococcus aureus*, *Pseudomonas aeruginosa*, *Escherichia coli*, and *Salmonella* by increasing cell permeability, altering the bacterial cell wall and membrane integrity, losing ATP, and inhibiting protein synthesis.

Compared to synthetic antimicrobials, medicinal plants are thought to have fewer side effects and exhibit varying degrees of efficacy against microbial infections [5, 10, 11]. Co-administration of antibiotics and non-antibiotic substances breaks down resistance and is a successful strategy for enhancing or restoring antibiotic efficacy [7]. This chapter describes the morphological description of *C. procera* and its phytochemical constituents or pharmacological properties described briefly.

2. *Calotropis procera*

The plants *Calotropis procera* referred to as "Raktha Arka", in traditional Ayurvedic medicine. It serves a variety of functions. The plant fibers are used to make baskets, ropes, bags, and nets. The wood serves as both fuel and building material. The leaves of the plant serve as the animal's food. The plant's latex is a crucial component of many folk medicines. The common names [12] of the plant are summarized in **Table 1**. The taxonomic classification of *C. procera* is tabulated in **Table 2**.

Country	Names
India	Sanskrit- Arka, Ganarupa, Mandara, Vasuka, Svetapushpa, Sadapushpa, Alarka, Pratapass
	Hindi- Aak, Madar
	Kannada- Ekka
	Tamil and Malayalam- Erukku
	Telugu- Jilledi, Puvvu
Malaysia	Remiga, Rembega, Kemengu
Indonesia	Sundanese and Madurese- Bidhuri
	Javanese- Sidaguri
	Aceh- Rubik
Philippines	Tagalog- Kapal-kapal

Country	Names
Laos	Kok May, Dok Kap, Dok Hak
Thailand	Northern- Po Thuean, Paan Thuean
	Central- Rak
Vietnam	Bootng, Lashen, Nam Tit Bat
French	Faux arbre de soie, Mercure vegetal
English	Giant Indian milked weed, Madar and Sodom apple
Turkey	Ipekag
Arab	Oshar or Ushar
Persia	Kharak
Pushto	Spalmai

Table 1.
Vernacular names of C. procera.

Taxonomic classification
Kingdom: Plantae
Subkingdom: Tracheobionta
Super division: Spermatophyta
Division: Magnoliophyta
Class: Magnoliopsida
Order: Gentianales
Family: Asclepiadaceae
Genus: Calotropis
Species: *C. procera*

Table 2.
Taxonomic classification of C. procera.

3. Distribution

Calotropis procera is a perennial plant belonging to the family Apocynaceae. The plant is abundant in Asia, America, Africa, Afghanistan, Algeria, Burkina Faso, Cameroon, Chad, Cote d'Ivoire, the Democratic Republic of the Congo, Egypt, Eritrea, Ethiopia, Gambia, Ghana, Guinea-Bissau, Pakistan, and India. It thrives as a wild shrub across Punjab, especially on plain pastures and roads [13]. Calotropis grows wild up to 900 meters (msl) throughout the nation [14] and is tolerant to salt, and likes disturbed environments. It readily establishes as a weed along deteriorated roadways, lagoon edges, and overgrazed native grasslands and is propagated by seeds spread by wind and animals. It prefers abandoned agriculture sites and frequently predominates there, especially in places with disturbed sandy soils and little rainfall. It is believed to be a sign of overcrowding. It is the first vegetation to grow on arid soil and is tolerant of drought [15]. The xerophytic adaptations include the presence of latex, a profoundly branching root system, and thick leaves covered with wax.

The vegetative characteristics of plant are summarized in **Table 3** [16].

Vegetative characters	Description
Habit	Shrub or a small tree up to 2.5 m (max.6 m) height.
Roots	Simple, branched, woody at base and covered with a fissured; corky bark; branches somewhat succulent and densely white tomentose; early glabrescent. All parts of the plant exude white latex when cut or broken.
Leaves	Opposite-decussate, simple, sub sessile, extipulate; blade-oblong obovate to broadly obovate, 5–30 × 2.5–15.5 cm, apex abruptly and shortly acuminate to apiculate, base cordate, margins entire, succulent, white tomentose when young, later glabrescent and glacouse.
Flowers	Bracteate, complete, bisexual, actinomorphic, pentamerous, hypogynous, pedicellate, pedicel 1–3 cm long.
Floral characteristics	Inflorescence- A dense, multiflowered, umbellate, peducled cymes, arising from the nodes and appearing axillary or terminal.
Calyx	Sepal five, Polysepalous, five lobed, shortly united at the base, glabrescent, quincuncial aestivation.
Corolla	Petals five, gamopetalous, five lobed, twisted aestivation.
Androecium	Stamens five, gynandrous, anther dithecous, coherent.
Gynoecium	Bicarpellary, apocarpus, styles are united at their apex, peltate stigma with five lateral stigmatic surfaces. Anthers adnate to the stigma forming a gynostegium.
Fruit	A simple, fleshy, inflated, subglobose to obliquely ovoid follicle up to 10 cm or more in diameter.
Seeds	Many, small, flat, obovate, 6 × 5 mm, compressed with silky white pappus, 3 cm or more long.

Table 3.
Vegetative characters of C. procera.

4. Phytochemistry of *C. procera*

C. procera contains cardenolide, triterpenoids, alkaloids, resins, anthocyanins, and other compounds. In addition to this it also contains hydrocarbons, saturated and un saturated fatty acids. Different phytoconstituents isolated from different parts of *C. procera* were tabulated in **Table 4**.

Plant part	Compounds present	References
Leaves	a-amyrin	[17]
	a-amyrin-acetate	–
	β-sitosterol	–
	Urosolic acid	–
	Cardenolide	[18]
	Calotropin	[19]
	Calotropagenin	[20]

Plant part	Compounds present	References
Latex	Caoutchouc	[18]
	Calotoxin	[18]
	Calactin	[21]
	Uscharin	–
	Trypsin	[21]
	Voruscharin	–
	Uzariginin	–
	Syrioginin	[21]
	Proceroside	[22]
Flower	Queretin-3-ratinoside	[22]
	Sterol	[23]
	D-arabinose	[22]
	Glucosamine	[23]
	L-rhamnose	[23]
Bark	Triterpenes	[24]
	Pentacyclic triterpinoides	[25]
	Calotropursenyl acetate	[26]
	Apundarol isovalerate	[24]
	Querecetin-3-rutinoside	[25]

Table 4.
Phytoconstituents of C. procera.

5. Traditional uses

Ancient Egyptians utilized *C. procera* as a medicinal plant throughout the Neolithic period in Egypt. The plant is a part of Greco-Arab medicine [27] and is traditionally used across 21 nations worldwide. The plant is also used in Ayurveda, Siddha, Unani, and Sudanese traditional systems of medicine. It is important to note that *C. procera* has been used more commonly to treat a variety of infectious disorders that may be generally divided into five categories:

1. Leprosy, boils, carbuncles, scabies, leishmaniasis, and infections of the skin, mouth, and teeth are examples of skin and dermal infections.

2. Respiratory infections include pneumonia, bronchitis, bronchial asthma, cough.

3. GIT infections include dysentery, diarrhea, cholera, gastritis, colitis, and worms.

4. GU infections include chronic renal failure and leucorrhea.

5. Systemic infections include malaria and both internal (oral) and external (topical) preparations have used *C. procera*.

Parts	Uses
Leaves	In leukoderma (skin disease); an antidote for rabies; prompt healing; applied for a poultice; treat migraine, fever, eczema, leprosy, elephantiasis, asthma, cough, and rheumatism).
Flower	Treat skin and gum infections; used in dysentery and antidote for scorpion bite.
Root	Used as a digestive; to treat body pain, malaria, eczema, leprosy elephantiasis, asthma, cough, and rheumatism.

Table 5.
Ethnomedicinal uses of C. procera.

However, given its increased usefulness in treating cutaneous infections, external or topical applications are more prevalent. Ethnomedicinal uses of *Calotropis procera* are summarized in **Table 5**.

6. Medicinal activity

This highly effective shrub is used in numerous widespread and traditional medicines to treat various illnesses like fever, leprosy, eczema, diarrhea, dysentery, and jaundice [28, 29]. The plant has reported anti-inflammatory, anti-tumor, anthelmintic, hepatoprotective, antioxidant, anticonvulsant, antibacterial, oestrogenic, antinociceptive, and antimalarial properties (**Figure 1**).

7. Antioxidant activity

The anti-inflammatory and anti-hyperglycemic effects of *Calotropis procera's* dry latex (DL) were demonstrated in rats that had been given an alloxan-induced diabetes model. In daily oral treatment of DL at dosages of 100 and 400 mg/kg, a dose-dependent drop in blood sugar and an increase in hepatic glycogen content were seen. Additionally, DL slowed the loss of body weight in diabetic animals and decreased their daily water intake to levels comparable to those of rodents without diabetes. Additionally, in rats with alloxan-induced diabetes, DL reduced the levels of thiobarbituric acid-reactive substances (TBARS) while increasing the levels of endogenous antioxidants like catalase, glutathione, and superoxide dismutase (SOD). Comparable to glibenclamide, a popular anti-diabetic drug, DL proved effective as an antioxidant and an anti-diabetic agent [16].

The antioxidant activity (free radical scavenging capacity) of the methanolic extract of *C. procera* roots was evaluated by the *in-vitro* DPPH scavenging assays. The IC50 value was found below 100 μg/ml, indicating the plant's potent antioxidant activity.

8. Antimicrobial activity

Calotropis procera seeds were extracted using chloroform and methanol and have been tested on a paper disc for possible in vitro antibacterial activity. The chloroform extract of the seeds showed superior antibacterial action [30]. The stems, fruit, leaves, and flowers of *C. procera*, as well as its n-hexane, ether, chloroform, and water fraction, were extracted with 70% methanol and water, and their antibacterial activity was investigated. The antibacterial and antifungal effects of the plant fractions were evaluated

Figure 1.
Health benefits of C. procera.

using *Klebsiella pneumoniae* and *Aspergillus niger*, respectively. The test extracts with a concentration of 10 mg/mL were used in the study. On Muller Hinton agar for bacteria and Yeast Peptone Glucose (YPG) agar for fungi, which had previously been seeded with the microbial inocula of 0.5 MacFarland density, a volume of 10 L of each examined extract sample was detected. The inhibition zones on the inoculation plates were measured in mm after a 24-hour period of 37°C incubation. The plant's flower extract showed the greatest antibacterial activity in the n-hexane and ether fractions.

9. Anti-inflammatory activity

The anti-inflammatory impact of *C. procera* was tested using the various acute and chronic models of inflammation. Oral administration of dried latex of *C. procera* significantly inhibited edema formation induced by carrageenan and Freund's Adjuvant [31]. The plant also has potent anti-inflammatory effects against cotton pellets and carrageenan-induced granulomas in albino Wistar rats. The methanolic extracts

(180 mg/kg, p.o.) of the roots of *C. Procera* can reduce subacute inflammation by interrupting the metabolism of arachidonic acid in both the cotton pellet and paw edema models [32].

10. Antipyretic activity

Calotropis procera may become a more widely available and effective antipyretic medication, according to the study. In contrast to aspirin, *C. procera's* ethanolic extract of the aerial parts, aqueous extract of the flower, and aqueous solution of the dry latex have all demonstrated potent antipyretic effects in animal models [33].

11. Anticancer activity

Cardenolide, a novel compound present in *C. procera*. According to Quaquebeke [34]. *C. procera* has strong anti-tumor properties in vitro and a high level of tolerance in vivo. Similar to this, di-(2-ethylhexyl) phthalate (DEHP) isolated from *C. procera* demonstrated anti-tumor activity, and copper nanoparticles synthesized using an aqueous extract of *C. procera* latex demonstrated cytotoxic [35] and cytostatic activity against tumor cells and cell lines [36].

12. Antimalarial activity

The alcoholic extract of *C. procera* flower extract exhibited a higher level of mosquito repellent activity against the female *Culex quinquefasciatus* mosquito as compared to the petroleum ether and chloroform extracts [37]. This study suggests the role of *C. procera* as a natural biocide for mosquito control. The aqueous extract of CG leaves at 125, 250, 500, and 1000 ppm exhibited larvicidal, mosquito-repellent, and ovicidal activity against *Culex gelidus* and *C. tritaeniorhynchus* mosquitoes. The extract showed dose-dependent larvicidal activity with a motility rate of 86 ± 1.42% (LC50 = 137.90) against *C. gelidus* and 94 ± 1.31% (LC50 = 110.05) against *C. tritaeniorhynchus*.

13. Anti-obesity activity (pancreatic lipase inhibitory activity)

The purified di-terpenoid fraction from the root extract of *C. procera* inhibits pancreatic lipase (PL) with an IC50 of 9.47 mg/mL. The purified di-terpenoid fraction was shown to have a considerably lower inhibition constant (Ki) than the positive control (Orlistat; IC50: 0.15 μM). The inhibition was determined to be competitive based on kinetic data. This explains the plant's antihyperlipidemic actions [38].

14. Antiviral activity

Globally, viral illnesses are regarded as one of the most significant hazards to people, animals, and plants. The outbreaks of deadly viral diseases like COVID-19, which pose a serious threat to human survival on a global scale, also call for the development of vaccines or other anti-toxin treatments. This is in addition to the challenges brought on by the emergence of antiviral resistance and the negative side effects of currently available antiviral drugs [39]. Research has shown the potential role of medicinal plants and their bioactive compounds as antiviral agents [40, 41].

15. Toxicity

In addition to its well-documented traditional uses across many nations, *C. procera* is categorized as a weed, a toxic plant and a poisonous plant [42–44]. The herb was formerly employed as an abortifacient. The plant leaves also cause ocular toxicity if splashed/entered accidentally. It causes ocular Keratouveitis accompanied by inflammations, corneal edema, irreversible endothelial cell damage and vision deterioration [30, 45, 46]. Ruminants have experienced harmful effects after consuming *C. procera* leaves (CPL) [47].

The leading cause of the plant's toxicity is the presence of poisonous substances like toxic cardenolides in its latex. Similar to those of Digitalis, the cardiac glycosides of *C. procera* severely increase heartbeat and finally result in animal mortality. CPL's pH is 5.2, which is harmful to the animal's mucous membranes [45]. Additionally, *C. procera* thrives in various soils, including those contaminated with heavy metals and found along roadsides. As a result of the plant's remarkable capacity to absorb diverse chemical components, such as heavy metals, it bioaccumulates more significant levels of dangerous heavy metals like lead (Pb), chromium (Cr), nickel (Ni) and cadmium (Cd) as well as other environmental contaminants which increase the plant's toxicity [48].

16. Conclusion

The *C. procera* is one of the globally distributed medicinal plants. Despite of having pharmacological and traditional uses this is the plant that is forgotten as the time passes. But now many scientists have worked to evaluate its phytochemicals and pharmacological property. The pharmacology, traditional uses, toxicology and use of secondary metabolites has been discussed in this chapter. *C. procera* is the richest source of phytochemicals and screening its phytoconstituents will give a new avenue to investigate its therapeutic role. *In vivo* and *in vitro* study of *C. procera* was well documented in literature but human safety and efficacy yet to be done and clinical trials need to be done to confirm its standard dosage.

Author details

Poonam Bansal[1*], Sunayna Choudhary[2], Tanvi Taneja[1], Sonali Sangwan[1], Bhupesh Gupta[3], Soniya Goyal[1], Raman Kumar[4] and Pooja Sharma[1]

1 Department of Biosciences and Technology, MM College of Engineering, MM (Deemed to be University), Ambala, Haryana, India

2 Department of Pharmacognosy, MM College of Pharmacy, MM (Deemed to be University), Ambala, Haryana, India

3 Computer Science and Engineering Department, MMEC, Maharishi Markandeshwar (Deemed to be University), Ambala, Haryana, India

4 Department of Biochemistry, School of Bioengineering and Bioscience, Lovely Professional University, Phagwara, Punjab, India

*Address all correspondence to: poonambansal@gmail.com

References

[1] El-Seedi HR, Khalifa SA, Taher EA, Farag MA, Saeed A, Gamal M, et al. Cardenolides: Insights from chemical structure and pharmacological utility. Pharmacological Research. 2019;**141**:123-175

[2] Fatima A, Ahmad M, Zafar M, Yaseen G, Khan MPZ, Butt MA, et al. Ethnopharmacological relevance of medicinal plants used for the treatment of oral diseases in Central Punjab-Pakistan. Journal of Herbal Medicine. 2018;**12**:88-110

[3] Khan IH, Javaid A. Antifungal activity and GC-MS analysis of n-butanol extract of quinoa leaves. Bangladesh Journal of Botany. 2020b;**49**(4):1045-1051

[4] Gupta A, Pandey AK. Antibacterial lead compounds and their targets for drug development. In: Phytochemicals as Lead Compounds for New Drug Discovery. Elsevier; 2020. pp. 275-292

[5] Vidyasagar GM. Plant-derived antifungal agents: Past and recent developments. Recent Trends in Antifungal Agents and Antifungal Therapy. 2016;**2016**:123-147

[6] Biharee A, Sharma A, Kumar A, Jaitak V. Antimicrobial flavonoids as a potential substitute for overcoming antimicrobial resistance. Fitoterapia. 2020;**146**:104720

[7] Khameneh B, Iranshahy M, Soheili V, Fazly Bazzaz BS. Review on plant antimicrobials: A mechanistic viewpoint. Antimicrobial Resistance & Infection Control. 2019;**8**(1):1-28

[8] Mulat M, Khan F, Muluneh G, Pandita A. Phytochemical profile and antimicrobial effects of different medicinal plant: Current knowledge and future perspectives. Current Traditional Medicine. 2020;**6**(1):24-42

[9] World Health Organization. 2019 Antibacterial Agents in Clinical Development: An Analysis of the Antibacterial Clinical Development Pipeline. 2019

[10] Pathania S, Bansal P, Gupta P, Rawal RK. Genus Calotropis: A hub of medicinally active phytoconstituents. Current Traditional Medicine. 2020;**6**(4):312-331

[11] Konaté K, Mavoungou JF, Lepengué AN, Aworet-Samseny RR, Hilou A, Souza A, et al. Antibacterial activity against β-lactamase producing Methicillin and Ampicillin-resistants *Staphylococcus aureus:* Fractional Inhibitory Concentration Index (FICI) determination. Annals of Clinical Microbiology and Antimicrobials. 2012;**11**:1-12

[12] Breckle SW, Hedge I, Rafiqpoor M. Vascular Plants of Afghanistan. Bonn: Scientia Bonnensis; 2013

[13] Azhar MF, Siddiqui MT, Ishaque M, Tanveer A. Study of ethnobotany and indigenous use of *Calotropis procera* (Ait.) in cholistan desert, Punjab, Pakistan. Journal of Agricultural Research. 2014;**52**(1):117-126

[14] Sharma AP, Tripathi BD. Assessment of atmospheric PAHs profile through Calotropis gigantea R. Br. leaves in the vicinity of an Indian coal-fired power plant. Environmental Monitoring and Assessment. 2009;**149**:477-482

[15] Smith NM. Weeds of the Wet/Dry Tropics of Australia-a Field Guide. USA:

Environment Centre NT. Inc Darwin North Territ; 2002

[16] Kumar VL, Padhy BM, Sehgal R, Roy S. Antioxidant and protective effect of latex of *Calotropis procera* against alloxan-induced diabetes in rats. Journal of Ethnopharmacology. 2005a;**102**:470-473

[17] Bryce TA, Eglinton G, Hamilton RJ, Martin-Smith M, Subramanian G. Triterpenoids from New Zealand plants-II.: The triterpene methyl ethers of Cortaderia toetoe Zotov. Phytochemistry. 1967;**6**(5):727-733

[18] Chopra RN, Nayar SL. Glossary of Indian Medicinal Plants. Vol. 46. New Delhi: Council of Scientific and Industrial Research; 1956. p. 29

[19] Yelne MB, Sharma PC, Dennis TJ. Database on medicinal plants used in ayurveda. Central Council for Research in Ayurveda & Siddha, New-Delhi. 2000;**2**:69-73

[20] Singh B, Rastogi RP. Structure of asclepin and some observations on the NMR spectra of *Calotropis* glycosides. Phytochemistry. 1972;**11**(2):757-762

[21] Baquar SR, Tasnif M. Medicinal plants of southern west Pakistan. Vivek Vihar, Delhi: Periodical Expert Book Agency, D-42; 1984. p. 30

[22] Grieve MM. A Modern Herbal. Vol. 10. London: Tigers Book International; 1994. p. 154

[23] Hanna AG, Elgamal MHA, Morsy NA, Duddeck H, Kovács J, Tóth G. Two cardenolides from *Calotropis procera*. Magnetic Resonance in Chemistry. 1999;**37**(10):754-757

[24] Ansari SH, Ali M. New oleanene triterpenes from root bark of *Calotropis procera*. Journal of Medicinal and Aromatic Plant Sciences. 1999;**21**(4):978-981

[25] Ansari SH, Ali M. Norditerpenic ester and pentacyclic triterpenoids from root bark of *Calotropis procera* (Ait) R. Br. Die Pharmazie. 2001;**56**(2):175-177

[26] Hamidi JA, Ismaili NH, Ahmadi FB, Lajisi NH. Antiviral and cytotoxic activities of some plants used in Malaysian indigenous medicine. Pertanika Journal of Tropical Agricultural Science. 1996;**19**(2/3):129-136

[27] Hassan LM, Galal TM, Farahat EA, El-Midany MM. The biology of *Calotropis procera* (Aiton) WT. Trees. 2015;**29**:311-320

[28] Meena AK, Yadav AK, Niranjan US, Singh B, Nagariya AK, Sharma K, et al. A review on *Calotropis procera* Linn and its ethnobotany, phytochemical, pharmacological profile. Drug Invent Today. 2010;**2**(2):185-190

[29] Verma SK, Singh SK, Mathur A. In vitro cytotoxicity of *Calotropis procera* and Trigonella foenum-graecum against human cancer cell lines. Journal of Chemical and Pharmaceutical Research (JOCPR). 2010;**2**(4):861-865

[30] Bhaskar VH, Ajay SS. Antimicrobial activity of *Calotropis procera* seeds. Asian Journal of Chemistry. 2009;**21**(7):5788-5790

[31] Kumar S, Dewan S, Sangraula H, Kumar VL. Anti-diarrhoeal activity of the latex of *Calotropis procera*. Journal of Ethnopharmacology. 2001;**76**(1):115-118

[32] Saba AB, Oguntoke PC, Oridupa OA. Anti-inflammatory and analgesic activities of ethanolic leaf extract of *Calotropis procera*. African Journal of Biomedical Research. 2011;**14**(3):203-208

[33] Dewan S, Kumar S, kumar VL. Antipyretic effect of latex of *Calotropis procera*. Indian Journal of Pharmacology. 2000;**32**(3):252

[34] Quaquebeke VE, Simon G, Andre A, Dewelle J, Yazidi ME, Bruyneel F. Identification of a novel cardenolide (2-oxovoruscharin) from *Calotropis procera* and the hemisynthesis of novel derivatives displaying potent in vitro antitumor activities and high in vivo tolerance: Structure-activity relationship analyses. Journal of Medicinal Chemistry. 2005;**48**:849-856

[35] Harne S, Sharma A, Dhaygude M, Joglekar S, Kodam K, Hudlikar M. Novel route for rapid biosynthesis of copper nanoparticles using aqueous extract of *Calotropis procera* L. latex and their cytotoxicity on tumor cells. Colloids and Surfaces. B, Biointerfaces. 2012;**15**:284-288

[36] Taylor P, Arsenak M, Abad MJ, Fernández A, Milano B, Gonto R. Screening of Venezuelan medicinal plant extracts for cytostatic and cytotoxic activity against tumor cell lines. Phytotherapy Research. 2013;**27**(4):530-539

[37] Larhsini M, Bousaid M, Lazrek HB, Jana M, Amarouch H. Evaluation of antifungal and molluscicidal properties of extracts of *Calotropis procera*. Fitoterapia. 1997;**68**(4):371-373

[38] Patil SG, Patil MP, Maheshwari VL, Patil RH. In vitro lipase inhibitory effect and kinetic properties of di-terpenoid fraction from *Calotropis procera* (Aiton). Biocatalysis and Agricultural Biotechnology. 2015;**4**(4):579-585

[39] Bagla VP, McGaw LJ, Eloff JN. The antiviral activity of six South African plants traditionally used against infections in ethnoveterinary medicine. Veterinary Microbiology. 2012;**155**(2-4):198-206

[40] Mohanraj R, Rakshit J, Nobre M. Anti HIV-1 and antimicrobial activity of the leaf extracts of *Calotropis procera*. International Journal of Green Pharmacy (IJGP). 2010;**4**(4):242-246

[41] Mukhtar M, Arshad M, Ahmad M, Pomerantz RJ, Wigdahl B, Parveen Z. Antiviral potentials of medicinal plants. Virus Research. 2008;**131**(2):111-120

[42] Gracia A, Rangel-Buitrago N, Castro-Barros JD. Non-native plant species in the Atlantico department coastal dune systems, caribbean of Colombia: A new management challenge. Marine Pollution Bulletin. 2019;**141**:603-610. DOI: 10.1016/j.marpolbul.2019.03.009

[43] Al-Zuhairi AH, Al-Ani JMK, Ibrahim SN. Toxicological effects of aqueous extract of *calotropis procera* leaves in experimentally poisoned rabbits. Iraqi Journal of Veterinary Medicine. 2020;**44**(1):46-56

[44] Tossou ML, Ballogou B, Maina J, Gicheha M. Effect of *Calotropis procera* on the proximate composition and potential toxicity of Wagashi (traditional cheese) in Benin. Food Science and Quality Management. 2018;**74**:30

[45] Al-Mezaine HS, Al-Amry MA, Al-Assiri A, Fadel TS, Tabbara KF, Al-Rajhi AA. Corneal endothelial cytotoxicity of the *Calotropis procera* (ushaar) plant. Cornea. 2008;**27**(4):504-506

[46] Lakhtakia S, Dwivedi PC, Choudhary P, Chalisgaonkar C, Rahud J. Ocular toxicity of Calotropis-missing links. Indian Journal of Ophthalmology. 2010;**58**(2):169

[47] Ahmed OM, Fahim HI, Boules MW, Ahmed HY. Cardiac and testicular toxicity effects of the latex and ethanolic leaf extract of *Calotropis procera* on male

albino rats in comparison to abamectin. Springerplus. 2016;**5**(1):1-21

[48] Naz A, Chowdhury A, Chandra R, Mishra BK. Potentialhuman health hazard due to bioavailable heavy metal exposure viaconsumption of plants with ethnobotanical usage at the largest chromite mine of India. Environmental Geochemistry and Health. 2020;**42**(12):4213-4231

Section 4

Essential Oils

Chapter 15

Melaleuca bracteata var. Revolution Gold (Myrtaceae) Essential Oil: Chemical Composition, Antibacterial, Membrane Damage, Antiplatelet Aggregation and Antiacetylcholinesterase Activities

Oladipupo A. Lawal, Kehinde O. Amisu, Rebamang A. Mosa, Foluso O. Osunsanmi and Andy R. Opoku

Abstract

Melaleuca bracteata var. Revolution Gold (a cultivar of *Melaleuca bracteata*) is an ornamental plant, which has been used in traditional medicine for the treatment of several diseases. Till moment, information is scanty on the biological activities of the essential oil from the plant. The water-distilled essential oil was analyzed by gas chromatography and gas chromatography-mass spectrometry. Antibacterial activity of the oil was evaluated by paper disc diffusion and micro-dilution methods. Cell membrane damage was assay using cytosolic lactate dehydrogenase released method. Platelet aggregation inhibitory activity was separately evaluated on Adenosine diphosphate, collagen, epinephrine and thrombin induced aggregation. Thirteen components representing 95.3% of the total oil were identified from the essential oil. Phenylpropanoids (82.9%) constitute the predominant class of compounds in the oil. On the whole, the oil displayed strong antibacterial action towards *Staphylococcus aureus*, moderate activity on *Bacillus cereus* and some strains of *Escherichia coli*. The lactate dehydrogenase released (0.78–47%) depicted the oil to exhibit low levels of membrane damage. The percentage platelet aggregation inhibition for the four platelet agonists was concentration dependent with thrombin > collagen > ADP > epi-nephrine. The acetylcholinesterase inhibitory activity (9.16%) indicated that the essential oil was not effective against the enzyme.

Keywords: *Melaleuca bracteata* var. revolution gold, Myrtaceae, essential oil, methyl eugenol, biological activity

IntechOpen

1. Introduction

Melaleuca bracteata L. (Syns: *Melaleuca daleana* Blakely, *Melaleuca glaucocalyx* Gand. or *Melaleuca monticola* J.M.Black) and commonly known as black tea tree, honey myrtle, golden bottle brush amid other names, belongs to the Myrtaceae family [1]. *Melaleuca bracteata* var. Revolution Gold (popularly known as *Melaleuca bracteata* var. "Johannesburg Gold") is a garden cultivar of *Melaleuca bracteata* and widely found in woodlands, open forests along watercourses and on the edges of swamps as well as garden and urban street ornamental plant in South Africa [1, 2]. *Melaleuca bracteata* var. Revolution Gold is a shrub or medium-size tree growing as tall as 5 m, with dark gray stem-bark. The leaves (*ca* 7 cm by 25 cm) with intact margin are evergreen, alternately arranged, ovate to lanceolate. The flowers in clusters vary from white to pink-red, pale yellow or greenish, with small petals and bundle of stamens. The fruits (2–3 mm) with numerous seeds of about 0.5–0.8 mm long aggregated into cylindrical stacks along the twigs [1, 2]. In traditional medicine, *M. bracteata* var. Revolution Gold has been reported used for treatment and prevention of numerous diseases [1, 2]. Previous studies on different extracts of *M. bracteata* var. Revolution Gold revealed the isolation of betulinic acid, oleanolic acid, maslinic acid and their derivatives, with many possessing antibacterial, anti-inflammatory, antiplatelet aggregation, antifungal, antiulcer antioxidant, anti-sickling and cytotoxic activities [3–7].

As a continuation of our studies on the flora of South African species [8–11], we reports the chemical composition, antibacterial, membrane damage, acetylcholinesterase and antiplatelet aggregation activities of essential oil of *Melaleuca bracteata* var. Revolution Gold collected from KwaDlangezwa area in uThungulu District Municipality, KwaZulu-Natal Province, South Africa.

2. Experimental

2.1 Chemicals and reagents

Analytical grade chemicals and reagents were purchased from Sigma-Aldrich Chemical Co. (St Louis, MO, USA).

2.2 Animals

Either sex of Sprague-Dawley rats (between 8 weeks and 220 to 250 kg) were collected from the Department of Biochemistry and Microbiology, University of Zululand animal house. The animals were preserved under standard temperature of 23 ± 2°C and 12 h light dark cycle and had free access to standard pellet feed and enough drinking water. Certificate of ethic clearance number: UZREC 171110–030 PGD 2014/53 was acquired from the Research Animal Ethical Clearance Committee (RAEC) of the University.

2.3 Plant material

Melaleuca bracteata var. Revolution Gold fresh plant materials were collected from the University of Zululand, KwaDlangezwa campus, South Africa. Dr. N. R. Ntuli, a plant taxonomist at the Department of Botany, University of Zululand, identified the plant material. Voucher specimen (VN 0256) was deposited in the Herbarium of the University.

2.4 Oil isolation

Air dried and squeezed leaves of *M. bracteata* var. Revolution Gold (300 g) were subjected to hydrodistillation in an all glassed Clevenger-type apparatus for 3 h according to an established procedure [12]. The distillate isolated was collected over water in the receiver arm of the apparatus into clean and previously weighed sample bottle, and refrigerated until further analyses.

2.5 Gas chromatography

Gas Chromatography analyses was carried out using an Agilent Gas Chromatography (7890A) equipped with Agilent 190,915 capillary column (30 m × 250 μmid; film thickness 0.25 μm) and FID detector. Oven temperature was programmed from 45°C (after 2 min) to 310°C at 5°C/min and final temperature was held for 10 min. Injection and detector temperatures were 200 and 240°C respectively. Helium was used as the carrier gas at a flow rate of 1 ml/min. Diluted oil (0.1 μl) was injected into the GC and peaks were measured by electronic integration method. *n*-Alkanes were runs at the same condition for retention indices determination.

2.6 Gas chromatography: mass spectrometry

Gas chromatography-mass spectrometry analyses was performed on an Agilent Gas Chromatography (7890A) equipped with an Agilent 190,915 capillary column (30 m × 250 μmid; film thickness 0.25 μm) interfaced with an Agilent mass spectrometer system (5975C VL MSD with Triple Axis Detector). Temperature oven was programmed from 70 to 240°C at the frequency of 5°C/min. Ion source was set at 240°C with electron ionization at 70 eV. Helium was used as the carrier gas at a flow rate of 1 ml/min. Diluted oil in hexane (1.0 μl) was injected into the GC/MS with the scanning ranges between 35 to 425 amu.

2.7 Identification of compounds

Constituents were identified on the basis of their retention times (RT) along with co-injection reference under identical experimental conditions. Comparison of their mass spectra was also check with those of NIST [13]. Furthermore, home-made MS library built up from pure substances and components of known essential oils was compared with literature [14].

3. Antimicrobial activity

3.1 Microorganisms

The acquired test microorganisms from the culture collection of the Applied and Environmental Microbiology Research Group (AEMREG), University of Fort Hare, Alice, South Africa were used in the antimicrobial activity. The microorganisms included referenced *Bacillus cereus* (ATCC 10702), *Staphylococcus aureus* (ATCC

25925), *Aeromonas hydrophila* (ATCC 7966) and seven species of *Escherichia coli*, as well as environmental isolates of *Vibrio vulnificus*, *Vibrio fluvialis* and *Vibrio parahaemolyticus*. The stock cultures were maintained in 20% glycerol at −80°C until use.

3.2 Disc diffusion method

The procedure of agar disc diffusion method [15] was used to determine the antibacterial activity of *M. bracteata* var. Revolution Gold essential oil. Briefly, the microorganisms were grown overnight at 37°C in 20 mL of Mueller-Hinton broth (Oxoid). The cultures were adjusted with sterile saline solution to obtain turbidity comparable to that of McFarland no. 5 standard (1.0×10^8) $CFU.mL^{-1}$. Standard (90 mm petri dishes (Merck, South Africa) containing 12 mL of sterilized Mueller-Hinton agar (Oxoid) were inoculated with the microbial suspensions. Whatman No.1 (6 mm) sterile discs paper was individually placed on the surface of the seeded agar plates and 10 μL of the oil (5 mg/mL) was applied to the filter paper disk. The plates were incubated at 37°C for 24 h and the diameter of the resulting zones of inhibition (mm) of growth was measured. All tests were performed in triplicates. Ciprofloxacin was used as positive control.

3.3 Minimum inhibitory concentrations (MIC)

The microbroth dilution method of EUCAST [16] as described by Penduka and Okoh [17] in 96 well microtiter plates was used to determine the MIC of the oil. The test organisms were standardized to match the 0.5 McFarland standard. A starting concentration of 5 mg/ml of the oil was serially diluted in double fold strength Mueller-Hinton Broth to make different test concentrations of the oil in the wells. A volume of 20 μL of the test organisms was introduced to 100 μL of the oil in broth. The plates were incubated at 37°C for 18–24 h, and the results were visually read by adding 40 μL of 0.2 mg/ml of ρ- iodonitrotetrazolium violet (INT) to all the wells. MIC was noted as the lowest concentration of the oil or antibiotic that prevented the growth of the organism after 18–24 h. Ciprofloxacin was used as positive control.

3.4 Minimum bactericidal concentrations (MBC)

The MBC was determined using the method described by Penduka et al. [18] with some minor modifications. Briefly, the oil and antibiotics were serially diluted in double fold strength Mueller-Hinton broth in 96 well microtitre plates to make different test concentrations starting with 8× MIC value of the test antibacterial agent up to its MIC value against each organism. The organisms were standardized to match the 0.5 McFarland standard and 20 μL of the organisms were inoculated into different well containing 100 μL of the antibacterial agent in broth. The plates were incubated for 18–24 h and 15 μL of the mixture from each well was sub cultured and inoculated onto fresh Mueller-Hinton agar plates. The plates were incubated for 18–24 h and MBC was taken as the minimum concentration of the antibacterial agent that barred the growth of viable colonies.

3.5 Cytosolic lactate dehydrogenase assay

The cytosolic lactate dehydrogenase assay of membrane damage was carried out according to the process of Mosman [19] as described by Soyingbe et al. [10] with some amendments. Standardized bacterial cultures similar 0.5 MacFarland standard were grown-up for 18–24 h in a concentration of 4× MIC value of the oil and the mixture was centrifuged at 5000 × g for 5 mins. About 50 μL of the supernatant was incubated with 50 μL mixed reaction solution of lactate dehydrogenase (LDH) release assay kit (Sigma Aldrich), at room temperature and incubated for 30 mins.

The absorbance of the mixture was determined at 492 nm using a 96 well microplate reader (Biotek Instrument ELx 808 UI). Cultures grown in 3% Triton X-100 were used as the positive control.

The percentage membrane damage (MD) was calculated using the formula:

$$\%MD = (\mathrm{E}-\mathrm{C})/(\mathrm{T}-\mathrm{C})\times 100. \quad (1)$$

where E = experimental absorbance of the cell cultures incubated with the test essential oil, C = control absorbance of the cell medium and T = 3% Triton X-100 treated cells supernatant.

4. Antiplatelet aggregation activity

4.1 Preparation of blood platelets

A method described previously [20] was used to prepared the blood platelets. The rats were sacrificed by a blow to the head and blood was instantly collected by cardiac puncture. The blood was mixed (5:1 $v.v^{-1}$) with an anticoagulant (acid-dextrose anticoagulant, 0.085 M Trisodium citrate, 0.065 citric acid, 2% dextrose). The platelets were gotten by runs of centrifugation at 1200 rpm for 15 min and at 2200 rpm for 3 min consecutively. The supernatant collected was centrifuged at 3200 rpm for 15 min and the resulting supernatant was discarded, and the platelets suspended in 5 ml washing buffer (pH 6.5). Re-centrifuged again at 3000 rpm for 15 min and finally re-suspended to a buffer solution of pH 7.4; containing 0.14 M NaCl, 15 mM Tris–HCl and 5 mM glucose. The platelets were further diluted with the re-suspending buffer (1:10) and the resulting solution was mixed with calcium chloride (0.4 ml: 10 μL $CaCl_2$).

4.2 Anti-platelet aggregation evaluation

Modified method [21] of anti-platelet aggregation activity was carried out to evaluate the oil action. The oil was solubilized in dimethyl sulfoxide (1:20), with 50 Mm Tris–HCl buffer to a final volume of 1% DMSO concentration. Different concentrations (1–10 $mg.mL^{-1}$) of the oil were used in the assay. The platelet aggregation inhibitory activity of the oil was separately evaluated on ADP (5 mM), collagen (5 mM), epinephrine (10 mM) and thrombin (5 $U.mL^{-1}$) induced aggregation. A 5 min. Pre-incubated platelets (100 μL) mixed with different concentrations of the oil and 20 μL of each platelets agonist was added to the mixture. Aggregation of the oil was determined using Biotek plate reader (Biotek Instrument ELx 808 UI) with Gen5 software following change in absorbance at 415 nm. DMSO (1%) was used as negative control and Aspirin was used as positive control.

4.3 Antiacetylcholinesterase (AChE) assay

Antiacetylcholinesterase activity of the essential oil was measured according to Ellman's method [22], using 96-well microplate reader. A mixture of 125 mL of 3 mM DTNB, 25 mL of 15 mM ATCI, and 50 mL of buffer, 25 mL of essential oil sample dissolved in a buffer containing than 10% methanol were added to the wells. The absorbance was measured using (Biotek Instrument ELx 808 UI with Gen5 software) at 405 nm for every 13 for 65 s. In addition, about 25 mL of 0.22 U/mL of AChE enzyme was added and the absorbance was again read at 415 nm for every 13 for 104 s. Inhibition was calculated by comparing the rates for the oil to the blank (10% MeOH in buffer).

5. Statistical analysis

Statistical analysis of the mean value obtained for experiments was calculated as mean ± standard deviation (SD) of three independent measurements using Microsoft excel program, 2016. One way analysis of variance (ANOVA) was used for the data analysis. While, P values ≤ 0.05 were regarded as significant and P values ≤ 0.01 as very significant.

6. Results and discussion

The hydrodistillation of dried leaves of *Mb*RG gave a colorless oil in a yield of 0.17% yield (w/w), calculated on dry weight basis. **Table 1** indicates the chemical compounds identified in the oil sample, their percentages and retention indices in order of their elution on DB-5 column. Thirteen components accounting for 95.3% of the total oil were identified in the oil. The oil was characterized by a high content of phenylpropanoids (82.9%) of which methyl eugenol (77.6%), a phenylpropene was the major constituent. The minor constituents of the oil were *p*-xylene (7.0%), *trans*-methyl cinnamate (4.4%) *m*-xylene (2.3%) and linalool (2.3%). The high content of methyl eugenol (77.6%) in the present sample is in agreement with the only previously reported samples of which the contents were 82.5 and 84.6% [23]. However, some compounds such as *p*-xylene, *m*-xylene and *trans*-methyl cinnamate that were identified in the present study were not detected in the previous study [23]. Additionally, methyl eugenol was described as a key volatile constituent of essential oils of some *M. bracteata* species [24–29]. There seems to be homogeneity in the main chemical compound of essential oils of *M. bracteata* species [23–29] and the present study.

The results indicate that *Mb*RG essential oil exhibited stronger antibacterial activity against *S. aureus* with ZI, MIC and MBC values of 12.3 mm, 0.63 and 2.5 mg/mL respectively. The essential oil also displayed stout action on the growth of *B. cereus* (ZI, 11.0 mm; MIC, 1.25 mg/mL, and MBC, 5 mg/mL); *E. coli* (DSM 1089), *E. coli* (DSM 10973) and *E. coli* (DSM 9025) with MIC of 1.25 mg/mL. In addition, mild antibacterial activities were observed towards *E. coli* (DSM 8695), *E. coli* (DSM 4618) and *E. coli* (ATCC 23922) with MIC of 1.25 mg/mL and MBC of 5 mg/mL, when compared with the standards. But, reduced antibacterial effects were recorded against other tested organisms such as *A. hydrophila, V. vulnificus, V.* fluvialis and *V. para-haemolyticus* with MIC and MBC values of 5 mg/mL. On the whole, the essential oil of *Mb*RG displayed antibacterial act towards *S. aurues, B. cereus* and some strains of *E. coli*. The most resistant action was recorded by *E. coli* (DSM 10974), *A. hydrophila* and the *Vibrio* species. It could therefore be postulated that the essential oil of *Mb*RG possessed reasonable antibacterial activity. This result is in agreement with the previous report on the antibacterial activities of *Mb*RG essential oil [23] and a number of species of the genus *Melaleuca* [27–30].

Compounds[a]	RI (Cal.)	RI (Lit.)	%Composition
p-Xylene	867	865	7.0
m-Xylene	871	870	2.3
α-Phellandrene	1002	1002	0.2
o-Cymene	1017	1016	0.8
1,8-Cineole	1028	1026	0.4
α-Terpinene	1015	1014	0.2
Z-Ocimene	1037	1034	0.1
Linalool	1099	1103	1.1
Terpinen-4-ol	1177	1179	0.1
α-Terpineol	1189	1191	0.2
Estragole	1197	1196	0.9
E-Methyl cinnamate	1357	1356	4.4
Methyl eugenol	1405	1407	77.6
Aromatic compounds			9.3
Monoterpene hydrocarbons			1.3
Oxygenated monoterpenes			1.8
Phenylpropanoids			82.9
Total identified			95.3

[a]*Elution order on DB-5 column; RI (Cal.) Experimental retention indices relative to* C_9-C_{24} n-*alkanes; RI (Lit.) literature retention indices.*

Table 1.
Chemical composition of Mb*RG essential oil.*

Lactate dehydrogenase (LDH) release is a suitable, dependable and non-radioactive colorimetric method of cytotoxicity to detect necrosis which is closely associated with inflammatory diseases, based on the leakage of cytosolic enzyme from the damaged cells [17]. **Table 2** displays the lactate dehyrogenase release from the bacterial cell exposed to *Mb*RG essential oil. The result shows the percentage lactate dehydrogenase release of 47, 43, 4–21, and 0.78% for *S. aureus*, *B. cereus, E. coli* species and *A. hydrophila*, respectively. When compared with the cells treated with 3% Triton X-100 (LDH of 96%) under the same experimental conditions, it could be concluded that *Mb*RG oil exhibited moderate membrane damage. However, the oil exhibited noticeable cell membrane disruption more on Gram-positive bacteria than Gram-negative bacteria. Essential oils and their constituents have been reported to exhibit antibacterial and anti-inflammatory activities, with many studies demonstrating membrane damage, removal of harmful stimuli and healing processes [31]. This finding also supports the fact that essential oils have greater activities against Gram-positive bacteria than Gram-negative bacteria [11].

The inhibitory actions of *Mb*RG essential oil against acetyl cholinesterase could be found in **Table 3**. The value of 9.16% was too low to achieve any significant inhibitory action against the enzyme, when compared with the standard drug Tacrine (96.9%). The platelet aggregation inhibitory activity of *Mb*RG essential oil against the four platelet stimulants (ADP, collagen, epinephrine and thrombin) and the standard

Microorganisms	Antibacterial					M.d
	*Mb*rgEO[a]			Ciprofloxacin		% LDH[e]
	ZI[b] (mm)	MIC[c]	MBC[d]	ZI (mm)	MIC	
B. cereus (ATCC 10702)	11.0 ± 0.0	1.25	5	29 ± 2.5	0.63	43
S. aureus (ATCC 25925)	12.3 ± 0.6	0.63	2.5	ND	NA	47
A. hydrophila (7966)	10.3 ± 0.6	5	5	ND	NA	0.78
E. coli (DSM 1089)	12.0 ± 0.0	1.25	5	24 ± 5.7	0.31	12
E. coli (DSM 8695)	10.7 ± 0.6	2.5	—	23 ± 3.5	0.31	4
E. coli (DSM 4618)	12.0 ± 1.0	2.5	5	29 ± 2.0	0.31	7
E. coli (DSM 10973)	12.0 ± 0.0	1.25	5	19 ± 1.2	0.63	ND
E. coli (DSM 10974)	11.0 ± 0.0	5	—	25 ± 3.0	0.63	ND
E. coli (DSM 9025)	11.7 ± 0.6	1.25	—	24 ± 3.5	0.31	ND
E. coli (ATCC 23922)	12.7 ± 0.6	2.5	5	25 ± 1.2	0.31	21
V. vulnificus	12.0 ± 0.0	5	5	26 ± 1.4	2.5	ND
V. fluvialis	11.0 ± 1.0	5	5	34 ± 0.7	0.16	ND
V. parahaemolyticus	11.3 ± 0.6	5	5	33 ± 1.4	0.78	ND

[a]Mb*EO -10 µg/mL.*
[b]*ZI: Inhibition zones diameter (mm) including diameter of sterile disc (6 mm), values are given as mean ± SD (3 replicates).*
[c]*MIC - minimum inhibitory concentration (mg/mL).*
[d]*MBC - minimum bactericidal concentration (mg/mL).*
[e]*%LDH releases in relation to Triton X-100.*
M.d - Membrane damage; ND - Not determined.

Table 2.
Antibacterial and membrane damaging activities of MbRG *essential oil.*

Sample		Acetyl cholinesterase[b]	Anti-platelet aggregation			
			ADP	Collagen	Epinephrine	Thrombin
*Mb*RG	1	9.16 ± 0.00	19.0 ± 3.3	37.0 ± 4.5	43.4 ± 1.9	32.1 ± 0.1
	3		44.0 ± 2.3	51.1 ± 3.1	45.0 ± 0.5	56.3 ± 2.5
	5		69.3 ± 0.7	62.0 ± 1.2	48.1 ± 3.4	73.0 ± 8.9
	10		79.0 ± 1.4	64.0 ± 1.1	59.6 ± 2.2	86.2 ± 6.0
	IC_{50}[c]		3.52	2.88	5.78	2.58
Aspirin	1	—	36.6 ± 0.4	27.0 ± 0.7	12.2 ± 0.6	46.1 ± 3.1
	3		55.2 ± 0.2	37.0 ± 3.2	37.1 ± 2.1	75.3 ± 2.0
	5		58.0 ± 0.7	59.0 ± 5.0	39.6 ± 1.6	77.6 ± 3.4
	10		61.0 ± 0.5	69.1 ± 1.1	57.6 ± 1.0	53.3 ± 1.2
	IC_{50}[c]		2.34	4.20	8.18	1.88
Tacrine[d]		96.90 ± 0.00	—	—	—	—

[a]*Values are given as mean ± SD (3 replicates).*
[b]*Percentage platelet aggregation inhibition.*
[c]*IC_{50} values (mg/mL).*
[d]*dcholinesterase inhibitor.*

Table 3.
Acetyl cholinesterase and antiplatelet aggregation activities of MbRG *essential oil.*[a]

are summarized in **Table 3** as percentage platelet aggregation inhibition and lethal concentration (IC_{50}). The results are concentration dependent. As concentrations of the oil increases, percentage platelet aggregation inhibition against the four platelet agonists was significantly. However, the lethal concentration (IC_{50}) values of the oil for the four platelet agonists showed the highest strength in the order thrombin (IC_{50}: 2.58 mg/mL) > collagen (IC_{50}: 2.88 mg/mL) > ADP (IC_{50}: 3.52 mg/mL) > epinephrine (IC_{50}: 5.78 mg/mL). It can be concluded that the essential oil of *Mb*RG showed potential anti-platelet aggregation inhibitory activity.

The observed biological activities displayed by the essential oil of *Mb*RG may be attributed to the effect of methyl eugenol. This compound was known to be an inhibitor of the enzyme acetylcholinesterase [31]. In addition, essential oils containing high contents of methyl eugenol have demonstrated cytotoxicity against cancer lines [32], antimicrobial and antioxidant [33] activities. Methyl eugenol is also most effective in terms of knockdown activity, as well as repelling and killing effects, apart from larvicidal activity against *Spodoptera litura* [31].

Acknowledgements

The authors (OAL, FOO, RAM and ARO) were grateful to Medicinal Research Council, South Africa, University of Zululand Research committee and National Research Fund, South Africa. OAL and KOA are also indebted to Lagos State University, Ojo, Lagos, Nigeria.

Authors' contribution

Author OAL designed the study and wrote the final draft of the manuscript. Authors OAL, RAM and FOO performed the experiments and analyzed the data. Authors KOA, RAM and FOO wrote part of the manuscript and manage the literature search. Authors OAL and ARO supervised the work. All authors read and approved the final manuscript.

Conflict of interest

Authors declared that there are no competing interests.

Author details

Oladipupo A. Lawal[1,2*], Kehinde O. Amisu[3], Rebamang A. Mosa[1], Foluso O. Osunsanmi[1] and Andy R. Opoku[1]

1 Department of Biochemistry and Microbiology, University of Zululand, South Africa

2 Department of Chemistry, Lagos State University, Lagos, Nigeria

3 Department of Microbiology, Lagos State University, Lagos, Nigeria

*Address all correspondence to: jumobi.lawal@lasu.edu.ng

References

[1] Craven LA, Lepschi BJ. Enumeration of the species and infraspecific taxa of *melaleuca* (*Myrtaceae*) occurring in Australia and Tasmania. Australian Systematic Botany. 1999;**12**:863-875

[2] Craven LY. Melaleuca (Myrtaceae) from Australia. Novon. 2009;**19**(4):444-453

[3] Adesanwo JK, Shode FO, Aiyelaagbe OO, Rabiu OO, Oyede RT, Oluwole FS. Antisecretory and antiulcerogenic activities of the stem bark extract of *Melaleuca bracteata* and isolation of principles. Journal of Medicinal Plant Research: Planta Medica. 2009;**3**:822-824

[4] Osunsanmi FO, Soyinbe OS, Ogunyinka IB, Mosa RA, Ikhile MI, Ngila JC, et al. Antiplatelet aggregation and cytotoxic activity of betulinic acid and its acetyl derivative from *Melaleuca bracteata*. Journal of Medicinal Plant Research: Planta Medica. 2015;**9**:647-654

[5] Penduka D, Gasa NP, Hlongwane MS, Mosa RA, Osunsanmi FO, Opoku AR. The antibacterial activities of some plant-derived triterpenes. African Journal of Traditional, Complementary, and Alternative Medicines. 2015;**12**:180-188

[6] Osunsanmi FO, Oyinloye BE, Mosa RA, Ikhile MI, Ngila JC, Shode FO, et al. Antiplatelet aggregation activity of betulinic acid, oleanolic acid, maslinic acid and derivatives from medicinal plants. Tropical Journal of Pharmaceutical Research. 2016;**15**:1613-1619

[7] Habila AJ, Habila JD, Shode FO, Opoku AR, Atawodi SE, Umar IA. Inhibitory effect of betulinic acid and 3β- acetoxybetulinic acid on rat platelet aggregation. African Journal of Pharmacy and Pharmacology. 2013;**7**:2881-2886

[8] Lawal OA, Ogunwande IA, Osunsanmi FO, Opoku AR, Oyedeji AO. *Croton gratissimus* leaf essential oil composition, antibacterial, antiplatelet aggregation and cytotoxic activities. Journal of Herbs Spices & Medicinal Plants. 2017;**23**:77-87

[9] Soyingbe OS, Myeni CB, Osunsanmi FO, Lawal OA, Opoku AR. Antimicrobial and efflux pumps inhibitory activities of *Eucalyptus grandis* essential oil against respiratory tract infectious bacteria. Journal of Medicinal Plant Research: Planta Medica. 2015;**9**:343-3484

[10] Soyingbe OS, Makhafola TJ, Mahlobo BP, Salahdeen HH, Lawal OA, Opoku AR. Antiasthma activity of *Eucalyptus grandis* essential oil and its main constituent: Vasorelaxant effect on aortic smooth muscle isolated from nomotensive rats. Journal of Experimental and Applied Animal Sciences. 2017;**2**(2):211-222

[11] Lawal OA, Ogunwande IA, Owolabi MS, Opoku AR, Oyedeji AO. Chemical composition, antibacterial activity and brine shrimp lethality test of essential oil from the leaves of *Eugenia natalitia* Sond. From South Africa. Chemistry of Natural Compounds. 2016;**52**(4):731-733

[12] British Pharmacopoeia. H.M. Stationery Office. Vol. II. London: British Pharmacopoeia; 1987. p. 109

[13] National Institute of Standards and Technology. Chemistry Web Book. Data from NIST Standard Reference,

Database. Washington DC, USA: National Institute of Standards and Technology; 2011. p. 69. Available from: http://www.nist.gov/

[14] Adams RP. Identification of Essential Oil Components by Ion Trap Mass Spectroscopy. New York. U.S.A: Academic Press; 1989

[15] Sudjana AN, D'Orazio C, Ryan V, Rasool N, Ng J, Islam N, et al. Antimicrobial activity of commercial *Olea europaea* (olive) leaf extract. International Journal of Antimicrobial Agents. 2009;**33**:461-463

[16] European Committee for Antimicrobial Susceptibility Testing. Determination of minimum inhibitory concentration (MICs) of antimicrobial agents by broth dilution. Clinical Microbiology and Infection. 2003;**9**:1-7

[17] Penduka D, Buwa L, Mayekiso B, Basson AK, Okoh AI. Identification of the anti-Listerial constituents in partially purified column chromatography fractions of *Garcinia kola* seeds and their interactions with standard antibiotics. African Journal of Traditional, Complementary, and Alternative Medicines. 2014;**2014**:1-8. DOI: 10.1155/2014/850347

[18] Penduka D, Mosa R, Simelane M, Basson A, Okoh A, Opoku A. Evaluation of the anti-*listeria* potentials of some plant-derived triterpenes. Annals of Clinical Microbiology and Antimicrobials. 2014;**13**:37. DOI: 10.1186/s12941-014-0037-1

[19] Tomita T, Umgegaki K, Hayashi E. Basic aggregation of properties of washed rat platelets: Correlation between aggregation, phospholipids degradation, malondialdehyde, and thromboxane formation. Journal of Pharmacological Methods. 1983;**10**:31-44

[20] Mekhfi H, Haouari EI, Legssyer A. Platelet anti-aggregate property of some Moroccan medicinal plants. Journal of Ethnopharmacology. 2004;**94**:317-322

[21] Marston A, Kissling J, Hostettmann K. A rapid TLC bioautographic method for the detection of acetylcholinesterase and butyrylcholinesterase inhibitors in plants. Phytochemical Analysis. 2002;**13**:51-54

[22] Oyedeji OO, Oyedeji AO, Shode FO. Compositional variations and antibacterial activities of the essential oils of three *melaleuca* species from South Africa. Journal of Essential Oil Bearing Plants. 2017;**17**:265-276

[23] Zhong CY, Huang ZQ, Liang ZY, Chen MY. Study on chemical components of essential oil from the branches and leaves of *Melaleuca bracteata*. Flavour Fragrance Cosmetics. 2009;**6**:8-10

[24] Ye ZM, Liu X, Shen MJ, Chen JY, Chen XD, Li YY. Optimization of process for extraction from *Melaleuca bracteata* essential oil with organic solvent and chemical composition identify. Chinese Journal of Tropical Crops. 2014;**2014**(35):992-998

[25] Aboutabl EA, Tohamy SFE, De Pooter HL, De Buyck LF. A comparative study of the essential oils from three *melaleuca* species growing in Egypt. Flavour and Fragrance Journal. 1991;**6**:139-141

[26] Almarie AMA, Rukunudi I. Chemical composition and herbicidal effects of *Melaleuca bracteata* F. Muell. Essential oil against some weedy species. International Journal of Scientific and Engineering Research. 2016;7:502-512

[27] Yatagai M, Ohira T, Kiyoshi N. Composition, miticidal activity and growth regulation effect on radish seeds of extracts from *melaleuca* species.

Biochemical Systematics and Ecology. 1998;**26**:713-722

[28] Siddique S, Parveen Z, Firdaus-e-Bareen MS. Chemical composition, antibacterial and antioxidant activities of essential oils from leaves of three *melaleuca* species of Pakistani flora. Arabian Journal of Chemistry. 2020;**13**:67-74

[29] Falci SPP, Teixeira MA, das Chagas PF, Martinez BB, ABAT L, Ferreira LM, et al. Antimicrobial activity of *melaleuca* sp. oil against clinical isolates of antibiotics resistant *Staphylococcus aureus*. Acta Cirúrgica Brasileira. 2015;**30**:401-406

[30] De França LKL, Martins ML, de Medeiros MMD, Iorio NLP, Fonseca-Gonçalves A, Cavalcanti YW, et al. Antibacterial activity of *Melaleuca alternifolia* (tea tree essential oil) on bacteria of the dental biofilm. Pesquisa Brasileira em Odontopediatria e Clínica Integrada. 2017;**17**:e3857

[31] Tripathi AK, Mishra S. Plant Monoterpenoids (prospective pesticides). In: Omkar A, editor. Ecofriendly Pest Management for Food Security. San Diego, USA: Elsevier; 2016. pp. 507-524. DOI: 10.1016/B978-0-12-803265-7.00011-7.ch16

[32] Noudogbessi JP, Gary-Bobo M, Adomou A, Adjalian E, Alitonou GA, Avlessi F, et al. Comparative chemical study and cytotoxic activity of *Uvariodendron angustifolium* essential oils from Benin. Natural Product Communications. 2014;**9**:261-264

[33] Joshi RK. Chemical composition, *in vitro* antimicrobial and antioxidant activities of the essential oils of *Ocimum gratissimum, O. Sanctum* and their major constituents. Indian Journal of Pharmaceutical Sciences. 2013;**75**:457-462

Chapter 16

Essential Oils from Medicinal Plants: Extraction Techniques, Biochemical Characterization, and Technical Analysis

Dhouha Alimi, Azhar Hajri, Slimen Selmi and Hichem Sebai

Abstract

Essential oils, called volatile oils or ethereal oils, are natural metabolic secretions widespread in the most varied organs of the plant. Some of these natural products are extremely potent and precise as action due to their complex molecular substances, in which mono- and sesquiterpenes constituents predominate, and contain aromatic compounds. The technological process of obtaining volatile oil intervenes decisively in its composition and its quality. In this chapter, the general overview of essential oils, their chemistry, extraction methods, and analyses are described. In detail, the chemical composition of essential oils is influenced by biotic, abiotic, and genetic factors, which are discussed in this chapter. In addition, extraction of EOs is one of the most effort-requiring and time-consuming processes. In this chapter, different methods used for the extraction are discussed. Furthermore, chemical structures in essential oil have been provided in detail. This chapter also discusses some of the developments in chromatography for essential oils analysis starting from gas chromatography to coupled techniques.

Keywords: essential oil, extraction, composition, terpenoids, biomolecules, analyses

1. Introduction

Several plant species have been recognized to possess volatile chemical compounds since antiquity, which can be extracted as "essential oils" using an appropriate lipid solvent. Despite only constitute a small portion of a plant, essential oils provide aromatic plants their distinctive qualities, which are employed in the food, fragrance, and pharmaceutical industries [1]. Today, it is evident that essential oils are mostly composed of highly complex, volatile organic compounds which are insoluble in water, mainly composed of monoterpenes and sesquiterpenes, representing one of the four main biological classes of natural compounds alongside polyphenols, alkaloids, and glycosides [2]. According to the French pharmacopoeia, essential oils or volatile oils are "products of generally quite complex composition containing the volatile principles contained in plants and more or less modified during preparation" [3].

These oils are separated from the various plant sources using different techniques. Even though it appears to be very simple to isolate such oils, the composition of the oil may vary considerably according to the method of extraction used. In 1998, AFNOR through its AFNOR NF T 75-006 standard defined an essential oil as a "Product obtained from a plant raw material, either by steam entrainment, either by mechanical processes from the citrus epicarp, or by dry distillation" [3]. Once the oils are obtained, the fundamental contribution of the organic chemistry to the industry resides in their characterization. In fact, the identification and determination of the components of the essential oil using a variety of techniques. We therefore thought that it was very appropriate to want to update the scientific knowledge regarding their indications, chemistry and toxicology thus making it possible to establish the links between these three concepts. We believe it's important to highlight to recent scientific studies regarding the utilization of essential oils. The aim of this chapter is to describe (i) the different extraction method, (ii) the chemical composition of essential oils, and (iii) the analytical techniques employed for the isolation and identification of phytoconstituents.

2. Techniques for obtaining essential oils

According to the definition provided by AFNOR, essential oils can be extracted either by cold expression (case of Hesperides) or by distillation which is available in hydro-distillation or steam entrainment. These three methods can be used in a continuous or discontinuous system, at atmospheric pressure, in overpressure or in depression [4]. These methods cannot be detailed without describing the processes taking place during the distillation and/or steam training:

2.1 The phenomena occurring during extraction

- First there is the actual extraction or hydro-diffusion step which consists of release of volatile compounds in the aqueous medium. This release is due to a phenomenon physical increase in the internal pressure of plant matter which has swollen by passive or osmotic water absorption, but also to a chemical phenomenon exerted by water [4, 5].

- Then comes the co-distillation of water and volatile elements [5, 6].

- Finally, the separation of the essential oil from the condensates involving the coalescence and settling [7].

2.1.1 Hydro-diffusion

This is under the influence of the osmotic exchanges that take place between the substrate plant and water phase, but also under the influence of physical forces [8].

Physical forces: When the plant mass is in a medium saturated with water, it follows that various hydraulic pressures build up. A plant particle submerged in water undergoes total hydraulic pressure, the components of which are as follows [9, 10]:

- Osmotic pressure

- Matrix pressure: between the particles and the water adsorbed on their surface

- Static pressure: exerted on the plant membranes, due to the conditions operative, it is zero at atmospheric pressure

- Gravity.

In addition to these different forces, there are different types of migration. Migrations of volatile compounds within the plant substrate

- Capillary diffusivity due to the porosity of the plant mass.

- Molecular diffusivity: the components of essential oils can migrate by simple molecular diffusion through plant tissues.

2.1.2 Co-distillation

At this stage, the gasoline is passed from the surface of the plant particle to the aqueous medium where it disperses. The entrainment of organic molecules during distillation is governed by two physical laws [11, 12].

- Dalton's law: the pressure of the vapor mixture is equal to the sum of the tension's vapors of the various constituents: PT = TH + TE.

- Raoul's law: the ratio of the quantities of products distilled simultaneously is function of the voltage and vapor densities at the distillation temperature chosen.

2.1.3 Coalescence and settling

$$\text{Mole H}/\text{Mole E} = \text{TH} \quad (1)$$

Isolation of volatile compounds largely depends on their solubility in water so that the distillate can be more or less rich in polar constituents, we then distinguish [13, 14]:

- part of the distilled oil is dissolved in water, this portion is of the order of 1%, rarely more than 2% and for some phenolic derivatives polar more than 5%.

- Another part is emulsified in water at the level of 10%.

- The last fraction is emulsified with water and organic molecules thirds playing the role of surfactants, it can exceed 10%.

2.2 Methods and equipment for obtaining essential oils

a. Extraction by entrainment with water vapor:

The vegetable mass is subjected to a stream of steam (without prior maceration), the vapor saturated with volatile components is condensed and then decanted.

b. Extraction by hydro-distillation:

The plant material is immersed in water, the whole is brought to the boil under pressure most often atmospheric.

c. Hydro-distillation under pressure:

It is strongly recommended for essential oils that are difficult to distill and/or with thermolabile compounds. Indeed, volatile compounds of high molecular mass like those of sandalwood, ginger and vetiver, cannot be pressure distilled atmospheric at acceptable temperature avoiding their degradation.

d. The heat pump system:

Based on the single or double effect still, the heat from the condenser is used to contribute to the formation of the vapor that will pass through the raw material. This is above all to save energy (60%) and cooling water (90%) [15].

e. Turbo-distillation:

This is an accelerated batch hydro-distillation, it is done under pressure atmospheric the only difference with conventional distillation is the presence of a turbine which shreds the plant material and agitates it. The latter increases the surface contact between the steam and the substrate and thus increases the yields energy and production. This device can be equipped with enrichment system vapors, most often it is a reflux system.

f. Microwave assisted distillation:

This is a laboratory process that has never been able to for technical reasons find its place in the industry. This is a particularly interesting technique for the gain of time it provides and by its performance.

g. Cold expression

Cold expression is reserved for the extraction of volatile compounds in the pericarps of Hesperides. This is a mechanical treatment that involves tearing the pericarps rich in secretory cells. The released gasoline is collected by a stream of water and receives all the usual product of entrainment with water vapor, hence the name of essential oil (AFNOR).

3. Biochemistry of essential oils

In the world of essential oils, terpenoids are by far the group of products most important natural, next to phenylpropanoids (C6-C3), C6-C1. The term terpene referred to the whole group in ancient literature but today it is limited to the designation of monoterpenoid hydrocarbons [16, 17]. Terpenoids are defined as substances composed of units of isoprene (2-methylbutadiene). This is not often present in essential oils and is not a synthetic intermediate either. But the 2-methylbutadiene backbone is easily recognizable in the structure of terpenoids. We distinguish between monoterpenes, and sesquiterpenes.

3.1 Terpenoids

3.1.1 Monoterpenoids

Geranyl pyrophosphate is the precursor of monoterpenoids it is formed by two five-carbon unit. Heterolysis of the bond between a pyrophosphate oxygen and distal carbon gives a cation (carbocation) of geranyl, the one that allows different pathways of monoterpenes biosynthesis. During the various biosynthesis described in the literature, other cations according to reactions under enzymatic control, which allows each plant according to its genetic material to produce specific terpenoids in kind or in quantity. Due to the high reactivity of the different cations observed and the diversity enzymatic, monoterpenes whether acyclic, monocyclic, or bicyclic exhibit a multiplicity of functionalization (**Figure 1**). We can then differentiate:

Aldehydes: Often acyclic such as geranial and citronellal

- Alcohols: Acyclic such as geraniol, linalool, and citronellol; monocyclic like menthol and α-terpineol, bicyclic like borneol and the fenchol
- Acyclic ketones like tagetone, monocyclic like menthone and carvone, bicyclics like fenchone, camphor, and thujone
- Esters especially of lynalyl acetates, citronellyl, menthyl, etc.
- Ethers such as eucalyptol (1.8-cineole)
- Peroxides such as ascaridole
- Phenols like thymol and carvacrol.

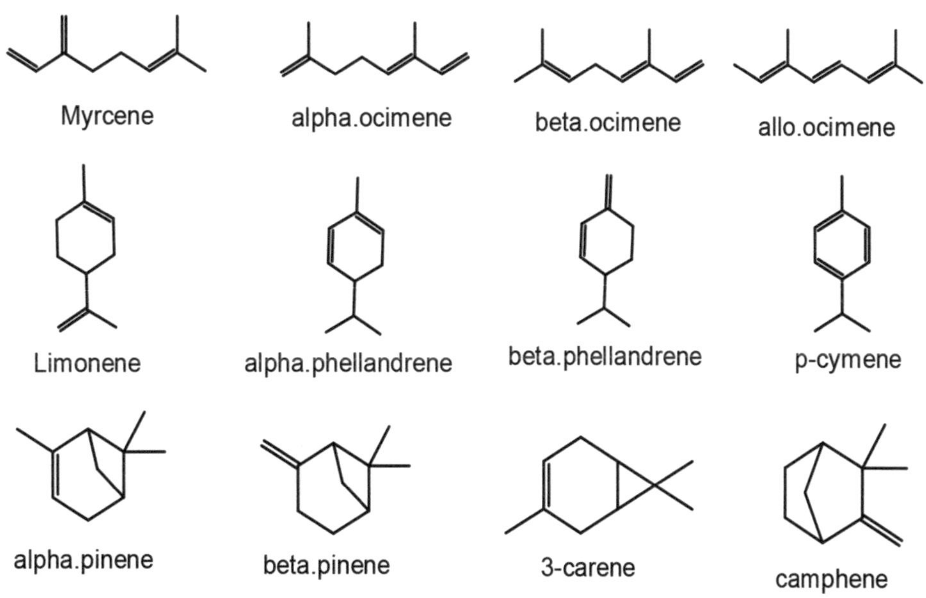

Figure 1.
Structures of some terpenoids.

Many of these products can be artifacts formed from dehydration of alcohol. Their presence in essential oil could well be due to the process extraction. Thus, the para-cymene being among the most stable can be an artifact obtained by various reactions (cyclization and/or isomerization and/or oxidation) from a number important of products.

From this brief description of the nature of monoterpenoids, we can cite the most common plants whose essential oils contain these products. Myrcene is a very common compound in hops, among others, and in some spices. Ocimene and alloocimene and their isomers are present in almost all essential oils, the most common isomer of β-ocimene is limonene.

Citronellol is a dihydrogeraniol quite common in nature in various forms enantiomeric rose, geranium, and lemongrass have the richest rates. The characteristic smell of roses is due to a mixture of geraniol, nerol, citronellol, and 2-phenylethanol. It should be noted that in essential oils there are also esters of these alcohols (**Figures 2** and **3**).

3.1.2 Sesquiterpenoids

Sesquiterpenoids contain 15 carbon atoms, which gives a high boiling point and therefore lower volatility. So, they will be a little less numerous in an essential oil and they will only rarely be responsible of its smell. In the same way as the geraniol, precursor of all the monoterpenoids, the farnesol is that of all sesquiterpenoids. Condensation between a pyrophosphate isopentenyl and geranyl pyrophosphate leads to farnesyl pyrophosphate (**Figure 4**).

Dehydration of farnesol gives a farnesyl cation which plays a role in formation of sesquiterpenes, similar to the role of the cation of geranyl for the synthesis of

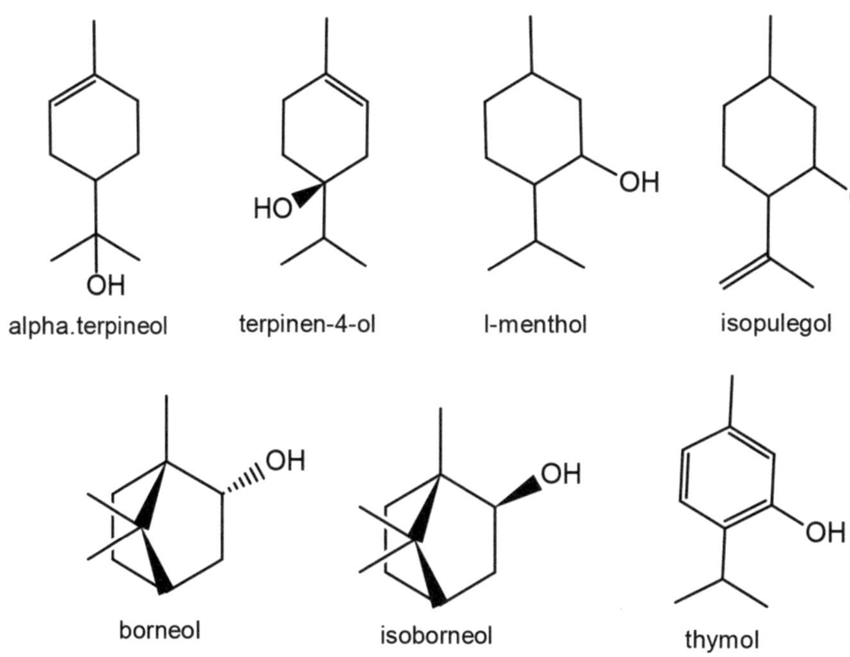

Figure 2.
Examples of cyclic alcohol-functional monoterpenoids.

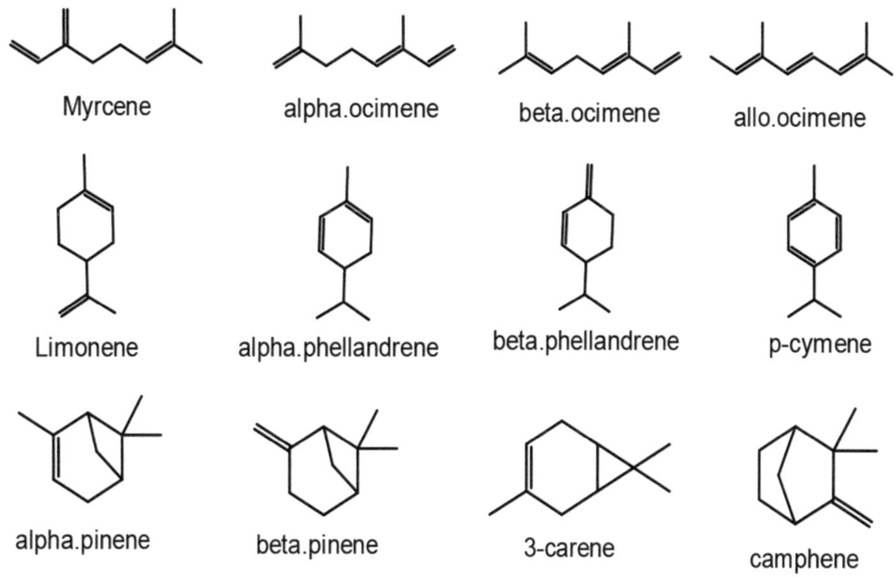

Figure 3.
Commonly encountered monoterpenoids.

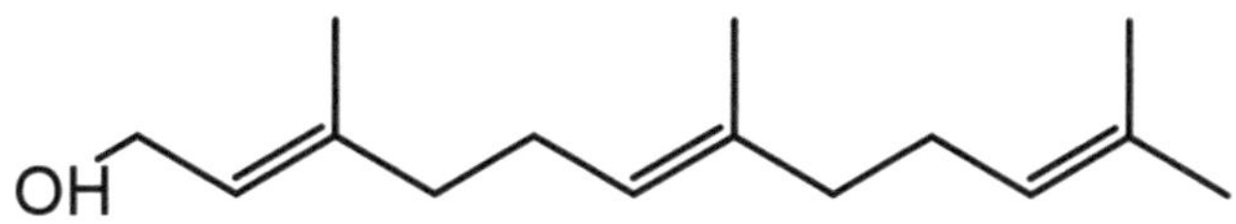

Figure 4.
Structure of farnesol.

monoterpenoids. A synthesis from farnesyl pyrophosphate allows a number greater potential of cyclic structures compared to a synthesis from the geranyl pyrophosphate. This is explained by the presence of three double bonds on the farnesyl molecule as well as by a greater variation in structure because there is a number significant rearrangements, oxidation of eliminations possible [18].

3.2 Shikimic acid derivatives: (phenylpropanoides, C6-C1, C6-C3)

Shikimic acid is a synthetic intermediate for plants, a precursor of lignin flavonoids [16]. Lignin is a structural constituent of plants, main component of wood. Flavonoids import to plants as agents' antioxidants and protection against ultraviolet radiation, they also give their colors to plants. With regard to the products found in essential oils, the key metabolite from shikimic acid is chorismic acid which can borrow several biosynthetic pathways. But the path that interests us the most is that of pre-phenic acid obtained by Claisen-type peri-cyclic rearrangement of the acid chorismic. This is the pathway which leads via phenylpyruvate to alanine and tyrosine to using amino transferase among others. This rearrangement is under the influence of chorismate-mutase. Phenylalanine allows us after reduction and elimination of nitrogen by phenyl-ammonia-lyase to go to cinnamic acid which, by hydroxylation of the nucleus aromatic, gives us o-coumaric acid and p-coumaric acid. From these. The latter are caffeic acid, then ferulic acid and then methylene caffeic acid (**Figure 5**).

Thus, in an essential oil we can find:

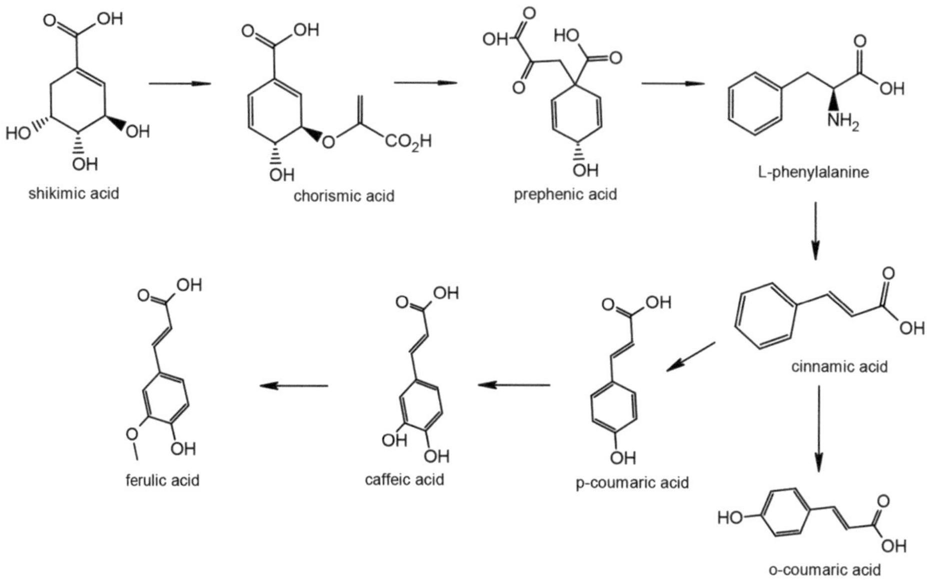

Figure 5.
Shikimic acid derivatives.

3.2.1 Benzoic acid derivatives

These are mainly C6-C1 phenol acids obtained by hydroxylation of the acid benzoic which itself comes from the aromatization of shikimic acid but without the addition of the three carbons of phosphoenol-pyruvate [19]. These derivatives also exist in free form than in combination with the ester or heterosides. In general, there are two categories: the pure benzyl acid-alcohols and the aldehydes corresponding to them and obtained by oxidative cleavage of the side chain (**Figure 6**).

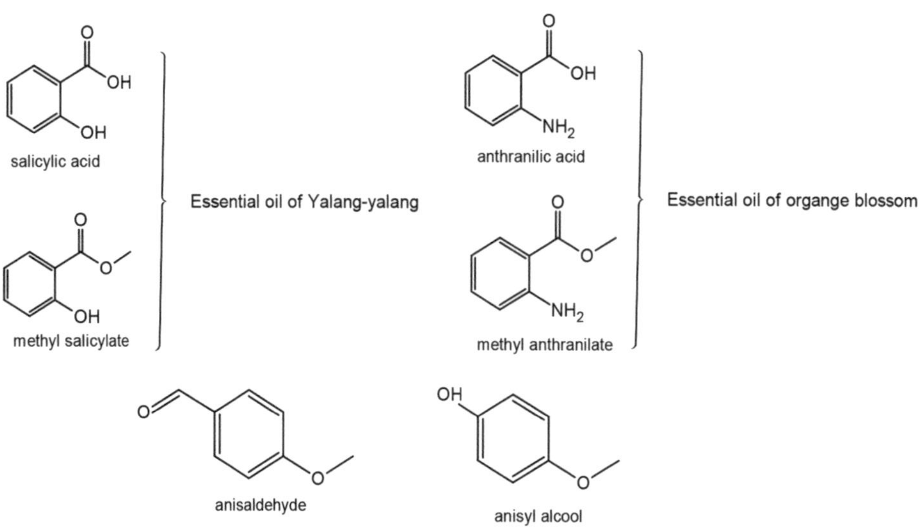

Figure 6.
Benzoic acid derivatives.

3.2.2 Ferulic acid derivatives

Ferulic acid itself is a derivative of benzoic acid, reducing its chain lateral leads to a very important family of essential oil components [20]. The key components are eugenol found in the essential oils of camphor tree, cinnamon, jasmine, basil among others, and isoeugenol present in essences of cassia, cloves, nutmeg. Eugenol methyl ether (methyl eugenol) is very common in nature, but however, raises questions about the toxicological safety of the use of oils essential oil comprising this compound, we can cite for example the essential oil of *Melaleuca alternifolia* which contains 98% methyleugenol (**Figure** 7).

3.2.3 Cinnamic acid derivatives

These are C6-C3 compounds most often designated by the term "Phenylpropanes" are the most numerous metabolites of shikimic acid and are universally distributed free or combined (esters, amides, glucosides). The pattern C6-C3 can polymerize to give lignin or cyclize to give coumarins or further lengthen its side chain to end up with flavonoids.

Cinnamic acid is obtained from chorismate via a phenylalanine which undergoes for this a stereospecific elimination of ammonia this reaction requires a phenyl ammonia-lyase. Most often the cinnamic acids encountered are esters or aldehydes. Cinnamic acid is present without modification in the essential oils of Cassia and Styrax. The corresponding aldehyde is in the essences of camphor, Cinnamyl alcohol and its esters in Daffodils and Lilacs. Lactonization of o-coumaric acid gives the nucleus of coumarins such as bergapten, oxygenated derivatives of the essential oils of Bergamot and Petitgrain. Still from cinnamic acid one can obtain estragol (methylchavicol) and by methylation of the phenol ring and reduction of the carboxylic acid function to alcohol function and elimination of the latter one obtains anethol. Estragol is naturally present in sage, rosemary, basil, and honeysuckle. Dillol is present in Apiaceae (fennel, anise, coriander) in lavender and Ylang-Ylang [20] (**Figure 8**).

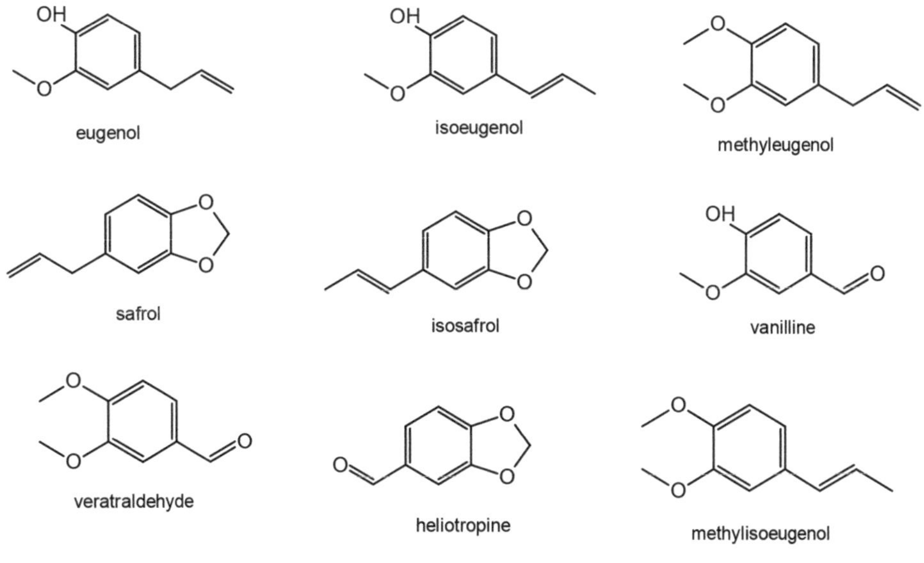

Figure 7.
Derivatives of ferulic acid.

Figure 8.
Structures of anethole and estragole.

4. Oil analysis techniques essential

For the analysis of essential oils, or at least for their chemical analysis, of many techniques have emerged during the second half of the previous century. We can then distinguish the techniques of separation of chemical components from the techniques of detection by proposing the following classification:

- Chromatographic separation techniques: CCM, HPLC, GIC, SFC
- Analysis techniques without separation/fragmentation: UV, IR, MS
- Coupling techniques: CPG-MS, CPG-UV, HPLC-CPG, HPLC-MS, CPG-IRTF, SFC-CPG

4.1 Chromatographic separation techniques

4.1.1 TLC or thin layer chromatography

This is the first and most widely used chromatographic technique, it provides simple information about the physicochemical characteristics of the components of a mixed. Many pharmacopoeias advocate the use of this technique given its simplicity for the characterization of essential oils in routine testing.

The foundations of TLC applied to essential oils were established by Stahl in 1969 [21, 22] and by Geiss in 1987 [20] who studied a significant number of metabolites secondary aromatic plants. Then in 2003, Shema and Fried [21] published the "CCM Handbook." Other approaches to CCM gave rise to the high CCM performance followed by overpressure TLC and phase rotation chromatography (RPC) which are forced flow techniques.

However, this technique should be indicated for the rapid determination of the different pathways and/or chemical families present in a given essential oil.

4.1.2 Gas chromatography

This is an analytical chemistry technique that separates compounds volatile or volatilizable without degradation (non-thermolabile). His power of separation exceeds that of all other techniques, at least for essential oils.

4.1.2.1 Principle and apparatus

This chromatographic method makes it possible to separate the compounds either by partition or by adsorption. This is a differential migration of the constituents of the mixture to be analyzed at through a chosen substrate. The best illustration of the evolution of CPG (**Figure 9**) applied to essential oils is found in the many works carried out for the determination of oil components essential of rue Ruta graveolens. Thus in 1961, only eight components were identified by Bruno during the first CPG of this oil; then in 1964, this number rose to 20 with the use of a Perkin-Elmer type chromatograph equipped with a thermal conductivity. Then with the introduction of programming systems temperature we went to 80 components and using a capillary column at high resolution with a flame ionization detector 150 components are obtained (1981) [23, 24].

4.1.2.2 The fast and ultra-fast GC

It was necessary, as part of a routine examination, to develop these two variants of the CPG. So, to significantly accelerate a CPG we act at the following levels.

- The dimensions of the column are reduced: the internal diameter and the length.
- The coatings are reduced: a thinner stationary phase.
- The flow of carrier gas is increased.
- Temperature transitions are accelerated during cycles.

Thus, the separation speed is considerably increased. This CPG technique rapid or ultrafast has been tested on lime essential oil with similar parameters to what was mentioned previously, that is to say a capillary column 5 m long, of 50 μm internal diameter, 0.05 μm coating, and a gas flow rate of 120 cm/min [25]. We arrive at a chromatography carried out in 1990s, i.e., 33 times faster than a traditional GIC, but generally a fast GIC lasts 13 min and instead of 60 for a classic GIC.

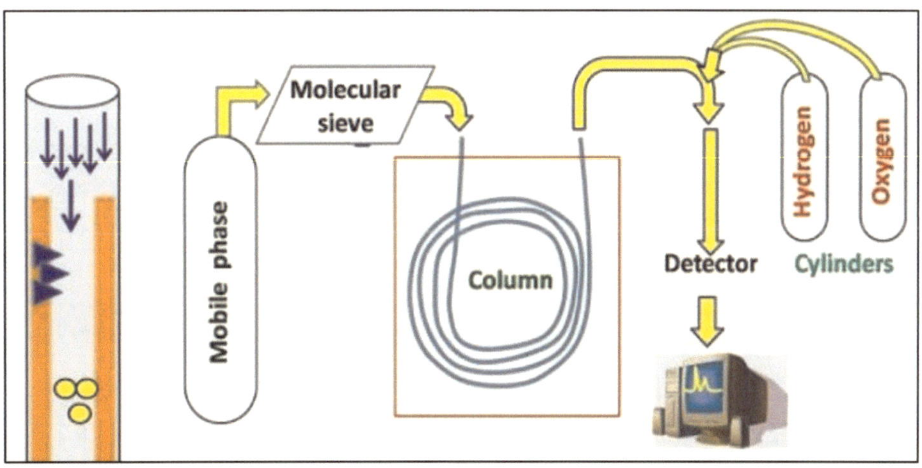

Figure 9.
Apparatus for a GC.

We can try to summarize the differences between the different types of chromatography in the gas phase by the following **Table 1**.

4.1.2.3 Chiral GC

This is an interesting evolution of CPG and consists of an enantioselectivity of the capillary column. The interest is even greater than essential oils are very often rich in mixture, racemic or not (depending on the botanical species and the chemotype), two enantiomers. This chiral chromatography then makes it possible to separate into using various stationary phases such as diamide phases which interact with chemical compounds through hydrogen bonds or as complex phases metal with low thermal stability. But most often we use derivative phases very selective cyclodextrin and used since their invention in particular in the determination of the enantiomeric composition of monoterpenoids and sesquiterpenoids of many essential oils [26].

4.1.2.4 The two-dimensional CPG CPGx CPG

This is a gas phase separation technique of course, in which all the compounds eluted from a first column are, directly after the latter, subjected to separation in a second column of different selectivity (Ecole supérieure de physics and industrial chemistry of Paris). The two columns are connected in series to the by means of a modulator which samples the effluent from the first column and transfers it, with or without concentrating it towards the second column. So the latter must be able to separate the different constituents in a time shorter than the duration of the modulation, and this is why a second column of the fast or ultra-fast CPG type is used.

The elution peak from column 1 is a first detector and each fraction is focused and continuously injected into column 2. In general, the detector1 and valve assembly constitutes a modulator. By joining the chromatograms of each column, one obtains a two-dimensional retention plane. This technique can be complicated by putting a first non-chiral column then a second chiral [27]. In the world of essential oils, this method has proven particularly effective in the study of the EO of Bergamot citrus bergamia in which only the L (−) enantiomers of linalool and linalyl acetate. So, this type of chromatography allows to update the adulterations of this essential oil.

		GIC conventional	GIC fast	GIC ultra-fast
Dimensions of the column	L	30 m	10 m	10–15 m
	Θ	0.25 mm	0.1 mm	0.1 mm
	Coating	0.25 μm	0.1 μm	0.1 μm
Temperature programming		50–350°C 3°C/min	50–350°C 14°C/min	45–325°C 45–200°C/min
Carrier gas Flow (flow)		H2 36 cm/s	H2 57 cm/s	H2 120 cm/s
Injection frequency (sampling or analysis)		10 HZ	20–50 HZ	50–250 HZ

Table 1.
The different types of gas chromatography.

4.1.3 Liquid chromatography

Due to the preponderant place of CPG in the analysis of essential oils, the liquid chromatography is most often only used for preparatory steps or semi-preparatory. Or at the limit for the individual isolation of a compound for its structural study.

4.1.3.1 HPLC or high-performance liquid chromatography

4.1.3.1.1 Principle and apparatus

The sample to be analyzed is pushed by a mobile phase into a packed column a stationary phase of fine particle size. The flow rate of the mobile phase is high which leads to an increase in pressure in the system. This high flow decreases the time required to separate the components along the stationary phase. The thin particle size of the stationary phase allows better separation of the components. Indeed, for the same volume of stationary phase, the exchange surface increases if the "grains" that compose it are of smaller diameter. The peaks obtained are therefore narrower the resolution is improved (the peaks are well separated, we can therefore differentiate them), the detection threshold is also lower (narrow and high peaks are easier to isolate from the background noise as wide and low peaks). The combination of these attributes speed and high-resolution leads to the term "high performance (**Figure 10**)."

The solvents used are miscible combinations of water and various liquids organic (alcohols, acetonitrile, dichloromethane, etc.).

Often the composition of the mobile phase is changed during analysis, this is the so-called "gradient" or "graduated elution" mode (in opposition to the "isocratic" mode, for which the composition of the mobile phase remains the same throughout the analysis). Through example, on an apolar column, using a water/methanol mixture as phase mobile, the more hydrophobic components are eluted with a high concentration of methanol, while the more hydrophilic components are preferentially eluted with a low methanol concentration. Depending on the nature of the stationary phase, we will start with a high methanol concentration or vice versa [28].

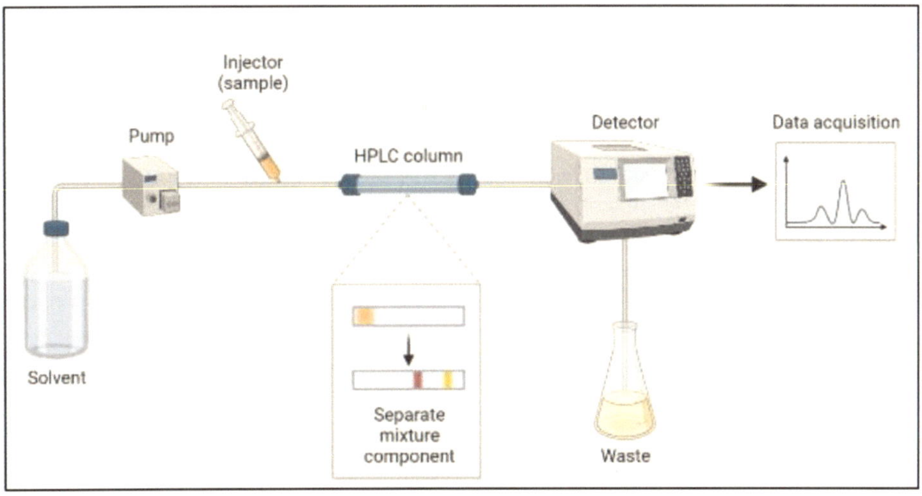

Figure 10.
Assembly for an HPLC.

4.1.3.1.2 HPLC and essential oils

The use of HPLC in the field of essential oils has been abandoned in favor of CPG with convincing results. However, HPLC analysis has a number of advantages, especially when looking for thermolabile compounds that is to say difficult to be analyzed by CPG. The main limitation of this technique lies in the methodology analysis of terpenoids that require a retention factor in an interval narrow. Hence the most common use of silica gel column, n-pentane as phase mobile, low temperature (−15°C in general) and a UV detection system at 220 nm. Under these conditions, the majority of mono and sesquiterpenoid hydrocarbons can be separated and analyzed [28].

Other chromatographic conditions were tested for the analysis of HE in acting on:

- The nature of the stationary phase (theory of complexation chromatography film)
- The temperature of the chromatography
- Setting up an acetonitrile-water gradient in the column

But despite everything, the separating power of HPLC is only remarkable for sesquiterpenoids and diterpenoids, with CPG doing better for monoterpenoids. He is to note however that researchers have succeeded in an enantiomeric separation by HPLC sesquiterpenes from an essential oil using a chiral stationary phase (chiralcel®) [28].

4.1.4 Supercritical fluid chromatography

The CFS uses as mobile phase a fluid or mixture of fluid brought to a point called critical by pressure and temperature control, i.e., at the point where the substance chemical, here the mobile phase, is in a hybrid state between a gas and a liquid. The diffusion coefficient of the fluid is then twice as important while the viscosity is two times less than the corresponding liquid, with a greater density than gas obviously. Most often carbon dioxide (CO_2) is used as the mobile phase, but since its polarity is low (comparable to hexane), it is incorporated in a small quantity polar solvent (methanol, ethanol, water) to be able to elute compounds endowed with a certain polarity [29].

This method also makes it possible to use a range of detection systems that are wider than that of HPLC. Apart from the use of this type of chromatography as a technique analysis, it is interesting to note that it can also be used for the extraction of oils essential. Indeed, several industries use it, most often when it is an EO with thermolabile compounds. We can then collect the majority of the components odorous, polar or non-polar, while avoiding the disadvantages of hydro-distillation or hydro-diffusion. Drawbacks such as hydrolysis reactions, solubilization of a non-negligible part of volatile products in water or the thermodegradation of these. There is no thermodegradation because, for the example of CO_2, the supercritical state is at 31.1°C and 74 bars. The use of this technique for the analysis of essential oils gives rise to a particular enthusiasm among researchers, especially over the past 10 years. We can then cite the establishment of a study protocol for the essential oil of Salvia angustifolia by CFSC [30].

4.1.5 Counter-current chromatography

Also called centrifugal partition chromatography, it consists of a liquid-liquid chromatography using no solid support but two non-liquid miscible (made from two or more solvents). The two phases (mobile and stationary) are liquids which prevents irreversible adsorption phenomena of the mobile phase.

This technique has two variants: 1. HSCCC: high speed counter current chromatography 2. DCCC: drop counter current chromatography.

4.1.5.1 HSCCC

Also called—rotation locular counter current chromatography|| and developed by Rikakikai, she uses a device made up of sixteen glass tubes, communicating with each other and arranged concentrically. Inside these tubes is the stationary phase which, by rotation and centrifugal force remains fixed in the tubes. It is crossed by a phase mobile under pressure (**Figure 11**).

This is a technique successfully used for the analysis of natural products such as EO [31, 32].

4.1.5.2 Countercurrent droplet chromatography

Moderately effective, it allows the separation of essential oils into fractions and rarely in pure compounds. Developed by Tanimura in 1970, it consists of a set of 300–600 glass tubes which are connected to each other with tubular connectors in Teflon and filled with the stationary phase (liquid). This is crossed by droplets mobile phase. However, since each compound of the essential oil has a partition coefficient which is specific to it (for a given couple of solvents), there is separation as the passages in the different tubes. It should be noted that for the fragmentation of oils essential you need a water-free solvent system [33] (**Figure 12**).

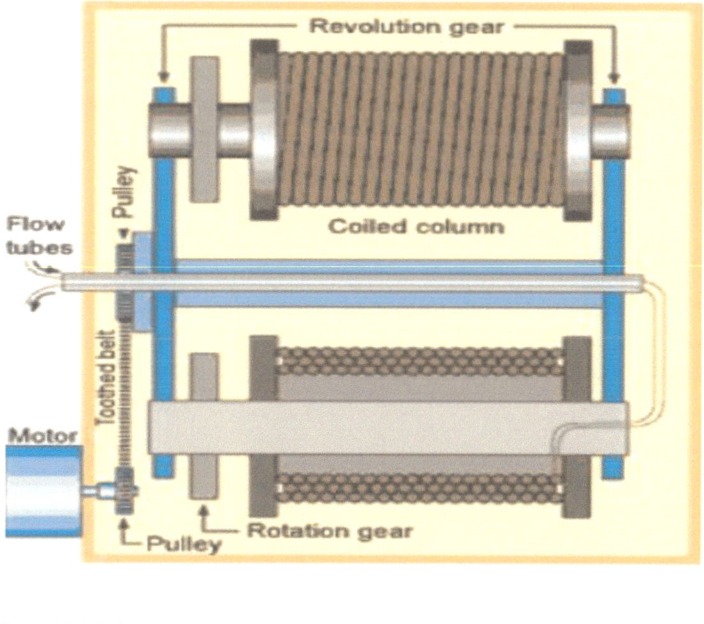

Figure 11.
Composition of an HSCCC device.

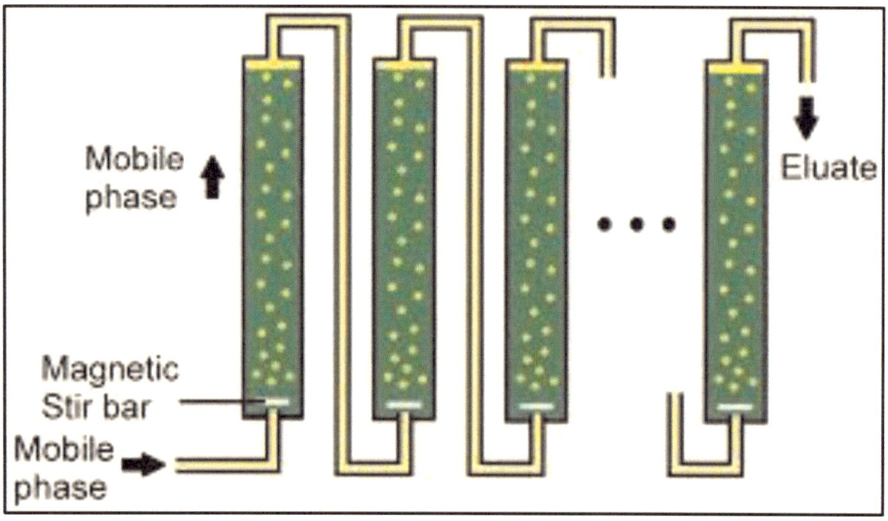

Figure 12.
Migration of the mobile phase of a DCCC device.

4.2 Analysis techniques without fragmentation

4.2.1 UV spectroscopy

4.2.1.1 Principle and apparatus

Ultraviolet-visible spectroscopy or ultraviolet-visible spectrometry is a spectroscopy technique involving photons whose wavelengths are in the range of ultraviolet (200–400 nm), visible, and up to near infrared (750–1400 nm). Subjected to radiation in this range of wavelengths, the molecules undergo an electronic transition. This technique is complementary to spectroscopy fluorescence in the sense that fluorescence involves transitions from the excited state down to the ground state while absorption spectroscopy deals with transitions between ground state and excited state.

4.2.1.2 Application to essential oils

The use of this technique for the analysis of essential oils is quite limited since it is not possible to obtain information on a chemical compound individually. But nevertheless, for the research of furanocoumarins, responsible for photo dermatoses in hesperidia essential oils, this is a method to privilege. It is recommended by the European Pharmacopoeia for the analysis of oil essential lime.

4.2.2 Infrared spectroscopy

4.2.2.1 Principle and apparatus

Infrared spectroscopy exploits the fact that molecules have frequencies specific for which they rotate or vibrate in correspondence with levels discrete energy (vibratory modes). These resonant frequencies are determined by the forms surfaces of potential energy, atomic masses and by vibronic coupling associate. For a vibrational mode in a molecule to be active in the infrared, it must be associated with changes in the permanent dipole. In particular, in the approximations of Born-Oppenheimer and harmonic, when the molecular Hamiltonian corresponding to the electronic ground state can be approximated by a harmonic oscillator at the neighborhood of the equilibrium molecular geometry, the resonance frequencies are determined by the normal modes corresponding to the potential energy surface of the state fundamental molecular electronics. However, the resonant frequencies can be in a first approach related to the strength of the bond, and to the atomic masses of termination. So, the frequency of the vibrations can be associated with a particular bond. Diatomic molecules have only one bond, which can be stretched. Molecules the more complex ones have a lot of bonds, and the vibrations can be conjugated, which leads to infrared absorptions at characteristic frequencies which can be related to chemical groups. For example, the atoms of a CH2 group, which we find commonly found in organic compounds can vibrate in six different ways: symmetrical and antisymmetric stretching, scissoring, rocking (rocking), wagging, and twisting [34].

4.2.2.2 Application to essential oils

Despite the low sensitivity and selectivity of this method in the case of mixtures with many components, in 1971 two hundred essential oils were analyzed for the publication of "Infrared analysis of essential oils" which is the reference of the genre in this domain [35]. On its own, conventional infrared spectroscopy does not allow measurements quantitative. For classic IR, new techniques such as spectroscopy attenuated reflection infrared or NIR-FT Raman spectroscopy, open a new method of analysis of essential oils since it is possible to identify the components of the oils essential using spectrographic references of pure chemical compounds. The definite advantage of these techniques lies in the ease of quality control of EO with the bonus of being able to quantify and analyze the components of an EO In-Situ, that is to say on living plant material (without prior isolation).

4.2.3 Mass spectrometry

4.2.3.1 Principle and apparatus

Mass spectrometry (MS) is a physical technique analysis to detect and identify molecules of interest by measuring their mass and characterize their chemical structure. Its principle lies in the gas phase separation of charged molecules (ions) in function of their mass/charge ratio (m/z).The mass spectrometer is often coupled with a chromatography system in gas phase, and this association, of a separating method and of an identification, allows the study of complex mixtures in trace amounts (a few nanograms of mixture).

The principle of mass spectrometry is as follows:

An organic compound introduced into the mass spectrometer is ionized by electronic bombardment at 70 eV. The ion thus obtained, called the molecular ion, allows the determination of the molar mass of the compound. There may be breaks in the chemical bonds within the molecular ion, forming thus characteristic fragment ions since this possible dissociation does not take place at the chance but according to well-defined mechanisms.

These fragment ions are then separated according to their mass/charge ratio by the application of a magnetic and/or electric field, then collected by a detector. All these fragment ions constitute the mass spectrum, the reading of which allows identification of the molecular structure [36].

4.2.3.2 Application to essential oils

This is the flagship technique for the determination of the molecular structures of isolated compounds. An essential oil mass spectrum always shows ions molecules corresponding to terpenoids with an m/z ratio of 136, 148, 152, and 154. And by focusing techniques without prior separation, the metastable ions are observed compounds such as anethol, fenchone, borneol, and cineol. And a variation is by the direct introduction of a part of the plant with EO (0.1–0.2 mg) which releases by heating the classically detected volatile compounds [37].

4.2.4 ^{13}C NMR spectroscopy

4.2.4.1 Principle and apparatus

NMR spectroscopy is based on the detection of the resonance phenomenon magnetic which occurs when atomic nuclei of non-zero spin are placed in a generally uniform external magnetic field and that they are excited by radio-frequency radiation tuned to the energy differences between different possible states of nuclear spin.

The resonant frequency $\nu 0$ is a first approximation directly proportional to the applied field B0: $\nu 0 = \gamma B0$, where $\gamma = 2\pi\gamma$ is the gyromagnetic ratio.

The fact that each isotope has a unique gyromagnetic ratio allows the NMR technique to be able to be tuned to a element. Just adjust the frequency of excitation and observation on the target nucleus.

The resonance frequency of nuclei also depends on their environment, the spins being in interaction with it. These interactions are called internal interactions by opposition to the external interactions of the spins with the external magnetic field and the radiofrequency radiation. These intra- or intermolecular interactions can be magnetic as is the case for chemical shift and dipole couplings, still or electric, which is the case with the dipole interaction. Interpretation and measurement of these interactions provide valuable information on:

- the nature and number of atoms close to the nuclei studied
- chemical bond
- molecular conformation
- interatomic distances
- molecular mobility

4.2.4.2 NMR and direct analysis of essential oils

In general, it allows the determination of the molecular structures of isolated compounds; but nevertheless, for the study of EOs and complex mixtures it presents a

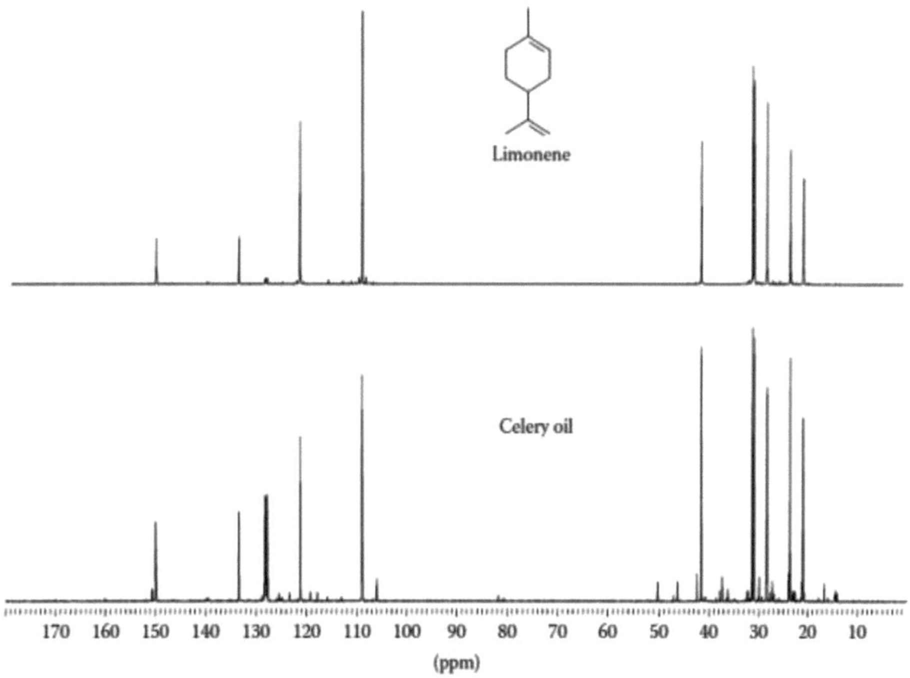

Figure 13.
Similarity between the ^{13}C NMR spectra of celery and limonene.

certain number of advantages in the presence of low volatile or unstable compounds thermodynamically. It has been established since the 1980s that it is possible to determine the constituents of an essential oil by comparing with spectra of pure products [23] (**Figure 13**).

Celery, or rather its essential oil, is extremely rich in limonene, hence a flagrant concordance between the two spectra. With an adequate computer system and a good database (bank of spectra), it is nowadays very easy to determine the composition of an EO by this NMR [38].

4.3 Analysis techniques with coupling

The advantage of a chain coupling of a chromatographic interface with a spectrometer is the ability to analyze the individual spectrum of a compound.

4.3.1 GC and mass spectrometry

This is the most widely used technique for EO analysis due in large part to the ease of handling efficient separation and detection systems, with a relatively low cost [39]. The first CPG-mass analysis dates from 1963.

For the analysis of essential oils, the most common equipment consists of a CPG capillary with an electron ionization quadrupole mass spectrometer. It is very easy to find mass spectrum databases due to very frequent use and worldwide of this technique.

For example, NIST/EPA/NIH 2005. WILEY REGISTERY 2006 and MASS FINDER 2007.

For the ionization of the compounds of essential oils, we can use two processes, electronic impact or chemical ionization which struggles to ionize alcohols and terpenic esters.

4.3.2 CPG and Fournier transform infrared

This is a complementary method to CPG-mass insofar as by TF infrared spectrometry distinguishes isomers of compounds eluted by GC not observable with mass spectrometry [40].

4.3.3 GPC and ultraviolet spectroscopy

The serial connection of the two devices, allows in the case of the use of a diode array detector, to detect and sometimes identify a certain number terpenoid hydrocarbon [41].

4.3.4 GPC and atomic emission spectrometry

This is only a complementary technique to CPG-Masse and CPG-FTIR.

4.3.5 GPC and isotope ratio mass spectrometry

A very interesting technique in the analysis of essential oils. It consists of the determination of the isotopic composition of the compounds eluted by carrying out combustion of these. We can then calculate the $^{13}C/^{12}C$ ratios: $^{18}O/^{16}O$ and $^{2}H/^{3}H$. This is the most sophisticated technique for judging the authenticity of oils essential.

4.3.6 HPLC-CPG

It is very simply a matter of putting an HPLC column and a CPG clone end to end. We can then have a modulation system to be able to choose the eluted fractions by HPLC and which it is desired to undergo CPG. The main thing is to choose a phase mobile volatilizable HPLC. A good separation of the esters, alcohols, and carbonyls of the compounds is then obtained from EO [42] (**Figure 14**).

4.3.7 HPLC-Mass and HPLC-NMR

The use of these two techniques is not widespread due to the relative HPLC ineffectiveness with respect to essential oils (see paragraph HPLC).

4.3.8 Extraction in supercritical fluid coupled to the CPG

It should be noted here that the supercritical fluid extraction is not a chromatographic technique since it involves extracting from the plant, at a critical point pressure and temperature specific to each chemical compound, which is then directly introduced into the chromatographic column, and which can therefore be identified. This is a fairly common technique in the world of essential oils since a many researchers use it for the analysis of herbal drugs such as Rosemary [41], *Thymus vulgaris* [42] orange, or cedarwood.

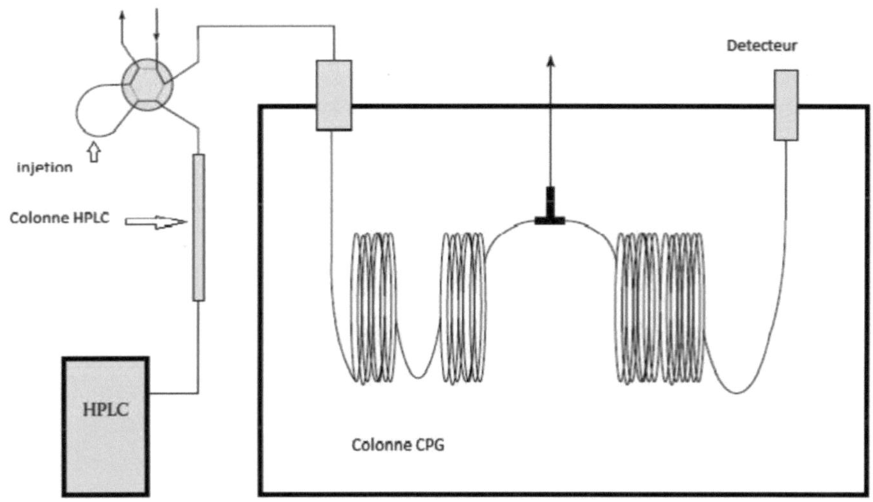

Figure 14.
HPLC-CPG coupling.

4.3.9 Super critical fluid chromatography (SFC) coupled with CPG

To be distinguished from the previous technique by the fact that the substance injected is oil essential and not the plant drug. In general, this is a common technique, used among other things for the analysis of the EO of sweet orange. Three hydrocarbon fractions are then obtained: aldehydes, alcohols, and esters [42].

4.3.10 SFC-Masse and SFC-FTIR

These are techniques based on the separating power of chromatography in supercritical fluid and on the complementarity between infrared spectrometry and mass spectrometry. They were applied during the characterization studies of Hops EO [43].

5. Conclusions

The chromatographic and the spectroscopic techniques fully changed the chemical analysis of the essential oils. The chemical composition of the essential oils was studied with the help of IR-spectroscopy, UV-Vis spectroscopy, gas chromatography, NMR spectroscopy. The enhanced demand for the essential oil in various fields of life provoked us to access the reliable methods for the essential oil analysis, and the techniques used are the GC-MS and GC analyses.

The characterization of the essential oil was carried out by using the gas chromatography. The compounds that are present in the essential oil was confirmed by using the GC and GC-MS analysis. The storage and handling of the essential oil also affect its yield and quality, ad essential oil was deposited in the oil glands that are present in the organization of the plant material. Essential oils are the natural volatile compounds having loveable odor. The essential oils are isolated mostly from the hydro-distillation method which is more suitable for this process and easy to carry. Whole parts of the plants are used for the extraction of plants. Steam distillation method is expensive than the hydro-distillation, so it is less preferred. Essential oils have good medicinal applications and used in the treatment of different diseases including the infectious diseases, depression, anxiety, act as the antifungal, antimicrobial, anticancer, and wound healing; they are also used in cosmetics and perfume industries. Researchers and industry professionals would surely benefit from this study's information as they choose the best extraction techniques for obtaining the highest yield and quality attributes.

Author details

Dhouha Alimi*, Azhar Hajri, Slimen Selmi and Hichem Sebai
Laboratory of Functional Physiology and Valorization of Bioresources (UR17ES27), Higher Institute of Biotechnology of Beja, University of Jendouba, Beja, Tunisia

*Address all correspondence to: dhouha.enmv@gmail.com

References

[1] Jugreet B, Sharmeen F, Mahomoodally M, Gokhan Z, Filippo M. Essential oils as natural sources of fragrance compounds for cosmetics and cosmeceuticals. Molecules. 2021;**26**(3):666

[2] Popa CL, Lupitu A, Daniela M, Copolovici L, Moisa C, Copolovici DM. Chemical and biochemical characterization of essential oils and their corresponding hydrolats from six species of the Lamiaceae family. Plants. 2021;**10**(11):2489

[3] Seyed M, Bagher H, Amin Mousavi K, de Souza A. Essential Oils in Food Processing: Chemistry, Safety and Applications. John Wiley & Sons Limited; 2018. p. 374

[4] Romdhane M, Tizaoui C. The kinetic modelling of a steam distillation unit for the extraction of aniseed (Pimpineliaanisum) essential oil. Journal of Chemical Technology and Biotechnology. 2005;**80**(7):759-766

[5] Perineau F, Ganou L, Vilarem G. Studying production of lovage essential oils in a hydrodistillation pilot unit equipped with a cohobation system. Journal of Chemical Technology and Biotechnology. 1992;**53**:165-171

[6] Bird RB, Stewart WE, Lightfoot EN. Fenómenos de transporte. Ediciones REPLA, S.A; 1987

[7] Rodríguez I, Gerbaud V, Joulia X. Feasibility of heterogeneous batch distillation processes. AICHE Journal. 2002;**48**(6):1168-1178

[8] Poulenat G. doctorat n° 2020. National Institut Polytechnic of Toulouse; 2003

[9] Kubátová A, Jansen B, Vaudoisot JF, Hawthorne SB. Thermodynamic and kinetic model for the extraction of essential oil from savory and polycyclic aromatic hydrocarbons from soil with hot (subcritical) water and supercritical CO2. Journal of Chromatography A. 2002;**975**:175-188

[10] Downes HR. Progress in essential oils. The chemistry of living cells. In: Handbook of Essentials Oils. Longmans; 1955

[11] Marshall TJ, Holmes JW. Plants and Soil Water in Soils Physics. London: Cambridge University Press; 1979. pp. 283-303 Google livre

[12] Bocchio E. Natural essentials oils. Parfums Cosmét. Arômes. 1985;**63**:61 in these marieluchese

[13] Garnero J. Semipreparative separation of terpenoids from essential oil. Phytotherapy. 1985;**15**:19

[14] Toulgoat K. Thèse de doctorat. Vol. n° 378. National Institute of Appliqued Sciences of Toulouse; 1996

[15] Boelens MH, Valverde F, Sequeiros L, Jimenez R. Perfume Flavor. 1990;**15**:11

[16] Bu'Lock JD. The Biosynthesis of Natural Products. New York: McGraw-Hill; 1965 Livre bibliothèque

[17] Baser KHC. In: Hüsnü Can Baser K, Buchbauer G, editors. Handbook of Essential Oils: Science, Technology, and Applications. Taylor & Francis Group; 2010. pp. 121-149

[18] Stahl E. A thermo micro procedure for rapid extraction and direct application in thin-layer chromatography. Analyst. 1969;**94**: 723-727

[19] Stahl E, editor. Thin-layer chromatography. In: A Laboratory Handbook. 2nd ed. Berlin: Springer; 1969

[20] Geiss F. Fundamentals of Thin-Layer Chromatography. Heidelberg: Hüthig Verlag; 1987

[21] Shema J, Fried B, editors. Handbook of Thin-Layer Chromatography. 3rd ed. New York: Marcel Dekker; 2003

[22] Hüsnü Can Baser K. Handbook of Essential Oils: Science, Technology, and Applications. Gerhard Buchbauer; 2010. p. 12

[23] König WA, Rieck A, Hardt I, Gehrcke B, Kubeczka KH, Muhle H. Enantiomeric composition of the chiral constituents of essential oils Part 2: Sesquiterpene hydrocarbons. Journal of High Resolution Chromatography. 1994; **17**:315-320

[24] Mondello-Mondello L, Shellie R, Casilli A, Marriott P, Dugo G. Ultra-fast essential oil characterization by capillary GC on a 50 μm ID column. Journal of Separation Science. 2004;**27**:699-702

[25] Hener U, Kreis P, Mosandl A. Enantiomeric distribution of a-pinene, b-pinene and limonene in essential oils and extracts. Part 2. Rutaceae and Gramineae. Flavour and Fragrance Journal. 1990;**5**:193

[26] Kubeczka KH. Application of HPLC for the separation of flavour compounds. In: Schreier P, editor. Flavour 81. Berlin, New York: Walter de Gruyter & Co. pp. 345-359. In: Handbook of EO from K. Hüsnü Can Baser Gerhard Buchbauer 2010 by Taylor and Francis Group, LLC

[27] Nishii Y, Yoshida T, Tanabe Y. Enantiomeric resolution of a germacrene-D derivative by chiral high-performance liquid chromatography. Bioscience, Biotechnology, and Biochemistry. 1997;**61**:547-548

[28] Chester TL, Innis DP. Separation of oligo-and polysaccharides by capillary supercritical fluid chromatography. Journal of High Resolution Chromatography. 1986;**9**:209-212

[29] Langa E, Della Porta G, Palavra AMF, Urieta JS, Mainar AM. Supercritical fluid extraction of Spanish sage essential oil: Optimization of the process parameters and modelling. The Journal of Supercritical Fluids. 2009; **49**(2):174-181

[30] Snyder JK, Nakanishi K, Hostettmann K, Hostettmann M. Application of rotation locular countercurrent chromatography in natural products isolation. Journal of Liquid Chromatography. 1984;**7**:243-256

[31] Kubeczka KH. Progress in isolation techniques for essential oil constituents. In: Vlietinck AJ, Dommisse RA, editors. Advances in Medicinal Plant Research. Stuttgart: Wissenschaftliche VerlagsgesellschaftmbH; 1985. pp. 197-224

[32] Becker H, Reichling J, Hsieh WC. Water-free solvent system for droplet counter-current chromatography and its suitability for the separation of non-polar substances. Journal of Chromatography. 1982;**237**:307-310

[33] Bellanato J, Hidalgo A. Infrared Analysis of Essential Oils. London: Heyden & Son Ltd; 1971

[34] Schultze W, Lange G, Heinrich G. Analysis of dried plant material directly introduced into a mass spectrometer. (Part I of investigations on medicinal plants by mass spectrometry). In: Baser KHC, Buchbauer G, editors. Handbook of Essential Oils: Science,

Technology, and Applications. Taylor and Francis Group; 1986. p. 2010

[35] Kubeczka KH. Studies on complex mixtures: Combined separation techniques versus unprocessed sample analysis. In: Hüsnü Can Baser K, Buchbauer G, editors. Handbook of Essential Oils: Science, Technology, and Applications. Taylor & Francis Group; 1989. p. 2010

[36] Ristorcelli, D., F. Tomi, and J. Casanova, 1997. Enantiomeric differentiation of oxygenated monoterpenes by carbon-13 NMR in the presence of a chiral lanthanide shift reagent. J. Magnet. Resonance Anal, 40–46. In Handbook of EO edited by K. Hüsnü Can Baser, Gerhard Buchbauer by Taylor and Francis Group, LLC

[37] Buttery RG, McFadden WH, Teranishi R, Kealy MP, Mon TR. Constituents of hop oil. Nature. 1963; **200**:435-436

[38] Herres W, Kubezka KH, Schultze W. HRGC-FTIR investigations on volatile terpenes. In: Brunke EJ, editor. Progress in Essential Oil Research. Berlin: W. de Gruyter; 1986. pp. 507-528 Handbook of Essential Oils: Science, Technology, and Applications, K. Hüsnü Can Baser, Gerhard Buchbauer. 2010

[39] Mondello L, Casilli A, Tranchida PQ, Cicero L, Dugo P, Dugo G. Comparison of fast and conventional GC analysis for citrus essential oils. Journal of Agricultural and Food Chemistry. 2003; **51**:5602

[40] Ibanez E, Lopez-Sebastian S, Ramos E, Tabera J, Reglero G. Analysis of highly volatile components of foods by off-line SFE/GC. Journal of Agricultural and Food Chemistry. 1997; **45**:3940-3943

[41] Hartonen K, Jussila M, Manninen P, Riekkola ML. Volatile oil analysis of Thymus vulgaris L. by directly coupled SFE/GC. Journal of Microcolumn Separations. 1992;**4**:3-7

[42] Yamauchi Y, Saito M. Fractionation of lemon-peel oil by semi-preparative supercritical fluid chromatography. Journal of Chromatography. 1990;**505**: 237-246

[43] Auerbach RH, Kenan D, Davidson G. Characterization of varietal differences in essential oil components of hops (Humulus lupulus) by SFC-FTIR spectroscopy. Journal of AOAC International. 2000;**83**:621-626

Section 5

Medicinal Plants and Phytopathology

Chapter 17

Genetic Engineering of Purslane (*Portulaca oleracea* L.)

Thalita Massaro Malheiros Ferreira, Fernanda Ferreira Salgado, Olga Costa Alves Souza, Rejane Valeriano Silva, Vivianny Nayse Belo Silva, Patrícia Abrão de Oliveira Molinari, Thales Lima Rocha and Manoel Teixeira Souza Junior

Abstract

Portulaca oleracea L., popularly known as purslane, is an herbaceous succulent plant classified as one of the most important invasive weeds in the world. Due to its high nutritional level and wide range of pharmacological effects, involving anti-inflammatory, antibacterial, antioxidant, and antiulcerogenic, purslane is one of the medicinal species listed by the World Health Organization. In addition, purslane produces several phytochemicals, including flavonoids, alkaloids, and terpenoids, which confer different pharmacological activities and make the plant highly attractive for use in the most diverse industries. It has high adaptability to extreme soil conditions, able to grow and spread in environments under drought stress, salinity, and poor nutrients; and has been presented as a potential model plant to study resistance to abiotic stresses. Among other purslane traits of interest to the agriculture sector, is worth to mention phytoremediation and allelopathy, thus being a sustainable alternative in organic agriculture. Here, we report a bibliometric analysis of purslane *in vitro* tissue culture and genetic modification/editing, and discuss opportunities and limitations to exploit the biotechnological potential of purslane as a source of valuable bio-molecules for many different industries.

Keywords: purslane, medicinal plant, multipurpose species, genetic transformation, tissue culture, biolistic, agrobacterium, abiotic stresses

1. Introduction

Portulaca oleracea L. (**Figure 1**), the most well-known species of the *Portulaca* genus, is commonly known as purslane, or common purslane, according to Ref. [1]. This genus belongs to the family Portulacaceae, order Caryophyllales, superorder Caryophyllanae, class Magnoliopsida, subdivision Spermatophytina, division Tracheophyta, superdivision Embryophyta, infrakingdom Streptophyta, and subkingdom Viridiplantae [2, 3].

Purslane is classified as a multipurpose plant species [4]. Plants cultivated for the purpose of providing more than one significant contribution to the production

Figure 1.
*Purslane (*Portulaca oleracea *L.).*

and/or service functions of a land use system are defined as multipurpose plants. They are classified according to the attributes of the plant species and the functional role of it in the technology under consideration, be it linked to the agricultural, pharmaceutical, chemical, or other economic sector.

2. Socioeconomic importance of purslane

2.1 A medicinal plant recommended by the World Health Organization

According to the World Health Organization (WHO), purslane is one of the most used medicinal plants. Known as a "Global Panacea"—a remedy supposed to heal all sicknesses, it is used extensively in folk medicine due to its wide array of health effects [5, 6]. The ethnobotanical importance of purslane led to various studies confirming its pharmacological properties. Those studies support its use as an antibacterial [7], anti-inflammatory, antioxidant [8], neuro- and hepatoprotective [6], antidiabetic [9], and antiulcerogenic agent [10], among other applications. In addition, it is reportedly a highly nutritious plant, being among the top terrestrial sources of essential fatty acids, tocopherol, ascorbic acid, glutathione, and other components, which suggests its nutraceutical potential [11, 12]. These valuable chemical constituents result from purslane's diverse set of chemical pathways.

All organisms have an integrated network of chemical reactions meticulously mediated and regulated by enzymes. It encompasses primary and secondary metabolic pathways synthesizing various organic compounds [13]. More specifically, primary metabolism involves generating components required for growth and development. Its products often serve as intermediates for the production of specialized chemicals that comprise the secondary metabolism, which plays a crucial part in a plant's interaction with the environment [14]. Processes that result in primary metabolites are highly conserved, while those of secondary metabolites are lineage-specific and continuously influenced by abiotic and biotic factors. That results in the formation or suppression of bioactive compounds that confer specific properties to the plant, intending to promote its survival and protection [15, 16].

Following a simple classification, secondary metabolites are divided into three main groups: phenolics, nitrogen-containing compounds, and terpenoids. Each is obtained through different biosynthetic pathways, resulting in chemicals with distinctive structures that confer valuable properties. These are building blocks to the manufacturing of high-added-value products focusing on health, nutrition, or agriculture, and thus are of enormous importance within the scientific and industrial framework [16, 17]. Purslane has a rich and unique content of these bioactive compounds that, individually or synergistically, provides beneficial effects and explains its extensive use in folk medicine [6, 12]. The following paragraphs will give insight into some highly important secondary metabolite groups and their known activities.

Flavonoids, which comprise the phenolic group, are among the main active ingredients from purslane—with kaempferol, luteolin, apigenin, myricetin, and quercetin as its major components. In addition, novel structures, namely portulacanones and oleracones, were first isolated from this plant. Studies have shown the anticancer [18, 19], anticholinesterase [20], anti-inflammatory, and antioxidant effects of these flavonoids [21, 22]. Furthermore, families belonging to Portulacaceae produce betalains, known as nitrogen-containing plant pigments with limited occurrence in nature [23, 24]. This subgroup is natural colorants in the food and cosmetic sectors, although studies have shown their neuroprotective [25], chemoprotective [26], and antimicrobial potential [24].

N-trans-feruloyltyramine, dopamine, noradrenaline, and oleraceins are alkaloids also identified in this plant species [27]. These nitrogen-containing compounds have reported immune-enhancing and neuroprotective effects, among others, and underwent studies for the prevention and treatment of neurodegenerative diseases [28–30]. The terpene content, which includes portulosides A-B, portulenes, and others, also contributes to potentializing antimicrobial and hepatoprotective effects of purslane extracts, and so on [27]. Other bioactive components include lignans, phenolic acids, and esters [31], and new molecules are constantly isolated from this plant through various methodologies [30, 32, 33].

Purslane is also a rich source of omega-3 and omega-6 fatty acids, thus contributing to its nutritional value [34]. These are the precursors of eicosapentaenoic and docosahexaenoic acids, which can reduce the risk of cardiovascular and cerebral diseases [12]. Studies on the development of functional food products from this plant are already available [35], as it also has considerable amounts of vitamins and dietary minerals [11]. Overall, each phytoconstituent mentioned contributes to establishing the ethnobotanical importance of purslane and supports the application of this plant in the pharmaceutical, food, and cosmetics industries.

2.2 A source of so many agricultural important traits

Besides being a source of many traits for the pharmaceutical and chemical industries, purslane is also a source of features of direct importance to the agricultural and agri-industrial sectors. Among the most important ones are phytoremediation, allelopathy, and tolerance to biotic and abiotic stress. Below we present some insights into some of those traits and then—more to the end of this chapter—report intensively on resistance to salinity stress.

Due to the purslane tolerance capacity for metal stress, it undergoes phytoremediation and biomonitoring studies in the field and closed conditions [36]. Phytoremediation is an economic process that exploits plants' capacity to accumulate heavy metals in polluted habitats by their harvestable parts [37], while biomonitoring is the capacity to monitor contaminated environments [36]. Mohammadzadeh and Hajiboland [38] reported a successful study using purslane in phytoremediation strategies to remove nitrate from nitrate-contaminated sites.

Allelopathy is the ability of a plant to suppress the germination, growth, survival, and reproduction of other plants in its surroundings. It produces and releases allelochemicals (secondary metabolites) that negatively affect other plants. Hamad [39] showed that aqueous extracts from purslane shoots and roots have allelopathic (inhibitor) effects on seed germination and the growth of monocots and dicots. Rashidi et al. [40] investigated the allelopathic effect of purslane on seed germination and growth of several plant species and demonstrated its allelopathic potential against *Phaseolus vulgaris* L. and *Allium cepa* L. as it reduced their seed germination rate.

After studying the chemical composition and yield of six purslane genotypes, Petropoulos et al. [34] reported that the biomass yield (fresh weight) in the open field was affected by genotype, with the highest yield of the tested genotypes being 33 tons/hectare, and the lowest being 11.5, with an average of about 22.5 among these genotypes. Kong and Zheng [41] evaluated the potential of producing purslane in a hydroponic system by testing two distinct cultivars—Green and Golden. Both cultivars performed similarly, generating a marketable yield of approximately 5.75 kg per m^2 on a bimonthly basis, which might yield 345 tons/hectare/year if cultivated in a bimestrial regime. Alu'datt et al. [42] evaluated the effect of different soil-less substrates on the fresh yield of purslane over five harvest cycles during the growing season under closed conditions and reported productivity of approximately 27 kg per m^2 when using Tuff: Peatmoss (2:1) substrates; what might yield 270 tons/hectare/year.

Purslane is a succulent herbaceous halophyte plant classified as invasive and considered the eighth most common weed in the world; it grows in warm moist places during the summer and spring seasons and can grow in almost any unshaded area, including gardens, crop fields, and waste places [43]. Because of that, its outdoor production in extensive areas faces several concerns. However, the above-mentioned high productivity of purslane in the context of controlled-environment agriculture [44–46] can open many doors of opportunities for the purslane industry. Many of those might take advantage of having a highly efficient protocol for engineering/editing purslane genome.

3. Genetic engineering/editing of purslane: state of the art

Genetically modified/edited plants are usually developed by *in vitro* regeneration from single transformed cells, and because of that, using *in vitro* plant tissue culture-based methods is required. However, that is not the only way to develop such types of plants. Some strategies of plant transformation that do not depend on *in vitro* regeneration are available and are known as "*in planta*" transformation methods. The floral dip transformation method is the most well known of them [47]; however, no report is available on its successful use in purslane.

3.1 Purslane *in vitro* tissue culture

Once the goal is the *in vitro* regeneration from single transformed cells, it is necessary to develop first a reliable and efficient purslane tissue culture protocol. Such a process may take advantage of the organogenesis or embryogenesis capability of the plant species in question and need to evaluate some factors such as the most appropriate type of explant, culture medium, growth regulators, and cultivation conditions, among others [48]. Unfortunately, there are not many reports on purslane *in vitro* tissue culture. The few ones available will be reported in the next paragraphs.

Safdari and Kazemitabar [49] was the first report on *in vitro* regeneration of purslane plants, intending to determine the best hormonal treatment for the induction of embryogenic callus from leaf tissue, the best type of explant and hormones for plant regeneration, and root induction from regenerated shoots. Later, Sharma et al. [50] reported an attempt to establish an efficient *in vitro* protocol for plant regeneration through organogenesis, using 1.5 cm long knots as explant, and achieving a stable efficiency of 70%.

Shekhawat et al. [51] reported an efficient *in vitro* regeneration method for purslane using a liquid medium, where the explants used were shoots with one and two nodes, obtaining a rooting efficiency rate of 96%. Oraibi et al. [52] reported success in efficiently inducing callus from purslane leaves, with subsequent production of extracts from the callus that presented antibacterial activity.

Purslane is sexually propagated, producing an enormous amount of seeds in a short period—within 60–90 days. Besides, purslane is also efficiently vegetatively propagated from cutting. The success of propagation (by seeds or cuttings) is probably one of the reasons that justify that there are not many reports on purslane *in vitro* tissue culture. The lack of demands for eradicating pathogens could be another reason to explain it.

The demonstrated capacity for producing over a hundred tons of biomass per hectare per year under closed conditions [42] makes purslane an ideal candidate as the crop to produce its bio-molecules, reducing the risk associated with the fact it is a weed [43]. However, one cannot forget that the growth of purslane cell suspension using bioreactors [53] is another way ahead to produce such bio-molecules under a controlled environment. In such case, there is the need to develop protocols to obtain and maintain purslane cell suspension.

Consequently, there is no doubt that for purslane to become a model plant for functional genomics research, aiming to advance on the exploitation of so many of its bio-molecules—whether in the pharmaceutical sector, in the agronomical sector, or in other sectors—the scientific community must expand and deeper the studies in many of the frontlines of plant tissue culture, such as haploid/di-haploid production, cell suspension production and maintenance, and, of course, genetic modification/editing. The results of the tissue culture survey on *Portulaca oleracea* are summarized in the table below (**Table 1**).

3.2 Genetic modification of purslane

The genetic transformation of plants involves the insertion, integration, and expression of exogenous genes into the genome of a plant species. One of the main focuses in obtaining transgenic cultures is incorporating new characteristics, studying primary biological processes, and producing bio-pharmaceutical proteins. Since the 1980s, different techniques became available for introducing heterologous genes into the genome, among which the transformation mediated by Agrobacterium and biolistics stands out [55]. The sonication-assisted Agrobacterium-mediated gene transfer system increases the transformation efficiency [56], and studies using sonication associated with vacuum infiltration proved to be efficient when applied to different cultivars of economic importance [57].

Sedaghati et al. [54] aimed to develop an Agrobacterium-mediated transformation and regeneration system using somatic embryogenesis in purslane, obtaining an efficiency of 72.22% from leaf explants. Studies carried out by the same group in 2021, seeking to optimize this transformation process, used sonication associated with

References	Explant	Multiplication technique	Efficiency	Objective
Safdari and Kazemitabar, [49]	Petioles, shoot tips, and leaves.	Callogenesis	70%	Test different hormones and explants.
Sharma et al. [50]	Nodal shoot segments	Organogenesis	70%	Development of an efficient regeneration protocol.
Shekhawat et al. [51].	Nodal shoot segments	Organogenesis	96%	Regeneration in liquid medium and implications of growth regulators.
Oraibi et al. [52]	Leaves	Callogenesis	80%	Antibacterial Activity.
Sedaghati et al. [54]	Seeds and leaves	Somatic embryogenesis	72.22%	Development of an efficient regeneration protocol.

Table 1.
Published findings on in vitro *tissue culture of purslane (*Portulaca oleracea *L.) - search done in December 2022.*

vacuum infiltration from seeds, obtaining a maximum efficiency of 39.25 ± 2.88%. Subsequently, Sedaghati et al. [58] carried a new study out where they applied purslane as a green bioreactor to evaluate the expression of the HSA human serum albumin gene.

Two reports describe success in transforming purslane using *Agrobacterium rhizogenes* [59, 60]. Tandon et al. [59] in an attempt to improve the phytoremediation of municipal waste water by increasing the root biomass, transformed four distinct plant species using *A. rhizogenes*, including *P. oleracea*. They used nodules and roots as explants for transformation, and achieved an increase in phytoremediation efficiency of transformed portulaca for the treatment of municipal waste water was observed over the non-transformed plants. The study by Ahmadi Moghadam et al. [60] aimed to optimize and evaluate the production of dopamine in hairy roots of purslane, using rolB as the gene of interest; and roots, stems, leaves, and cotyledons as explants. It was found that the most suitable explants for the induction of hairy roots were the cotyledons, with an efficiency of 53.3%. The effect of methyl jasmonate and salicylic acid on

References	Explant	Technique	Efficiency	Gene
Tandon et al. [59]	Nodal segments and roots	*Agrobacterium rhizogenes*	42–68%	–
Moghadam et al. [60]	Cotyledon leaves	*Agrobacterium rhizogenes*	53.30%	*rolB*
Sedaghati et al. [54]	Stem and leaf explants	Agrobacterium-mediated	72.22%	*GUS*
Sedaghati et al. [57]	Seeds	Agrobacterium-mediated transformation using sonication and vacuum infiltration.	39.25%	uidA
Sedaghati et al. [58]	Seeds and leaves	Agrobacterium-mediated transformation.	100%	HSA

Table 2.
*Published findings on genetic modification in purslane (*Portulaca oleracea *L.) - search done in December 2022.*

the accumulation and biosynthesis of dopamine in hairy roots culture of purslane was investigated, showing that elicitation with the former resulted in about 4.3-fold higher dopamine yield compared to the control hairy root cultures. The results of the genetic modification survey on *Portulaca oleracea* are summarized in the table above (**Table 2**).

4. Purslane as a model plant to study resistance to abiotic stresses

Purslane is a vegetable highly adaptable to various climates and adverse conditions (such as heat, drought, and salinity), giving it a competitive advantage over many other plant species produced [48]. It can grow in arid and saline soils, and it is considered both a halophyte (adapted to salty environments) and a xerophyte (adapted to dry ones); it is also listed as a halophyte plant in the eHaloph database [3, 61, 62].

Ren et al. [63] characterized the responses of 10 different purslane accessions to drought stress, and two African accessions were the most drought tolerant ones, Tokombiya and Egypitum. Drought tolerance was highly variable among the tested accessions at seed germination, seedling development, and adult stages. Jin et al. [64] evaluated the effect of drought and heat stress on purslane, individually and simultaneously. They demonstrated different survival strategies to physiological and metabolic stress, expressed as an increase in malondialdehyde, electrolyte leakage, O_2 and superoxide dismutase, and peroxidase activities, and a decrease in chlorophyll content. Furthermore, they suggested that combined stresses have a more severe effect on purslane plants compared to individual stresses. Another indication is that purslane is a promising candidate to be used in the recovery of ecosystems in arid and semi-arid regions [64].

Another very important fact about purslane is that, depending on the environmental condition, it can switch between the C4 photosystem and Crassulacean acid metabolism (CAM). Ferrari et al. [65] used purslane as a model system to investigate the involvement of the circadian clock, transcription factors, and plant hormones in coordinating the expression of C4 and CAM genes. They showed that, in general, the endogenous circadian clock coordinates and optimizes the daily time of regulation of the C4 and CAM genes in purslane leaves that are well-irrigated and under water stress [65].

There are studies of taxonomy, ecology, physiology, biochemistry, and genetics of purslane, including those characterizing its genetic variability. That information already available regarding the development of purslane, together with characteristics such as a short life cycle, small size, and relatively easy handle, reinforces the idea of turning purslane a research model plant for a better understanding of plant resistance to those two abiotic stresses [66, 67]. It is necessary to use different model plants, instead of the regular ones—Arabidopsis, for instance—as responses to those stresses are different between species [68].

Our research group developed a robust salt stress protocol (**Figure 2**) to characterize the morphophysiological responses of young purslane plants to salinity stress, along with multi-omics analyses, where it was possible to observe three distinct levels of salt stress by electrical conductivity gradients and water potential in substrate saturation extract [69]. The multi-omics integration studies showed 51 pathways having at least one enzyme and one metabolite differentially expressed in the leaves of purslane as a result of salt stress [69]. The characterization of the metabolomics, transcriptomics, and proteomics profiles on the leaves and roots of adult purslane plants were generated, and their analyses allowed further insights on the resistance of purslane to salinity stress (Rodrigues-Neto et al., unpublished).

Figure 2.
*Purslane (*Portulaca oleracea *L.) submitted to salinity stress studies at Embrapa Agroenergia during seed germination, seedling development, and different adult stages. Source: Souza Jr., M. T.*

Salinity tolerance is an important agronomic feature due to the great problem of salinized soil, which corresponds to 20% of the irrigated land in the world [48]. Studies carried out by Hamidov et al. [70] suggest the use of purslane to promote the rehabilitation of saline soils in the northern part of Uzbekistan, in addition to highlighting its great nutritional quality and its efficiency in tolerating chloride salinity, which makes this species a potential candidate for biosaline agriculture [70].

Yang et al. [30] applied physiological and comparative proteomics to investigate the mechanisms underlying purslane's tolerance to high-temperature and high-humidity stresses, demonstrating that this plant species deploys multiple strategies to cope with these combined stressors. Several strategies, from the activation of several metabolisms to the transient development of a CAM-like metabolism, were responsible for increased adaptation to water stress and combined stress in purslane [71].

To become a reliable model plant in functional genomics studies aiming to characterize the genetic basis of the resistances to those different abiotic stresses, it would be helpful to avail a reference whole genome sequence, as well as an efficient and reliable genetic modification/editing protocol.

5. Purslane as a source of bio-molecules of biotechnological importance

5.1 Screening for bio-molecules in purslane biomass

The final composition of bioactive compounds in purslane is affected by harvesting period, soil characteristics, cultivation practices, biotic/abiotic stresses, and other environmental factors, besides the genetic background. Temperature conditions during storage also play an essential role in this plant's quality, as it may lead to the degradation of particular components. Moreover, the efficiency of the extraction process is highly dependent on the plant's material, target metabolites, the temperature set, and the type of solvent used. Understanding these aspects that result in purslane's and its extract variability can allow for the management of specialized chemicals, thus leading to increased added-value products [12, 34, 72].

Novel bioactive compounds are continuously being isolated and identified from purslane through different extraction methods, including alkaloids [32, 73], flavonoids [22, 31], lignans [31, 74], and so on. Furthermore, changes in the metabolism—caused by biotic and abiotic factors—can be analyzed through various analytical methods. Metabolomic profiling techniques can determine and compare the chemical content of cultivars grown in varying conditions [12]. For instance, studies have shown that certain levels of salinity treatment result in increased concentration of valuable metabolites in purslane. Salt, as abiotic stress, triggers the plant to potentialize specific chemical pathways. However, it is essential to consider that these effects may also depend on the cultivar and geographic distribution, among other factors. Still, this indicates the potential for purslane's accessions to be explored for stress-induced augmented production of bioactive compounds [12, 48].

Moreover, fermentation can enrich the profile of active compounds found in purslane. During this specific process, enzymes derived from the metabolism of microorganisms promote the degradation of antinutritional factors and organic complexes. That gives rise to low-molecular-weight compounds—such as free amino acids [75] that can serve as substrates in the biosynthetic routes of secondary metabolites [76]. Lactic acid fermentation, for instance, is a suitable tool that explores the functional potential of purslane's biomass. It promoted increased bio-availability and profile modification of its specialized constituents and supplemented the matrix with active metabolites from the bacteria. Based on this, choosing a microbial group with high metabolic potential and versatility could lead to a selection of suitable strains for effective biotechnological processes [77, 78].

In summary, purslane is a rich source of specialized chemicals, and abiotic and biotic factors strongly influence their final content. Post-harvest treatments and extraction methods may also result in compound variability. Based on that, this species is the source of novel phytochemicals, and many techniques are employed to understand those aspects and compare cultivars. Stress-induced methods can lead to obtaining higher concentrations of bioactive compounds. Additionally, fermentation and biotechnological processes are promising strategies for exploring its functional potential.

5.2 Bio-molecules to control plant pathogens

The intensification of agricultural activity through the implementation of extensive areas of monoculture led to the emergence of pests at high levels, capable of reaching the threshold of economic damage. Groups of pathogens such as fungi, bacteria, viruses, nematodes, and insects are biotic agents whose infection to cultivated plants drastically reduces productivity. In the case of phytonematodes, global losses reach values above US$ 100 billion annually [79], with emphasis on the species *Meloidogyne incognita*. Several methods can be used to control phytopathogens, and the integration of these different strategies results in the most effective form of management. However, under high population densities, the suppression of these pathogens is largely carried out by making use of synthetic chemical pesticides, products that represent a risk of intoxication for humans and animals, as well as contamination of the environment. Chemical pest control substantially burdens the production process and also incurs the risk of selecting resistant organisms [80]. In view of all the problems involved in the adoption of this control method, it is necessary to search for more sustainable measures as an alternative way to contain the attack of agricultural pests. A trend that has shown promise and is gaining space in the field of research is the investigation of the activity of botanical extracts against plant parasites [81].

A vast number of species belonging to different botanical families can produce, through their secondary metabolism, several compounds from different chemical classes with the potential to suppress pest attacks. Chemical groups, such as alkaloids, terpenes, coumarins, flavonoids, among others, represent some of the elements responsible for inhibiting the action of various phytoparasites and, therefore, have competence to serve as a basis for the development of more eco-friendly pesticides, and mitigate the negative effects caused by pesticides [82]. Brazil has a mega diversity of plant resources with potential to be exploited in the prospection of biopesticides [83].

Botanical extracts can be produced from different structures, such as roots, flowers, seeds, and leaves, among others, and even from industrial co-products using different extracting solutions [84]. In this context, the use of these solutions with different physical-chemical characteristics allows obtaining a greater number of compounds, of different chemical classes with multiple applicability related to the medical, cosmetics, livestock, and agricultural areas [85]. Purslane is a plant of wide geographic distribution and highly attractive as an object of study for having a rich variety of phytochemicals with biocidal action. Furthermore, the use of plant residues as a green technology in pest control is a bi-sustainable possibility. This measure is characterized by the removal of materials from the industrialization process, for the production of sustainable pesticides, replacing conventional agrochemicals, whose molecules are generally classified as having a high toxicological degree [86].

Among the advantages of using phytochemicals in pest management is the possibility of associating extracts from different species, aiming at obtaining multifunctional technologies without generating incompatibility between the different molecules that participate in the compounds, which commonly happens in mixtures of synthetic chemicals [87]. Besides, the miscellany of extracts from different botanical families might result in the development of technologies that control two or more pathogens simultaneously, without negatively affecting the community of beneficial microorganisms in the soil.

Identified and characterized botanical compounds with biocidal effect can be used as tools in activities devoted to the discovery of genes and gene products involved in their synthesis routes. Such information is later used in metabolic engineering studies, which will allow the synthesis of the compound of interest in different parts of the plant, such as the root, for the control of soil pathogens. In addition, extracts and botanical compounds can also be used as basic constituents in the generation of green nanoparticles [88, 89]. Brazil has one of the greatest biodiversity on the planet and represents a relevant source of natural plant-based products. This fraction of biodiversity constitutes the so-called plant genetic resources. At the Brazilian Agricultural Research Corporation—Embrapa, a significant inter- and intraspecific genetic variability is conserved in active Germplasm Banks. These banks represent a source of diverse materials with numerous physicochemical characteristics and, therefore, an important and interesting reservoir for the prospection of extracts, fractions, and biocidal compounds. These materials might be used in the formulation of new eco-pesticides, following the global market trend that arises from society's growing demand for safer food and environmentally friendly practices [90].

The use of plant extracts, as a source of bioactive compounds in the suppression of phytopathogens, represents another possibility to be used in the management of agricultural pests [91]. However, some points deserve attention regarding the use of plant extracts as a basis for the development of natural pesticides. They are (i) the availability of plant material considering the seasonality of the crop in question, (ii) the search for biocidal action standardization showed by plants of the same species, but coming from different regions, (iii) the obstacles in the production of extracts on a large scale, (iv) the low residual characterizing a shorter time of action and, therefore, the demand for a greater number of applications, and (v) the restrictions that come up against the legislation concerning the registration of these products as they fall under the pesticide law.

In short, the implementation of natural pesticides in the production process, replacing synthetic agrochemicals whenever possible, is highly required. The planning to reduce the population of plant parasites must be prioritized and carried out rationally. Furthermore, the biology of the pathogen, the different methods of control, and their respective tactics must still be considered and could contribute to the minimization of the environmental imbalance. The joint implementation of these actions may strengthen the implantation of preventive measures and restrict the use of curative agricultural practices.

Purslane extract can be used as a pesticide as a pest control measure [4, 92]. Studies have shown that the methanolic extract of *Portulaca oleracea* showed high contact toxicity and dichloromethane exhibited antifood toxicity against the *Aphis gossypii* pest that affects cotton [4]. Tayyab et al. [93] carried out a study with 40 species of indigenous plants, including purslane, evaluating the insecticidal potential of acetone extracts. About 41% death rate was observed for the cotton aphid using 10% acetone extract obtained from purslane leaves [93].

6. Conclusions and future perspectives

Purslane is a multipurpose plant species source of many known bio-molecules and features of interest to several distinct industries—pharmaceutical, chemical, food, and feed, just to name a few. It is one of the most used medicinal plants, according to the WHO; but, it is also the eighth most common weed in the world. As an invasive weed, purslane raises several concerns when considered to be grown outdoors, in open fields. Because of that, it has been subject of increasing interest in the context of controlled-environment agriculture, where it deliveries very high biomass productivity. Within the scope of controlled-environment, purslane could be produced in greenhouses, in containers (vertical farming systems), or in bioreactors (cell suspensions).

After performing a bibliometric analysis of purslane *in vitro* tissue culture and genetic modification/editing, we came across with very little scientific data available. To exploit the biotechnological potential of purslane, as a source of valuable bio-molecules for the many different industries, it is imperative to change this scenario. Among the key initiatives necessary to do so there are: (a) a Whole Genome Project to generate and make available a reference genome for this plant species; (b) a reliable and efficient protocol for genetic modification/editing, able to produce a large number (hundreds or thousands) of mutants (genetically modified/edited) per period (a few months or a year); and (c) to build up a multi-omics database harboring information regarding the response of this plant species to many environmental conditions.

Other initiatives, such as the generation of haploid/di-haploid individuals; the production and maintenance of cell suspensions; the production, maintenance, and characterization of purslane's extract library, would be also of great value to exploit the biotechnological potential of this important plant species.

Acknowledgements

The author(s) disclosed receipt of the following financial support for the research, authorship, and/or publication of this article: The grant (01.13.0315.00 - DendêPalm Project) for this study was awarded by the Brazilian Innovation Agency - FINEP.

Conflict of interest

The authors declare no conflict of interest.

Author details

Thalita Massaro Malheiros Ferreira[1], Fernanda Ferreira Salgado[1],
Olga Costa Alves Souza[2], Rejane Valeriano Silva[2], Vivianny Nayse Belo Silva[1],
Patrícia Abrão de Oliveira Molinari[3], Thales Lima Rocha[4]
and Manoel Teixeira Souza Junior[3]*

1 Universidade Federal de Lavras (UFLA), Lavras, MG, Brazil

2 Universidade de Brasília (UnB), Brasília, DF, Brazil

3 Embrapa Agroenergia, Brasília, DF, Brazil

4 Embrapa Recursos Genéticos e Biotecnologia, Brasília, DF, Brazil

*Address all correspondence to: manoel.souza@embrapa.br

References

[1] Integrated Taxonomic Information System – IT IS [Internet]. 2022. Available from: www.itis.gov. [Accessed: May 1, 2023]

[2] WFO - Portulaca oleracea L. [Internet]. 2023. Available from: http://www.worldfloraonline.org/taxon/wfo-0000484425. [Accessed: May 1, 2023]

[3] Portulacaceae in Flora e Funga do Brasil [Internet]. 2023. Available from: https://floradobrasil.jbrj.gov.br/FB601987. [Accessed: May 1, 2023]

[4] Srivastava R. Multipurpose bene fi ts of an underexplored species purslane (Portulaca oleracea L.): A critical review. Environmental Management. 2021;**72**(2):309-320. DOI: 10.1007/s00267-021-01456-z

[5] Dehghan F, Soori R, Gholami K, Abolmaesoomi M, Yusof A, Muniandy S, et al. Purslane (Portulaca oleracea) seed consumption and aerobic training improves biomarkers associated with atherosclerosis in women with type 2 diabetes (T2D). Scientific Reports. 2016;**6**(1):37819. DOI: 10.1038/srep37819

[6] Zhou Y-X, Xin H-L, Rahman K, Wang S-J, Peng C, Zhang H. Portulaca oleracea L.: A review of Phytochemistry and pharmacological effects. BioMed Research International. 2015;**2015**:1-11. DOI: 10.1155/2015/925631

[7] Lei X, Li J, Liu B, Zhang N, Liu H. Separation and identification of four new compounds with antibacterial activity from Portulaca oleracea L. Molecules. 2015;**20**(9):16375-16387. DOI: 10.3390/molecules200916375

[8] Baradaran Rahimi V, Rakhshandeh H, Raucci F, Buono B, Shirazinia R, Samzadeh Kermani A, et al. Anti-inflammatory and anti-oxidant activity of Portulaca oleracea extract on LPS-induced rat lung injury. Molecules. 2019;**24**(1):139. DOI: 10.3390/molecules24010139

[9] Bao M, Hou K, Xin C, Zeng D, Cheng C, Zhao H, et al. *Portulaca oleracea* L. extract alleviated type 2 diabetes via modulating the gut microbiota and serum branched-chain amino acid metabolism. Molecular Nutrition & Food Research. 2022;**66**(11):2101030. DOI: 10.1002/mnfr.202101030

[10] Karimi G, Hosseinzadeh H, Ettehad N. Evaluation of the gastric antiulcerogenic effects of Portulaca oleracea L. extracts in mice. Phytotherapy Research. 2004;**18**(6):484-487. DOI: 10.1002/ptr.1463

[11] Kumar A, Sreedharan S, Kashyap AK, Singh P, Ramchiary N. A review on bioactive phytochemicals and ethnopharmacological potential of purslane (Portulaca oleracea L.). Heliyon. 2022;**8**(1):e08669. DOI: 10.1016/j.heliyon.2021.e08669

[12] Montoya-García CO, García-Mateos R, Becerra-Martínez E, Toledo-Aguilar R, Volke-Haller VH, Jesús Magdaleno-Villar J. Bioactive compounds of purslane (Portulaca oleracea L.) according to the production system: A review. Scientia Horticulturae. 2023;**308**:111584. DOI: 10.1016/j.scienta.2022.111584

[13] Dewick PM. Medicinal Natural Products: A Biosynthetic Approach. 3rd ed. New Jersey: John Wiley & Sons; 2002

[14] Pott DM, Osorio S, Vallarino JG. From central to specialized metabolism:

An overview of some secondary compounds derived from the primary metabolism for their role in conferring nutritional and organoleptic characteristics to fruit. Frontiers in Plant Science. 2019;**10**:835. DOI: 10.3389/fpls.2019.00835

[15] Erb M, Kliebenstein DJ. Plant secondary metabolites as defenses, regulators, and primary metabolites: The blurred functional trichotomy. Plant Physiology. 2020;**184**(1):39-52. DOI: 10.1104/pp.20.00433

[16] Guerriero G, Berni R, Muñoz-Sanchez J, Apone F, Abdel-Salam E, Qahtan A, et al. Production of plant secondary metabolites: Examples, tips and suggestions for biotechnologists. Genes (Basel). 2018;**9**(6):309. DOI: 10.3390/genes9060309

[17] TFRRS A-C, DHAM S. Secondary metabolites. In: Chromatography and its Applications. London, UK: InTech; 2012. DOI: 10.5772/35705

[18] Tavsan Z, Kayali HA. Flavonoids showed anticancer effects on the ovarian cancer cells: Involvement of reactive oxygen species, apoptosis, cell cycle and invasion. Biomedicine & Pharmacotherapy. 2019;**116**:109004. DOI: 10.1016/j.biopha.2019.109004

[19] Yan J, Sun L-R, Zhou Z-Y, Chen Y-C, Zhang W-M, Dai H-F, et al. Homoisoflavonoids from the medicinal plant Portulaca oleracea. Phytochemistry. 2012;**80**:37-41. DOI: 10.1016/j.phytochem.2012.05.014

[20] Yener I, Kocakaya SO, Ertas A, Erhan B, Kaplaner E, Oral EV, et al. Selective in vitro and in silico enzymes inhibitory activities of phenolic acids and flavonoids of food plants: Relations with oxidative stress. Food Chemistry. 2020;**327**:127045. DOI: 10.1016/j.foodchem.2020.127045

[21] Chagas MSS, Behrens MD, Moragas-Tellis CJ, Penedo GXM, Silva AR, Gonçalves-de-Albuquerque CF. Flavonols and flavones as potential anti-inflammatory, antioxidant, and antibacterial compounds. Oxidative Medicine and Cellular Longevity. 2022;**2022**:9966750. DOI: 10.1155/2022/9966750

[22] Yang X, Zhang W, Ying X, Stien D. New flavonoids from Portulaca oleracea L. and their activities. Fitoterapia. 2018;**127**:257-262. DOI: 10.1016/j.fitote.2018.02.032

[23] Iwashina T. Flavonoid properties in plant families synthesizing Betalain pigments (review). Natural Product Communications. 2015;**10**(6):1103-1114. DOI: 10.1177/1934578X1501000675

[24] Wijesinghe VN, Choo WS. Antimicrobial betalains. Journal of Applied Microbiology. 2022;**133**(6):3347-3367. DOI: 10.1111/jam.15798

[25] Wang C-Q, Yang G-Q. Betacyanins from Portulaca oleracea L. ameliorate cognition deficits and attenuate oxidative damage induced by D-galactose in the brains of senescent mice. Phytomedicine. 2010;**17**(7):527-532. DOI: 10.1016/j.phymed.2009.09.006

[26] Gandía-Herrero F, Escribano J, García-Carmona F. Biological activities of plant pigments Betalains. Critical Reviews in Food Science and Nutrition. 2016;**56**(6):937-945. DOI: 10.1080/10408398.2012.740103

[27] Iranshahy M, Javadi B, Iranshahi M, Jahanbakhsh SP, Mahyari S, Hassani FV, et al. A review of traditional uses, phytochemistry and pharmacology of Portulaca oleracea L. Journal of Ethnopharmacology. 2017;**205**:158-172. DOI: 10.1016/j.jep.2017.05.004

[28] Liu K, Tan J-N, Wei Y, Li C, Dou Y, Zhang Z. Application of choline chloride-based deep eutectic solvents for the extraction of dopamine from purslane (Portulaca oleracea L.). Journal of Materials Chemistry C. 2022;**4**:100299. DOI: 10.1016/j.rechem.2022.100299

[29] Sun H, He X, Liu C, Li L, Zhou R, Jin T, et al. Effect of Oleracein E, a neuroprotective Tetrahydroisoquinoline, on rotenone-induced Parkinson's disease cell and animal models. ACS Chemical Neuroscience. 2017;**8**(1):155-164. DOI: 10.1021/acschemneuro.6b00291

[30] Yang Y, Chen J, Liu Q, Todd CD, Shi J, Yang Y, et al. Comparative proteomic analysis of the Thermotolerant plant Portulaca oleracea acclimation to combined high temperature and humidity stress. Journal of Proteome Research. 2012;**11**(7):3605-3623. DOI: 10.1021/pr300027a

[31] Duan Y, Ying Z, He F, Ying X, Jia L, Yang G. A new skeleton flavonoid and a new lignan from Portulaca oleracea L. and their activities. Fitoterapia. 2021;**153**:104993. DOI: 10.1016/j.fitote.2021.104993

[32] Song M, Ying Z, Ying X, Jia L, Yang G. Three novel alkaloids from Portulaca oleracea L. and their anti-inflammatory bioactivities. Fitoterapia. 2022;**156**:105087. DOI: 10.1016/j.fitote.2021.105087

[33] Xu W, Wang J, Ju B, Lan X, Ying X, Stien D. Seven compounds from Portulaca oleracea L. and their anticholinesterase activities. Natural Product Research. 2022;**36**(10):2547-2553. DOI: 10.1080/14786419.2021.1916928

[34] Petropoulos S, Karkanis A, Martins N, Ferreira ICFR. Phytochemical composition and bioactive compounds of common purslane (Portulaca oleracea L.) as affected by crop management practices. Trends in Food Science & Technology. 2016;**55**:1-10. DOI: 10.1016/j.tifs.2016.06.010

[35] Souza PG, Rosenthal A, Ayres EMM, Teodoro AJ. Potential functional food products and molecular mechanisms of Portulaca Oleracea L. on anticancer activity: A review. Oxidative Medicine and Cellular Longevity. 2022;**2022**:7235412. DOI: 10.1155/2022/7235412

[36] Subpiramaniyam S, Portulaca oleracea L. For phytoremediation and biomonitoring in metal-contaminated environments. Chemosphere. 2021;**280**:130784. DOI: 10.1016/j.chemosphere.2021.130784

[37] Elshamy MM, Heikal YM, Bonanomi G. Phytoremediation efficiency of Portulaca oleracea L. naturally growing in some industrial sites, Dakahlia District, Egypt. Chemosphere. 2019;**225**:678-677. DOI: 10.1016/j.chemosphere.2019.03.099

[38] Mohammadzadeh P, Hajiboland R. Phytoremediation of nitrate contamination using two halophytic species, Portulaca oleracea and Salicornia europaea. Environmental Science and Pollution Research. 2022;**29**:46127-46144. DOI: 10.1007/s11356-022-19139-5

[39] Hamad SW. Bioherbicidal actions of common purslane on seed germination and growth of some crop and weed species. Proceedings of the IOP Conf Ser: Earth Environ Sci. 2021;**910**:012107. DOI: 10.1088/1755-1315/910/1/012107

[40] Rashidi S, Reza Yousefi A, Goicoechea N, Pouryousef M, Moradi P, Vitalini S, et al. Allelopathic interactions between seeds of Portulaca oleracea

L. and crop species. Applied Sciences. 2021;**11**(8):3539. DOI: 10.3390/app11083539

[41] Kong Y, Zheng Y. Hydroponic production of purslane as a sodium-removing vegetable in NaCl-rich nutrient solution. HortScience. 2014;**49**(2):201-206

[42] Alu'datt MH, Rababah T, Alhamad MN, Al-Tawaha A, Al-Tawaha AR, Gammoh S, et al. Herbal yield, nutritive composition, phenolic contents and antioxidant activity of purslane (Portulaca oleracea L.) grown in different soilless media in a closed system. Industrial Crops and Products. 2019;**141**:111746. DOI: 10.1016/j.indcrop.2019.111746

[43] Ozturk M, Altay V, Güvensen A. Portulaca oleracea: A vegetable from saline habitats. In: Grigore MN, editor. Handbook of Halophytes: From Molecules to Ecosystems towards Biosaline Agriculture. Cham: Springer Nature Switzerland AG; 2020. pp. 1-14. DOI: 10.1007/978-3-030-57635-6_96

[44] van Delden SH, SharathKumar M, Butturini M, Graamans LJA, Heuvelink E, Kacira M, et al. Current status and future challenges in implementing and upscaling vertical farming systems. Nat Food. 2021;**2**(12):944-956. DOI: 10.1038/s43016-021-00402-w

[45] Stein EW. The transformative environmental effects large-scale indoor farming may have on air, water, and soil. Air, Soil and Water Research. 2021:14. DOI: 10.1177/1178622121995819

[46] Vatistas C, Avgoustaki DD, Bartzanas T. A systematic literature review on controlled-environment agriculture: How vertical farms and greenhouses can influence the sustainability and footprint of urban microclimate with local food production. Atmosphere. 2022;**13**(8):1258. DOI: 10.3390/atmos130812581

[47] Zlobin NE et al. CRISPR/Cas9 genome editing through in planta transformation. Critical Reviews in Biotechnology. 2020;**40**:153-168. DOI: /10.1080/07388551.2019.1709795

[48] Amirul Alam M, Juraimi AS, Rafii MY, Hamid AA, Aslani F, Alam MZ. Effects of salinity and salinity-induced augmented bioactive compounds in purslane (Portulaca oleracea L.) for possible economical use. Food Chemistry. 2015;**169**:439-447. DOI: 10.1016/j.foodchem.2014.08.019

[49] Safdari Y, Kazemitabar SK. Plant tissue culture study on two different races of purslane (Portulaca oleracea L.). African Journal of Biotechnology. 2009;**8**(21):5906-5912

[50] Sharma MM, Abhijeet S, Verma RN, Ali DZ, Amla B. Influence of PGRS for the in vitro plant regeneration and flowering in Portulaca oleracea (L.): A medicinal and ornamental plant. International Journal of Botany. 2011;**7**(1):103-107

[51] Shekhawat MS, Kannan N, Manokari M. Propagation OF Portulaca oleracea L. in liquid medium: Implications of plant growth regulators in culture. Journal of Microbiology, Biotechnology and Food Sciences. 2021;**4**(4):332-335. DOI: 10.15414/jmbfs.2015.4.4.332-335

[52] Oraibi AG, AlShammari AA, Mohsien RA, Obaid W. Investigation the antibacterial activity of Portulaca oleracea L. tissue cultures in vitro. Journal of Pharmaceutical Research International. 2017;**18**:1-7

[53] Su R, Sujarani M, Shalini P, Prabhu N. A review on bioreactor technology

assisted plant suspension culture. Asian Journal of Biotechnology and Bioresource Technology. 2019;**5**(3):1-13. DOI: 10.9734/ajb2t/2019/v5i330062

[54] Sedaghati B, Haddad R, Bandehpour M. Efficient plant regeneration and agrobacterium-mediated transformation via somatic embryogenesis in purslane (Portulaca oleracea L.): An important medicinal plant. PCTOC. 2019;**136**(2):231-245

[55] Joung YH, Choi PS, Kwon SY, Harn CH. Plant transformation methods and applications. In: Koh H-J, Kwon S-Y, Thomson M, editors. Current Technologies in Plant Molecular Breeding. California: Springer, Dordrecht; 2015. pp. 297-343

[56] Khemkladngoen N, Cartagena JA, Fukui K. Physical wounding-assisted agrobacterium-mediated transformation of juvenile cotyledons of a biodiesel-producing plant. Jatropha curcas L. Plant Biotechnology Reports. 2011;**5**(3):235-243

[57] Sedaghati B, Haddad R, Bandehpour M. Development of an efficient in-planta agrobacterium-mediated transformation method for Iranian purslane (Portulaca oleracea L.) using sonication and vacuum infiltration. Acta Physiologiae Plantarum. 2021;**43**(2):1-9

[58] Sedaghati B, Haddad R, Bandehpour M. Purslane (Portulaca oleracea L.) as a novel green-bioreactor for expression of human serum albumin (HSA) gene. Transgenic Research. 2022;**31**(3):369-380. DOI: 10.1007/s11248-022-00296-9

[59] Tandon SA, Badrike H, Kumar R, Kakde U. Enhancement in phytoremediation of municipal wastewater by genetically modifying wetland and non-wetland plant species (non-edible). Global Journal of Bio-Science and BioTechnology. 2014;**3**:359-364

[60] Ahmadi Moghadam Y, Piri K, Bahramnejad B, Ghiyasvand T. Dopamine production in hairy root cultures of Portulaca oleracea (purslane) using agrobacterium rhizogenes. Journal of Agricultural Science and Technology. 2014;**16**(2):409-420

[61] Aronson JA. Haloph: A Database of Salt Tolerance Plants of the World. Tucson, AZ: Office of Arid Land Studies, University of Arizona; 1989. p. 77

[62] Kafi M, Rahimi Z. Effect of salinity and silicon on root characteristics, growth, water status, proline content and ion accumulation of purslane (Portulaca oleracea L.). Soil Science and Plant Nutrition. 2011;**57**(2):341-347. DOI: 10.1080/00380768.2011.567398

[63] Ren S, Weeda S, Akande O, Guo Y, Rutto L, Mebrahtu T. Drought tolerance and AFLP-based genetic diversity in purslane (Portulaca oleracea L.). Journal of Biotech Research. 2011;**3**:51-61

[64] Jin R, Wang Y, Liu R, Gou J, Chan Z. Physiological and metabolic changes of purslane (Portulaca oleracea L.) in response to drought, heat, and combined stresses. Frontiers in Plant Science. 2016;**6**:1123. DOI: 10.3389/fpls.2015.01123

[65] Ferrari RC, Kawabata AB, Ferreira SS, Hartwell J, Freschi L. A matter of time: Regulatory events behind the synchronization of C4and crassulacean acid metabolism in Portulaca oleracea. Journal of Experimental Botany. 2022;**73**(14):4867-4885. DOI: 10.1093/jxb/erac163

[66] López-González D, Ledo D, Cabeiras-Freijanes L, Verdeguer M,

Reigosa MJ, Sánchez-Moreiras AM. Phytotoxic activity of the natural compound norharmane on crops, weeds and model plants. Plants. 2020;**9**(10):1-23. DOI: 10.3390/plants9101328

[67] Saheri F, Barzin G, Pishkar L, Boojar MMA, Babaeekhou L. Foliar spray of salicylic acid induces physiological and biochemical changes in purslane (Portulaca oleracea L.) under drought stress. Biologia. 2020;**75**(12):2189-2200. DOI: 10.2478/s11756-020-00571-2

[68] Borsai O, Al Hassan M, Boscaiu M, Sestras R, Vicente O. The genus Portulaca as a suitable model to study the mechanisms of plant tolerance to drought and salinity. The EuroBiotech Journal. 2018;**2**:1-10. DOI: 10.2478/ebtj-2018-0014

[69] Silva VNB, Silva TLC, Ferreira TMM, Rodrigues Neto JC, Leão AP, Aquino Ribeiro JA, et al. Multi-omics analysis of young Portulaca oleracea L. plants' responses to high NaCl doses reveals insights into pathways and genes responsive to salinity stress in this halophyte species. Phenomics. 2022;**3**:1-21. DOI: 10.1007/s43657-022-00061-2

[70] Hamidov A, Beltrao J, Costa C, Khaydarova V, Sharipova S. Environmentally useful technique - Portulaca Oleracea Golden purslane as salt removal techniques. WSEAS Transactions on Environment and Development. 2007;**7**(3):117-122

[71] Andrea RMD, Andreo CS, Lara MV. Deciphering the mechanisms involved in Portulaca oleracea (C 4) response to drought: Metabolic changes including crassulacean acid-like metabolism induction and reversal upon re-watering. Physiologia Plantarum. 2014;**152**(3):414-430. DOI: 10.1111/ppl.12194

[72] Chen W-C, Wang S-W, Li C-W, Lin H-R, Yang C-S, Chu Y-C, et al. Comparison of various solvent extracts and major bioactive components from Portulaca oleracea for antioxidant, anti-Tyrosinase, and anti-α-glucosidase activities. Antioxidants (Basel). 2022;**11**(2):398. DOI: 10.3390/antiox11020398

[73] Lan X, Ying Z, Guo S, Duan Y, Cui X, Leng A, et al. Two novel amide alkaloids from Portulaca oleracea L. and their anti-inflammatory activities. Natural Product Research. 2022;**36**(21):5567-5574. DOI: 10.1080/14786419.2021.2021519

[74] Ma Y, Bao Y, Zhang W, Ying X, Stien D. Four lignans from Portulaca oleracea L. and its antioxidant activities. Natural Product Research. 2020;**34**(16):2276-2282. DOI: 10.1080/14786419.2018.1534852

[75] Olukomaiya OO, Adiamo OQ, Fernando WC, Mereddy R, Li X, Sultanbawa Y. Effect of solid-state fermentation on proximate composition, anti-nutritional factor, microbiological and functional properties of lupin flour. Food Chemistry. 2020;**315**:126238. DOI: 10.1016/j.foodchem.2020.126238

[76] Nakagawa A, Minami H, Kim J-S, Koyanagi T, Katayama T, Sato F, Kumagai H. A bacterial platform for fermentative production of plant alkaloids. Nature Communications 2011;**2**(1):326. DOI: 10.1038/ncomms1327

[77] Di Cagno R, Filannino P, Vincentini O, Cantatore V, Cavoski I, Gobbetti M. Fermented Portulaca oleracea L. juice: A novel functional beverage with potential ameliorating effects on the intestinal inflammation and epithelial injury. Nutrients. 2019;**11**(2):248. DOI: 10.3390/nu11020248

[78] Filannino P, di Cagno R, Trani A, Cantatore V, Gambacorta G, Gobbetti M.

Lactic acid fermentation enriches the profile of biogenic compounds and enhances the functional features of common purslane (Portulaca oleracea L.). Journal of Functional Foods. 2017;**39**:175-185. DOI: 10.1016/j.jff.2017.10.022

[79] El-Ashry RM, El-Saadony MT, El-Sobki AEA, El-Tahan AM, Al-Otaibi S, El-Shehawi AM, et al. Biological silicon nanoparticles maximize the efficiency of nematicides against biotic stress induced by Meloidogyne incognita in eggplant. Saudi Journal of Biological Sciences. 2022;**29**:920-932

[80] Bhandari S, Thakuri LS, Rimal S, Bhatta T. Management of okra jassid (Amrasca biguttula biguttula) through the use of botanicals and chemical pesticides under field conditions in Chitwan, Nepal. Journal of Agriculture and Food Research. 2022;**10**:100403

[81] Dolma SK, Reddy SGE. Characterization of Triadica sebifera (L.) small extracts, Antifeedant activities of extracts, fractions, seed oil and isolated compounds against Plutella xylostella (L.) and their effect on detoxification enzymes. Molecules. 2022;**27**:6239. DOI: 10.3390/molecules27196239

[82] Khursheed A, Rather MA, Jain V, Wani AR, Rasool S, Nazir R, et al. Plant based natural products as potential ecofriendly and safer biopesticides: A comprehensive overview of their advantages over conventional pesticides, limitations and regulatory aspects. Microbial Pathogenesis. 2022;**173**:105854

[83] Rocha TL, Polez VLP, Viol LCS, Pimentel RR, Biscaia D, Pinheiro JB. Use of natural and residual resources for the sustainable Management of Phytonematodes: Challenges and future trends. In: Chaudhary KK, Meghvansi MK, editors. Sustainable Management of Nematodes in Agriculture, Vol. 1: Organic Management. Cham: Springer International Publishing; 2022. pp. 3-37. DOI: 10.1007/978-3-031-09943-4_1

[84] Lengai GMW, Muthomi JW, Mbega ER. Phytochemical activity and role of botanical pesticides in pest management for sustainable agricultural crop production. Scientific African. 2020;**7**:1-13

[85] Giunti G, Benelli G, Palmeri V, Laudani F, Ricupero M, Ricciardi R, et al. Non-target effects of essential oil-based biopesticides for crop protection: Impact on natural enemies, pollinators, and soil invertebrates. Biological Control. 2022;**176**:105071

[86] Hanif K, Zubair M, Hussain D, et al. Biopesticides and insect management. International Journal of Tropical Insect Science. 2022;**42**:3631-3637. DOI: 10.1007/s42690-022-00898-0

[87] Udo I A, Uko AE, Etim DO. Management of root-knot disease in okra with poultry manure and leaf extracts of Senna alata. South Asian Journal of Parasitology. 2020;**4**:1-10

[88] Silva LP, Bonatto CC. Green nanotechnology for sustained release of eco-friendly agrochemicals. In: Vaz S Jr, editor. Sustainable Agrochemistry. Cham: Springer; 2019. pp. 137-164. DOI: 10.1007/978-3-030-17891-8_4

[89] Hernández-Díaz JA, Garza-García JJ, Zamudio-Ojeda A, León-Morales JM, López-Velázquez JC, García-Morales S. Plant-mediated synthesis of nanoparticles and their antimicrobial activity against phytopathogens. Journal of the Science of Food and Agriculture. 2020;**101**(4):1270-1287. DOI: 10.1002/jsfa.10767

[90] Bélanger J, Pilling D. The State of the World's Biodiversity for Food

and Agriculture. Rome, Italy: FAO Commission on Genetic Resources for Food and Agriculture Assessments; 2019. 572 p. DOI: 10.4060/ca3129en. ISBN: 978-92-5-131270-4

[91] Sivasubramaniam N, Hariharan G, Zakeel MCM. Sustainable management of plant-parasitic nematodes: An overview from conventional practices to modern techniques. In: Ansari R, Rizvi R, Mahmood I, editors. Management of Phytonematodes: Recent Advances and Future Challenges. Singapore: Springer; 2020. pp. 353-399. DOI: 10.1007/978-981-15-4087-5_16

[92] Isman MB. Botanical insecticides in the twenty-first century — Fulfilling their promise ? Annual Review of Entomology. 2020;**65**:233-249. DOI: 10.1146/annurev-ento-011019-025010

[93] Tayyab MB, Majeed MZ, Riaz MA, Anjum M, Ouedraogo SN, Luqman M, et al. Insecticidal potential of indigenous Flora of soon valley against Asian citrus psyllid. Sarhad Journal of Agriculture. 2018;**38**:1

Chapter 18

Biomolecules Produced by *Trichoderma* Species as Eco-Friendly Alternative Suppressing Phytopathogens and Biofertilizer Enhancing Plant Growth

Abdenaceur Reghmit, Farida Benzina-tihar and Fatma Sahir-Halouane

Abstract

Olive (*Olea europeae* L.) is one of the most important fruit trees of the Mediterranean regions. Biotic factors such as phytopathogenic diseases have a significant negative impact on olive productivity in the Mediterranean Basin including Algeria. Currently, phytopathogens management is focus mainly on the use of chemical pesticides which is not recommended because it leads to environmental pollution, development of chemical resistance, and its low cost-efficiency. Eco-friendly methods and alternative disease control measures such as the use of biocontrol agents and biofertilizer should be opted as alternatives to the use of synthetic chemicals. *Trichoderma* species associated with olive roots are known for their ability to produce antimicrobial compounds, such as antibiotics, volatile organic compounds and lytic enzymes that restrict phytopathogenic strain growth. Besides, they are considered as plant growth promoting fungi (PGPF). This genus colonize the root systems of plants and promote their growth; it can increase nutrient availability and uptake in plants by fixing nitrogen, solubilizing phosphorus, producing several biomolecules and phytohormones. Moreover, it helps plants tolerate environmental stresses such as drought, salinity and diseases. In this work, we review pionnering and recent developments on several important biomolecules and functions that *Trichoderma* species isolated from olive rhizosphere soil exhibit to enhance plant growth and control phytopathogen diseases. Therefore, the use of highly competitive strains in open field in order to obtain consistent and better results in agricultural production activities.

Keywords: *Trichoderma* spp., biocontrol, pesticides, biomolecules, phytohormones, biofertilizer

1. Introduction

Olive is affected by a wide range of biotic constraints such as soil-borne diseases which can cause significant damage and economic losses. Currently, plants diseases are managed mainly through the use of chemical pesticides, which can generate negative effects, such as health problems, loss of ecological diversity, and the bioaccumulation of toxic substances [1]. Nowadays, a key practice to deal with plant pathogens in sustainable agriculture is the biological control, which is based on managing natural resources and developing antagonistic activities against harmful microorganisms [2] which make it an effective and eco-friendly approach against plant diseases [3]. Many microorganisms with antagonistic activity such as *Trichoderma* spp. offer an environment-friendly alternative to get out of chemical pesticides damages [4–7]. *Trichoderma* spp. are known as promising fungal for the management of plant diseases, especially against soil-borne pathogens [8]. Therefore, *Trichoderma* spp. are most investigated and employed as biopesticide [9–11]. This genus has antagonistic activities and can act by various mechanisms against a wide range of soil-borne phytopathogenic fungi including competition for nutrients and the systemic activation of plant defense responses [12–14]. Thus, *Trichoderma* spp. are used as biopesticides in management of plant diseases worldwide [15]. Furthermore, they act by different modes of action against plant pathogens, including, mycoparasitism through the production of the cell wall degrading enzymes such as chitinases, glucanases, and proteases [16, 17], production of antibiotics and volatile organic compounds (VOCs) [18]. *Trichoderma* spp. produce wide spectrum of VOCs which are part from several chemical groups such as monoterpenes, sesquiterpenes, alcohols, aldehydes, aromatic compounds, esters, furans, hydrocarbons, ketones, and compounds containing S and N elements [19, 20]. These volatile compounds can diffuse through pores in the soil, move long distance, and affect pathogen without direct contact [21], which makes it more efficient at microbial interactions compared to non-volatile compounds [22]. Hence, it could be responsible for several biological processes such as biocontrol or communication between microorganisms [23]. Importantly, *Trichoderma* spp. are considered as plant growth-promoting fungi (PGPF) which can colonize and proliferate within the rhizosphere environment and enhancers of plant defense mechanisms. They display stimulation of plant growth because of their capacity to produce plant growth promoters [24–26]. *Trichoderma* species are able to promote plant growth through various mechanisms such as solubilizing insoluble phosphate, production of siderophore and plant hormone such as indole-3-acetic acid (IAA) [27].

2. Characteristics of *Trichoderma* spp.

Trichoderma is a genus of a heterogeneous group of fungal species. They are considered as anamorphic Hypocreales [28]. *Trichoderma* species are free-living and/or endophytic fungi that grow vigorously in soil and plant root ecosystems, they are known as ubiquitous saprophytic fungi [12, 29–31] as well as aboveground such as on rotting wood and other organic materials [17, 32–35]. Further, *Trichoderma* strains produce a few pigments, ranging from a greenish-yellow up to a reddish tinge and sometimes colorless strains might likewise be available. The conidia can have different hues, going from drab to various tints of green or dim or earthy colored hints [28]. Microscopic identification criteria of *Trichoderma* are as follows: septate and translucent hyphae;

conidiophores are short, translucent, branched often giving the pyramidal appearance, not verticillate, the phialides are attached at right angles to the conidiophores. Spores produced are conidia which are translucent, ovoid in shape, borne in small terminal clusters at the tips of phialides, some species can produce globose chlamydospores, which are intercalary or terminal. These chlamydospores are usually unicellular but can be multicellular for some species such as *Trichoderma stromaticum* [36].

3. *Trichoderma* spp. with biocontrol potentials against plant diseases

In recent years, *Trichoderma* species are considered as good alternatives for substituting chemical usages. They act by various mechanisms against a wide range of phytopathogenic fungi. Biocontrol by *Trichoderma* including myco-parasitism, antibiosis as well as competition for nutrients and space with plant pathogens [29, 32, 37–39]. Furthermore, their interaction with root induces the plant's resistance to pathogens, and destruction of the root-knot nematode at the different stages of its growth phases [40].

3.1 Volatile organic compounds

Volatile organic compounds (VOCs) are low-molecular-weight molecules that have contain fewer than 12 carbon atoms, and may be associated with other elements such as nitrogen, sulfur, bromine, oxygen, fluorine, and chlorine [41]. These compounds exhibit antimicrobial activities, promote plant development, the induction of systemic resistance, and considered as chemical signaling in plants [42, 43]. Most of volatile compounds produced by *Trichoderma* species exerted antifungal activity against plant pathogens. It has been reported that [44] were identified more than 278 volatile compounds from liquid cultures of *Trichoderma harzianum* using GC–MS; these compounds are part of different chemical groups such as normal saturated hydrocarbons (C7–C30), cyclohexane, fatty acids, alcohols, cyclo-pentane, esters, sulfur-containing compounds, simple pyrane, and benzene derivatives. Similar compounds were detected by [45], suggesting that *Trichoderma* spp. strains isolated from the rhizosphere of healthy olive trees have antagonistic activity against plant pathogen by production of VOCs. Among compounds detected in this study, eicosane which has antifungal activity [46, 47]. On the other hand, several volatile compounds with antifungal activities were detected by [45] such as benzeneethanol, toluene, alcohols, phenols, cyclohexane. Palmitic acid, alkanes, octadecenoic acid, palmitic acid, and limonene. Many works revealed the antifungal effect of these compounds such as the finding of [48] who reported that Fatty acids (e.g., palmitic acid and octadecenoic acid) are known to possess antibacterial and antifungal activities. Furthermore, [49] report the production of palmitic acid by *Trichoderma virens* and *T. harzianum*. Compounds such as methoxy-acetic acid and benzene were also detected; these compounds have been demonstrated to exhibit antimicrobial activities [50]. Moreover, in a previous study, [51] reported that the terpenoid and limonene were the main components which were observed as effective biological control compounds. Moreover, limonene is considered as a mediator of plant growth that leads to a change in the concentration of chlorophyll and the size of plants. In similar studies, alkanes with antifungal activity were also detected such as cyclohexane and cyclopentane, other alkanes were identified such as dodecane which has a role as antifungal agent [52].

3.2 Production of antimicrobial compounds

Plants and microorganisms are in constant competition for nutrients, microorganisms produce various antimicrobial compounds as a strategy to compete with other microorganisms for establishment in an ecological niche [53]. These compounds have bactericidal or bacteriostatic effect. Antimicrobial compounds produced by *Trichoderma* species can act by various mechanisms against a wide range of soil-borne phytopathogenic fungi including the production of antibiotics and/or hydrolytic enzymes, as well as competition for nutrients [12–14].

3.2.1 Lytic enzymes

The cell walls of fungi contain chitin, cellulase, and glucan. Therefore, phytopathogenic fungi are affected by some lytic enzymes, including 1,3-glucanases, lipases, cellulases, and chitinases [54]. *Trichoderma* species have been widely recognized for the production of extracellular enzymes with mycoparasitic effect such as glucanases, cellulases, and chitinases. Mycelium lysis was observed in the confrontation zone between the pathogen *Verticillium dahliae* and *Trichoderma* species suggesting the ability of these isolates isolated from the rhizosphere of healthy olive to produce enzymes involved on cell wall degradation process and lysis of the mycelium during the mycoparasitic activity [45]. Moreover, [12] suggested that *Trichoderma* isolates are able to produce cell wall degrading enzymes such as cellulase, xylanase, pectinase, glucanase, lipase, amylase, arabinase, protease, and chitinases that are the most important lytic enzymes playing a key role in the degradation of cell walls of other plant pathogenic fungi (**Figure 1**) [12].

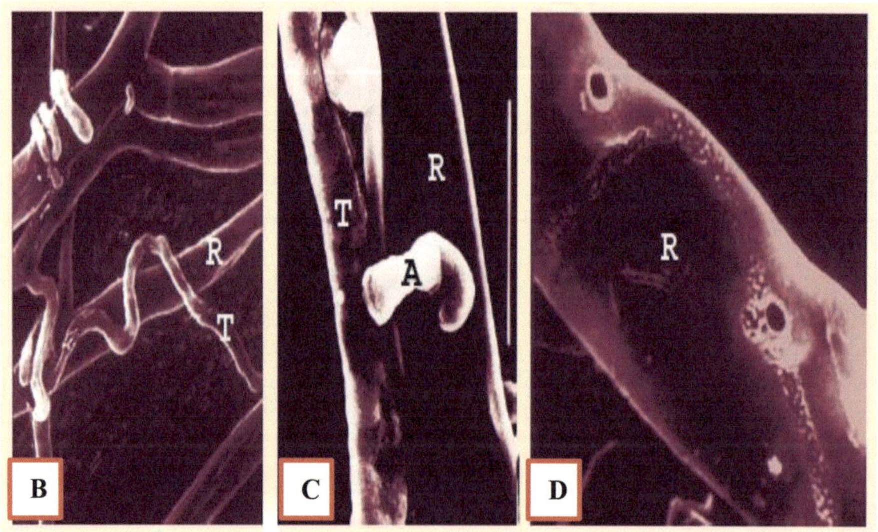

Figure 1.
The different action stages of Trichoderma *against* Rhizoctonia solani *through mycoparasitism are as follows: (A) appressorium-like structures, (B)* Trichoderma *wrapping around the hyphae of* R. solani, *(C) an enlargement of the interaction between* Trichoderma *and* Rhizoctonia *in which appressorium-like structures are observed. The scale bar equals 10 μm. (D) The hypha of* R. solani, *from which the* Trichoderma *hypha has been removed, shows the pores caused by the mycoparasite at the junction points between the two hyphae. R: hyphae of* R. solani. *T: hyphae of* Trichoderma *spp. [12].*

3.2.2 Antibiotics

Fungi are able to produce compounds with antibiotic properties. They have low molecular weight and can interfere with the development of various microorganisms through inhibition [28]. This inhibition action called antibiosis. It is another mechanism found in *Trichoderma* which can restrict phytopathogens growth. There are more than 180 secondary metabolites of *Trichoderma* that have been identified into various classes of compounds [55]. These compounds have different effects against pathogens. Some secondary metabolites affect plant metabolism and growth. For example *T. viride*, *T. harzianum*, and *Trichoderma koningii*, are capable in the production and secretion of a volatile compound, 6-pentyl-α-pyrone (6-PAP) which exhibit antifungal activity against several pathogenic species such as *Botrytis cinerea*, *R. solani*, and *Fusarium oxysporum* [28].

Antibiotics produced by *Trichoderma* species inhibit the growth of other microorganisms. Most *Trichoderma* strains produce metabolites with antibiotic properties that prevent colonization by antagonized microorganisms; among these metabolites, the production of harzianic acid, alamethicins, tricholin, peptaibols, antibiotics, 6-PAP, massoilactone, viridin, gliovirin, glisoprenins, heptelidic acid, and others have been described [56].

3.2.3 Competition

Competition between fungi and pathogens is another mechanism of biocontrol. *Trichoderma* can compete for nutrients and restrict the growth of pathogens especially when nutrients become a limiting factor. They can easily compete the rhizosphere of different plants and cause changes in plant metabolism. Moreover, *Trichoderma* may colonize also space around infection sites [57]. Through chelators production such as siderophores which increase the absorption and concentration of certain nutrients (copper, iron, phosphorus, manganese, and sodium). As a result, iron will be less available for the pathogen. For this reason, competition in the soil between microorganisms is considered an indirect control mechanism for pathogens [58]. Studies conducted by [59] showed the important role of siderophores produced by *Trichoderma asperellum* in antagonism against *F. oxysporum*. They also play a role in stimulating plant growth by reducing oxidative stress. Besides, studies conducted by [59] showed the role of siderophores produced by *T. asperellum* T34 in controlling *F. oxysporum*, with a reduction in tomato infestation and stimulation of root plant growth.

3.2.4 Siderophores and the acquisition of iron

Siderophores are low-molecular-weight secondary metabolites produced by a wide range of plant and fungal species such as *Trichoderma* spp. [60]. They have the ability to capture metal ions with a high affinity for Fe (III) than Fe (II). Depending on the functional group that acts as the sequestrant, they can be classified into catecholates, hydroximates, and hydroxycarboxylates [61]. In the rhizosphere, crops may obtain iron through microbially produced siderophores [62]. There are more than 500 biomolecules that are classified as siderophores [62]. Iron deficiency can lead to severe biological inhibition for organisms by depriving them of this element because it is essential in cellular processes such as DNA synthesis, respiration, and free-radical detoxification [63]. Siderophores produced by *Trichoderma* spp. demonstrate various functions in the rhizosphere. In addition to conferring an advantage to take iron into the rhizosphere, under limiting conditions, siderophores may also inhibit the growth of pathogens that could potentially cause damage to the plant [64]. *Trichoderma* spp. producing siderophores in rhizospheres can restrict iron and make it less available to pathogens, indirectly promoting plant growth [65]. Studies conducted by [59] showed the role of siderophores produced by *T. asperellum* T34 in controlling *F. oxysporum*, reducing tomato infestation and stimulating plant root growth.

4. Biomolecules enhancing plant growth

4.1 Production of phytohormones

Phytohormones play an important role in agriculture [66]; they are synthesized by many rhizosphere microorganisms including *Trichoderma* spp. They have various roles such as modification of the physiological functions of plants to accelerate their growth by intensive cell division in callus tissue, promotion of phloem development, enhance lateral root development, plant growth stimulation and prevention of leaf aging by slowing down the breakdown of chlorophyll pigments in plants as well as improving metabolism even at low concentrations [67–69]. IAA and gibberellins (GAs) are among the most important phytohormones that regulate the plant's development and enhance plant growth through several processes [70–74].

4.2 Acquisition and nutrients solubilization

Various fungi such as *Trichoderma* spp. are associated to the plant roots' rhizosphere, they provide nutrients, protection against biotic and abiotic stresses, and stimulate plant growth [75, 76]. *Trichoderma* species have the ability to acquire nutrients in the rhizosphere through various mechanisms.

4.2.1 Phosphate solubilization

Phosphorus is important elements for plant growth. It can be found in two forms: organic phosphorus and inorganic phosphorus, which usually forms insoluble mineral compounds with calcium, aluminum, or manganese [77, 78]. The distribution of these forms in soils is influenced by several factors such as microbial activity, pH, soil type, and organic matter availability [1]. Recently, phosphate solubilizing microorganisms have attracted the attention of agronomists; these microorganisms were used as soil inoculum to improve plant growth [79]. Plants and fungi including *Trichoderma* spp. compete for the limited available phosphorus through various processes, such as solubilization, precipitation, absorption, and desorption. Inorganic phosphate and organic phosphorus can be mineralized through enzymatic action [1]. *Trichoderma* species have the ability to solubilize insoluble phosphate into soluble phosphate [80, 81]. In previous studies; [82] reported that *Trichoderma atroviride* LBM 112 and *T. stilbohypoxyli* LBM 120 revealed positive results for phosphate solubilization with formation of halo-zone on the solid medium containing insoluble inorganic phosphorus source. In addition, *T. harzianum* T11 (OL587563) isolated from rhizosphere soil of olive trees has several plant growth-promoting traits, such as the phosphate-solubilizing ability and the production of siderophores [74].

4.2.2 Nitrification and nitrogen fixation

Nitrogen fixation processes have significant ecological importance in various ecosystems, including those of agricultural interest. Nitrogen plays a critical role in plant synthesis as it is a component of important biomolecules such as nucleic acids, peptides, organic acids, and fatty acids, which are necessary for the structure and activity of all organisms. Nitrogen-fixing by microorganisms play a key role on growth-promoting plant. It has been suggested that the promotion effect on plant growth might be mediated by providing nitrogen through biological nitrogen fixation and hormones [83, 84]. Production of ammonia and nitrogen-fixing ability by *Trichoderma* strains are reported in previous findings. Ahemad and Kibret [85] reported that ammonia is useful for plants as directly or indirectly. Ammonia production by the *Trichoderma* isolates may influence plant growth indirectly;

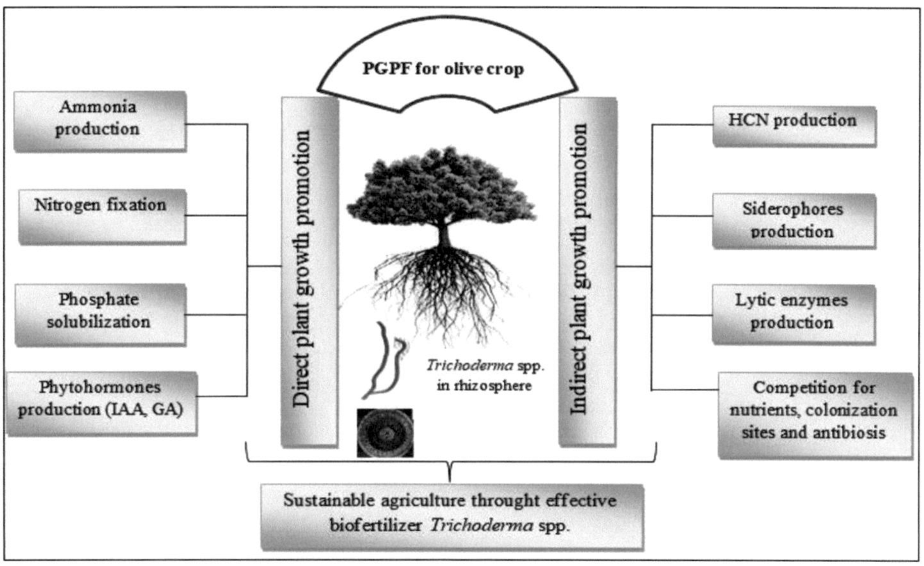

Figure 2.
Schematic description of the main mechanisms used by Trichoderma *spp. to competitively colonize the rhizosphere of host plants [74].*

Chemical nature	Secondary metabolites	*Trichoderma* species	Bio-activity observed	Reference
Alcohol	2-Phenylethanol	*T. harzianum*	Reduces the growth of *Aspergillus flavus* and aflatoxin production	[87, 88]
Anthraquinone	Pachybasin	*T. harzianum*	Increases the number of coils of the biocontrol agent against *R. solani*	[89]
	Emodin	*T. viride*	Antimicrobial and antineoplasic agent	[90–92]

Chemical nature	Secondary metabolites	*Trichoderma* species	Bio-activity observed	Reference
Azaphilone	T22azaphilone	*T. harzianum*	Inhibits the growth of *R. solani*, *Pythium ultimum* and *Gaeuman nomyces graminis*	[93]
Bisorbicillinoid	Bisvertinolone	*T. longibrachiatum*	Antifungal properties via inhibition of β- (1,6)-glucan biosynthesis	[94]
Butenolide	Dehydro derivative of harzianolide	*T. harzianum*	Antifungal activity against *Gaeumannomy ces graminis* var. *tritici*	[95]
Hydrolytic enzymes	Cellulases	*T. reesei*	Degrades cellulase during root colonization to penetrate the plant tissue	[96]
	β-1,6- Glucanases	*Trichoderma* sp.	Hydrolyses fungal pathogen cell walls of *B. cinerea*, *R. solani*, *Phytophthora citrophthora*	[17]
	Chitinases	*Trichoderma* sp.	Hydrolytic enzymes of the fungal cell wall	[97, 98]
Indolic compound	Indole-3- acetic acid (IAA)	*T. atroviride*, *T. virens*	Controls a number of growth and development processes in plants	[99]
	Indole-3- acetaldehyde	*T. atroviride*, *T. virens*	Controls root growth in *Arabidopsis thaliana*	[99]
	Indole-3-carboxaldehyde	*T. atroviride*, *T. virens*	Induces adventitious root formation in *A. thaliana*	[100]
Koninginins	Koninginins A–E	*T. koningii* *T. harzianum*	Antifungal activity against *F. oxysporum*, *Fusarium solani*, and *Alternaria panax*	[101, 102]
Monoterpene	β-Myrcene	*T. virens*	Regulates the expression of genes	[22, 103]

Chemical nature	Secondary metabolites	*Trichoderma* species	Bio-activity observed	Reference
Nitrogen heterocyclic compound	Harzianic acid	*T. arundinaceum; T. harzianum*	Antimicrobial metabolite, siderophore and plant growth regulator	[104–106]
	Harzianopyridone	*T. harzianum*	Antifungal activity against *B. cinerea*, *R. solani* and inhibitor of the protein phosphatase type 2A (PP2A)	[107]
	Melanoxadin	*T.* sp. strain ATF-451	Inhibits melanin formation in the larval hemolymph of the silkworm, *Bombyx mori*	[108]
Peptide	Trichokonin VI (Tk VI)	*T. longibrachiatum*	Inhibits primary root growth in *A. thaliana*	[109]
Pyrane	Koninginin A	*T. koningii*	Plant growth regulator	[110]
	Koninginin D	*T. koningii*	Alters pathogen fungal growth of *R. solani*, *Phytophthora cinnamomi*, *Pythium middletonii*, *F. oxysporum* and *Bipola ris sorokiniana*	[111]
Pyridones	Hharzianopyridone	*T. harzianum*	antifungal activity against plant pathogenic fungi, such as *P. ultimum*, *G. graminis* var. *tritici*, *R. solani*, and *B. cinerea*	[112]
Pyrones	6-Pentyl-2H- pyran—2-one	*Trichoderma viride* *T. atroviride*	Antifungal activity against *R. solani*, *F.*	[113]
Siderophore	Fusarinine C	*Trichoderma* sp.	Fe-chelated, can be available to plants	[114]
	Ferricrocin	*T. atroviride*[a], *T. virens*[a], *T. reesei*[a]	Key metabolite in the competition for iron in the rhizophere	[115]
	Coprogen B	*Trichoderma* spp.	Solubilizes iron unavailable to the plant	[116]

Chemical nature	Secondary metabolites	*Trichoderma* species	Bio-activity observed	Reference
Steroidal compound	Viridin	*T. koningii*, *T. virens*, *T. viride*	Antifungal metabolite that alter the spore germination of *Botrytis allii*, *Colletotrichum lini* and *Fusarium caeruleum*	[55]

[a]*http://genomebiology.com/2011/12/4/R40*

Table 1.
Secondary metabolites secreted by Trichoderma *sp. and their bio-active role.*

ACC synthesized in plant tissues by ACC synthase is released from plant roots and taken up by neighboring micro-organisms. Then, *Trichodrema* may hydrolyze ACC (1-aminocyclopropane-1-carboxylic acid) to ammonia. Besides, [74] reported that production of ammonia by *Trichoderma* species isolated from rhizosphere soil of

Compound	Strain	Crops	Application mode	Beneficial outcome	References
Biofertilizer	*Trichoderma azevedoi*	Lettuce	Simple exposure	Increases carotenoids and chlorophyll with reduction in the white mold attack to about 78.83%	[117]
	Trichoderma afroharzianum	Tomato	Seed inoculation or treatment	Helps in the secretion of Phytohormones like homeostasis, antioxidant activity, phenylpropanoid biosynthesis and glutathione metabolism	[118]
	T. harzianum, *T. asperellum*, *Trichoderma hamatum*, *T. atroviride*	Chinese cabbage	Irrigation	Increases soil enzyme activity, yield by 37%, and increases the concentration of inorganic nitrogen and phosphorus content of the soil	[119]
	Trichoderma brevicompactum, *Trichoderma gamsii*, *T. harzianum*	Tomato	Seedling drenching	Improved growth and yield due to the production of IAA	[35, 120]
	T. harzianum *T. asperellum*	Tomato	Seed treatment	Improves phosphorus uptake	[121]
	T. brevicompactm, *T. gamsii*, *T. harzianum*	Tomato	Seed drenching	Improves phosphorus solubilization	[120]

Table 2.
Trichoderma *sp. as bio-fertilizers and their role in promoting plant growth and yield.*

olive is sustained with the results obtained by [86] who reported that among 20 *Trichoderma* spp. isolated from chili rhizosphere, 13 isolates were able to produce ammonia (**Figure 2**; **Tables 1** and 2).

5. Conclusion

This report reviews the importance of *Trichoderma* spp. as a biocontrol agent suppressing the growth of the fungal pathogens and as biofertilizer enhancing plant growth. Therefore, the increase use of *Trichoderma* spp.as commercial mycofungicides and biofertilizers offers promising prospects for sustainable and environmentally friendly agriculture. These eco-friendly alternatives can substitute the excessive use of chemical products that can cause problems in the long term. The biotechnological advances from these microorganisms such as fungi are immense and yet to be explored. Thus, more studies need to be explored to elucidate the development of sustainable biotechnological applications of the *Trichoderma* species on soil–plant system.

Conflict of interest

The authors declare no competing interests.

Author details

Abdenaceur Reghmit*, Farida Benzina-tihar and Fatma Sahir-Halouane
Laboratory of Valorization and Conservation of Biological Resources, Department of Biology, University of M'hamed Bougara, Boumerdes, Algeria

*Address all correspondence to: a.reghmit@univ-boumerdes.dz

References

[1] Khatoon Z, Huang S, Rafique M, Fakhar A, Kamran MA, Santoyo G. Unlocking the potential of plant growth-promoting rhizobacteria on soil health and the sustainability of agricultural systems. Journal of Environmental Management. 2020;**273**:111118

[2] Jie WG, Bai L, Yu WJ, Cai BY. Analysis of interspecific relationships between *Funneliformis mosseae* and *Fusarium oxysporum* in the continuous cropping of soybean rhizosphere soil during the branching period. Biocontrol Science and Technology. 2015;**25**:1036-1051

[3] Huang XQ, Zhang N, Yong XY, Yang XM, Shen QR. Biocontrol of *Rhizoctonia solani* damping-off disease in cucumber with Bacillus pumilus SQR-N43. Microbiological Research. 2012;**167**:135-143

[4] Naing KW, Anees M, Kim SJ, Nam Y, Kim YC, Kim KY. Characterization of antifungal activity of Paenibacillus ehimensis KWN38 against soilborne phytopathogenic fungi belonging to various taxonomic groups. Annales de Microbiologie. 2014;**64**(1):5563

[5] Anees MR, Azim S, Ur Rehman M, Jamil S, El Hendawy NA, Al Suhaiban NA. Antifungal potential of *Trichoderma* strains originated from North Western regions of Pakistan against the plant pathogens. Pakistan Journal of Botany. 2018;**50**(5):2031-2040

[6] Li YT, Hwang SG, Huang YM, Huang CH. Effects of Trichoderma asperellum on nutrient uptake and Fusarium wilt of tomato. Crop Protection. 2018a;**110**:275-282

[7] Li Z, Wang T, Luo X, Li X, Xia C, Zhao Y, et al. Biocontrol potential of Myxococcus sp. strain BS against bacterial soft rot of calla lily caused by Pectobacterium *carotovorum*. Biological Control. 2018b;**126**:36-44

[8] Elad Y, Williamson B, Tudzynski P, Delen N. *Botrytis* spp. and diseases they cause in agricultural systems- an introduction. In: Elad Y, Williamson B, Tudzynski P, Delen N, editors. Botrytis, Biology, Pathology and Control. Netherland: Kluwer Academic Publishers; 2004. pp. 1-8

[9] Whipps JM, Lumsden RD. Commercial use of fungi as plant disease biological control agents: status and prospects. In: Butt T M, Jackson C, Magan N, editors. Fungal Biocontrol Agents: Progress, Problems and Potential. Wallingford, UK: CABI Publishing; 2001. pp. 9-22

[10] Singh H, Singh BN, Singh S, Sarma B. Exploring different avenues of Trichoderma as a potent bio-fungicidal and plant growth promoting candidate-an overview. Annual Reviews in Plant Pathology. 2013;**5**:315-321

[11] Keswani C, Mishra S, Sarma BK. Unraveling the efficient application of secondary metabolites of various *Trichoderma*. Applied Microbiology and Biotechnology. 2014;**98**:533-544

[12] Harman GE, Howell CR, Viterbo A, Chet I, Lorito M. *Trichoderma* species – opportunistic, avirulent plant symbionts. Nature Reviews Microbiology. 2004;**2**: 43-56. DOI: 10.1038/nrmicro797

[13] Hermosa R, Viterbo A, Chet I, Monte E. Plant-beneficial effects of Trichoderma and of its genes. Microbiology. 2012;**158**:17-25

[14] Shoresh M, Harman GE, Mastouri F. Induced systemic resistance and plant

responses to fungal biocontrol agents. Annual Review of Phytopathology. 2010;**48**:21-43

[15] Ghisalberti EL, Sivasithamparam K. Antifungal antibiotics produced by Trichoderma spp. Soil Biology and Biochemistry. 2011;**23**:1011-1020

[16] Khatri DK, Tiwari DN, Bariya HS. Chitinolytic efficacy and secretion of cell wall-degrading enzymes from Trichoderma spp. In response to phytopathological fungi. Journal of Applied Biology and Biotechnology. 2017;**5**(6):1-8. DOI: 10.7324/JABB.2017.50601

[17] Druzhinina IS, Seidl-Seiboth V, Herrera-Estrella A, Horwitz BA, Kenerley CM, Monte E, et al. Trichoderma: the genomics of opportunistic success. Nature Reviews Microbiology. 2011;**9**:749-759. DOI: 10.1038/nrmicro2637

[18] Santos SS, Augusto DG, Alves PAC, Pereira JS, Duarte LM, Melo PC, et al. *Trichoderma asperelloides* ethanolic extracts efficiently inhibit Staphylococcus growth and biofilm formation. PLoS One. 2018;**13**(8):0202828. DOI: 10.1371/journal.pone.0202828

[19] Korpi A, Järnberg J, Pasanen AL. Microbial volatile organic compounds. Critical Reviews in Toxicology. 2009;**39**:139-193

[20] Schnürer J, Olsson J, Börjesson T. Fungal volatiles as indicators of food and feeds spoilage. Fungal Genetics and Biology. 1999;**27**:209-217

[21] Tirranen LS, Gitelson II. The role of volatile metabolites in microbial communities of the LSS higher plant link. Advances in Space Research. 2006;**38**(6):1227-1232

[22] Crutcher FK, Parich A, Schuhmacher R, Mukherjee PK, Zeilinger S, Kenerley CM. A putative terpene cyclase, vir4, is responsible for the biosynthesis of volatile terpene compounds in the biocontrol fungus Trichoderma virens. Fungal Genetics and Biology. 2013;**56**:67-77

[23] Bitas V, Kim HS, Bennett JW, Kang S. Sniffing on microbes: diverse roles of microbial volatile organic compounds in plant health. Molecular Plant-Microbe Intractions. 2013;**26**:835-843

[24] Saber WIA, Abd El-Hai KM, Ghoneem KM. Synergistic effect of *Trichoderma* and *Rhizobium* on both biocontrol of chocolate spot disease and induction of nodulation, physiological activities and productivity of Vicia faba. Research Journal of Microbiology. 2009;**4**(8):286-300

[25] Vinale F, Sivasithamparam K, Ghisalberti EL, Woo SL, Nigro M, Marra R, et al. Trichoderma secondary metabolites active on plants and fungal pathogens. The Open Mycology Journal. 2014;**8**:127-139

[26] Ezzat AS, Ghoneem KM, Saber WIA, Alaskar AA. Control of wilt, stalk and tuber rots diseases using Arbuscular mycorrhizal fungi, Trichoderma species and hydroquinone enhances yield quality and storability of Jerusalem artichoke (Helianthus tuberosus L.). Egyptian Journal of Biological Pest Control. 2015;**25**(1):11-22

[27] Napitupulu, TP, Kanti A, Sudiana IM. Evaluation of the environmental factors modulating indole-3-acetic acid (IAA) production by *Trichoderma harzianum* InaCC F88. In: IOP conference series: earth and environmental science (Vol. 308, No. 1, *of bacterial infections*. Clin Microbiol Open Acc 3:15. 2019

[28] Rincón AM, Benítez T, Codón AC, Moreno-Mateos MA. Biotechnological aspects of *Trichoderma* spp. In: Rai M, Bridge PD, editors. Applied Mycology. London, UK: CAB International; 2009. pp. 216-223

[29] Vinale FK, Sivasithamparam LE, Ghisalberti R, Marra LS, Lorito M. Trichoderma-plant-pathogen interactions. Soil Biology and Biochemistry. 2008;**40**:1-10. DOI: 10.1016/j.soilbio.2007.07.002

[30] Kodsueb R, McKenzie EHC, Lumyong S, Hyde KD. Diversity of saprobic fungi on Magnoliaceae. Fungal Diversity. 2008;**30**:37-53. Available from: http://cmuir.cmu.ac.th/jspui/handle/6653943832/60072

[31] Kaewchai S, Soytong K, Hyde KD. Mycofungicides and fungal biofertilizers. Fungal Diversity. 2009;**38**:25-50

[32] Howell CR. Mechanisms employed by *Trichoderma* species in the biological control of plant diseases; the history and evolution of current concepts. Plant Disease. 2003;**87**(1):4-10. DOI: 10.1094/pdis. 2003.87.1.4

[33] Mukherjee AK, Sampath Kumar A, Kranthi S, Mukherjee PK. Biocontrol potential of three novel *Trichoderma* strains: isolation, evaluation and formulation. 3 Biotech. 2014;**4**(3):275-281. DOI: 10.1007%2Fs13205-013-0150-4

[34] Swain H, Adak T, Mukherjee AK, Mukherjee PK, Bhattacharyya P, Behera S, et al. Novel Trichoderma strains isolated from tree barks as potential biocontrol agents and biofertilizers for direct seeded rice. Microbiological Research. 2018;**214**:83-90. DOI: 10.1016/j.micres.2018.05.015

[35] Kamal RK, Athisayam V, Gusain YS, Kumar V. Trichoderma: a most common biofertilizer with multiple roles in agriculture. Biomedical Journal of Scientific & Technical Research. 2018;**4**:4136-4137. DOI: 10.26717/BJSTR.2018.04.001107

[36] Yuri. Fusarium oxysporum. Retrieved from the Web site: http://thunderhouse4-yuriblogspot.com/2012/06/fusarium-oxysporum.html; 2012

[37] Benítez T, Rincón AM, Limón MC, Codón AC. Biocontrol mechanisms of Trichoderma strains. International Microbiology. 2004;**7**:249-260

[38] Mayo S, Gutiérrez S, Malmierca MG, Lorenzana A, Campelo MP, Hermosa R, et al. Influence of *Rhizoctonia solani* and *Trichoderma* spp. in growth of bean (Phaseolus vulgaris L.) and in the induction of plant defense-related genes. Frontiers in Plant Science. 2015;**6**:685. DOI: 10.3389/fpls.2015.00685

[39] Zin NA, Badaluddin NA. Biological functions of Trichoderma spp. for agriculture applications. Annals of Agricultural Science. 2020;**65**(2):168-178. DOI: 10.1016/j.aoas.2020.09.003

[40] Kumar A, Patel A, Singh S, Tiwari R. Effect of Trichoderma sp. in plant growth promotion in Chilli. International Journal of Current Microbiology and Applied Science. 2019;**8**(3):1574-1581

[41] Ueda H, Kikuta Y, Matsuda K. Plant communication. Plant Signaling & Behavior. 2012;**7**:222-226

[42] Pieterse CMJ, Zamioudis C, Berendsen RL, Weller DM, Van Wees SCM, Bakker PAHM. Induced systemic resistance by beneficial microbes. Annual Review of Phytopathology. 2014;**52**:347-375

[43] Bakker PAHM, Pieterse CMJ, Van Loon LC. Induced systemic resistance

by fluorescent Pseudomonas spp. Phytopathology. 2007;**97**:239-243

[44] Siddiquee S, Cheong BE, Taslima K, Kausar H, Hasan MdM. Separation and identification of volatile compounds from liquid cultures of Trichoderma harzianum by GC-MS using three different capillary columns. Journal of Chromatographic Science. 2012;**50**:358-367

[45] Reghmit A, Benzina-tihar F, López Escudero FJ, Halouane-Sahir F, Oukali Z, Bensmail S, et al. Trichoderma spp. isolates from the rhizosphere of healthy olive trees in northern Algeria and their biocontrol potentials against the olive wilt pathogen, Verticillium dahliae. Organic Agriculture. 2021;**11**:639-657

[46] Karanja E, Boga H, Muigai A, Wamunyokoli F, Kinyua J, Nonoh J. Growth characteristics and production of secondary metabolites from selected novel *Streptomyces* species isolated from selected Kenyan national parks. In: Scientific conference proceeding. 2012

[47] Nandhini SU. Gas chromatography–mass spectrometry analysis of bioactive constituents from the marine Streptomyces. Asi J Pharm Clin Res 8:244-246.Ng LC, Ngadin A, Azhari M, Zahari NA (2015) Potential of Trichoderma spp. as biological control agents against bakanae pathogen (*Fusarium fujikuroi*) in rice. Asian J Plant Pathol. 2015;**9**:46-58

[48] Pohl CH, Kock JL, Thibane VS. Antifungal free fatty acids: a review. Science Against Microbial Pathogens: Communicating Current Research and Technological Advances. 2011;**1**:61-71

[49] Dubey SC, Tripathi A, Dureja P, Grover A. Characterization of secondary metabolites and enzymes produced by *Trichoderma* species and their efficacy against plant pathogenic fungi. Indian Journal of Agricultural Research. 2011;**81**(5):455-461

[50] Sohrabi M, Zhang L, Zhang K, Ahmetagic A, Wei MQ. Volatile organic compounds as novel markers for the detection of bacterial infections. Clinical Microbiology Peer Reviewed Open Access Journals. 2014;**3**:151

[51] Khethr FBH, Ammar S, Saidana D, Daami M, Chriaa J, Liouane K, et al. Chemical composition, antibacterial and antifungal activities of Trichoderma sp. growing in Tunisia. Annales de Microbiologie. 2008;**58**:303. DOI: 10.1007/BF03175334

[52] Hsouna AB, Trigie M, Mansour RB, Jarraya RM, Damak M, Jaoua S. Chemical composition, cytotoxicity effect and antimicrobial activity of *Ceratonia siliqua* essential oil with preservative effects against listeria inoculated in minced beef meat. International Journal of Food Microbiology. 2011;**148**(1):66-72

[53] Santoyo G, del Orozco-Mosqueda MC, Govindappa M. Mechanisms of biocontrol and plant growth-promoting activity in soil bacterial species of Bacillus and Pseudomonas: a review. Biocontrol Science and Technology. 2012;**22**:855-872

[54] Bhagwat A, Collins CH, Dordick JS. Selective antimicrobial activity of cell lytic enzymes in a bacterial consortium. Applied Microbiology and Biotechnology. 2019;**103**:7041-7054

[55] Reino JL, Guerrero RF, Hernandez-Gal R, Collado IG. Secondary metabolites from species of the biocontrol agent Trichoderma. Phytochemistry Reviews. 2008;**7**(1):89-123. DOI: 10.1007/s11101-006-9032-2

[56] Vey A, Hoagland RE, Butt TM. Toxic metabolites of fungal biocontrol

agents. In: Butt TM, Jackson C, Magan N, editors. Fungi as Biocontrol Agents: Progress, Problems and Potential. Bristol: CAB International; 2001. pp. 311-346

[57] Gosling P, Hodge A, Goodlass G, Bending GD. Arbuscular mycorrhizal fungi and organic farming. Agriculture, Ecosystems & Environment. 2006;**113**:17-35. DOI: 10.1016/j.agee.2005.09.009

[58] Eisendle M, Oberegger H, Buttinger R, Illmer P, Haas H. Biosynthesis and uptake of siderophores is controlled by the Pae C-mediated ambient-pH regulatory system in *Aspergillus nidulans*. Eukaryotic Cell. 2004;**3**:561-563

[59] Segarra G, Casanova E, Avilés M, Trillas I. *Trichoderma asperellum* Strain T34 controls fusarium wilt disease in tomato plants in soilless culture through competition for iron. Microbial Ecology. 2010;**59**:141-149

[60] Ortiz-Galeana MA, Hernández-Salmerón JE, Valenzuela-Aragón B, De Los Santos-VillalobosS, Rocha-GranadosMDC, Santoyo G. Diversity of cultivable endophytic bacteria associated with blueberry plants (Vaccinium corymbosum L.) cv. Biloxi with plant growth-promoting traits. Chilean Journal of Agricultural & Animal Sciences. 2018;**34**:140-151

[61] Hider RC, Kong X. Chemistry and biology of siderophores. Natural Product Reports. 2010;**27**:637-657

[62] Dimkpa C. Microbial siderophores: Production, detection and application in agriculture and environment. Endocytobiosis and Cell Research. 2016;**27**:7-16

[63] Aguado-Santacruz GA, Moreno-Gómez B, Jiménez-Francisco B, García-Moya E, Preciado-Ortiz RE. Impacto de los sideróforos microbianos y fitosideróforos en la asimilación de hierro por las plantas: Una síntesis. Revista Fitotecnia Mexicana. 2012;**35**:9-21

[64] Crowley DE. Microbial siderophores in the plant rhizosphere. In: Barton LL, Abadia J, editors. Iron nutrition in plants and rhizospheric microorganisms. Dordrecht: Springer; 2006. pp. 169-198

[65] Budzikiewicz H. Siderophores of the Pseudomonadaceae sensu stricto (Fluorescent and Non-*Fluorescent Pseudomonas* spp.). In: Budzikiewicz H, Flessner T, Jautelat R, Scholz U, Winterfeldt E, Herz W, Falk H, Kirby GW, editors. Progress in the Chemistry of Organic Natural Products. Progress in the Chemistry of Organic Natural Products. Vienna: Springer Vienna; 2004. pp. 81-237

[66] Jaroszuk-Ściseł J, Tyśkiewicz R, Nowak A, Ozimek E, Majewska M, Hanaka A, et al. Phytohormones (auxin, gibberellin) and ACC deaminase in vitro synthesized by the mycoparasitic *Trichoderma* DEMTkZ3A0 strain and changes in the level of auxin and plant resistance markers in wheat seedlings inoculated with this strain conidia. International Journal of Molecular Sciences. 2019;**20**(19):4923

[67] Karadeniz A, Topcuoğlu ŞF, İnan S. Auxin, gibberellin, cytokinin and abscisic acid production in some bacteria. World Journal of Microbiology and Biotechnology. 2006;**22**(10):1061-1064. DOI: 10.1007/s11274-005-4561-1

[68] Chanclud E, Morel JB. Plant hormones: a fungal point of view. Molecular Plant Pathology. 2016;**17**(8):1289-1297. DOI: 10.1111/mpp.12393

[69] Fayziev V, Jovlieva D, Juraeva U, Shavkiev J, Eshboev F. Effects of

PVXN-UZ 915 necrotic isolate of potato virus X on amount of pigments of *Datura stramonium* leaves. Journal of Critical Reviews. 2020;7(9):400-403. DOI: 10.31838/jcr.07.09.82

[70] Pallardy SG. Plant hormones and other signaling molecules. In: Pallardy SG, editor. Physiology of Woody Plants. 3rd edition. Columbia, Mossouri: Academic Press; 2008:367-377. DOI: 10.1016/B978-012088765 1.50014-2

[71] Tsavkelova EA, Klimova SY, Cherdyntseva TA, Netrusov AI. Microbial producers of plant growth stimulators and their practical use: a review. Applied Biochemistry and Microbiology. 2006;**42**(2):117-126. DOI: 10.1134/S0003683806020013

[72] Hamayun M, Sumera A, Ilyas I, Bashir A, In-Jung L. Isolation of a Gibberellin producing fungus (Penicillium sp.MH7) and growth promotion of crown daisy (Chrysanthemum coronarium). Journal of Microbiology and Biotechnology. 2010;**20**(1):202-207. DOI: 10.4014/jmb.0905.05040

[73] Jaroszuk-Ściseł J, Kurek E, Trytek M. Efficiency of indoleacetic acid, gibberellic acid and ethylene synthesized in vitro by Fusarium culmorum strains with different effects on cereal growth. Biologia. 2014;**69**(3):281-292. DOI: 10.2478/s11756-013-0328-6

[74] Abdenaceur R, Farida BT, Mourad D, Rima H, Zahia O, Fatma SH. Effective biofertilizer Trichoderma spp. isolates with enzymatic activity and metabolites enhancing plant growth. International Microbiology. 2022;**25**(4):817-829

[75] Brimecombe MJ, De Leij FA, Lynch JM. The effect of root exudates on rhizosphere microbial populations. In: Pinton R, Varanini Z, Nanipieri P, editors. The Rhizosphere. New York: Marcel Dekker Inc.; 2001. pp. 95-140

[76] Filiz O, Takil E, Kayan N. The role of plant growth promoting rhizobacteria (Pgpr) and phosphorus fertilization in improving phenology and physiology of bean (*phaseolus vulgaris* l.). Applied Ecology and Environmental Research. 2021;**19**(3):2507-2517. DOI: 10.15666/aeer/1903_25072517

[77] Rodríguez H, Fraga R. Phosphate solubilizing bacteria and their role in plant growth promotion. Biotechnology Advances. 1999;**17**:319-339 [CrossRef]

[78] Alori ET, Glick BR, Babalola OO. Microbial phosphorus solubilization and its potential for use in sustainable agriculture. Frontiers in Microbiology. 2017;**8**:971

[79] Fasim F, Ahmed N, Parson R, Gadd GM. Solubilization of zinc salts by a bacterium isolated from air environment of a tannery. FEMS Microbiology Letters. 2002;**213**:1-6. DOI: 10.1111/j.1574-6968.2002.tb11277.x

[80] Akintokun AK, Akande GA, Akintokun PO, Popoola TOS, Babalola AO. Solubilization on insoluble phosphate by organic acid-producing fungi isolated from Nigerian soil. International Journal of Soil Science. 2007;**2**(4):301-307. DOI: 10.3923/ijss.2007.301.307

[81] Saravanakumar K, Arasu VS, Kathiresan K. Effect of *Trichoderma* on soil phosphate solubilization and growth improvement of *Avicennia marina*. Aquatic Botany. 2013;**104**:101-105. DOI: 10.1016/j.aquab ot.2012.09.001

[82] Lopez AC, Alvarenga AE, Zapata PD, Luna MF, Villalba LL. Trichoderma spp. from Misiones, Argentina: effective fungi to promote plant growth of the

regional crop Ilex paraguariensis St. Hil. Mycology. 2019;**10**(4):210-221

[83] Bach E, dos Santos Seger GD, de Carvalho FG, Lisboa BB, Passaglia LMP. Evaluation of biological control and rhizosphere competence of plant growth promoting bacteria. Applied Soil Ecology. 2016;**99**:141-149. DOI: 10.1016/j.apsoil.2015.11.002

[84] Hemerly A. Genetic controls of biomass increase in sugarcane by association with beneficial nitrogen-fixing bacteria. Plant and Animal Genome Conference XXIV. Universidade Federal do Rio de Janeiro, Rio de Janeiro, Brazil. 2016

[85] Ahemad M, Kibret M. Mechanisms and applications of plant growth promoting rhizobacteria: current perspective. Journal of King Saud University - Science. 2014;**26**(1):1-20. DOI: 10.1016/j.jksus.2013.05.001

[86] Mohiddin FA, Bashir I, Padder SA, Hamid B. Evaluation of different substrates for mass multiplication of *Trichoderma* species. Journal of Pharmacognosy and Phytochemistry. 2017;**6**(6):563-569

[87] Tarus PK, Langat-Thoruwa CC, Wanyonyi AW, Chhabra SC. Bioactive metabolites from Trichoderma harzianum and Trichoderma longibrachiatum. Bulletin of the Chemical Society of Ethiopia. 2003;**17**:185-190

[88] Chang PK, Hua SS, Sarreal SB, Li RW. Suppression of aflatoxin biosynthesis in Aspergillus flavus by 2 phenylethanol is associated with stimulated growth and decreased degradation of branched-chain amino acids. Toxins (Basel). 2015;**7**:3887-3902

[89] Lin YR, Lo CT, Liu SY, Peng KC. Involvement of pachybasin and emodin in self-regulation of *Trichoderma harzianum* mycoparasitic coiling. Journal of Agricultural and Food Chemistry. 2012;**60**:2123-2128

[90] Ali S, Watson MS, Osborne RH. The stimulant cathartic, emodin, contracts the rat isolated ileum by triggering release of endogenous acetylcholine. Autonomic & Autacoid Pharmacology. 2004;**24**:103-105

[91] Huang Q, Shen HM, Shui G, Wenk MR, Ong C-N. Emodin inhibits tumor cell adhesion through disruption of the membrane lipid raft-associated integrin signaling pathway. Cancer Research. 2006;**66**:5807-5815

[92] Wu YW, Ouyang J, Xiao XH, Gao WY, Liu Y. Antimicrobial properties and toxicity of anthraquinones by microcalorimetric bioassay. Chinese Journal of Chemistry. 2006;**24**:45-50

[93] Vinale F, Marra R, Scala F, Ghisalberti EL, Lorito M, Sivasithamparam K. Major secondary metabolites produced by two commercial Trichoderma strains active against different phytopathogens. Letters in Applied Microbiology. 2006;**43**:143-148. DOI: 10.1111/j.1472-765X.2006.01939.x

[94] Kontani M, Sakagami Y, Marumo S. First β-1,6- glucan biosynthesis inhibitor, bisvertinolone isolated from fungus, Acremonium strictum and its absolute stereochemistry. Tetrahedron Letters. 1994;**35**:2577-2580

[95] Ordentlich A, Wiesman Z, Gottlieb HE, Cojocaru M, Chet I. Inhibitory furanone produced by the biocontrol agent *Trichoderma harzianum*. Phytochemistry. 1992;**31**:485-486. DOI: 10.1016/0031-9422(92)90021-H

[96] Payne CM, Knott BC, Mayes HB, Hansson H, Himmel ME,

Sandgren M, et al. Fungal cellulases. Chemical Reviews. 2015;**115**:1308-1448

[97] Gruber S, Seidl-Seiboth V. Self versus non-self: fungal cell wall degradation in Trichoderma. Microbiology. 2012;**158**:26-34

[98] Tzelepis G, Dubey M, Jensen DF, Karlsson M. Identifying glycoside hydrolase family 18 genes in the mycoparasitic fungal species Clonostachys rosea. Microbiology. 2015;**161**:1407-1419

[99] Contreras-Cornejo HA, Macias-Rodriguez L, Cortes-Penagos C, Lopez-Bucio J. Trichoderma virens, a plant beneficial fungus enhances biomass production and promotes lateral root growth through an auxin dependent mechanism in Arabidopsis. Plant Physiology. 2009;**149**:1579-1592

[100] Contreras-Cornejo HA, Macias-Rodriguez L, Beltran-Peña E, Herrera-Estrella A, Lopez-Bucio J. Trichoderma-induced plant immunity likely involves both hormonal and camalexin dependent mechanisms in Arabidopsis thaliana and confers resistance against necrotrophic fungi Botrytis cinerea. Plant Signaling & Behavior. 2011;**6**:1554-1563. DOI: 10.4161/psb.6.10.17443

[101] Cutler HG, Himmelsbach DS, Yagen B, Arrendale RF, Jacyno JM, Cole PD, et al. Koninginin B: a biologically active congener of koninginin A from Trichoderma koningii. Journal of Agricultural and Food Chemistry. 1991;**39**:977-980. DOI: 10.1021/jf00005a035

[102] Chen JL, Liu K, Miao CP, Guan HL, Zhao LX, Sun SZ. Chemical constituents with siderophores activities from Trichoderma koningiopsis YIM PH30002. Natural Product Research and Development. 2015;**27**:1878-1883

[103] Godard K, White R, Bohlmann J. Monoterpene-induced molecular responses in Arabidopsis thaliana. Phytochemistry. 2008;**69**:1838-1849

[104] Vinale F, Flematti G, Sivasithamparam K, Lorito M, Marra R, Skelton BW, et al. Harzianic acid, an antifungal and plant growth promoting metabolite from Trichoderma harzianum. Journal of Natural Products. 2009;**72**(11):2032-2035. DOI: 10.1021/np900548p

[105] Vinale F, Nigro M, Sivasithamparam K, Flematti G, Ghisalberti EL, Ruocco M, et al. Harzianic acid: a novel siderophore from Trichoderma harzianum. FEMS Microbiology Letters. 2013;**347**(2):123-129. DOI: 10.1111/1574-6968.12231

[106] Malmierca MG, Cardoza RE, Alexander NJ, McCormick SP, Collado IG, Hermosa R, et al. Relevance of trichothecenes in fungal physiology: disruption of tri5 in Trichoderma arundinaceum. Fungal Genetics and Biology. 2013;**53**:22-33

[107] Kawada M, Yoshimoto Y, Kumagai H, Someno T, Momose I, Kawamura N, et al. PP2A inhibitors, harzianic acid and related compounds produced by fungal strain F-1531. Journal of Antibiotics. 2004;**57**:235-237

[108] Hashimoto R, Takahashi S, Hamano K, Nakagawa A. A new melanin biosynthesis inhibitor, melanoxadin from fungal metabolite by using the larval haemolymph of the silkworm, Bombyx mori. Journal of Antibiotics. 1995;**48**:1052-1054

[109] Shi WL, Chen XL, Wang LX, Gong ZT, Li S, Li CL, et al. Cellular and molecular insight into the inhibition of primary root growth of Arabidopsis induced by peptaibols, a class of linear

peptide antibiotics mainly produced by Trichoderma spp. Journal of Experimental Botany. 2016;**67**(8):2191-2205. DOI: 10.1093/jxb/erw023

[110] Cutler HG, Himmelsbach DS, Arrendale RF, Cole PD, Cox RH. Koninginin A: a novel plant growth regulator from Trichoderma koningii. Agricultural and Biological Chemistry. 1989;**53**:2605-2611. DOI: 10.1271/bbb1961.53.2605

[111] Dunlop RW, Simon A, Sivasithamparam K, Ghisalberti EL. An antibiotic from Trichoderma koningii active against soilborne plant pathogens. Journal of Natural Products. 1989;**52**:67-74

[112] Dickinson JM, Hanson JR, Hitchcock PB, Claydon N. Structure and biosynthesis of harzianopyridone, an antifungal metabolite of Trichoderma harzianum. Journal of the Chemical Society, Perkin Transactions. 1989;**1**:1885-1887. DOI: 10.1039/P19890001885

[113] Garnica-Vergara A, Barrera-Ortiz S, Munoz-Parra E, Raya-Gonzalez J, Mendez-Bravo A, Macias-Rodriguez L, et al. The volatile 6-pentyl-2H-pyran-2-one from Trichoderma atroviride regulates Arabidopsis thaliana root morphogenesis via auxin signalling and ethylene insensitive 2 functioning. The New Phytologist. 2015;**209**:1496-1512

[114] Anke H, Kinn J, Bergquist KE, Sterner O. Production of siderophores by strains of the genus Trichoderma isolation and characterization of the new lipophilic coprogen derivative, palmitoylcoprogen. Biometals. 1991;**4**:176-180

[115] Kubicek CP, Herrera-Estrella A, Seidl-Seiboth V, Martinez DA, Druzhinina IS, Thon M, et al. Comparative genome sequence analysis underscores mycoparasitism as the ancestral life style of Trichoderma. Genome Biology. 2011;**12**:R40

[116] Vinale F, Sivasithamparam K, Ghisalberti EL, Ruocco M, Wood S, Lorito M. Trichoderma secondary metabolites that affect plant metabolism. Natural Product Communications. 2012;7(11):1545-1550. PMID: 23285827

[117] Rafael L, Valadares-inglis MC, Henrique G, Peixoto S, Eliza B, de Lucas G, et al. Volatile organic compounds emitted by *Trichoderma azevedoi* promote the growth of lettuce plants and delay the symptoms of white mold. Biological Control. 2021;**152**:104447. DOI: 10.1016/j.biocontrol.2020.104447

[118] Juan Z, Ting LIU, Wei-cheng LIU, Dian-peng Z, Dan D, Hui-ling WU, et al. Transcriptomic insights into growth promotion effect of Trichoderma afroharzianum TM2-4 microbial agent on tomato plants. Journal of integrative. Agriculture. 2021;**20**(5):1266-1276. DOI: 10.1016/S2095-3119(20)63415-3

[119] Ji S, Liu Z, Liu B, Wang Y, Wang J. The effect of Trichoderma biofertilizer on the quality of flowering Chinese cabbage and the soil environment. Scientia Horticulturae. 2020;**262**:109069. DOI: 10.1016/j. scienta. 2019.109069

[120] Bader AN, Salerno GL, Covacevich F, Consolo VF. Native *Trichoderma harzianum* strains from Argentina produce indole-3 acetic acid and phosphorus solubilization, promote growth and control wilt disease on tomato (*Solanum lycopersicum* L.). Journal of King Saud University-Science. 2020;**32**(1):867-873. DOI: 10.1016/j.jksus.2019.04.002

[121] Yu Z, Wang Z, Zhang Y, Wang Y, Liu Z. Biocontrol and growth-promoting effect of Trichoderma asperellum TaspHu1 isolate from Juglans mandshurica rhizosphere soil. Microbiological Research. 2021;**242**:126596. DOI: 10.1016/j.micres.2020.126596

Section 6

Ethnomedicine

Chapter 19

Use of Medicinal Plants: Interindividual Variability of Their Effects from a Genetic and Anthropological Perspective

Alda Pereira da Silva Oliveira, Maria do Céu Costa and Manuel Pires Bicho

Abstract

The use of plants for nutritional and therapeutic purposes has been constant over the centuries. The variability of enzymatic activity between individuals and populations in an attempt to adapt has been a conditioning mechanism, reflected in the incidence and prevalence of certain diseases, possible adverse effects of plant-derived nutrients and their interaction with medications, in addition to interference in natural selection and consequent geographical distribution of specific genetic polymorphisms in harmony with indigenous medicinal plants. The metabolizer type may influence the anticancer protective effect of certain plant-derived constituents, with interindividual variability to be considered. This chapter will deepen and develop the role of using plants in different geographic areas and populations over the centuries in producing the genetic variability of the metabolism of plant constituents in the context of environmental adaptation and ecogenetics. Possible therapeutic/adverse effects due to this variability will be discussed.

Keywords: medicinal plants, nutrigenetics, pharmacogenetics, ecogenetics, genetic variability, anthropology

1. Introduction

Since time immemorial, medicinal plants have been a fundamental aspect of human health and continue to play a vitally important role in different cultures worldwide. Primitive medicine before the Christian era was based from a therapeutic point of view, on a powerful psychological component supported by magical beliefs and rites combined with medicinal plants.

Today, however, it is known that medicinal plants' effects can vary significantly between individuals and interfere with medicinal substances. This variability involves aspects ranging from inherent to the medicinal plant to complex genetic and anthropological factors.

The interindividual variability of the effects of medicinal plants arises from the complex interaction between the plant phenotype and genetic and anthropological factors specific to each individual and community. It is essential to recognize and respect this variability in the use of medicinal plants for health purposes.

Nutrigenetics and pharmacogenomics make it possible to identify genetic markers associated with responses to specific food or medicinal plants. The patient's genetic background, cultural environment, and lifestyle must be considered when recommending medicinal plants or herbal medicines.

Furthermore, the importance of collaboration between therapists from alternative or traditional approaches and modern healthcare providers stands out for a holistic and personalized approach to recommending herbal medicines, within integrative medicine programs.

It is currently recognized as imperative to understand the modes of interaction between different medicines from conventional and traditional healthcare systems when used in treatment combinations. Both synthetic and natural medicinal chemical entities are metabolized by the same enzyme systems in the human body, resulting in pharmacokinetic and pharmacodynamic interactions, the properties of which are still largely unknown/unquantified.

This chapter will address these three aspects, plant, individual, and anthropological, which lead to interindividual variability and its effects, highlighting the growing importance of medicine that respects variability and, increasingly, is centered on the person.

2. Medicinal plant variability

The variability of the response to therapeutically beneficial plants begins with its natural variability. The plant has variability depending on its phenotype, the seed quality, the climatic conditions, and the terrain where it grows.

Chemical variation in a plant sample can influence the effectiveness of medicines formulated against a specific disease. Therefore, selecting raw materials based on their chemical composition is a prerequisite [1].

Preparations based on medicinal plants still require detailed scientific analytical studies for quantification of markers and active ingredients or just for chemical standardization purposes, so that they can guarantee the reproducibility of their effects in in vitro biological tests and in pre-clinical animal models. For the clinical evaluation stage, quality control is a completely indispensable practice in accordance with international standards.

The already validated quality control methods for some medicinal plants are present in monographs found in all European Pharmacopoeia: United States Pharmacopoeia, Chinese Pharmacopoeia, WHO Monographs, Japanese Pharmacopoeia, Brazilian Pharmacopoeia—they are universal reference works, updated in all countries on different continents.

Geographical origin and climatic conditions are the notable factors that affect the metabolome of a plant. Plants are adapted to different geographic, climatic, and soil conditions through genotypic and phenotypic changes. Genotypic change also influences plants' production and accumulation of secondary metabolites [2, 3].

Although the specialized metabolic profile is unique to individuals within a species or a closely related taxonomic group, it can be altered if its biosynthetic pathways are influenced by environmental conditions such as climate, soil, pathogen infection, and pest infestation. Therefore, regional variation may be due to different mixtures or proportions of active compounds, which links the geography and climate of the medicinal plant habitat.

Genetic diversity can help evaluate the evolution and conservation of varieties [4]. Genetic diversity is generally estimated through DNA sequences (polymorphisms between varieties) and cytological and morphological markers. However, morphological characteristics are often influenced by the environment. Therefore, molecular markers are relatively more stable and popular than morphological markers [5]. Inbreeding and evolution events can alter allele frequency and reduce genetic diversity [6]. Therefore, it is vital to accurately estimate the correlation between different germplasm resources to ensure high-efficiency utilization and management and to maintain adequate genetic variability for breeding diverse plant varieties [7].

Genetic diversity and population structure analysis have examined various plant species. An analysis of 1151 ramie germplasms using SSR and phenotypic markers reveals that the genetic diversity of wild germplasms is greater than that of domesticated germplasms. This finding of diversity and subpopulations [8] has been observed in several plants such as cannabis [9], sunflower from Iran [10], beans from Brazil [11], allowing technological advances. This wealth of variability is substantial and needs to be preserved by this observation of genetic diversity and the population structure of plants, whether they are sources of medicines, nutrition, or fiber.

2.1 The case for turmeric (*Curcuma longa* L.)

Curcuma longa L., rhizoma (turmeric root; **Figure 1**) with long-standing use, was approved in Europe as a traditional herbal medicinal product for the relief of digestive disturbances, such as feelings of fullness, slow digestion, and flatulence [12]. However, there are also studies showing a potential role as an immune modulator and anti-inflammatory [13–15].

The characteristic compounds are curcuminoids, of which curcumin makes up approximately 90% of the curcuminoid content in turmeric [16]. Chemically, curcumin is a diferuloylmethane, i.e., a beta-diketone derived from methane in which two of the hydrogens are substituted by feruloyl groups (**Figure 2**). These phenolic groups in the structure of curcumin explain the ability of curcumin to eliminate oxygen-derived free radicals [17]. However, as generally observed in medicinal plants' bioactive markers, the curcumin content of the *Curcuma longa* rhizome is very low, as it varies from 0.6 to 5% of the dry mass [18].

Recently, Chen et al. [19] studied the genetic and chemical variability among five Curcuma species, and the results showed that the similarity of the chemical composition of medicinal plants was the primary evidence for the selection of the original plants of Curcuma medicinal materials [19]. In this study, the ITS2 and trnK intron gene sequences were used to analyze the genetic distance between different Curcuma species—chemical composition by HPLC. The authors found that the correlation between genetic distance based on finite genetic sequence and chemical variability showed a relatively low level. The pharmacodynamic potential of new species can be predicted by analyzing the genetic distance between them of the same genus and known medicinal plants.

According to this research, genetic distance data could provide some reference clues for finding new medicinal plant resources.

The huge variety of secondary metabolites produced by plants used to treat various diseases and illnesses are often difficult to obtain in large quantities, limiting their industrial use.

Figure 1.
Curcuma plant (hand-drawn illustration: In Franz Eugen Köhler, Köhler's Medizinal-Pflanzen (1887)).

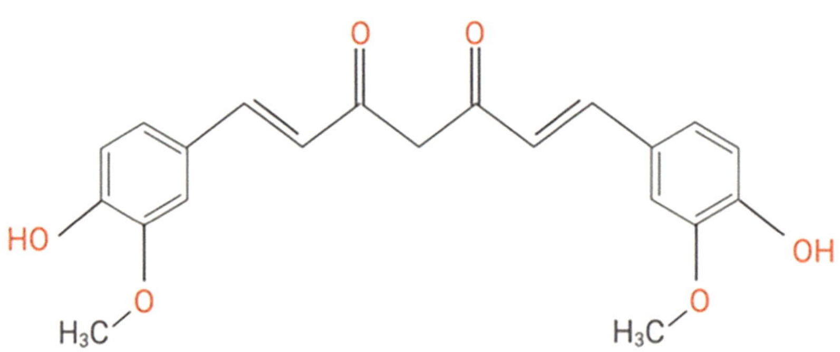

Figure 2.
Chemical structure of curcumin, (1E,6E)-1,7-bis(4-hydroxy-3-methoxyphenyl)hepta-1,6-diene-3,5-dione.

Cytochrome P450 enzymes (CYPs) are fundamental catalysts in the biosynthesis and metabolism of highly valued active metabolites. The technological development of high-throughput sequencing and high-resolution mass spectrometry has allowed new biosynthetic pathways and identified associated CYPs.

Current challenges and possible strategies to overcome limitations associated with CYP engineering to improve the biosynthesis of target secondary metabolites were highlighted [20].

2.2 Variability, scientific research, and use: example of cannabis

Cannabis contains several secondary metabolites belonging to different chemical classes including cannabinoids, terpenoids, flavonoids, and steroids among 545 identified compounds [21–28].

The term "variety" is the adaptation of a species resulting from changes in its habitat due to accidental factors such as climate change, soil changes, diseases, insect attacks, nematodes, and other similar influences [29]. The term "cultivar" is a combination of "cultivated variety," abbreviated to "cultivar" [29].

Unlike varieties, cultivars are not products of natural evolutionary processes. Instead, they are bred through deliberate breeding or agricultural techniques for improved, uniform characteristics [30]. This distinction is crucial as it highlights the human intervention in developing specific plant traits and characteristics.

Cannabis is typically classified into three species: *Cannabis sativa*, *Cannabis indica*, and *Cannabis ruderalis* (**Figure 3**). These species display genetic differences in growth characteristics, cannabinoid profiles, and terpene compositions. Thus, patients, growers, and dispensaries differentiate the three species. However, poly hybrids between these species have been developed worldwide with varying percentage contributions from each species and are currently commonly purchased as "*Cannabis sativa*". Each of the three species has a wide range of cultivars and varieties, each with its unique genetic makeup. These genetic differences result in variations in plant morphology, cannabinoid content (e.g., THC and CBD levels), and terpene profiles, leading to different effects and uses.

For medical applications, researchers largely adopt a chemotaxonomic perspective that describes three chemical phenotypes or chemotypes based on the content

Figure 3.
Cannabis sativa, Cannabis indica, *and* Cannabis ruderalis *have different heights, shapes, leaf structures, content of psychoactive molecules, and geographic origins (source: [31]).*

of two main cannabinoids (**Figure 4**): psychoactive tetrahydrocannabinol (THC) and non-psychoactive cannabidiol (CBD; [32]). THC-dominant strains have a THC/CBD ratio > 1, intermediate strains have THC/CBD ≈ 1, and CBD-dominant strains have THC/CBD < 1. Although most clinical research studies focus on THC and CBD, increasing evidence shows that whole plant extract has additional benefits compared to individual cannabinoids.

As early as 2020, Reimann-Philipp et al. highlighted that medical cannabis patients receive clinical benefits from the diverse plant's secondary metabolites, which contain a variety of other cannabinoids besides THC and CBD, several different combinations of cannabinoids and terpenoids that can be used to classify chemovars [33].

In state-regulated medical cannabis programs, no conventional naming system correlates breeder-reported names with their active ingredient profiles, and these "strain" names are invalid (as well as appropriate for microbial agents) as chemical differences are referred to as chemovars. The taxonomy of cannabis, a versatile plant with a long history of human use, has been the subject of constant debate and review, especially with the advancement of molecular studies. New taxonomic developments are expected.

Cannabis sativa and *Cannabis indica* were recognized as distinct species within the *Cannabis* genus, with different morphological and chemical characteristics, and have different cannabinoid profiles and terpene compositions [34].

Classification of cannabis is a fundamental requirement for future research and medical applications. This approach is facilitated by obtaining an overview of the class and secondary metabolites with potentially therapeutic properties associated with each part of the plant.

Currently, researchers have attempted to discriminate and identify chemical differences between the categories of "Sativa" (narrow leaflet drug, NLD) and "Indica" (broad leaflet drug, WLD; [35]). The results of the chemotaxonomic separation of

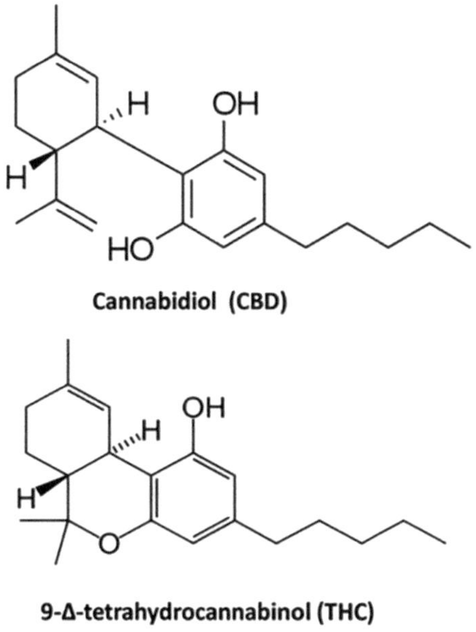

Figure 4.
THC and CBD are the more abundant markers of cannabis from a chemotaxonomic perspective.

"Sativa" and "Indica" were mixed, and the concentrations of THC and CBD seemed to have no differentiating value. However, specific terpenoids were more prominent in some varieties than others [25, 35].

An analysis of 81 marijuana samples and 43 hemp samples using single nucleotide polymorphisms (SNPs) revealed that marijuana and hemp were significantly different at the genome level, and that hemp was genetically more similar to the *Cannabis indica* type than to the *Cannabis sativa* [36].

Cannabis breeding has been actively pursued to develop varieties with specific characteristics, such as high cannabinoid content, improved terpene profiles, and resistance to pests and diseases [9].

These are exciting times for medical research into cannabis and its dozens of cannabinoids. After almost four millennia of its documented medical use in treating convulsive, spastic disorders and numerous severe syndromes, we are very close to obtaining conclusive proof of its effectiveness. We can foresee the era of evidence-based prescription of cannabis-based medicines in serious pathologies.

These examples show how plants are a source of variability, highlighting the importance of their control, from their production to handling of food or phytomedicine, to guarantee safety and quality in their use.

3. Genetic variability of the person

Talking about interindividual variability is a huge challenge, as it is a very vast topic, mainly because we are all different people from each other, and around 8 billion humans inhabit the planet.

The human genome is highly diverse, and genetic polymorphisms can influence how individuals metabolize and respond to medicinal plants.

The action of the plant on people who ingest it depends on variables ranging from ingestion to excretion and, above all, the metabolizer type.

The enzymes involved in the metabolism of medicinal plants and medicines, such as the enzymes that together constitute the so-called Cytochrome P450, can vary in terms of activity due to genetic differences, leading to variations in the way in which medicinal compounds are metabolized and resulting in its efficacy and toxicity.

In terms of pharmacogenomics, it is known that medicinal plants are metabolized by specific enzymes that metabolize drugs. This fact has several implications for its therapeutic efficacy. First, these enzymes are encoded by polymorphic genes that can affect how these compounds are metabolized. Variation in these respective genes results in variant enzymes exhibiting altered or abolished activity [37, 38].

Thus, in a population with individuals carrying a myriad of polymorphic drug-metabolizer genes, there is enormous variability in how individuals respond to herbal medicines. It is, therefore, necessary to consider the pharmacogenetic effects of these enzymes when medicinal plants and products derived from them are used as therapeutic means. In addition to being metabolized, herbal medicines can also affect the expression of some of this extensive group of enzymes through inhibition or induction [39, 40].

Notably, we can focus our attention on the detoxification systems of the human body, addressing Phase 0, I, II, and III systems, their variability, and their importance in the metabolization of herbal medicines and, consequently, their effectiveness and toxicity as well as possible interactions with other medicinal substances, which are processed by the same enzymatic protein systems.

Phase 0 system consists of membrane receptors (ex., OATs, OATPs) that function as small entry openings for substances to be metabolized, that is, to become active or, on the contrary, to make them water-soluble so that they can be excreted and eliminated [41].

Phase I system consists of a set of enzymatic proteins called Cytochromes (CYPs), known, as a whole, as CYP 450 that involves an enzyme superfamily that, to date, has 18 different families and 44 subfamilies of enzymes [42]. Approximately 50–80 genes support these families, which encode all the necessary enzyme structures that makeup Citrochrome P450. Cytochrome P450 enzymes are called "CYP," a term that means proteins linked to heme, a prosthetic group, formed by around 500 amino acids [42] and composed of 57 isoenzymes, grouped into families and subfamilies. The identification of CYPs is presented with the prefix "CYP," followed by a number that represents the family, a letter that indicates the subfamily, and, finally, a number that indicates the isoform: CYP 1 (family) A (subfamily) 2 (isoform).

These hemic proteins contain the chemical element iron in their constitution, linked to a heme group. In addition to the animal kingdom, they exist in bacteria, fungi, and plants [42].

The name was adopted because this structure has an optical absorption capacity of around 450 nanometers when complexed with carbon monoxide [43]. This characteristic is due to the ferrous content of the constituent hemoproteins. CYT P450 proteins are widely disseminated in body tissues, with high concentrations in tissues such as the liver and small intestine [42]. They are anchored explicitly beyond the endoplasmic reticulum and in mitochondria membranes, predominantly in liver and bowel cells [42].

In addition to interacting with membranes, CYT P450 proteins interact with each other and other proteins, such as NADPH-cytochrome P450 reductase and cytochrome b5, which may contribute to controlling the detoxification process [44].

Cytochrome P450 enzymes are involved in around 95% of the redox reactions [45, 46] of the chemicals they metabolize, performing mainly in Phase I metabolism, essential functions. In this phase, the functions of detoxification/deactivation of xenobiotics (any substance foreign to the body, namely drugs, toxicants) predominate, with the most prominent CYP enzymes for these functions being those belonging to families 1–3 [42], being responsible by the metabolism of around 80% of medications, and contributing approximately 50% of the work of CYPs [47].

CYPs are responsible for several reactions, with monooxygenation [43] being the predominant chemical reaction, which is why they are also called enzymes with "monooxygenase" activity [48]. CYPs are directly involved in metabolic pathways that process not only endogenous substances (steroids, fatty acids, vitamins, etc.) but also exogenous substances (drug medicines, environmental pollutants, and carcinogens), making the molecules of these compounds more soluble in water and facilitating its excretion on the one hand, or promoting its activation, as is the case with some drugs or carcinogenic substances [47].

At a cellular level and as an example with a liver cell where much of the metabolism takes place, variability begins in the protein transporters responsible for the entry of the substance into the cell (Phase 0) then, in the variability of the enzymes responsible for the chemical modification of the ingredient to be detoxified (Phase I), predominantly in the endoplasmic reticulum, then in the enzymes involved in the conjugation process (Phase II) in the cytosol and, finally, the variability in the membrane transporters ABC, ATP-binding cassette transport (Phase III), responsible for forwarding the products already metabolically transformed, for the body's excretory pathways, mainly urinary and intestinal.

Considering Phase 0, it is essential to highlight that uptake transporters deemed specific to the liver, such as OATP1B1 and 1B3, as well as OATP2B1 and 1A2, were also found to be expressed in the intestine [49]. Grapefruit juice can inhibit the OATP1A2 transporter and thus compete with the bioavailability of certain medications [49].

Regarding CYPs (Phase I), there are different drug response phenotypes, which include poor metabolizers, extensive metabolizers, and ultra-rapid metabolizers, which influence the physiological effects of medications [50]. Particularizing and considering a Phase I enzymatic protein, CYP 2D6, for example, some people are ultra-fast, extensive, intermediate, and slow metabolizers depending on the active and inactive genes they possess, resulting from the phenotype of the person in question. Genetic variations and individual differences can affect the excretion rate of metabolites and influence the response to medications and other substances, including medicinal plants, and determine their possible side effects.

Regarding Phase II, there is also variability in interindividual enzymatic activity, which can condition changes in the metabolism of endogenous and exogenous substances with consequent repercussions on individual health.

For example, catechol O-methyltransferase (COMT) is an essential enzyme in Phase II of metabolism, deactivating endogenous or exogenous catechols, such as catecholamines and catechol estrogens, as well as in the metabolism of some medications. The catechol O-methyltransferase transfers a methyl group from SAM (S-adenosylmethionine) to a catechol-containing substrate molecule. A Val158Met genetic variant in the COMT gene leads to a several-fold decrease in enzymatic activity, accumulating potentially carcinogenic endogenous catechol estrogens and their reactive intermediates, thus increasing the risk of carcinogenesis [51]. The variation in COMT activity can also explain the effect of certain medicinal plants, such as green tea (**Figure 5**).

3.1 The case for green tea (*Camellia sinensis* (L.) Kuntze)

Although from herbal medicine perspective, *Camellia sinensis* (L.) Kuntze, *non fermentatum folium*, has been recognized in Europe as a traditional herbal medicinal product for the relief of fatigue and the sensation of weakness [52]; some studies have highlighted the antineoplastic potential of green tea polyphenols, quercetin, fisetin, or luteolin. These are common phytochemicals that can be markedly altered (either decreased or increased) by COMT-mediated O-methylation of these exogenous

Figure 5.
*Green tea (*Camellia sinensis *(L.) Kuntze). Source: Jardim Botânico UTAD, Flora Digital de Portugal.*

substrates; flavonoids can also behave as potent inhibitor compounds of the COMT enzyme, delaying the detoxification of endogenous catechol estrogens, potentially carcinogenic [51].

The human COMT gene contains a functional polymorphism, with a G → A substitution in nucleotide 1947 of exon 4 (COMTG1947A) altering the amino acid codon at position 108 (Val → Met) in the COMT protein, which is associated with a variation in the activity of COMT enzyme. Individuals with the G/G genotype have three to four times greater COMT enzyme activity than those with the A/A genotype, while heterozygotes have intermediate enzyme activity [53].

Green tea, a traditional drink in Asian countries such as Japan and China, is also rich in polyphenols such as catechins and gallocatechins, including epigallocatechin-3 gallate (EGCG), which have also been shown to exhibit antiproliferative and antiangiogenic effects in breast cancer cell lines. However, this protective effect of green tea is observed mainly among women with the genotype of low catechol-O-methyltransferase COMT activity. The inverse association between tea intake and breast cancer risk was observed only among individuals with at least one low-activity COMT allele [54].

Similarly, it was found that green tea consumers with the highest activity of the COMT genotype, in which polyphenols are effectively excluded, will obtain less protective benefits against the development of lung cancer [55].

Another Phase II metabolism enzyme is arylamine N-acetyltransferase 2 (NAT2), which is involved in physiological responses to xenobiotics, including medicines and exogenous chemicals in the diet and the environment.

The extensive polymorphism in NAT2 gives rise to wide interindividual variation in acetylation capacity, influencing individual susceptibility to various drug-induced adverse reactions [56] and even the risk of malignant neoplasms [57].

As mentioned, interindividual variability also occurs in Phase III transport proteins. There are several families of ABC genes and multiple encoded proteins, each with different specificities.

Genetic variation influences the effects of plant-medicine interactions, demonstrating pharmacogenomic studies that this influence may involve pharmacokinetic and pharmacodynamic pathways, with this knowledge being essential in contributing to the safe use of herbal medicines in clinical practice [58].

Thus, the dosage of herbal-based supplements can be adjusted to improve efficacy and reduce toxicity according to pharmacogenetic knowledge whose development leads to the discovery and identification of the targets/mechanisms of pharmacological effects and therapeutic responses of natural products effectively and efficiently at the complete genome level, allowing the rational development of herbal medicine as part of an accurate, practical medicine [59].

Plants and vegetable foods such as turmeric, thistle, apple, and green tea, in addition to grapefruit and broccoli, can interfere with the metabolization processes by activating or inhibiting the enzymes of the different phases, with interindividual variability.

Many plants are CYP activators, such as cruciferous vegetables (broccoli, brussels sprouts, cabbage, cauliflower, radish, and watercress) and St. John wort, which activate CYP 1A2; others are inhibitors, such as grapefruit, coffee (caffeine), and echinacea with CYP 1A2 inhibitor properties. Equivalently, grapefruit inhibits and St. John's wort activates the CYP 3A4.

Echinacea (**Figure 6**) root (*Echinacea purpurea* root), popularly used for conditions such as common cold, coughs, bronchitis, influenza, and inflammation of the

Figure 6.
Echinacea purpurea. *Source: Jardim Botânico UTAD, Flora Digital de Portugal.*

mouth and pharynx, reduces the oral clearance of CYP1A2 substrates and selectively modulates the catalytic activity of CYP3A in hepatic and intestinal sites. Care must be taken when co-administered with medications dependent on CYP3A or CYP1A2 for their elimination [60].

The intestine's high level of expression of CYP3A4 may condition CYP3A4 susceptibility to dietary modulation. Numerous food-drug interactions involving CYP1A2, CYP2E1, glucuronosyltransferases, and glutathione S-transferases have been documented, in addition to interactions involving transporters such as P-glycoprotein (ABC-transporter of Phase III) and organic anion transporting polypeptide [61].

The binding affinity and response to active compounds in medicinal plants are related to the variability of receptors and their expression; for example, variations in opioid receptors can influence an individual's response to analgesic plants, such as opium poppy.

The variability of response to medicinal plants is more complex regarding plants with psychotropic action. For example, considering cannabis, several factors seem to contribute to this variability, from purely genetic characteristics of the individual but also the cannabinoid profile, individual tolerance, route of administration, dose, mental state and physical environment of an individual, previous experience of consumption, one's health and metabolism, and age; all of them in addition to the purity and quality of the herbal material such as contaminants or impurities that can introduce unexpected or side effects. Its properties are mainly related to its chemical composition, which depends on the manufacturing method, hemp variety, and seeds used [22].

Another aspect of great importance in interindividual variability in metabolism concerns the intestinal microbiota. Gut microbial communities represent a source of human genetic and metabolic diversity. Gut microbiomes differ between people when viewed from the perspective of microbial lineage components, encoded metabolic functions, postnatal developmental stage, and environmental exposures [62].

Finally, it is also worth noting that, within the individual, variability also involves temporality. For example, human glucuronidation (Phase II metabolism) begins after birth and is scarce during fetal life. This fact may justify indirect hyperbilirubinemia in children in the neonatal period.

4. Anthropological variability

Anthropological factors such as diet, lifestyle, and exposure to environmental toxins can influence an individual's response to medicinal plants. For example, people from different regions may have different tolerances or sensitivities to specific herbal remedies based on their environmental exposures. Ethnic differences were found in the enzymatic activities of CYP3A4 enzymes [61] and in the pharmacodynamic response to cyclosporine between healthy African American men and White men [63].

However, it was noted that these differences could have resulted from non-genetic factors, such as diet or drug therapy, with the type of menu being another factor that adds to the variability of response to medicinal plants. There were found ethnic differences concerning CYP3A4 activity; however, these differences could be partially explained by different dietary patterns that can modulate this enzyme with a high level of expression in the intestine and specific for a wide range of substrates [61].

Ethnic aspects are to be considered. For example, *Moringa pteridosperms* and *Moringa oleifera* are widely used in sub-Saharan Africa, and polymorphisms in drug-metabolizing enzymes have been found to affect activities, with differences between racial and ethnic populations. An example of this is CYP2D6, where certain variants are found only in specific people; for instance, CYP2D6*17 among Black Africans, CYP2D6*10 among Asians, and CYP2D6*2 N reported in most populations but at different frequencies [64].

Ethnological variability is also associated with Phase II enzymes, for example, in the case of COMT and NAT2.

Regarding COMT, the frequency of the homozygous A/A genotype in the Fujian Han Chinese population was similar to that of the Kenyan, Japanese, Korean, and Taiwanese Han populations but much lower than in Caucasians and southwest Asians demonstrating differences and variability between groups—ethnicities in COMT enzyme activity [53].

The diverse functioning of COMT and its complex regulation by several genetic and environmental factors, including plant-based food ingredients, emphasizes the need to stratify further association studies between COMT genotype and cancer risk from product consumption containing catecho l [51].

The variation in the protective activity of green tea concerning cancer may have to do with variants in COMT activity between populations and the different distributions of phenotypic frequencies, being the result of the selection of phenotypes over the years, according to dietary patterns, which may have an impact and implications for the prevalence of the disease.

Regarding arylamine N-acetyltransferase 2 or NAT2, many questions remain about the evolutionary mechanisms that led to the high prevalence of NAT2 slow acetylators between humans. Recent research studies demonstrate some evidence about NAT2 gene variation, suggesting that slow-acting NAT2 variants may have become targets of a natural positive selection due to changing livelihoods and lifestyles in human populations over the past 10,000 years [65].

A higher prevalence of the slow acetylation phenotype was observed in populations that practice agriculture (45.4%) and herding (48.2%) as compared to people that rely primarily on hunting and gathering (22.4%) ($P = 0.0007$). This fact began to be seen in the frequency of the slow variant 590A, which occurred three times more frequently in food producers and farmers (25%) compared to hunter-gatherers (8%) [65].

These findings are consistent with the hypothesis that the Neolithic transition to subsistence economies based on agricultural and pastoral resources modified the selective regime that affects NAT2 in the acetylation pathway, with evidence of a correlation between the prevalence of slow acetylators in humans may have been a subsistence strategy adopted by past populations over time in the last 10,000 years; it appears that a slower rate of acetylation may represent a selective advantage in people that change from foraging and hunting-based food to pastoralism/agriculture in the Neolithic period [65].

Understanding how NAT2 genetic diversity is structured in humans is of anthropological importance and medical relevance for pharmacogenetics and epidemiological applications. Genetic heterogeneity is observed between populations from different parts of Asia and between people from Africa and America, and differences in allele frequencies between populations and individuals of different ethnic or geographic origins must be considered, as they may respond differently to acetylated drugs [65].

Impressive patterns of geographic differentiation were described for the slow acetylation variants of the NAT2 gene, suggesting that this genetic locus has been subject to the action of natural selection over time. The correlation of the allele associated with the enzyme's slow activity may have conferred a selective advantage in populations switching from food gathering to agricultural activities in the Neolithic period in an adaptive evolution of the NAT2 gene. The rs1799930 A allele has been associated with slower acetylation capacity in vivo and is much more frequent in farmers and pastoralists compared to hunter-gatherers, highlighting the functional importance of this polymorphism in human adaptation to environmental fluctuations in xenobiotics [66].

Another example related to gut microbiome variability is that Native Hawaiian and Pacific Islander (NHPI) populations demonstrate a disproportionately higher rate of diabetes mellitus type 2, a chronic disease that arises from metabolic dysfunction and is often associated with obesity and inflammation. Reversible lifestyle habits, such as diet, may protect against or contribute to the increased prevalence of health inequities in these populations through the gut microbiome-immunogenetic axis, i.e., the connection between diet, epigenetics, microbiome composition, immune function, and response to infections [67].

Different dietary patterns and eating habits have contributed to variability from an anthropological perspective and can condition different disease prevalences between other communities. A diet rich in fiber, found in whole grains and some fruits and vegetables, facilitates a favorable composition of the intestinal microbiome and increases the production of butyrate, acetate, and propionate, which are short-chain fatty acids that act in metabolic and immunological pathways, protecting against the metabolic syndrome and chronic inflammatory states associated with dysbiosis. Native Hawaiians and Pacific Islanders who once thrived on healthy traditional diets may be more sensitive than non-Indigenous peoples to the metabolic disruption of Westernized diets that affect the immunogenetic-gut microbiome axis [67].

Another example was pronounced differences in bacterial species assemblages and functional gene repertoires observed between individuals residing in the United States compared to other countries. These distinctive characteristics are evident in early childhood as well as adulthood. Furthermore, the similarity of fecal microbiomes between member families extends across cultures. These findings highlight the need to consider the microbiome when evaluating human development, nutritional deficiencies, physiological variations, and the impact of westernization, with sustainable agriculture policies and better nutrition having to be adapted to different cultural conditions but also to different intestinal microbiomes [62].

Anthropological factors can also influence medicinal plants' use in other health practices. For example, some individuals may combine herbal remedies with modern medicine, while others may rely solely on traditional herbal treatments. Among patients using conventional and traditional medicine systems, issues to be addressed include interactions between drug-drug, herb-drug, and herb-herb, and genetic polymorphisms in genes coding for drug-metabolizing enzymes [68].

4.1 Plant/diet-drug interactions

P450 is a cytochrome with an essential role in metabolism, being susceptible to induction or inhibition caused by substances found in plants [47, 69–72] with consequent repercussions on the expected effects of drugs.

Patients should be trained to avoid certain plant-drug combinations that are clinically relevant [73].

Meta-analyses demonstrated a significant effect on CYP1A2 and glutathione S-transferase-alpha (GST-α), with Cruciferae consumption increasing the activities of these enzymes by 20–40% and 15–35% respectively, suggesting that patients undergoing pharmacotherapy with CYP1A2 or GST-α substrates may have altered drug exposure profiles if they concomitantly consume large quantities of cruciferous vegetables [74].

The interactions of grapefruit juice with cyclosporine and felodipine, St. John's wort with cyclosporine and indinavir, and red wine with cyclosporine have the potential to require dosage adjustment to maintain drug concentrations within their therapeutic windows [61].

There is still some controversy regarding the clinical significance of potential interactions between diet and medications. For example, regarding St. John's wort (*Hypericum perforatum*), some results suggest that it is unlikely to inhibit the activity of CYP 2D6 or CYP 3A4 when taken at doses recommended for depression [75].

Food-drug interactions involving Phase I CYP1A2, CYP2E1, and Phase II glucuronosyltransferases and glutathione S-transferases have also been reported. However, most of these interactions are modest in magnitude and clinically relevant only for drugs with a narrow therapeutic range. Recently, interactions involving drug transporters, including P-glycoprotein and organic anion-transporting polypeptide, have also been identified. More research is needed to determine the scope, clinical relevance, and magnitude of the effects of food on drug metabolism and transport.

Another aspect linked to variability in the use of medicinal plants is interference in pharmacodynamics, which is also a result of individual phenotypic variability. For example, cranberry (*Vaccinium macrocarpon*), commonly used as an aid in the treatment and prevention of urinary infections [76], can interfere with warfarin without altering the binding to plasma proteins of S- or R-warfarin. The interaction depends on the VKORC1 1173 T > C polymorphism, an epoxide reductase essential for activating vitamin K, a cofactor of clotting factors. It was found that individuals with CT and TT genotypes of VKORC1 present a reduction in warfarin activity when administered with cranberry extract by 22% and 11%, respectively [77]. This case is an example of genetic polymorphisms in the pharmacodynamic pathway that may also be involved in plant-drug interactions.

Numerous interactions between plants and medicines have been described and explained by the pharmacodynamic and metabolization process [78–83] where CYP play a role (**Tables 1–3**).

Pharmacodynamics (PD) and pharmacokinetics (PK) are hard to predict in all patients, and best practice involves the use of standard dosing based on weight and therapeutic drug monitoring (TDM).

Pharmacodynamics (PD) and pharmacokinetics (PK) are hard to predict in all patients, and best practice involves the use of standard dosing based on weight and therapeutic drug monitoring (TDM). Pharmacogenetics (PG) is the use of genetic screening to predict metabolic responses to different drugs and enables more accurate predictions of PD and PK to be made. The biggest challenge in reducing metabolic

Pharmacodynamic and/or pharmacokinetic changes	Pharmacokinetic mechanism	Examples of influenced drugs	Probable outcome
Increased concentration of ARV*	Inhibition of CYP3A4 (intestinal)	Indinavir	Increased ADR risk**
	Inhibition of CYP2D6	Ritonavir	
		Cobicistat	
Concentration reduction	Alteration of GPP	Celiprolol	Therapeutic failure
Concentration reduction	Inhibition of CYP1A2	Antipsychotics and antidepressants	Therapeutic failure
Increased concentration	Unknown	Sulfasalazine	Increased ADR risk
Synergistic effect	Unknown	Warfarin	Increased hemorrhagic risk
Increased concentration	Unknown	Palbociclib	Increased ADR risk
		Capecitabine	
		Enzalutamide	
Concentration reduction		Talinolol	Therapeutic failure
Increased concentration	Inhibition of CYPA4/5	Tacrolimus	Increased ADR risk
Concentration reduction	Inhibition of GPP	Azatioprin, ciclosporin	Hepatotoxicity
Increased concentration	Inhibition of CYP3A4	Everolimus	Increased ADR risk
	Inhibition of GPP		
Reduced bioavailability (dose dependant)	Inhibition of GPP	Digoxin	Therapeutic failure

References: [78, 80–86].
**ARV: antirretrovirals.*
***ADR: adverse drug reaction.*

Table 1.
Curcuma: approved use for dyspeptic problems [12]. Some studies show a potential role as an immune modulator and anti-inflammatory [73, 82].

Pharmacodynamic and/or pharmacokinetic changes	Pharmacokinetic mechanism	Examples of influenced drugs	Probable outcome
Increased bioavailability	Unknown	Antirretrovirals	Increased ADR risk**
		Tacrolimus, ciclosporin	Toxicity/ therapeutic failure
Increased AUC	Inhibition of GPP	Simvastatin	Increased ADR risk
Inhibition of folic acid absortion			
Antagonism	Unknown	Warfarin	Therapeutic failure
Increased AUC	Inhibition of OATP1A1 and OATP1A2	Beta-blockers	Increased ADR risk
		Fluoroquinolones	Increased ADR risk
		Statins	Increased ADR risk
Reduced bioavailability		Nadolol	Therapeutic failure
Decreased absorption	Inhibition of OATP1B1, OATP1B3	Atorvastatin	Increased ADR risk
Decreased absorption	Inhibition of MATE1, MATE 2	Atorvastatin	Increased ADR risk
	Inhibition of OCT1, OCT2* (when in interaction with Metformin)		
Antagonism	Unknown	Lisinopril	Therapeutic failure
Increased bioavailability	Inhibition of GPP	Tacrolimus, ciclosporin	Increased ADR risk
Decreased bioavailability	Inhibition of OATP1B1, OATP2B1, OATP1A2 (gut) Activation of OATP1B3	Rosuvastatin	Therapeutic failure

References: [78, 83–85, 88–92].
**OCT1, OCT2: organic cation influx transporters. They are responsible for the hepatic and renal transport of Metformin.*
***ADR: adverse reaction to the drug.*

Table 2.
Green tea: approved use [52] to combat asthenia and also described as having anti-inflammatory, antibacterial, and antiviral potential [87].

instability arises when we face different crops, plant selection, preparation methods and dosage that can vary and contribute to variations in the effectiveness of herbal medicines.

Anthropological studies can help preserve, understand, and respect the cultural context of practices in the use of medicinal plants and understand aspects linked to the variability of responses associated with the environment, adaptability, and the evolution of the human species.

Pharmacodynamic and/or pharmacokinetic changes	Pharmacokinetic mechanism	Examples of influenced drugs	Probable outcome
Risk of increased viral load	Inhibition of CYP3A4	ARV*	Therapeutic failure
	Inhibition of CYP1A2, CYP2C9	Caffeinne, tolbutamid	ADR risk**
Concentration reduction	CYP3A4 induction	Midazolam	Therapeutic failure
		Warfarin	Therapeutic failure
Concentration reduction	Unknown	Imunossupressors	Therapeutic failure

References: [80, 82–84, 89, 91, 92, 94].
**ARV: antirretrovirals.*
***ARD: adverse reaction to the drug.*

Table 3.
Echinacea-approved use for common cold and acne [93].

4.2 The anthropological case of cannabis

Cannabis has been used for various purposes across cultures and throughout history. It has been used not only for medicinal but also for recreational, spiritual, and industrial purposes.

Different cultures have developed unique traditions and practices related to cannabis consumption. Over time, and in different rules, cannabis has been used as an analgesic, for pain relief, as an anti-inflammatory, and to treat various medical conditions. Traditional knowledge about the properties and aspects linked to the uses of the plant has been observed to vary between different cultures, often linked to beliefs and practices that are not only cultural but also religious and spiritual. For example, Rastafarians use it as a sacrament and certain Native American tribes use it in sacred rituals. Industrial uses such as hemp fiber production are historically known for manufacturing textiles, ropes, paper, and other products [95]. However, recreational use for its psychoactive properties in many parts of the world, both historically and in contemporary times, is recognized as the "anthropological marker of cannabis," although different varieties and methods of consumption have emerged based on cultural preferences. The legal status and social acceptance of cannabis vary widely from one region to another. Some countries and states have legalized its recreational and medicinal use, while others maintain strict prohibitions [96].

Gathering data to develop medicinal cannabis use may lead to two types of products: herbal medicines based on padronized and/or quantified extracts on one side and medicines based on specific cannabinoids or combinations of the cannabinoids on the other. In any case, an integrated systematic approach shall consider the interindividual variability of their effects from a genetic and anthropological perspective (**Figure** 7).

Finally, high-quality evidence on the short- and long-term safety of medicinal cannabis is still lacking. Although there is no known level of cannabinoid ingestion that will result in a toxic or lethal dose in humans, it is reported that the median lethal dose of THC in animal models ranges from 800 to >9000 mg/kg (depending on the species). Thus, the estimates of a lethal dose of THC for a 70 kg human range up to >15 g, and for CBD, doses of ca. 1000 mg/kg have been tolerated in humans [97, 98].

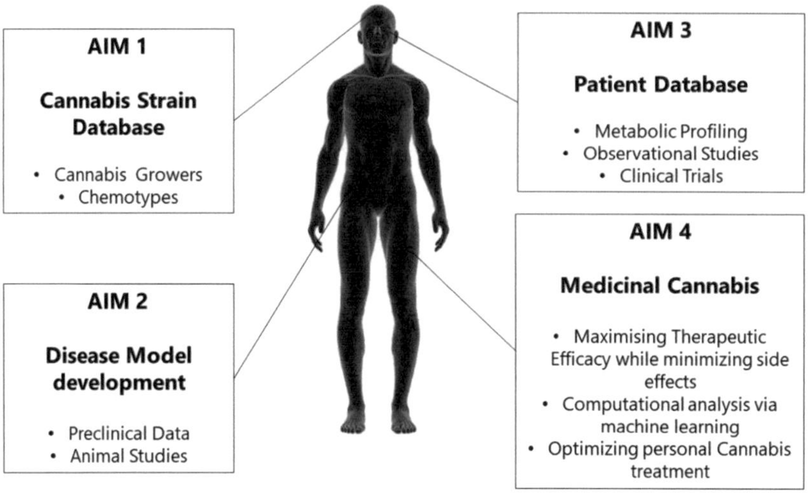

Figure 7.
Research on the therapeutical potential of specific cannabis strains.

5. Conclusions

In conclusion, understanding the interindividual variability of the effects of medicinal plants from a genetic and anthropological perspective is crucial to maximizing their benefits while minimizing potential risks. This approach can contribute to developing personalized and culturally sensitive herbal medicine practices based on nutrigenetics and ecogenetics aspects.

In the food supplements market, numerous plants are freely available for sale that people use, either by prescription or self-medication, intending to treat a health problem and often in combination with chemically synthesized medicines. Due to their multifactorial variability, this aspect denotes the concern and attention necessary to guarantee herbal supplements' standardization and quality control.

Furthermore, this guarantee is fundamental for the safety of consumers and healthcare providers, who already have the factor of interindividual variability to consider in responding to the use of medicinal plants and possible interactions with medications.

The domain of pharmacogenetics and pharmacogenomics, through the understanding of mechanisms of genetic variations and associations of differences in physiological actions of medicinal plants, allows us to understand how the interindividual variability, in part due to genetic composition, added to the genotypic biodiversity inherent to plants, can influence their physiological effects on the human body. This information can help personalize herbal remedies in a person-centered, holistic healthcare professional approach.

Author details

Alda Pereira da Silva Oliveira[1,2,3]*, Maria do Céu Costa[3,4,5] and Manuel Pires Bicho[2,6]

1 University Clinic of General and Family Medicine, Lisbon School of Medicine, University of Lisbon, Portugal

2 Ecogenetics and Human Health Research Unity, Institute for Environmental Health, ISAMB and Lab Associate TERRA, Lisbon School of Medicine, University of Lisbon, Portugal

3 Research Center for Biosciences and Health Technologies, CBIOS, Portugal

4 School of Health Sciences and Technologies, ECTS, Grupo Lusófona, Campo Grande, Lisboa, Portugal

5 NICiTeS—Núcleo de Investigação em Ciências e Tecnologias da Saúde, ERISA—Escola Superior de Saúde Ribeiro Sanches, Polytechnic Institute of Lusophony, Lisboa, Portugal

6 Instituto de Investigação Bento e Rocha Cabral, Cç. Bento Rocha Cabral, Lisboa, Portugal

*Address all correspondence to: aldapsilva@medicina.ulisboa.pt

References

[1] Raskar S et al. Assessing the impact of geographical distribution and genetic diversity on metabolic profiles of a medicinal plant, Embelia ribes Burm. f. Plants. 2022;**11**(2861):1-19. DOI: 10.3390/plants11212861

[2] Albert A et al. Temperature is the key to altitudinal variation of phenolics in *Arnica montana* L. cv. ARBO. Oecologia. 2009;**160**(1):1-8. DOI: 10.1007/s00442-009-1277-1

[3] Karimi A et al. Metabolomics approaches for analyzing effects of geographic and environmental factors on the variation of root essential oils of *Ferula assa-foetida* L. Journal of Agricultural and Food Chemistry. 2020;**68**(37):9940-9952. DOI: 10.1021/acs.jafc.0c03681

[4] Ellegren H, Galtier N. Determinants of genetic diversity. Nature Reviews Genetics. 2016;**17**(7):422-433. DOI: 10.1038/nrg.2016.58

[5] Nadeem MA et al. DNA molecular markers in plant breeding: Current status and recent advancements in genomic selection and genome editing. Biotechnology and Biotechnological Equipment. 2018;**32**(2):261-285. DOI: 10.1080/13102818.2017.1400401

[6] Hufbauer RA. Population genetics of invasions: Can we link neutral markers to management? Weed Technology. 2004;**18**(sp1):1522-1527. DOI: 10.1614/0890-037X(2004)018[1522:PGOICW]2.0.CO;2

[7] Sanchez D et al. Improving the use of plant genetic resources to sustain breeding programs' efficiency. Proceedings of the National Academy of Sciences. 2023;**120**(14):1-9. DOI: 10.1073/pnas

[8] Feng Y, Ryan UM, Xiao L. Genetic diversity and population structure of cryptosporidium. Trends in Parasitology. 2018;**34**(11):997-1011. DOI: 10.1016/j.pt.2018.07.009

[9] Zhang J et al. Genetic diversity and population structure of cannabis based on the genome-wide development of simple sequence repeat markers. Frontiers in Genetics. 2020;**11**(September):1-12. DOI: 10.3389/fgene.2020.00958

[10] Jannatdoust M et al. Analysis of genetic diversity and population structure of confectionery sunflower (*Helianthus annuus* L.) native to Iran. Journal of Crop Science and Biotechnology. 2016;**19**(1):37-44. DOI: 10.1007/s12892-015-0052-6

[11] Delfini J et al. Population structure, genetic diversity and genomic selection signatures among a Brazilian common bean germplasm. Scientific Reports. 2021;**11**(1):1-12. DOI: 10.1038/s41598-021-82437-4

[12] European Medicines Agency. European Union herbal monograph on *Curcuma longa* L., rhizoma final. Committee on Herbal Medicinal Products (HMPC). 2018;**44**(September 2018):1-7. Available from: https://www.ema.europa.eu/en/documents/herbal-monograph/final-european-union-herbal-monograph-curcuma-longa-l-rhizoma-revision-1_en.pdf

[13] Akhter M. Herbal drug interactions. Research Anthology on Recent Advancements in Ethnopharmacology and Nutraceuticals. 2021;**2**(10):120-141. DOI: 10.4018/978-1-6684-3546-5.ch008

[14] Chainani-Wu N. Safety and anti-inflammatory activity of curcumin:

A component of tumeric (*Curcuma longa*). Journal of Alternative and Complementary Medicine (New York, N.Y.). United States. 2003;**9**(1):161-168. DOI: 10.1089/107555303321223035

[15] Kocaadam B, Şanlier N. Curcumin, an active component of turmeric (*Curcuma longa*), and its effects on health. Critical Reviews in Food Science and Nutrition. 2017;**57**(13):2889-2895. DOI: 10.1080/10408398.2015.1077195

[16] Ruby AJ et al. Anti-tumour and antioxidant activity of natural curcuminoids. Cancer Letters. 1995;**94**(1):79-83. DOI: 10.1016/0304-3835(95)03827-J

[17] Indira KIP. Free radical reactions of curcumin in membrane models. Free Radical Biology and Medicine. 1997;**23**(6):838-843. DOI: 10.1016/S0891-5849(97)00026-9

[18] Cronin JR. Curcumin: Old spice is a new medicine. Alternative and Complementary Therapies. 2003;**9**(1):34-38. DOI: 10.1089/10762800360520776

[19] Chen M et al. Analysis of genetic and chemical variability of five curcuma species based on DNA barcoding and HPLC fingerprints. Frontiers in Plant Science. 2023;**14**:1-14, 1229041. DOI: 10.3389/fpls.2023.1229041

[20] Sethi A, Bhandawat A, Pati PK. Engineering medicinal plant-derived CYPs: A promising strategy for production of high-valued secondary metabolites. Planta. 2022;**256**(6):1-14. DOI: 10.1007/s00425-022-04024-9

[21] Chandra S, Lata H, ElSohly MA. *Cannabis Sativa* L.-Botany and Biotechnology. Berlin/Heidelberg, Germany: Springer; 2017

[22] Citti C et al. Cannabinoid profiling of hemp seed oil by liquid chromatography coupled to high-resolution mass spectrometry. Frontiers in Plant Science. 2019;**10**(February):1-17. DOI: 10.3389/fpls.2019.00120

[23] Gill EW, Paton WDM, Pertwee RG. Preliminary experiments on the chemistry and pharmacology of cannabis. Nature. 1970;**228**(5267):134-136. DOI: 10.1038/228134a0

[24] Hanuš LO et al. Phytocannabinoids: A unified critical inventory. Natural Product Reports. 2016;**33**:1357-1392. DOI: 10.1039/c6np00074f

[25] Jin D et al. Secondary metabolites profiled in cannabis inflorescences, leaves, stem barks, and roots for medicinal purposes. Scientific Reports. 2020;**10**(1):1-14. DOI: 10.1038/s41598-020-60172-6

[26] McPartland JM, Russo EB. Cannabis and cannabis extracts. Journal of Cannabis Therapeutics. 2001;**1**(3-4):103-132. DOI: 10.1300/J175v01n03_08

[27] Mechoulam R, Gaoni Y. Recent advances in the chemistry of hashish. Fortschritte der Chemie organischer Naturstoffe = Progress in the chemistry of organic natural products. Progres dans la chimie des substances organiques naturelles. 1967;**25**:175-213. DOI: 10.1007/978-3-7091-8164-5_6

[28] Pavlovic R et al. Phytochemical and ecological analysis of two varieties of hemp (*Cannabis sativa* L.) grown in a mountain environment of Italian Alps. Frontiers in Plant Science. 2019;**10**(October):1-20. DOI: 10.3389/fpls.2019.01265

[29] Arévalo RA et al. Los términos cultivar o variedad de caña de azúcar (*Saccharum* spp.). Revista Chapingo

Serie Horticultura. 2006;**XII**(1):5-9. DOI: 10.5154/r.rchsh.2004.04.027

[30] Tooker JF, Frank SD. Genotypically diverse cultivar mixtures for insect pest management and increased crop yields. Journal of Applied Ecology. 2012;**49**(5):974-985. DOI: 10.1111/j.1365-2664.2012.02173.x

[31] McPartland JM. Cannabis systematics at the levels of family, genus, and species. Cannabis and Cannabinoid Research. 2018;**3**(1):203-212. DOI: 10.1089/can.2018.0039

[32] de Meijer EPM et al. The inheritance of chemical phenotype in *Cannabis sativa* L. Genetics. 2003;**163**:335-346. DOI: 10.1300/J237v08n02_04

[33] Reimann-Philipp U et al. Cannabis chemovar nomenclature misrepresents chemical and genetic diversity; survey of variations in chemical profiles and genetic markers in Nevada medical cannabis samples. Cannabis and Cannabinoid Research. 2020;**5**(3):215-230. DOI: 10.1089/can.2018.0063

[34] Zandkarimi F et al. Comparison of the cannabinoid and terpene profiles in commercial cannabis from natural and artificial cultivation. Molecules. 2023;**28**(2):1-15. DOI: 10.3390/molecules28020833

[35] Hazekamp A, Tejkalová K, Papadimitriou S. Cannabis: From cultivar to chemovar II - A metabolomics approach to cannabis classification. Cannabis and Cannabinoid Research. 2016;**1**(1):202-215. DOI: 10.1089/can.2016.0017

[36] Sawler J et al. The genetic structure of marijuana and hemp. PLoS One. 2015;**10**(8):1-9. DOI: 10.1371/journal.pone.0133292

[37] Li J, Bluth MH. Pharmacogenomics of drug metabolizing enzymes and transporters: Implications for cancer therapy. Pharmacogenomics and Personalized Medicine. 2011;**4**(1):11-33. DOI: 10.2147/PGPM.S18861

[38] Sim SC, Kacevska M, Ingelman-Sundberg M. Pharmacogenomics of drug-metabolizing enzymes: A recent update on clinical implications and endogenous effects. Pharmacogenomics Journal. 2013;**13**(1):1-11. DOI: 10.1038/tpj.2012.45

[39] Fasinu PS et al. The potential of Sutherlandia frutescens for herb-drug interaction. Drug Metabolism and Disposition. 2013;**41**(2):488-497. DOI: 10.1124/dmd.112.049593

[40] Taesotikul T et al. Effects of Phyllanthus amarus on the pharmacokinetics of midazolam and cytochrome P450 activities in rats. Xenobiotica. 2012;**42**(7):641-648. DOI: 10.3109/00498254.2012.655703

[41] Roth M, Obaidat A, Hagenbuch B. OATPs, OATs and OCTs: The organic anion and cation transporters of the SLCO and SLC22A gene superfamilies. British Journal of Pharmacology. 2012;**165**(5):1260-1287. DOI: 10.1111/j.1476-5381.2011.01724.x

[42] Deodhar M et al. Mechanisms of cyp450 inhibition: Understanding drug-drug interactions due to mechanism-based inhibition in clinical practice. Pharmaceutics. 2020;**12**(9):1-18. DOI: 10.3390/pharmaceutics12090846

[43] Omura T. Forty years of cytochrome P450. Biochemical and Biophysical Research Communications. 1999;**266**(3):690-698. DOI: 10.1006/bbrc.1999.1887

[44] Johnson EF et al. Correlating structure and function of

drug-metabolizing enzymes: Progress and ongoing challenges. Drug Metabolism and Disposition. 2014;**42**(1):9-22. DOI: 10.1124/dmd.113.054627

[45] Guengerich FP, Waterman MR, Egli M. Recent structural insights into cytochrome P450 function. Trends in Pharmacological Sciences. 2016;**37**(8):625-640. DOI: 10.1016/j.tips.2016.05.006

[46] Rendic S, Guengerich FP. Survey of human oxidoreductases and cytochrome P450 enzymes involved in the metabolism of xenobiotic and natural chemicals. Chemical Research in Toxicology. 2015;**28**(1):38-42. DOI: 10.1021/tx500444e

[47] Zhao M et al. Cytochrome p450 enzymes and drug metabolism in humans. International Journal of Molecular Sciences. 2021;**22**(23):1-16. DOI: 10.3390/ijms222312808

[48] José A, Lemos G, Trindade EJ. Interferências no Efeito Farmacológico Mediadas pelas Biotransformações dos Citocromos P450. Revista Científica do ITPAC. 2014;**7**(2):1-11

[49] Glaeser H et al. Intestinal drug transporter expression and the impact of grapefruit juice in humans. Clinical Pharmacology and Therapeutics. 2007;**81**(3):362-370. DOI: 10.1038/sj.clpt.6100056

[50] Guengerich FP. Cytochrome P450s and other enzymes in drug metabolism and toxicity. AAPS Journal. 2006;**8**(1):E101-E111. DOI: 10.1208/aapsj080112

[51] Sak K. The Val158Met polymorphism in COMT gene and cancer risk: Role of endogenous and exogenous catechols. Drug Metabolism Reviews. 2017;**49**(1):56-83. DOI: 10.1080/03602532.2016.1258075

[52] EMA Monograph. Community herbal monograph on *Camellia sinensis* (L.) Kuntze, non fermentatum folium final discussion in working party on community monographs and community list (MLWP). EMA. 2013;**283630**(November 2013):1-5. Available from: www.ema.europa.eu

[53] Lin CH et al. Genetic polymorphism of catechol O-methyltransferase and pharmacokinetics of levodopa in healthy Chinese subjects. Methods and Findings in Experimental and Clinical Pharmacology. 2009;**31**(6):389-395. DOI: 10.1358/mf.2009.31.6.1386990

[54] Wu AH et al. Tea intake, COMT genotype, and breast cancer in Asian-American women. Cancer Research. 2003;**63**(21):7526-7529

[55] Lai C-Y et al. Genetic polymorphism of catechol-O-methyltransferase modulates the association of green tea consumption and lung cancer. European Journal of Cancer Prevention. 2019;**28**(4):316-322. Available from: https://journals.lww.com/eurjcancerprev/fulltext/2019/07000/genetic_polymorphism_of.10.aspx

[56] Ladero JM. Influence of polymorphic N-acetyltransferases on non-malignant spontaneous disorders and on response to drugs. Current Drug Metabolism. 2008;**9**(6):532-537. DOI: 10.2174/138920008784892038

[57] Agúndez JAG. Polymorphisms of human N-acetyltransferases and cancer risk. Current Drug Metabolism. Netherlands. 2008;**9**(6):520-531. DOI: 10.2174/138920008784892083

[58] Liu MZ et al. Pharmacogenomics and herb-drug interactions: Merge of

future and tradition. Evidence-based Complementary and Alternative Medicine. 2015;**2015**:8, Article ID 321091. DOI: 10.1155/2015/321091

[59] Rao T et al. The pharmacogenetics of natural products: A pharmacokinetic and pharmacodynamic perspective. Pharmacological Research. 2019;**146**:104283. DOI: 10.1016/j.phrs.2019.104283

[60] Gorski JC et al. The effect of echinacea (*Echinacea purpurea* root) on cytochrome P450 activity in vivo. Clinical Pharmacology and Therapeutics. 2004;**75**(1):89-100. DOI: 10.1016/j.clpt.2003.09.013

[61] Harris RZ, Jang GR, Tsunoda S. Dietary effects on drug metabolism and transport: Clinical pharmacokinetics. Clinical Pharmacokinetics. 2003;**42**(13):1071-1088

[62] Yatsunenko T et al. Human gut microbiome viewed across age and geography. Nature. 2012;**486**(7402):222-227. DOI: 10.1038/nature11053

[63] Stein CM et al. Cyclosporine pharmacokinetics and pharmacodynamics in African American and white subjects. 2001;**69**(5):317-323. DOI: 10.1067/mcp.2001.115073

[64] Dandara C et al. Cytochrome p450 pharmacogenetics in african populations: Implications for public health. Expert Opinion on Drug Metabolism and Toxicology. 2014;**10**(6):769-785. DOI: 10.1517/17425255.2014.894020

[65] Sabbagh A et al. Arylamine N-acetyltransferase 2 (NAT2) genetic diversity and traditional subsistence: A worldwide population survey. PLoS One. 2011;**6**:1-10, e18507. DOI: 10.1371/journal.pone.0018507

[66] Patillon B et al. A homogenizing process of selection has maintained an “ultra-slow” acetylation NAT2 variant in humans. Human Biology. 2014;**86**(3):185-214. DOI: 10.13110/humanbiology.86.3.0185

[67] Rubas NC, Maunakea A. Immunoepigenetic-microbiome Axis: Implications for health disparities research in native Hawaiians and Pacific islanders. Hawaii Journal of Health and Social Welfare. 2021;**80**(8):195-198

[68] Thomford NE et al. Pharmacogenomics implications of using herbal medicinal plants on African populations in health transition. Pharmaceuticals. 2015;**8**(3):637-663. DOI: 10.3390/ph8030637

[69] Gurley BJ et al. Clinical assessment of CYP2D6-mediated herb-drug interactions in humans: Effects of milk thistle, black cohosh, goldenseal, kava kava, St. John’s wort, and Echinacea. Molecular Nutrition & Food Research. 2008;**52**(7):755-763. DOI: 10.1002/mnfr.200600300

[70] Husain I et al. Screening of medicinal plants for possible herb-drug interactions through modulating nuclear receptors, drug-metabolizing enzymes and transporters. Journal of Ethnopharmacology. 2023;**301**(August 2022):115822. DOI: 10.1016/j.jep.2022.115822

[71] Paul P et al. Interactions-readdressing the issue. Journal of Current Medical Research and Opinion. 2021;**4**(04):895-919. DOI: 10.15520/jcmro.v4i04.414

[72] Sharma AK, Kapoor VK, Kaur G. Herb–drug interactions: A mechanistic approach. Drug and Chemical Toxicology. 2022;**45**(2):594-603. DOI: 10.1080/01480545.2020.1738454

[73] Spanakis M et al. PharmActa: Empowering patients to avoid clinical

significant drug–herb interactions. Medicine. 2019;**6**(1):26. DOI: 10.3390/medicines6010026

[74] Eagles SK, Gross AS, McLachlan AJ. The effects of cruciferous vegetable-enriched diets on drug metabolism: A systematic review and meta-analysis of dietary intervention trials in humans. Clinical Pharmacology and Therapeutics. 2020;**108**(2):212-227. DOI: 10.1002/cpt.1811

[75] Markowitz JS et al. Effect of St. John's wort (*Hypericum perforatum*) on cytochrome P-450 2D6 and 3A4 activity in healthy volunteers. Life Sciences. 2000;**66**(9):133-139. DOI: 10.1016/s0024-3205(99)00659-1

[76] Bruyère F et al. A multicenter, randomized, placebo-controlled study evaluating the efficacy of a combination of propolis and cranberry (*Vaccinium macrocarpon*) (DUAB®) in preventing low urinary tract infection recurrence in women complaining of recurrent cystitis. Urologia Internationalis. 2019;**103**(1):41-48. DOI: 10.1159/000496695

[77] Mohammed Abdul MI et al. Pharmacodynamic interaction of warfarin with cranberry but not with garlic in healthy subjects. British Journal of Pharmacology. 2008;**154**(8):1691-1700. DOI: 10.1038/bjp.2008.210

[78] Ali Y et al. The involvement of human organic anion transporting polypeptides (OATPs) in drug-herb/food interactions. Chinese Medicine (United Kingdom). 2020;**15**(1):1-10. DOI: 10.1186/s13020-020-00351-9

[79] Choi JG et al. A comprehensive review of recent studies on herb-drug interaction: A focus on pharmacodynamic interaction. Journal of Alternative and Complementary Medicine. New York, N.Y, United States. 2016;**22**(4):262-279. DOI: 10.1089/acm.2015.0235

[80] Clairet al et al. Interaction between phytotherapy and oral anticancer agents: Prospective study and literature review. Medical Oncology. 2019;**36**(5):1-18. DOI: 10.1007/s12032-019-1267-z

[81] Coimbra University. OIPM-Observatório de Interações Planta-medicamento. 2022. Available from: http://www.oipm.uc.pt/home [Accessed: 8 December 2022]

[82] Mukadam M et al. Herbal drug interactions. Herbal Drug Interactions. International of Recent Advances in Multidisciplinary Topics. 2021;**2**(10):111-114

[83] Orellana-Paucar A, Vintimilla-Rojas D. Interactions of clinical relevance associated with concurrent administration of prescription drug and food or medicinal plants: A systematic review protocol. Systematic Reviews. 2020;**9**(1):4-9. DOI: 10.1186/s13643-019-1259-2

[84] Bordes C et al. Interactions between antiretroviral therapy and complementary and alternative medicine: A narrative review. Clinical Microbiology and Infection. 2020;**26**(9):1161-1170. DOI: 10.1016/j.cmi.2020.04.019

[85] Pochet S et al. Herb-anticancer drug interactions in real life based on VigiBase, the WHO global database. Scientific Reports. 2022;**12**(1):1-13. DOI: 10.1038/s41598-022-17704-z

[86] Babos MB et al. Herb–drug interactions: Worlds intersect with the patient at the center. Medicine. 2021;**8**(8):44. DOI: 10.3390/medicines8080044

[87] Proença da Cunha A, Pereira da Silva A, Roque OR. In: Gulbenkian FC,

editor. Plantas e Produtos Vegetais em Fitoterapia. 1a ed. Lisboa: Fundação Calouste Gulbenkian; 2003

[88] Amadi CN, Mgbahurike AA. Selected food/herb-drug interactions: Mechanisms and clinical relevance. American Journal of Therapeutics. United States. 2018;**25**(4):e423-e433. DOI: 10.1097/MJT.0000000000000705

[89] Asher GN, Corbett AH, Hawke RL. Common herbal dietary supplement-drug interactions. American Family Physician. 2017;**96**(2):101-107

[90] Loretz C et al. Application of cryopreserved human intestinal mucosa and cryopreserved human enterocytes in the evaluation of herb-drug interactions: Evaluation of CYP3A inhibitory potential of grapefruit juice and commercial formulations of twenty-nine herbal supplement. Drug Metabolism and Disposition. 2020;**48**(10):1084-1091. DOI: 10.1124/dmd.120.000033

[91] Surana AR et al. Current perspectives in herbal and conventional drug interactions based on clinical manifestations. Future Journal of Pharmaceutical Sciences. 2021;7:1-12. Article ID 103. DOI: 10.1186/s43094-021-00256-w

[92] Tan CSS, Lee SWH. Warfarin and food, herbal or dietary supplement interactions: A systematic review. British Journal of Clinical Pharmacology. 2021;**87**(2):352-374. DOI: 10.1111/bcp.14404

[93] EMA/HMPC. European Union herbal monograph on *Echinacea purpurea* (L.) Moench, herba recens. Vol. 44(May). 2017. pp. 1-7. Available from: http://www.ema.europa.eu/docs/en_GB/document_library/Herbal_-_Community_herbal_monograph/2015/04/WC500185437.pdf

[94] Mukadam MS et al. Herbal drug interactions. International of Recent Advances in Multidisciplinary Topics. 1997;**2**(10):2582-7839

[95] High N. The History of Cannabis: Origin, Spread, and Cultural Significance. HighThailand; 2023. Available from: https://www.highthailand.com/the-history-of-cannabis/

[96] MacCallum CA, Russo EB. Practical considerations in medical cannabis administration and dosing. European Journal of Internal Medicine. 2018;**49**(October):12-19. DOI: 10.1016/j.ejim.2018.01.004

[97] Gable RS. Comparison of acute lethal toxicity of commonly abused psychoactive substances. Addiction. 2004;**99**(6):686-696. DOI: 10.1111/j.1360-0443.2004.00744.x

[98] Queensland Government. Clinical Guidance: For the Use of Medicinal Cannabis Products. Queensland, Australia: Queensland Health, Department of Health Medicinal Cannabis; 2017. pp. 1-27

Chapter 20

Complementary or Alternative Plant Based Medicines and Its Active Constituents Responsible for Overall Therapeutic Efficacy

Rakhi Mishra and Binit Dwivedi

Abstract

Complementary or Alternative Medicine, like Homeopathic medicine, is made from plant, animal, and mineral kingdoms and sometimes from biochemical substances. Most of the Homeopathic remedies come from plant-based drugs. The presences of the bioactive compound in the plants are responsible for the overall therapeutic efficacy of Homeopathic medicines. The presence of bioactive compounds such as alkaloids, flavonoids, and phenols in plant drugs acts as a natural source of antioxidant substances of high importance. The concentration of these bioactive compounds and their antioxidant activity indicates that these compounds contribute to the intense antioxidant activity of Homeopathic drugs. The scope of the present research is to provide detailed information on plant-based Homeopathic medicines containing specific active compounds, which justify their typical medicinal usage in Homeopathy. It is one of the big reasons for the cure and healing properties of Complementary or Alternative Medicine medicines.

Keywords: homeopathy, antioxidant, digitoxin, reserpine, potency

1. Introduction

One of the Complementary or Alternative medicine-based systems like Homeopathy is a belief that the body can cure itself [1]. Homeopathy developed in the late 1700s in Germany, and the two primary principles of Homeopathy are [2, 3] "Like cure likes" principles. According to this principle, patients with particular signs and symptoms are treated with homeopathy remedies that produce signs and symptoms in healthy individuals [4]. Homeopathy is a medical system devised by German physician Samuel Hahnemann (1755–1843). The first edition in which he summarized homeopathy is Organon [5]. In his first fifteen years as a physician, Hahnemann struggled a lot. One day, however, he discovered that he started taking regular doses of cinchona or the bark "which contains quinine, a medicine to treat malaria. The results produced symptoms like intermittent fever (malaria) and mild degree characteristic rigors diseases [6]. This article was first published in 1796 as an essay on a new principle for ascertaining the curative power of Drugs which was included in his famous work "The Organon of the Healing Art" [7]. Homeopathy is based on using highly diluted solutions of substances selected by matching the patient's symptoms with the symptoms these substances produced in healthy individuals [8].

Homeopathy preparations are termed remedies and are made using homeopathic dilution, where selected substances are repeatedly diluted until the final product is chemically indistinguishable from the diluents. Homeopaths hit or shake the product in each dilution, claiming this makes the diluents remember the original substance after its removal [9]. Homeopathic medicine is made from plant, animal, and mineral kingdoms. Homeopathic medicines are made from plants such as belladonna, arnica, and chamomile, minerals such as sulfur and mercury, animal products such as Sepia, Lachesis, and sometimes biochemical substances such as histamine or human growth factor. Remedies are prepared by the process of serial dilution and succession. The greater the succession, the greater will be the potency of the remedy.

2. Homeopathic mother tincture and potencies

In homeopathy, Mother Tincture (Q) is defined as the original drug substance prepared with the aid of ethyl alcohol and water directly from the crude drug. They are the precursors of the corresponding potencies of the respective drug and the starting point for the preparation of homeopathic medicines. The original drug substance is used in extremely minute quantities to prepare a given homeopathic medicine, and the method of preparing homeopathic medicines is called potentization; in this method, one part of the original drug substance is mixed with nine parts of a vehicle (ethyl alcohol and water) and shaken vigorously by a special device that converts the preparation into the one tenth potency (1X) one tenth. On the other hand, one part of the original drug substances is mixed with 99 parts of the carrier (ethyl alcohol) is denoted as 1C potency (1:100). This process continued until the required potency is reached and homeopathic medicines are available in certain standard potencies such as 3X, 6X, 12X, 30X, 200X, 30C, 100C, 200C and 1000C etc. The potencies are nothing but energized dilutions (or attenuations) of the Mother Tinctures of homeopathic remedies [10].

In homeopathy, practitioners select a drug that would, if given to a healthy volunteer, cause the presenting symptoms of the patient example *Allium cepa* is derived from the common onion. While in contact with raw onion typically causes lacrimation, stinging and irritation around the eyes and nose, and clear nasal discharge. *Allium cepa* might be prescribed to patients with hay fever, significantly if both nose and eyes are affected.

Phytochemicals are biologically active naturally occurring chemical compounds found in plants. These phytochemical compounds are highly beneficial for human health. Generally, phytochemical compounds are classified as primary constituents and secondary constituents. The phytochemicals are generally present in the roots, stems, flowers, leaves, fruits, and seeds like various parts of the plant. This is the reason all complementary medicines are made from different parts of the plants depending upon the phytochemicals present therein to get maximum efficacy.

Chlorophylls, nucleic acid, common sugar, amino acid, proteins, and purines are known as primary constituents present in the plant. Chlorophyll forms during the photosynthesis process, responsible for giving green color to the plant. It also plays a beneficial role for humans and helps in the binding to the carcinogenic compounds in the body and inhibit their absorption in your intestine which helps to prevent their reach to our liver tissues which can cause further harm to the body.

Alkaloids, phenols, flavonoids, terpenes, lignans, steroids, glucosides, and saponins are known as secondary constituents present in the plant. Phenolic compounds are the widely distributed phytochemicals present in various parts of the plant. Flavonoids, phenolic acids, and polyphenols are the largest group of phenolic phytochemicals present in the plant kingdom. They act as strong antioxidant agents which help the human body to fight against free radical groups and protect it from free–radical mediated disease processes.

2.1 Flavonoids

Flavonoids are the polyphenolic phytochemical compounds present in vegetables, fruits, and beverages like tea, coffee, and fruit drinks [11]. Generally, flavonoids are found in conjugated form in nature and are characterized as monoglycosidic and diglycoscidic [12]. Flavonoids have many biological properties including anti-inflammatory, anti-microbial, cytotoxic, and anti-tumor activities [13].

2.2 Phenolic acids

Phenolic acids are a diverse group of hydroxyl benzoic and hydroxycinnamic acids and esters with glucose and carboxylic acid, hydroxycinnamic acid-like molecules. Phenolic acids act as a strong antioxidant agents and prevent the human body from degenerative diseases [14]. The degenerative diseases are cardiovascular disease, inflammation, and cancer. Phenolic acids have properties to act as strong antidiabetic agents [15]. Phenolic acids can influence the role of insulin and glucose receptors [16]. Phenolic compounds also have properties to inhibit the activities of α-glucosidase and α-amylase which are the main component of the conversion of dietary carbohydrates into glucose hydroxybenzoic and hydroxyl cinnamic acids [17]. Phenolic acids known for the prevention and treatment of cancer [18].

2.3 Tannins

Tannins are a group of polyphenolic compounds with a high molecular weight that can form complexes with polysaccharides, alkaloids, proteins, nucleic acid, and minerals [19]. As per epidemiological studies [20] tannins are beneficial for human health as they decrease the chance of creating chronic diseases. The tannins present in plants are used as astringent, used against diarrhea, used as diuretics against stomach problems, and duodenal tumors. Tannins have many biological properties including anti-inflammatory [21], antiseptic, and antioxidant properties. Major phenolic compounds found in various plants are quercetin, oleanolic acid, ursolic acid, kaempferol, luteolin, chlorogenic acid, curcumin, ascorbic acid, tannic acid, gallic acid [22] and, rutin etc.

2.4 Alkaloids

Alkaloids are heterocyclic nitrogen atoms naturally synthesized by a large number of organisms, including plants, animals, bacteria, and fungi [23]. Based on the type of heterocyclic ring system present in the molecule, they are pyrrolidine alkaloids, pyridine alkaloids, pyrrolidine-pyridine alkaloids, pyridine-piperidine alkaloids, quinoline alkaloids and isoquinoline alkaloids [24].

2.5 Essential oils

Essential oils are extracted from plants by a distillation process i.e. (liquid extracts), also obtained by the physically squeezed method in which essential oil squeeze out of some plants (like orange peels) in the case of flowers, extracted with a nonpolar solvent i.e. lemon oil [25].

Lemon contains essential oil like limonene, sesquiterpene, limonene, sesquiterpenes, aldehydes (citral, and citronellal), and geranyl acetate). Eucalyptusoil contains 1,8-cineol and α-pinene, and chamomile oil contains (Bisabolol oxide, Chamazulene, and α-Terpineol).

Essential oil like lemon oil is used as flavoring agent and are primarily used for adding fragrance to cosmetics, foods, or your home via a diffuser, clove oil used as an antimicrobial agent. Clove oil is used as a pain reliever such as toothache and muscle pain, kills bacteria, and is also recommended for digestive upset. Eucalyptus oil is commonly used in cold remedies like vapor rub.

2.6 Triterpenes

Triterpenes are six isoprene units like Lanosterol and squalene. Triterpenes are found in wheat germ and olives. Acyclic triterpene hydrocarbon squalene constitutes more than half of the liver oil of various species which are widely distributed in nature. These are found in fish liver oil, vegetable oil, fungi, human earwax, and sebaceous secretions.

2.7 Saponin

Saponins are secondary metabolites of the plant kingdom and are stable in aqueous solutions such as soap, hence the name "saponin" is given to these compounds. Saponins are a group of compounds that includes glycosylated steroids, triterpenoids, and steroid alkaloids. Spirostan and furostan derivatives are the two main types of steroid aglycones. The main triterpene aglycone is a derivative of oleanane. The steroidal saponins are essential precursors for steroid drugs. These steroid drugs act as strong anti-inflammatory effects for example androgens, oestrogens, and progestins [26].

2.8 Sterols

Sterols are a group of fat-like substances which occur naturally in the animal and plant kingdom. Animal sterol is known as cholesterol whereas plant sterols are known as ß-sitosterol, campesterol, and stigmasterol [27]. They have a similar structure to cholesterol. The basis of their molecular framework is a double bond. If this double bond is hydrolyzed, then the saturated plant sterols are formed from the unsaturated plant sterols.

3. Some important Homeopathic plant-based drugs and their active constituents responsible for the overall therapeutic efficacy of Homeopathic drugs

3.1 *Leucas aspera*

Leucas aspera (Wild) Linn, family Lamiaceae [28], is commonly called "Thumbai" [29, 30]. It is a common weed in India and is widely distributed throughout the country. *L. aspera* whole plant is traditionally used as an antipyretic [31] and insecticide [32, 33] drug. In homeopathy, *L. aspera,* mother tincture is used to treat dysentery, jaundice, fever, a bite from venomous animals [34], and enlargement of the liver and spleen. The main bioactive constituents of *L. aspera* are ursolic and oleanolic acid [35]. Due to the presence of ursolic [36] and oleanolic acid-like triterpenes [37], it possesses many pharmacological [38, 39] properties. *L. aspera* plant possesses anti-inflammatory, hepatoprotective [40], antitumor [41], anticancer [42], antimicrobial anti-HIV, antifungal, gastroprotective, hypoglycemic, and antihyperlipidemic properties (**Table 1**).

S. no.	Part used	Different pharmacological activity
1.	Aerial Part	Arthritis, Ulcer protective effect, Diabetes, Anticancer activity, antipsoriatic activity
2.	Leaf	Hepatoprotective activity, Cytotoxic activity, Antidiabetic activity, antivenom activity
3.	Roots	Analgesic activity, antinociceptive activity, Central nervous system depressant activity
4.	Whole plant	Cytotoxic activity, anti-inflammatory activity, Hepatoprotective activity, Antihyperglycemic activity, Antimutagenic activity, anthelminthic activity
5.	Shoot system including stem, leaves, and flower	anti-inflammatory activity

Table 1.
Leucas aspera *plant parts responsible for different pharmacological activity [43, 44].*

3.2 *Rauwolfia serpentina*

Rauwolfia serpentina (Family: Apocynaceae) is a medicinally famous herb in Homeopathy [45, 46]. *Rauvolfia serpentina* is therapeutically used as a sedative, a hypnotic drug, and in hypertension. Reserpine (indole alkaloid) was isolated in 1952 from the dried root of *R. serpentina* (Indian snakeroot) [47], which had been known as sarpagandha [48], and used for the treatment of insanity, fever, snakebites anxiety, and neuropsychiatric conditions. Reserpine has ability to deplete catecholamines [49]. Reserpine reduces glycemia in some cases, but the effect is short-lived. In some patients, it has a stimulating effect on prothrombin activity. Reserpine produces sedation and a lowering of blood pressure if administered orally, in hypertension; the impact of Reserpine is slow, seldom appearing before 3–6 days of administration and continuing for some time after the withdrawal of the drug a cumulative effect. It is most valuable in young patients with mild labile hypertension and tachycardia. In long-established hypertension, it is best used in conjunction with more potent hypertensive drugs such as hexamethonium or hydralazine. Combined with polythiazide; it is a useful hypertensive in mild to moderate thiazide, it is a useful hypotensive in mild to moderate conditions. The response to Reserpine varies in patients, and the dosage must be adjusted to individual requirements. In severe hypertension, it may be given by intravenous or intramuscular injection when the effect begins within a few hours. Parenteral therapy of Reserpine is indicated in the treatment of hypertension only when oral administration is impracticable [50]. Different alkaloids are present in Rauwolfia, viz. ajmaline, ajmalinine, ajmalicine, scrpentine, and serpentinine [51]. Reserpine, yohimbine [52], recinnamine, reserpinine, rauwolfinine, renoxidine, rescinnamine, reserpiline, sarpagine, serpentinine, tetraphyllicine, 3-epi-a-yohimbine. It also contains small amounts of phytosterol and fatty substances [53]. *R. serpentina* is one of the essential drugs for various disorders, including hypertension (**Table 2**).

S. no.	Alkaloid	Different pharmacological activity
1.	Reserpine	Antipsychotic activity and antihypertensive activity
2.	Reserpiline	Antihypertensive activity
3.	Rescinnamine	Antihypertensive activity
4.	Ajmaline	Antiarrhythmic activity
5.	Ajmalicine	Vasodilator
6.	Serpentine	Tranquilizer
7.	Alstonine	Antipsychotic activity

Table 2.
Rauwolfia serpentina *major alkaloids responsible for different pharmacological activity [54].*

3.3 *Digitalis purpurea*

Digitalis purpurea, commonly known as Foxglove [55], is a biennial plant belonging to the family Plantaginaceae. It is indigenous to part of western and southwestern Europe. In India, *Digitalis purpurea* is found in the Nilgiri hills of Tamil Nadu, southern Sikkim, and the eastern Himalayan region. Its leaves contain both primary and secondary glycosides. Digitoxin is the main active constituent of the *D. purpurea* plant, which is used as cardiac glycoside in medicines. Digitoxin is generally known as a highly toxic by-product [56]. In Homeopathy, *D. purpurea* is used for treating heart-related diseases where the heart is primarily involved where the pulse is weak, irregular, intermittent, abnormally slow, and dropsy of external and internal parts. In females, during labor-pain in the abdomen and back before menses, and for uterine hemorrhage [57] *D. purpurea* homeopathic medicine is recommended in Homeopathy. As per the information given in the homeopathic book Materia Medica Pura, *D. purpurea* is mainly used for diseases where the heart is primarily involved [58], such as atrial flutter, atrial fibrillation, and in case of congestive heart failure conditions. Cardiac glycosides [59], such as digitoxin in the leaves of *D. purpurea* [60], help to prevent congestive heart failure by increasing the force of contractions of the heart in the body. In previous suggested studies, digitoxin present in the leaves of Digitalis has the highest gastrointestinal (GI) absorption of 90–100% with a half-life of 4–5 days [61] which is more significant than other commercially available cardiac steroids such as Digoxin, Deslanoside, Ouabain [62]. *D. purpurea*, being rich in active constituents, shows cardiovascular, cytotoxic [63], antioxidant, anti-diabetic [64], insecticidal, immunological, cardioprotective, hepatoprotective, and neuroprotective effects and has greater importance in Homeopathy.

3.4 *Hydrocotyle asiatica*

Hydrocotyle asiatica, synonym *Centella asiatica* commonly known as brahami, belongs to the family of perennial plants in the flowering plant Umbelliferae (Apiaceae) [65]. *H. asiatica* is an essential Homeopathic medicinal plant widely used as a medicine due to the benefit of its bioactive compounds such as asiatic acid [66], rutin, kaempferol, quercetin, gallic acid, luteolin, and catechin [67]. Hydrocotyle asiatica possess diverse pharmacological activities such as neuroprotective, nerve regenerative, immunomodulatory, anti-depressive, memory enhancing [68], gastro-protective, cardioprotective, radioprotective, anti-cancer, antimicrobial, anti-inflammatory, anti-diabetic and antioxidative properties. *H. asiatica* contains the most abundant triterpene glycoside asiatic acid, which shows cytotoxic activity on cancer cells. Asiatic acid is a triterpene glycoside, commonly used for wound healing. Asiatic acid has antioxidant, anti-inflammatory, and neuroprotective properties. *H. asiatica* has been used in folk herbal medicine for centuries for memory enhancement, anti-depressants [69], wound healing [70] and psoriasis remedy [71], and chronic disease treatments [72]. In homeopathy, *H. asiatica* acts as a curative in disorders that exhibit interstitial inflammation, cellular proliferation, leprosy, lupus, granular ulceration of the womb, profuse leucorrhœa, psoriasis gyrates, and syphilitic affections [73].

3.5 *Nux vomica*

Nux vomica, commonly known as Kuchla, belongs to the family Loganiaceae. In homeopathy, *N. vomica* seed part is used. *N. vomica* is one of the most prescribed medicines in homeopathy [74]. The main bioactive compounds of *N. vomica* seeds are strychnine and brucine. Brucine in *N. vomica* is responsible for its anti-tumor, anti-inflammatory, analgesic, and effects on the cardiovascular and nervous systems. The presence of brucine makes *N. vomica* an analgesic agent and can be used to relieve arthritis and traumatic pain [75]. The alkaloids in *N. vomica* seeds are protostrychinine, isostrychinine, vomicine, n-oxystrychinine, pseudostrycheinine, chlorogenic acid, and glycoside. In the homeopathy system of medicine, *N. vomica* is prescribed for anger effect, colic, constipation, dyspepsia, gastrodynia, hemorrhoids, tobacco habit, insomnia, nightmares, lumbago, diabetes, asthma, aphrodisiac and to improve appetite [76]. *N. vomica* has an anti-alcoholism effect. In homeopathy, *Nux vomica* 30C, 200C, and 1000C are recommended for anti-alcoholism effects [77]. In recent years brucine displayed an excellent anti-tumor effect on various tumors. For hepatocellular carcinoma, brucine inhibits the proliferation of Hep G2 cells by regulating calcium concentration and depolarization of mitochondria [78].

3.6 *Matricaria chamomilla*

Matricaria chamomilla is, commonly known as chamomilla and belongs to the Asteraceae family which is referred to as the star among medicinal species. *M. chamomilla* was introduced in India during the Mughal period. In India, *M. chamomilla* is grown in Punjab, Uttar Pradesh, Maharashtra, Jammu, and Kashmir [79]. In Homeopathy, *M. chamomilla* part used is the whole plant. The whole *chamomilla* plant contains over 120 constituents [80]. *M. chamomilla* acts as a natural antioxidant and antimicrobial agent. The latest research has proved the presence of 52 active components in the *M. chamomilla* plant in which the highest contents are β-farnesene α-farnesene, α-bisabolol, and its oxide chamazulene, germacrene, and spiroether [81]. *M. chamomilla* homeopathic extracts contains active groups of phenol, flavonoids, and coumarins. It act as strong antioxidant, antibacterial, antifungal, antiparasitic agents [82]. In general, *M. chamomilla* is used for many ailments in humans such as fever, inflammation, menstrual disorder, insomnia, and ulcers; very good in wound healing, muscle spasms, gastrointestinal upset, rheumatic pain, and hemorrhoids [83]. This plant has immense curative properties. In Homeopathy, *M. chamomilla* is used for tooth problems in infants [84]. The symptoms are twitching, convulsion during teething, intolerant pain, frantic irritability, ugly, cross, uncivil, quarrelsome, and colic after anger [85].

3.7 *Aconitum napellus*

Aconitum napellus belongs to the family *Ranunculaceae,* commonly known as Monkshood. The main active constituent present in *A. napellus* is "aconitine", which is present throughout the plant, with incredibly high levels in the leaves and roots. In homeopathy *A. napellus* plant part used is the whole plant. *A. napellus* plant contains various diterpenes alkaloids, isonapelline, luciculine, and napelline [86]. Aconitine present in *A. napellus* acts as a voltage-gated sodium channel activator. These sodium channel act as transmembrane proteins that are responsible for the rapid depolarization that underlies the upstroke of action potentials in neurons and is responsible for crucial nerve impulse conduction [87]. *A. napellus* has solid neuroprotective properties. *A. napellus* is a constituent of Ayurvedic, Unani medical preparation, and polyherbal formulation to treat diabetes, as a nerve tonic with potent antioxidant properties [88]. *A. napellus* is very useful in numbness [89], a sensation of cold, and pain in the extremities, associated with diabetic neuropathy, and is also used for the treatment of rheumatoid and joint pain [90]. *A. napellus* is the first homeopathic medicine that is recommended for the onset of acute croup [91]. In homeopathy, *A. napellus* is used in case of shock, sudden or violent onset ailments from shock, fright or fear, intense fear, terror-struck, restlessness with fear of death and diseases from exposure to cold and dry wind [92].

3.8 *Rhus toxicodendron*

Rhus toxicodendron synonym *Toxicodendron pubescens* is commonly known as poison ivy and belongs to the Anacardiaceae family [93]. This plant family produces the chemical substance pentadecylcatechols. The main active constituent in *R. toxicodendron* is fisetin, rhamnose, and gallic acid [94]. This plant's main habitat is in the forest of the United States. In Homeopathy, the leaves part is used. In Homeopathy, *R. toxicodendron* is used for joint pains [95], stiffness worse in damp weather, irritability, restlessness [96], back pains, and asthma alternating with skin eruptions [97]. *R. toxicodendron* is used for anti-inflammatory activities for arthritis [98] and rheumatism [99]. In homeopathy, Rhus tox 6C, 12C, 30C, and 200C are recommended [100]. *R. toxicodendron* retains its anti-arthritis properties at 1 M (potency 1000C), 10 M (potency 10,000C), and CM (potency 100,000C) homeopathic dilution [101]. *R. toxicodendron al*so possesses immune stimulatory activity in its crude form and in homeopathically diluted forms [102]. It is commonly used for strains [103]. This medicine is recommended for patients with high fever and body aches and restlessness problems. A person experiencing this type of fever might want to remain in motion throughout to avoid restlessness. This medicine is recommended in case a person catches a fever due to getting wet in the rain [104].

4. Conclusion

Homeopathic medicines show high therapeutic efficacy and synergetic effects due to the presence of many active constituents in the plants, which altogether cause synergetic effect and are responsible for the overall therapeutic efficacy of the homeopathic drug. There is an essential need for physicochemical, phytochemical, and High-Performance Thin Layer Chromatography studies of homeopathic medicines to laid down the standard for a monograph of pharmacopeia standards. In the future, these studies may open the area of vast research opportunities in medicinal plant biochemistry. The quantitative estimation of homeopathic drugs should be carefully evaluated and quantified to determine the presence of active compounds in these drugs, responsible for the overall pharmacological activity of the homeopathic medicine. The presence of alkaloids, flavonoids, and phenols in plant drugs act as a natural source of antioxidant substances of high importance in homeopathic medicine. The concentration of these active compounds and their antioxidant activity indicate that these compounds contribute to the intense antioxidant activity of Homeopathic medicine. The presence of active constituents in homeopathic mother tincture suggests that the mother tincture of homeopathic plant drugs contained specific active compounds which justify typical medicinal usage in Homeopathy and the reason for a cure and healing properties for homeopathic medicines.

Author details

Rakhi Mishra* and Binit Dwivedi
Department of Chemistry, DDPR CRI(H), Ministry of AYUSH, Government of India, Drug Standardization Unit, Noida, Uttar Pradesh, India

*Address all correspondence to: rakhimishra020@gmail.com

References

[1] Ratini M. What is Homeopathy? 2021. Available at: https://www.webmd.com/balance/what-is-homeopathy [Accessed 7 June 2022]

[2] Ernst E. Homoeopathy: Past, present and future. British Journal of Clinical Pharmacology. 1997;**44**(5):435-437. DOI: 10.1046/j.1365-2125.1997.t01-1-00611.x

[3] Linde K, Clausius N, Ramirez G, Melchart D, Eitel F, Hedges LV, et al. Are the clinical effects of homoeopathy placebo effects? A meta-analysis of placebo-controlled trials. The Lancet. 1997;**350**(9081):834-843. DOI: 10.1016/s0140-6736(97)02293-9

[4] Ernst E. A systematic review of systematic reviews of homeopathy. British Journal of Clinical Pharmacology. 2002;**54**(6):577-582. DOI: 10.1046/j.1365-2125.2002.01699.x

[5] Grams N. Homeopathy—Where is the science? EMBO Reports. 2019;**2019**:e47761. DOI: 10.15252/embr.201947761

[6] Loudon I. A brief history of homeopathy. Journal of the Royal Society of Medicine. 2006;**99**(12):607-610. DOI: 10.1258/jrsm.99.12.607

[7] Bynum WF, Porter R, editors. Companion Encyclopedia of the History of Medicine. London. 1st ed. 1993. DOI: 10.4324/9781315002514

[8] Piltan D, Rist L, Simões-Wüst P, Saller R. Test of a homeopathic dilution of *Aconitum napellus*. Forschende Komplementärmedizin/Research in Complementary Medicine. 2009;**16**(3):168-173. DOI: 10.1159/000219316

[9] Rastogi RP, Mehrotra BN. Compendium of Indian Medicinal Plants I: 1960-1969: CDRI. New Delhi: Lucknow and Publications and Information Directorate; 1993. pp. 338-343

[10] Vickers A, Zollman C. ABC of complementary medicine: Homoeopathy. BMJ. 1999;**319**(7217):1115-1118. DOI: 10.1136/bmj.319.7217.1115

[11] Panche AN, Diwan AD, Chandra SR. Flavonoids: An overview. Journal of Nutrition Science. 2016;**5**:e47. DOI: 10.1017/jns.2016.41

[12] Santos EL, Maia BHLNS, Ferriani AP, Teixeira SD. Flavonoids: Classification, Biosynthesis and Chemical Ecology. London, United Kingdom: InTech; 2017. DOI: 10.5772/67861

[13] Ullah A, Munir S, Badshah SL, Khan N, Ghani L, Poulson BG, et al. Important flavonoids and their role as a therapeutic agent. Molecules. 2020;**25**(22):5243. DOI: 10.3390/molecules25225243

[14] Kumar N, Goel N. Phenolic acids: Natural versatile molecules with promising therapeutic applications. Biotechnology Reports (Amsterdam, Netherlands). 2019;**24**:e00370. DOI: 10.1016/j.btre.2019.e00370

[15] Sun C, Zhao C, Guven EC, et al. Dietary polyphenols as antidiabetic agents: Advances and opportunities. Food Frontiers. 2020;**1**:18-44. DOI: 10.1002/fft2.15

[16] Williamson G, Sheedy K. Effects of polyphenols on insulin resistance. Nutrients. 2020;**12**(10):3135. DOI: 10.3390/nu12103135

[17] Lin D, Xiao M, Zhao J, Li Z, Xing B, Li X, et al. An overview of plant phenolic compounds and their importance in human nutrition and management of Type 2 diabetes. Molecules. 2016;**21**(10):1374. DOI: 10.3390/molecules21101374

[18] Abotaleb M, Liskova A, Kubatka P, Büsselberg D. Therapeutic potential of plant phenolic acids in the treatment of cancer. Biomolecules. 2020;**10**(2):221. DOI: 10.3390/biom10020221

[19] Frutos P, Hervás G, Giráldez FJ, Mantecón AR. Review: Tannins and ruminant nutrition. Spanish Journal of Agricultural Research. 2004;**2**(2):191-202. DOI: 10.5424/sjar/2004022-73

[20] Serrano P et al. Nuclear magnetic resonance structure of the nucleic acid-binding domain of severe acute respiratory syndrome coronavirus nonstructural protein 3. Journal of Virology. 2009;**83**(24):12998-13008

[21] de Bruyne M, Clyne PJ, Carlson JR. Odor coding in a model olfactory organ: The Drosophila maxillary palp. The Journal of Neuroscience. 1999;**19**(11):4520-4532

[22] Tong Z, He W, Fan X, Guo A. Biological function of plant tannin and its application in animal health. Frontier in Veterinary Science. 2022;**8**:803657. DOI: 10.3389/fvets.2021.803657

[23] Cornu A, Jm Besle P, Mosoni EG. Lignin-carbohydrate complexes in forages: Structure and consequences in the ruminal degradation of cell-wall carbohydrates. Reproduction Nutrition Development. 1994;**34**(5):385-398

[24] Casu B, Molyneux P. A comparative study of efficiency in European banking. Applied Economics. 2003;**35**(17):1865-1876. DOI: 10.1080/0003684032000158109

[25] Golmohammadi M, Borghei A, Zenouzi A, Ashrafi N, Taherzadeh MJ. Optimization of essential oil extraction from orange peels using steam explosion. Heliyon. 2018;**4**(11):e00893. DOI: 10.1016/j.heliyon.2018.e00893

[26] Takeuchi O, Hoshino K, Kawai T, Sanjo H, Takada H, Ogawa T, et al. Differential roles of TLR2 and TLR4 in recognition of gram-negative and gram-positive bacterial cell wall components. Immunity. 1999;**11**(4):443-451. DOI: 10.1016/s1074-7613(00)80119-3

[27] Sterol, Sterol regulatory element binding proteins (SREBPs) are membrane-bound transcription factors that activate genes involved in cholesterol synthesis. Clinical Biochemistry: Metabolic and Clinical Aspects (Third Edition). 2014

[28] Islam AK, Ohno O, Suenaga K, Kato-Noguchi H. Two novel phytotoxic substances from Leucas aspera. Journal of Plant Physiology. 2014;**171**(11):877-883. DOI: 10.1016/j.jplph.2014.03.003

[29] Latha B, Rumaisa Y, Soumya C, Shahul S, Sadhiya N. Phytochemical studies on *Leucas aspera*. Journal of Chemical and Pharmaceutical Research. 2013;**5**(4):222-228

[30] Enjamoori VK, Nampalli A, Vasudha B, Gangarapu K, Boggula N. A review on *Leucas aspera* for phytopharmacological studies. INNOSC Theranostics and Pharmacological Sciences. 2019;**2**(1). DOI: 10.26689/itps.v2i1.436

[31] Gupta N, Subhramanyam E, Jha S, Bhatia V, Narang E. A comparative antipyretic activity of the crude extracts of the plant Leucas aspera

and Glycosmis pentaphylla. Journal of Chemical and Pharmaceutical Research. 2011;**3**(1):320-323

[32] Devi MS, Vinothini K, Arjun P, Sekar S, Mavumengwana V. In vitro biomass accumulation of calli and root enhancement of Leucas aspera (Willed) Linn. Under stress conditions. African Journal of Science, Technology, Innovation and Development. 2015;**7**(6):395-400. DOI: 10.1080/20421338.2015.1096509

[33] Jayashree D. Phytochemicals analysis and thin layer chromatography finger printing of methanolic extracts of three medicinal plants. International Research Journal of Pharmacy. 2013;**4**(6):123-1236. DOI: 10.7897/2230-8407.04627

[34] Evans WC, Evans D. Chapter 5 – A taxonomic approach to the study of medicinal plants and animal-derived drugs. In: Evans WC, Evans D, editors. Trease and Evans' Pharmacognosy. Sixteenth ed. London: W.B. Saunders; 2009. pp. 18-44

[35] Shukla M, Agarwal M, Singh J, Tripathi A, Srivastava A, Singh V. Predictors of depression among people living with HIV/AIDS on antiretroviral therapy attending tertiary care hospitals in the Capital of Uttar Pradesh: A cross-sectional study. Pharmacognosy Magazine. 2016;**12**(46):159-164. DOI: 10.4103/0971-9962.209200

[36] Hussain H, Green IR, Ali I, Khan IA, Ali Z, Al-Sadi AM, et al. Ursolic acid derivatives for pharmaceutical use: A patent review (2012-2016). Expert Opinion on Therapeutic Patents. 2017;**27**(9):1061-1072. DOI: 10.1080/13543776.2017.1344219

[37] Sultana N, Ata A. Oleanolic acid and related derivatives as medicinally important compounds. Journal of Enzyme Inhibition and Medicinal Chemistry. 2008;**23**(6):739-756. DOI: 10.1080/14756360701633187

[38] Asadi-Samani M, Moradi MT, Bahmani M, Shahrani M. Antiviral medicinal plants of Iran: A review of ethnobotanical evidence. International Journal of PharmTech Research. 2016;**9**(5):427-434

[39] Mučaji P, Nagy M. Contribution to the TLC separation of ursolic and oleanolic acid mixture. European Pharmaceutical Journal. 2011;**58**(1): 56-61. DOI: 10.2478/v10219-011-0006-0

[40] Banu S, Bhaskar B, Balasekar P. Hepatoprotective and antioxidant activity of Leucas aspera against D-galactosamine induced liver damage in rats. Pharmaceutical Biology. 2012;**50**(12):1592-1595. DOI: 10.3109/13880209.2012.685130

[41] Patel K, Patel DK. The beneficial role of Rutin, A naturally occurring flavonoid in health promotion and disease prevention: A systematic review and update. In: Bioactive Food as Dietary Interventions for Arthritis and Related Inflammatory Diseases. Second ed. Academic Press; 2019. DOI: 10.1016/B978-0-12-813820-5.00026-X

[42] Yin R, Li T, Tian JX, Xi P, Liu RH. Ursolic acid, a potential anticancer compound for breast cancer therapy. Critical Reviews in Food Science and Nutrition. 2018; **58**(4):568-574. DOI: 10.1080/10408398.2016.1203755

[43] Rao P, Prasanna MN. Immunological studies on rheumatoid arthritis treated with homeopathic drugs: Results of the Pilot Study. Indian Journal of Research Homoeopathy. 2008;**2**:42-49. Available from: https://www.ijrh.org/text.asp?2008/2/4/42/130659

[44] Nirmala and Kanchana. Antimicrobial and pharmacological value of *Leucas aspera*. Systematic Reviews in Pharmacy. 2018;**9**(1):41-44

[45] Ajay IA, Ajibade O, Oderinde RA. Preliminary phytochemical analysis of some plant seeds. Research Journal of Chemical Science. 2011;**1**(3):58-62

[46] Qureshi SA, Nawaz A, Udani SK, Anmi B. Hypoglycaemic and Hypolipidemic activities of Rauwolfia serpentina in Alloxan- induced diabetic rats. International Journal of Pharmacology. 2009;**2009**:1-4

[47] Henry JP, Scherman D. Radioligands of the vesicular monoamine transporter and their use as markers of monoamine storage vesicles. Biochemical Pharmacology. 1989;**38**(15):2395-2404. DOI: 10.1016/0006-2952(89)90082-8

[48] Lobay D. Rauwolfia in the treatment of hypertension. Integrated Medicine (Encinitas). 2015;**14**(3):40-46

[49] Paravati S, Rosani A, Warrington SJ. Physiology, Catecholamines. Treasure Island (FL): StatPearls Publishing; 2022. Available from: https://www.ncbi.nlm.nih.gov/books/NBK507716/

[50] Baumeister AA, Hawkins MF, Uzelac SM. The myth of reserpine-induced depression: Role in the historical development of the monoamine hypothesis. Journal of the History of the Neurosciences. 2003;**12**(2):207-220. DOI: 10.1076/jhin.12.2.207.15535

[51] Wilkins RW, Judson WE. The use of Rauwolif serpentina in hypertensive patients. The New England Journal of Medicine. 1953;**248**:48-53. DOI: 10.1056/NEJM195301082480202

[52] Siddiqui S. Some contributions to researches on medicinal plants in a historical context. In: Rahman A, editor. Natural Product Chemistry. Berlin, Heidelberg: Springer; 1986. DOI: 10.1007/978-3-642-71425-2_22

[53] Deshmukh SR, Dhanashree S, Patil Bhausaheb A. Extraction and evaluation of indole alkaloids from rauwolfia serpentina for their antimicrobial and antiproliferative activities. International Journal of Pharmacy and Pharmaceutical Sciences. 2012;**4**(5):329-334

[54] Singh M, Kaur R, Rajput R, Mathur G. Evaluating the therapeutic efficiency and drug targeting ability of alkaloids present in Rauwolfia serpentina. International Journal of Green Pharmacy. 2023;**11**(3):132-142

[55] Ali M, Ravinder E, Ramachandram R. A new flavonoid from the aerial parts of *Tridax procumbens*. Fitoterapia. 2001;**2001**:313-315

[56] Chowdhury R, Ramond A, O'Keeffe LM, Shahzad S, Kunutsor SK, Muka T, et al. Environmental toxic metal contaminants and risk of cardiovascular disease: Systematic review and meta-analysis. BMJ. 2018;**362**:k3310. DOI: 10.1136/bmj.k3310

[57] Ali M, Ravinder E, Ramachandram R. A new flavonoid from the aerial parts of Tridax procumbens. Fitoterapia. 2001;**72**(3):313-315. DOI: 10.1016/s0367-326x(00)00296-3

[58] Boericke W. Pocket Manual of Homoeopathic Materia Medica Comprising the Characteristic and Guiding Symptoms of All Remedies Clinical and Pathogenetic, presented by Médi-T *Digitalis purpurea* Foxglove Digitalis. Ninth ed. India: B Jain Publishers Pvt. Ltd.; 1927. pp. 252-253

[59] Boericke W. Homoeopathic Materia Medica; Presented by Médi-T *Digitalis*

purpurea Foxglove Digitalis. 1990. Available at: http://www.homeoint.org/books/boericmm/d/dig.htm [Accessed 4 August 2022]

[60] Fujii Y, Ikeda Y, Yamazaki M. High-performance liquid chromatographic determination of secondary cardiac glycosides in digitalis purpurea leaves. Journal of Chromatography A. 1989;**479**:319-325. DOI: 10.1016/s0021-9673(01)83346-x

[61] Smith TW. Pharmacokinetics, bioavailability and serum levels of cardiac glycosides. Journal of the American College of Cardiology. 1985;**5**(5 Suppl. A):43A-50A. DOI: 10.1016/s0735-1097(85)80462-9

[62] Bhagwat DA, Killedar SG, Adnaik RS. Anti-diabetic activity of leaf extract of *Tridax procumbens*. International Journal of Green Pharmacy. 2008;**2**(2):126-128. Available from: https://eurekamag.com/research/022/159/022159330.php

[63] Ali Esmail Al-Snafi. Phytochemical Constituents and Medicinal Properties of Digitalis lanata and Digitalis Purpurea: A Review. Iajps: CSK Publications; 2017. DOI: 10.5281/zenodo.344926

[64] Jadhav M, Ghanghav S, Singh N. Digitalis purpurea: An Overview on Phytochemical and Pharmacological Profile. IJP. 2018;**5**(9):563-570. DOI: 10.13040/IJPSR.0975-8232.IJP.5(9).563-70

[65] Menglan S, Fading P, Zehui P, Watson MF, et al. Apiaceae, Umbelliferae, Flora of China. 2005;14: pp. 1-205. Available at: http://flora.huh.harvard.edu/china/mss/volume14/APIACEAE.pdf [Accessed 4 August 2022]

[66] Joseph GV, Chaturvedi S, Deokule SS. Standardisation and quality evaluation of Centella asiatica Linn. Ancient Science of Life. 2001;**20**(4):99-110

[67] Borhan MZ, Ahmad R, Rusop M, Abdullah S. Green extraction: Enhanced extraction yield of Asiatic acid from Centella asiatica (L.) nanopowders. Journal of Applied Chemistry. 2013:2356-7171. DOI: 10.1155/2013/460168

[68] Orhan IE. *Centella asiatica* (L.) Urban: From traditional medicine to modern medicine with neuroprotective potential. Evidence-based Complementary and Alternative Medicine. 2012;**2012**:946259. DOI: 10.1155/2012/946259

[69] Puttarak P, Dilokthornsakul P, Saokaew S, Dhippayom T, Kongkaew C, Sruamsiri R, et al. Effects of Centella asiatica (L.) Urb. On cognitive function and mood related outcomes: A systematic review and Meta-analysis. Scientific Reports. 2017;**7**(1):10646. DOI: 10.1038/s41598-017-09823-9

[70] Yao CH, Yeh JY, Chen YS, Li MH, Huang CH. Wound-healing effect of electrospun gelatin nanofibres containing *Centella asiatica* extract in a rat model. Journal of Tissue Engineering and Regenerative Medicine. 2017;**11**(3):905-915. DOI: 10.1002/term.1992

[71] Ravi A, Khan S, Suklabaidya M, Priyadurairaj P, Sudarsanan PS, Chandrasekar K. Antioxidant activity and In Silico analysis of *Centella asiatica* and Indigofera aspalathoides in psoriasis. Biomedical Pharmacology Journal. 2018;**11**:3. DOI: 10.13005/bpj/1504

[72] Gohil KJ, Patel JA, Gajjar AK. Pharmacological review on *Centella asiatica*: A potential herbal cure-all. Indian Journal of Pharmaceutical Sciences. 2010;**72**(5):546-556. DOI: 10.4103/0250-474X.78519

[73] Homoeopathic Materia Medica by William Boericke. Ninth edition, Hydrocotyle asiatica Indian Pennywort. 2000. Available at: http://www.homeoint.org/books/boericmm/h/hydrc.htm [Accessed 04 August 2022]

[74] Rehman T. A brief review on Nux vomica : A panacea homoeopathic remedy. Journal of Integrated Standardized Homoeopathy. 2021;**4**:59-63. DOI: 10.25259/JISH_3_2021. DOI:10.25259/JISH_3_2021

[75] Lu L, Huang Rui W, Ye JJ-M, Hong-Zhuan C, Li-Jun Z, Xin L. Brucine: A review of Phytochemistry, pharmacology, and toxicology. Frontiers in Pharmacology. 2020;**11**:1663-9812. DOI: 10.3389/fphar.2020.00377

[76] A. Strychnos nux-vomica seeds: Pharmacognostical standardization, extraction, and antidiabetic activity. Journal of Ayurveda Integrated Medicine. 2012;**3**(2):80-84. DOI: 10.4103/0975-9476.96523

[77] Coelho C, Valvassoura F. Use of ultradiluted medications, Nux vomica, and Papaver somniferum as an aid to the Anesthesia recovery of cats submitted to elective ovariohysterectomy. Homeopathy. 2018;**107**(S 01):55-78. DOI: 10.1055/s-0038-1632417

[78] Lu L, Huang R, Wu Y, et al. Brucine: A review of phytochemistry, pharmacology, and toxicology. Frontiers in Pharmacology. 2020;**11**:377. DOI: 10.3389/fphar.2020.00377

[79] Singh O, Khanam Z, Misra N, Srivastava MK. Chamomile (*Matricaria chamomilla* L.): An overview. Pharmacognosy Reviews. 2011;**5**(9):82-95. DOI: 10.4103/0973-7847.79103

[80] El Joumaa MM, Borjac JM. *Matricaria chamomilla*: A valuable insight into recent advances in medicinal uses and pharmacological activities. Phytochemistry Reviews. 2022;**21**:1913-1940. DOI: 10.1007/s11101-022-09817-0

[81] Stanojevic LP, Marjanovic-Balaban ZR, Kalaba VD, Stanojevic JS, Cvetkovic DJ. Chemical composition, antioxidant and antimicrobial activity of chamomile flowers essential oil (*Matricaria chamomilla* L.). Journal of Essential Oil Bearing Plants. 2016;**19**(8):2017-2028. DOI: 10.1080/0972060X.2016.1224689

[82] El Mihyaoui A, Esteves da Silva JCG, Charfi S, Candela Castillo ME, Lamarti A, Arnao MB. Chamomile (Matricaria chamomilla L.): A review of ethnomedicinal use, phytochemistry and pharmacological uses. Life. 2022;**12**(4):479. DOI: 10.3390/life12040479

[83] Srivastava JK et al. Chamomile: A herbal medicine of the past with bright future. Molecular Medicine Reports. 2010;**3**(6):895-901

[84] Singh O, Khanam Z, Misra N, Srivastava MK. Chamomile (*Matricaria chamomilla* L.): An overview. Pharmacognosy Reviews. 2011;**5**(9):82-95. DOI: 10.4103/0973-7847.79103

[85] Vickers A, Zollman C. ABC of complementary medicine. BMJ. 1999;**319**(7217):1115-1118. DOI: 10.1136/bmj.319.7217.1115

[86] Artal FJC. Chapter 23 – Adverse neurological effects caused by the ingestion of plants, seeds, and fruits. In: Watson RR, Preedy VR, editors. Bioactive Nutraceuticals and Dietary Supplements in Neurological and Brain Disease. United States: Academic Press; 2015. pp. 215-219. DOI: 10.1016/B978-0-12-411462-3.00023-

[87] Yamakawa K. Chapter 15 – Mutations of voltage-gated sodium channel genes

SCN1A and SCN2A in epilepsy, intellectual disability, and autism. In: Sala C, Verpelli C, editors. Neuronal and Synaptic Dysfunction in Autism Spectrum Disorder and Intellectual Disability. United States: Academic Press; 2016. pp. 233-251. DOI: 10.1016/B978-0-12-800109-7.00015-7

[88] Shoaib A, Siddiqui HH, Dixit RK, Siddiqui S, Deen B, Khan A, et al. Neuroprotective effects of dried tubers of *Aconitum napellus*. Plants. 2020;**9**(3):356. DOI: 10.3390/plants9030356

[89] Narayan Das Prajapati, U. Kumar, Agro's Dictionary of Medicinal Plants, Agrobios: Jodhpur, India, 2003. Available online at https://content.kopykitab.com/ebooks/2014/04/3140/sample/sample_3140.pdf [Accessed 22 July 2022]

[90] Shoaib A, Siddiqui HH, Dixit RK, Akhtar J. *Aconitum Napellus*: Detoxification and acute toxicity investigation followed by sub-acute toxicity and bioavailability assessment of highest and lowest LD extract. Journal of Biologically Active Products from Nature. 2019;**9**(2):108-119. DOI: 10.1080/22311866.2019.1605931

[91] Loo M. Chapter 15 – Bronchiolitis (Croup). In: Loo M, editor. Integrative Medicine for Children. United States: W.B. Saunders; 2009. pp. 207-213. DOI: 10.1016/B978-141602299-2.10015-5

[92] Vickers A, Zollman C. ABC of complementary medicine. BMJ. 1999;**319**(7217):1115-1118. DOI: 10.1136/bmj.319.7217.1115

[93] Government of India, Ministry of Health, India. 1979; p. 180. Available online at: http://www.hplism.nic.in/sites/default/files/rhus-toxicodendron.pdf [Accessed 25 July 2022]

[94] William Anderson Syme. Some constituents of the Poison Ivy Plant: (Rhus Toxicodendron). 2010; 18-23. Available at: https://www.gutenberg.org/files/34510/34510-h/34510-h.htm [Accessed 03 August 2022]

[95] Lee KJ, Yeo MG. Homeopathic Rhus toxicodendron has dual effects on the inflammatory response in the mouse preosteoblastic cell line MC3T3-e1. Homeopathy. 2016;**105**(1):42-47. DOI: 10.1016/j.homp.2015.09.004

[96] Kayne SB. Chapter 13 – Materia medica. In: Kayne SB, editor. Homeopathic Pharmacy. Second ed. USA: Philadelphia: Churchill Livingstone; 2006. pp. 337-354. DOI: 10.1016/B978-044310160-1.50018-2

[97] Castro M. Homeopathy A theoretical framework and clinical application. Journal of Nurse-Midwifery. 1999;**44**(3):280-290. DOI: 10.1016/s0091-2182(99)00040-3

[98] Huh Y, Kim M, Yeo M. Homeopathic Rhus toxicodendron treatment increased the expression of cyclooxygenase-2 in primary cultured mouse chondrocytes. Homeopathy: The Journal of the Faculty of Homeopathy. 2013;**102**:248-253. DOI: 10.1016/j.homp.2013.07.001

[99] Rajgurav AB et al. To study the efficacy of Rhus tox in management of cases of osteoarthritis of knee joint. International Journal of Research Orthopedics. 2017;**3**(1):54-60. DOI: 10.18203/issn.2455-4510.IntJResOrthop20164832

[100] Magar S, Nayak D, Mahajan UB, et al. Retracted article: Ultra-diluted Toxicodendron pubescens attenuates pro-inflammatory cytokines and ROS-mediated neuropathic pain in rats. Scientific Reports. 2018;**8**:13562. DOI: 10.1038/s41598-018-31971-9

[101] Patel DR, Ansari IA, Kachchhi YN, Patel RB, Patil KR, Jadhav RB, et al.

Toxicodendron pubescens retains its anti-arthritic efficacy at 1M, 10M and CM homeopathic dilutions. Homeopathy. 2012;**101**(3):165-170. DOI: 10.1016/j.homp.2012.02.007

[102] Magar S, Nayak D, Mahajan UB, Patil KR, Shinde SD, Goyal SN, et al. Ultra-diluted toxicodendron pubescens attenuates pro-inflammatory cytokines and ROS- mediated neuropathic pain in rats. Scientific Reports. 2018;**8**(1):13562. DOI: 10.1038/s41598-018-31971-9

[103] Dana Ullman MPH. Ten common homeopathic medicines. Available at: https://homeopathic.com/ten-common-homeopathic-medicines/ [Accessed 10 June 2022]

[104] Avinashi S. 7 best homeopathic medicines for fever in child & adults. Available at: https://www.lybrate.com/topic/homeopathic-medicines-for-fever bbe8/02ca98eee4317ca693ca2dd92 5ed0075 [Accessed 10 June 2022]

Chapter 21

Use of Indigenous Knowledge Systems for Managing Cattle Health in Zimbabwe: Challenges and Opportunities

Vimbai Gobvu, Kudakwashe C. Chirigo, Takudzwa L. Charakupa and Clarice P. Mudzengi

Abstract

Cattle play a pivotal role especially for the rural farmer by providing milk, draught power, meat and serving as an indication of wealth among other roles. Research and development of cattle production especially in communal areas can be a sustainable way to improve the livelihoods of the rural population. Major constraints to communal cattle production include high prevalence of diseases, limited forage and poor marketing linkages. For reasons that include; lack of veterinary clinics and extension services, high costs of drugs and potency of the ethnoveterinary medicines, many farmers have resorted to the use of their indigenous knowledge systems (IKS) in the management of cattle diseases. Generally, these practices are cheap, locally available, and sustainable especially in times of climate change and variability. One of the challenges in the use of (IKS) is the lack of scientific evidence on their efficacy and the lack of precise dosages, which could lead to toxicity. There is need therefore for documentation, research and scientific validation of IKS to increase their sustainable use and adoption in livestock health management.

Keywords: ethnoveterinary medicine, livestock, conventional, research, scientific validation

1. Introduction

Livestock production is important in the livelihoods of people living in developing countries like Zimbabwe. It has an important role in food security and nutrition, and the general economy [1, 2]. For instance, cattle provide food, manure for crop production and soil fertility management, raw material for industry, cash income as well as in promoting saving, fuel, social functions, and employment [3]. Cattle are an important livestock species in Zimbabwe, as they contribute about 25% to gross domestic product (GDP). Generally, cattle rearing is significant to the socio-economic profile of rural households [4].

IntechOpen

Cattle production is hindered by several challenges in the rural areas which include nutrition-related problems, market linkages problems, lack of technical know-how and cattle health related problems. High costs of veterinary medicines, dilapidated government veterinary services that have since been incapacitated to supply veterinary drugs, poor road, and other communication networks affect information dissemination, hence the availability of conventional drugs to treat livestock diseases in most rural areas [5]. With farmers being aware of the need to keep their cattle in a state of well being they then resort to the use of indigenous knowledge systems (IKS) through the use of ethnoveterinary medicines to treat their livestock.

Indigenous knowledge systems are an adhesive that binds society, being durable local knowledge [6], hence the cornerstone for building of our identity and ensuring coherence of social structures within communities [7]. Mapara [8] mentions that IKS have traditionally been passed from generation to generation orally; hence people possess little or no knowledge of invaluable practices such as ethnoveterinary medicine which depend only on historical evidence of use as proof of safety and effectiveness [5]. Ethnoveterinary medicine (EVM) is the study of people's folk beliefs, knowledge, skills, methods, and practices on the healthcare of animals, including the use of medicinal plants, surgery techniques, and management practices for the prevention and treatment of livestock diseases [9]. Generally, ethnoveterinary practices are cheap, locally available, and sustainable especially in times of climate change and variability [5]. EVM have controlled a wide spectrum of common livestock diseases successfully, including diarrhoea, wounds, coccidiosis and reproductive disorders [10–12]. This review focuses on the use of IKS in managing cattle diseases in Zimbabwe including the methods of preparing and administering the remedies. The review will serve as a baseline for the use of these IKS remedies for potential veterinary drugs development.

2. Cattle production trends post Land Reform Programme in Zimbabwe

Figure 1 highlights trends in cattle numbers from 2001 to 2021. As depicted in **Figure 1**, from 2001 to 2008, cattle production in Zimbabwe has been massively underperforming as the total national herd declined from between 6,270,000 to 5,012,000. The decline in cattle production was evidenced by various challenges such

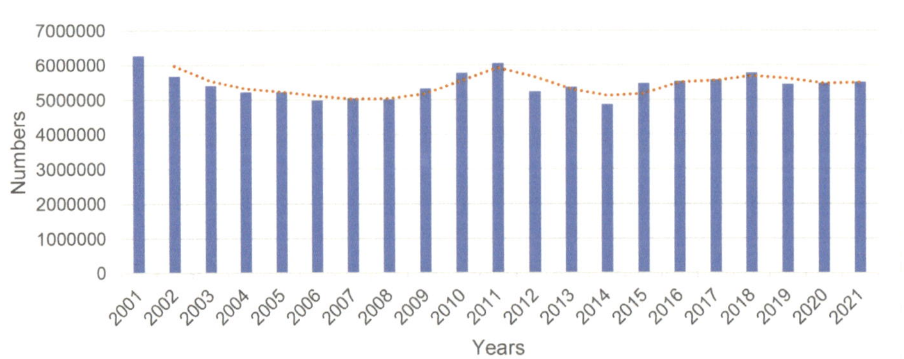

Figure 1.
Cattle populations, (2001–2021). Source: Ministry of Agriculture, Water and Rural Resettlement Reports [13–19].

as the Fast Track Land Reform Programme which saw several large-scale cattle farmers selling their cattle and some losing their farms [20, 21]. Nevertheless, the effects of hyperinflation, lack of access to credit, foreign currency shortages, unfavourable and cost of doing business conditions, and the high cost of critical utilities such as electricity, inappropriate breeds, inadequate feed supply, lack of government support, cost of production, poor marketing channels, poor disease control methods, inadequate infrastructure, weak extension support, among other factors [22–24].

From 2009 to 2011, there was a slight increase in the national herd from 5,331,000 to 6,058,388. Unexpectedly, growth in national herd increased from 5,241,192 to 4,868,357 between 2012 to 2014. Various cases of diseases emanated from foot and mouth, anthrax, black leg, lumpy skin disease and tick borne [13, 25]. Also, in 2015–2018 growth in national herd increased from 5,477,338 to 5,774,525. The year 2018 started with the continuation of the South-Eastern Lowveld cluster disease outbreak which spread mainly due to illegal movements and movements in search of grazing eventually covering twenty districts. Foot and mouth disease originating from Mozambique was precipitated by movement of Zimbabwean cattle deep inside the Mozambican territory to access water at time when Mozambique was experiencing a serious outbreak of foot and mouth disease in the area [14, 26]. From 2019 to 2021, national herd increased from 5,443,770 to 5,509,983. Though the national herd increased, calf mortality across provinces ranged from 2 to 31% which is against the recommended 2% and this was due to poor calf management, predation, poor housing and poor nutrition. The national average calving rates remain very low ranging from 35% in communal areas to 48% in large scale commercial farming sector, against a national target of above 60% [18].

3. Constraints to cattle production

In view of anticipated increases in temperature, decreases in rainfall and subsequent shortening of the growing season due to climate change, livestock production is expected to become a more sustainable livelihood source compared to cropping. Actually, there already are projections of increases in the global demand for cattle products by 2050, mainly due to improvements in the worldwide standard of living [27]. However, there are various constraints to cattle production such as nutrition, diseases and markets (**Table 1**). An improved comprehension of these constraints is necessary for development of strategies to increase sustainability of livestock production.

Nutrition and health are generally regarded as the main challenges of livestock production and productivity [3, 23]. Animal feeds are expensive, accounting for more than 70% of livestock production costs. During the dry season, there are limitations in both the quality and quantity of feed. The grazing resource is more available in the wet season than the dry season in which poor quality cereal stover constitute the bulk of the feed resource. Moreover, there is competitive use of the common feed ingredients like maize and soya bean between humans and animals. The productivity of the rangeland, which is the main feed resource in extensive livestock production has also been decreasing, due to unsustainable utilisation which can be partly attributed to increasing population growth, mismanagement, and adverse effects of climate change. Management factors which cause reductions in livestock production include uncontrolled breeding which causes inbreeding, poor housing and low adoption of routine health management practices such as vaccination and dosing [25]. For

Constraint	References
Feeds shortages and quality	[28–30]
Diseases and parasites	[5, 25, 31]
Markets and prices	[32, 33]
Breeds and breeding	[34–38]
Management	[25, 39]
Water access and quality	[23]
Extension service	[5, 23]
Capital and other financial resources	[40]

Table 1.
Constraints to livestock production.

instance, vector associated dermatophilosis, tick borne diseases, and parafilariosis are some of the common cattle health constraints.

Climate change has also caused a geographical shift in the occurrence of livestock diseases and parasites. Efforts to address the emerging diseases are hampered by increasing resistance to antibiotics. Resultantly, growth and reproductive performance of breeds that do not adapt to the prevailing climatic conditions are reduced. Similar to the grazing resource, water availability is also seasonal, with challenges of access experienced more in the dry season. The IPCC [41] anticipates even more water challenges due to decreases in precipitation resulting from climate change. Other limitations to livestock production include limited financial resources, e.g. capital, and extension services that offer veterinary or technical assistance. Market challenges in livestock production include low prices, high transport costs and shortages of buyers [36].

4. Use of indigenous knowledge systems in livestock health management in Zimbabwe

In the absence of funds, farmers face problems of scarcity, erratic supply and prohibitive costs of synthetic drugs or veterinary services and they usually revert back to sustainable traditional systems of animal health care [42]. In Zimbabwe, there is evidence that ethnoveterinary medicines are gaining recognition at the expense of conventional drugs especially because of greater accessibility, lower costs and apparent effectiveness [5, 12, 42, 43]. A study by [5] in Masvingo province has shown that farmers use plants that occur in their respective districts of the province. Farmers justify the potency of the ethno-veterinary remedies in relation to livestock's health and production performance in terms of feed intake, body weight, carcass size and quality [12, 44]. **Table 2** shows indigenous plants and remedies that are used to manage different cattle ailments in Zimbabwe.

There are several challenges downplaying the use of IKS in Zimbabwe. One of the main disadvantages of the use of herbal plants is the lack of scientific evidence on their efficacy and the lack of precise dosages, which could lead to toxicity [42]. Indigenous ethnoveterinary practices were carried out essentially based on private practice. The information reserved by traditional healers due to high secrecy is relatively less susceptible to distortion but ends up being less accessible to the public [12]. The fact that

Disease/ symptom	Local and scientific names of remedies	Method of treatment	References
Septic wounds	Muvengahonye (*Canthium* spp.)	Fresh leaves are ground and applied to the wound. Fresh leaves are ground and applied to the wound as a powder on the wound	[11, 42]
	Gavakava (*Aloe* spp.)	Dry leaves are crushed and the powder applied	[11, 42]
	Murenja (*Cissus quandrangularis*)	Fruit is crushed and the fluid applied to wound	[11, 42, 45, 46]
	Muvheva (*Kigelia africana*)	The inner core of dried fruit is applied as a powder on the wound	[11, 42]
Eye problems	Nhundurwa (*Solanum indicum*)	Fruit is crushed and the fluid is applied to the eye	[11, 42, 47]
	Snail's shell	Shell is ground to powder and applied to the eye	[11, 42, 47]
	Tomato leaves (*Lycopersicon esculentum*)	Animal made to drink crushed leaves and water mixture	[47]
Gastrointestinal worms	Muzhozho (*Venonia amygdalina*)	Add water to ground fresh leaves; animal made to swallow mixture	[11, 42]
	Banana (*Musa paradisiacal*)	Add water to crushed fresh roots; animal made to swallow mixture	[11, 42, 47]
	Gavakava (*Aloe* spp.)	Add water to crushed fresh leaves; animal made to swallow mixture	[11, 42, 47]
Bloat	Munhanzva (*Pauzzozi amixta*)	Leaves crushed and water added; animal made to swallow mixture	[11, 42, 46, 47]
	Chin'ai (*Phlegmostomium*)	Mix with table salt, add water; animal made to swallow mixture	[11, 42]
	Muhumbakumba (*Bridellia mollis*)	Soak bark and administer orally	[48]
Goitre	Mukwakwa (*Strychnos madagascariensis*)	Boil roots in water, cool and administer orally	[48]
Coccidiosis	Gavakava (*Aloe* spp.)	Grind fresh leaves and add to drinking water	[10–12, 42, 43]
	Mhiripiri (*Capsicum annum*)	Animal made to drink crushed fruit and water mixture	[47]
	Mucherenje/ Muwora (*Albizia gummisera*)	Animal made to drink suspension of powdered bark in water	[47]
	Muchakata (*Parinaria curatellifolia*)	Animal made to drink bark powder and water mixture	[47]
Retained afterbirth	Munhanzva (*Pauzzoziamixta*)	Fresh leaves are crushed and the slippery paste inserted into the vagina	[11, 42, 46, 47]
Lumpy skin disease	Muhumbakumba (*Bridellia mollis*)	Boil leaves in water and administer orally	[48]
Snake bite	Munyoka (*Amaranthusgneizaus*)	Add water to crushed fresh roots; animal made to swallow mixture	[11, 42]
	Mubhanana (*Musa paradisiacal*)	Add water to crushed dried roots; animal made to swallow mixture	[11, 42, 47]

Disease/ symptom	Local and scientific names of remedies	Method of treatment	References
Foot and mouth	Nhengeni (*Annona senegalensis*)	Administer sap from fruit orally	[48]
Dystocia	Munanzva (*Pouzolzia mixta*)	Crush bark, soak and administer orally	[48]
Fertility	Gomarara (*Loranthus* spp.)	Feeding fresh leaves to cows improves calving rate	[11, 42]
Delayed parturition	Murenja (*Cassius quandrangularis*)	Crush fresh stem and leaves, place in the vagina to hasten parturition	[11]
Fleas	Rutapatsikidzi (*Aneilema hockii*)	Branches of plant are placed near sleeping animals. Fleas are attracted by the herb and leave the animal.	[11]
Poor milk flow	Baobab (*Adanonsia digitata*)	Inner core of dried fruit is removed, added to water; animal made to swallow mixture	[11]
Diarrhoea/ Gastrointestinal problems	Gavakava (Aloe spp.)	Grind fresh leaves and add to drinking water	[10, 43, 46, 47]
	Mufandichimuka (*Myrothamnus flabellifoilius*)	Animal made to drink root powder and water mixture	[47]
	Rusungwe/ Nyakadombo (*Sarcostemma viminale*)	Animal made to drink stem powder and water mixture	[47]
	Mhiripiri (*Capsicum annum*)	Animal made to drink crushed fruit and water mixture	[47]
	Muchakata (*Parinaria curatellifolia*)	Animal made to drink bark powder and water mixture	[47]
	Murumanyama (*Xeroderris stuhlmannii*)	Animal made to drink crushed bark and water mixture	[47]
Fractures	Batanai (*Bulbophylum spp.*)	Bark is tied around fracture as supporting pad	[47]
Foot rot	Mushozhiwa (*Pseudolachnostylis maprouneifolia*)	Leaves or bark infusion taken through the mouth	[47]
External parasites	Murunjurunju (*Cissus quadrangularis*)	Crush stems and mix with water to spray	[49]
	Mopane (*Colophospermum mopane*)	Burn branches and twigs and apply ashes on animal skin	[49]
	Gavakava (*Aloe excelsa*)	Crush leaves, mix with water for 24 hours and spray	[49]
	Croton gratissimu	Use leaves and twigs	[46]
	Lippia javanica	Water leaf extracts sprayed on cattle	[50]

Table 2.
IKS remedies used to manage cattle health in Zimbabwe.

some herbs are available only in certain seasons often limits the use of herbal plants. Moreover, some of the preparations are mixtures of many kinds of plants which may be difficult to find at the same time.

Ethnoveterinary medicines are often not as fast working and potent as allopathic medicines and their use is time consuming in their preparation and use. The ethnoveterinary medicines may therefore be less suitable to control and treat epidemic and endemic infectious diseases and acute life-threatening bacterial infections. Paucity of treatment against the infectious epidemic diseases is another limitation of ethnoveterinary medicines. The preparation and use of ethnoveterinary medicines is often difficult and has inconveniences [5].

There is still need for the validation, documentation and acknowledgement of EVM in Zimbabwe among other tropical countries [42]. Additionally, socio-economic factors as population pressure, agricultural expansion into the veld, and other stress on the land may cause unsustainable consumption [5] It is not easy to standardise herbal therapies as the concentration of active ingredients varies in different parts of the plants [9]. Cases of toxicity and underdosing are more as there is no exact dosage in relation to body weight. That some herbs are available only in certain seasons often limits the application of ethnoveterinary medicines. In the absence of regulatory control, product quality becomes variable.

Research and scientific validation of IKS, including EVM are therefore important to increase their adoption in livestock health management. The knowledge of traditional healers and experienced elderly people should be tapped to gather information on these practices so that the future generations can enjoy the same benefits [5]. The knowledge of traditional healers, stockmen, hunters, and other experienced elderly people may be the only ones with appropriate information regarding, for instance, the forms in which the drug has to be given for a particular disease: this information needs to be documented for the benefit of future generations [5, 42].

5. Conclusion

Conventional veterinary services have played a paramount role in the control of livestock diseases. The conventional veterinary services cannot yet deliver complete coverage in preventive and curative health care practices because of inadequate labor, logistical problems, erratic supply of drugs, and the high cost of drugs and equipment. In Zimbabwe, there is evidence that ethnoveterinary medicines are gaining recognition at the expense of conventional drugs especially because of greater accessibility, lower costs and apparent effectiveness. A practical solution to cutbacks in veterinary services is to develop socially acceptable and effective remedies from reasonably inexpensive sources that can complement modern medicine like ethnoveterinary medicines.

Author details

Vimbai Gobvu*, Kudakwashe C. Chirigo, Takudzwa L. Charakupa
and Clarice P. Mudzengi
Gary Magadzire School of Agriculture and Engineering, Great Zimbabwe University,
Masvingo, Zimbabwe

*Address all correspondence to: vgobvu@gmail.com

References

[1] Hatab AA, Cavinato MER, Lagerkvist CJ. Urbanization, livestock systems and food security in developing countries: A systematic review of the literature. Food Security. 2019;**11**(2):1-21

[2] Mapiye O, Chikwanha OC, Makombe G, Dzama K, Mapiye C. Livelihood, food and nutrition security in southern Africa: What role do indigenous cattle genetic resources play? Diversity. 2020;**12**(2):74

[3] Ma'alin A, Abdimahad K, Hassen G, Mahamed A, Hassen M. Management practices and production constraints of indigenous Somali cattle breed in Shabelle zone, Somali regional state, Ethiopia. Open Journal of Animal Sciences. 2022;**12**:103-117. DOI: 10.4236/ojas.2022.121008

[4] Nyamushamba GB, Mapiye C, Tada O, Halimani TE, Muchenje V. Conservation of indigenous cattle genetic resources in Southern Africa's smallholder areas: turning threats into opportunities—A review. Asian-Australasian Journal of Animal Sciences. 2016;**30**(5):603

[5] Mudzengi CP, Dahwa E, Skosana JLN, Murungweni C. Promoting the use of ethnoveterinary practices in livestock health Management in Masvingo Province, Zimbabwe. Ethnobotany Research and Applications. 2014;**12**(12):398-406. DOI: 10.17348/era.12.0.397-405

[6] Sithole PM. Indigenous knowledge Systems in Crop Management and Grain Storage in Chimanimani District of Zimbabwe. Southern African Journal of Environmental Education. 2020;**36**:21-32

[7] Chiwanza K, Musingafi M, Mupa P. Challenges in preserving indigenous knowledge systems: Learning from past experiences. Information and Knowledge Management. 2013;**3**(2):19-25

[8] Mapara J. Indigenous Knowledge Systems in Zimbabwe: Juxtaposing Postcolonial Theory. Journal of Pan African Studies. 1 Sep 2009;**3**(1)

[9] Temeche MA, Asnakew AT. A review on status of ethnoveterinary medicine and challenges it faces in Ethiopia. International Journal of Veterinary Sciences and Animal Husbandry. 2020;**5**:39-48

[10] Masimba ES, Mbiriri DT, Kashangura MT, Mutibvu T. Indigenous practices for the control and treatment of ailments in Zimbabwe's village poultry. Livestock Research for Rural Development. 2011;**23**(12):2-9

[11] Matekaire T, Bwakura TM. Ethnoveterinary medicine: A potential alternative to orthodox animal health delivery in Zimbabwe. International Journal of Applied Research in Veterinary Medicine. 2004;**2**(4):269-273

[12] Mwale M, Bhebhe E, Chimonyo M, Halimani TE. Use of herbal plants in poultry health management in the Mushagashe small-scale commercial farming area in Zimbabwe. International Journal of Applied Research in Veterinary Medicine. 2005;**3**(2):163-170

[13] Ministry of Agriculture, Mechanisation and Irrigation Development. Second Round Crop and Livestock Assessment Report 2014/2015 Season. 2015

[14] Ministry of Lands, Agriculture, Water and Rural Resettlement. Second Round Crop and Livestock Assessment Report 2018/2019 Season. 2019

[15] Ministry of Lands, Agriculture, Water and Rural Resettlement. Second Round Crop and Livestock Assessment Report 2017/2018 Season. 2018

[16] Ministry of Lands, Agriculture, Water and Rural Resettlement. Second Round Crop and Livestock Assessment Report 2016/2017 Season. 2017

[17] Ministry of Lands, Agriculture, Water and Rural Resettlement. Second Round Crop and Livestock Assessment Report 2015/2016 Season. 2016

[18] Ministry of Lands, Agriculture, Water and Rural Resettlement. Second Round Crop and Livestock Assessment Report 2020/2021 Season. 2021

[19] Ministry of Lands, Agriculture, Water and Rural Resettlement. Second Round Crop and Livestock Assessment Report 2019/2020 Season. 2020

[20] Chatikobo P, Choga T, Ncube C, Mutambara J. Participatory diagnosis and prioritization of constraints to cattle production in some smallholder farming areas of Zimbabwe. Preventive Veterinary Medicine. 2013;**109**(3-4):327-333

[21] Mavedzenge BZ, Mahenehene J, Murimbarimba F, Scoones I, Wolmer W. The dynamics of real markets: Cattle in southern Zimbabwe following land reform. Development and Change. Wiley Online Library. 2008;**39**(4):613-639

[22] Chisango FF, Deliwe T, Prince N, Saziso M. Challenges and opportunities on beef cattle marketing and off take rates in Zimbabwe's small holder farming sector: A case of A1 resettlement farmers in Umzingwane District of Matabeleland South Province. International Journal of Innovative Research & Development. 2015;**4**(4):221-226

[23] Mutibvu T, Maburutse BE, Mbiriri DT, Kashangura MT. Constraints and opportunities for increased livestock production in communal areas: A case study of Simbe, Zimbabwe. Livestock Research for Rural Development. 2012;**24**(9):2012. Available from: http://www.lrrd.cipav.org.co/lrrd24/9/cont2409.htm

[24] Tembachako DS, Ndlovu P, Mukomana S. Challenges and opportunities on beef cattle marketing and off take rates in Zimbabwe's small holder farming sector: A case of A1 resettlement farmers in Umzingwane District of Matabeleland South Province. International Journal of Innovative Research & Development. 2015;**4**(4):221-226

[25] Tavirimirwa B, Mwembe R, Ngulube B, Banana NYD, Nyamushamba GB, Ncube S, et al. Communal cattle production in Zimbabwe: A review. Livestock Research for Rural Development. 2013;**25**(12):217

[26] Ndlovu T, Belle J, Silengo M. Participation of communal cattle farmers in drought risk reduction in southern Zimbabwe. Jàmbá: Journal of Disaster Risk Studies. 2021;**13**(1):1-10

[27] Rojas-DowningMM,NejadhashemiAP, Harrigan T, Woznicki SA. Climate change and livestock: Impacts, adaptation, and mitigation. Climate Risk Management. 2017;**16**:145-163. DOI: 10.1016/j.crm.2017.02.001

[28] Mavhura E, Manatsa D, Mushore T. Adaptation to drought in arid and semi-arid environments: Case of the Zambezi Valley, Zimbabwe. Journal of Disaster Risk Studies. 2015;7:1-7

[29] Chakoma I, Manyawu G, Gwiriri LC, Moyo S, Dube S, Imbayarwo-Chikosi VE, et al. Promoting the use of home-mixed

supplements as alternatives to commercial supplements in smallholder beef production systems in the sub-humid region of Zimbabwe. African Journal of Range and Forage Science. 2016;**33**:165-171

[30] Dzavo T, Zindove TJ, Dhliwayo M, Chimonyo M. Effects of drought on cattle production in sub-tropical environments. Tropical Animal Health and Production. 2018;**51**:669-675

[31] Sungirai M, Moyo ZD, De Clercq P, Madder M. Communal farmers' perceptions of tick-borne diseases affecting cattle and investigation of tick control methods practiced in Zimbabwe. Tick and Tick-borne Diseases. 2016;**7**(1):1-9

[32] Paenda O, Musemwa L, Ndhleve S, Sibanda M. Determinants of farmers' marketing choices and preferences under communal cattle farming: Evidence from Mwenezi District in Zimbabwe. Journal of Human Ecololgy. 2020;**72**(1-3):13-23. DOI: 10.31901/24566608.2020/72.1-3.3263

[33] Mujeyi A, Mutenje M, Manyawu GJ, Gwiriri L, Chakoma I. Spearheading development through empowering smallholder farmers along beef cattle value chains: A case of Goromonzi and Murehwa districts, Zimbabwe. International Journal of Managing Value and Supply Chains. 2015;**6**(4):31-44

[34] Mapiye C, Chimonyo M, Dzama K, Raats JG, Mapekula M. Opportunities for improving Nguni cattle production in the smallholder farming systems of South Africa. Livestock Science. 2009;**124**:196-204

[35] Scholtz MM, Theunissen A. The use of indigenous cattle in terminal cross-breeding to improve beef cattle production in sub-Saharan Africa. Animal Genetic Resources. 2010;**46**:33-39

[36] Bidi NT, Dube AB, Khombe CT, Assan N. Community based small scale commercial cattle breeding programme in Mangwe district of Zimbabwe. Agricultural Advances. 2015;**4**(3):22-33. DOI: 10.14196/aa.v4i3.1845

[37] Gororo E, Makuza SM, Chatiza FP, Gwatibaya S, Gahadzikwa P, Chidzwondo F. The potential of reproductive technologies in breeding smallholder cattle populations in Zimbabwe. International Journal of Livestock Production. 2017;**8**:168-179

[38] Nyamushamba GB et al. Conservation of indigenous cattle genetic resources in southern Africa's smallholder areas: Turning threats into opportunities—A review. Asian-Australasian Journal of Animal Sciences. 2017;**30**(5):603-621. DOI: 10.5713/ajas.16.0024

[39] Homann S, Van Rooyen A. Unexploited agricultural growth: The case of crop–livestock production Systems in Zimbabwe. In: 2nd African Association of Agricultural Economists Conference, Accra, Ghana. 2007;**22**(40):503-506. Available from: www.aaae-africa.org

[40] Tambi MD, Anyah FJ. Constraints and challenges in livestock production in Cameroon. South Asian Research Journal of Business and Management. 2019;**1**:10-17

[41] IPCC. IPCC fourth assessment report. The Physical Science Basis. 2007;**2**:580-595

[42] Marandure T. Concepts and key issues of ethnoveterinary medicine in Africa: A review of its application in Zimbabwe. African Journal of Agricultural Research. 2016;**11**(20):1836-1841

[43] Jambwa P, Katsande S, Matope G, McGaw LJ. Ethnoveterinary remedies

used in avian complementary medicine in selected communal areas in Zimbabwe. Planta Medica. 2021;**88**(03/04):313-323

[44] Muchadeyi FC, Sibanda S, Kusina NT, Kusina JF, Makuza SM. Village chicken flock dynamics and the contribution of chickens to household livelihoods in a smallholder farming area in Zimbabwe. Tropical Animal Health and Production. 2005;**37**(4):333-344

[45] Marume A, Matope G, Katsande S, Khoza S, Mutingwende I, Mduluza T, et al. Wound healing properties of selected plants used in ethnoveterinary medicine. Frontiers in Pharmacology. 2017;**8**:544

[46] Maroyi A. Use of ethnoveterinary medicine by small-scale farmers to treat livestock diseases: An alternative to orthodox livestock health delivery system in southern Africa. International Journal of Scientific and Technology Research. 2021;**10**(07):32-40

[47] Maroyi A. Use of traditional veterinary medicine in Nhema communal area of the midlands province, Zimbabwe. African Journal of Traditional, Complementary, and Alternative Medicines. 2012;**9**(3):315-322

[48] Gumbochuma G, Hamandishe VR, Nyahangare ET, Imbayarwo-Chikosi VE, Ncube S. Ethnoveterinary practices for poultry and cattle in Zimbabwe: A case study of Takavarasha village. South African Journal of Animal Science. 2013;**2**(12):355-359

[49] Nyahangare ET, Mvumi BM, Mutibvu T. Ethnoveterinary plants and practices used for ecto-parasite control in semi-arid smallholder farming areas of Zimbabwe. Journal of Ethnobiology and Ethnomedicine. 2015;**11**(1):1-6

[50] Madzimure J, Nyahangare ET, Hamudikuwanda H, Hove T, Stevenson PC, Belmain SR, et al. Acaricidal efficacy against cattle ticks and acute oral toxicity of *Lippia javanica* (Burm F.) Spreng. Tropical Animal Health and Production. 2011;**43**(2):481-489